Environmental Science and Engineering

The ultimate goal of this series is to contribute to the protection of our environment, which calls for both profound research and the ongoing development of solutions and measurements by experts in the field. Accordingly, the series promotes not only a deeper understanding of environmental processes and the evaluation of management strategies, but also design and technology aimed at improving environmental quality. Books focusing on the former are published in the subseries Environmental Science, those focusing on the latter in the subseries Environmental Engineering.

Ipsita Dipamitra Behera · Alok Prasad Das
Editors

Environmental Hydrocarbon Pollution and Zero Waste Approach Towards a Sustainable Waste Management

Springer

Editors
Ipsita Dipamitra Behera
Department of Chemical Engineering
Indira Gandhi Institute of Technology
Dhenkanal, Odisha, India

Alok Prasad Das
Department of Life Sciences
RDW University
Bhubaneswar, Odisha, India

ISSN 1863-5520 ISSN 1863-5539 (electronic)
Environmental Science and Engineering
ISBN 978-3-031-81614-7 ISBN 978-3-031-81615-4 (eBook)
https://doi.org/10.1007/978-3-031-81615-4

This Springer imprint is published by the registered company Springer Nature Switzerland AG
The registered company address is: Gewerbestrasse 11, 6330 Cham, Switzerland

Contents

Chapter 1
Hydrocarbon Waste: The Unseen Threat to Our Biosphere—Origins, Traits, Fate, and Environmental Impact

Debshankar Paramanik and Manisha Bal

Abstract As a result of industrialization, urbanization, and human activity, hydrocarbon waste is a ubiquitous and oftentimes overlooked hazard to the biosphere. With an emphasis on its pervasive existence in ecosystems across the world, this study investigates the origins, characteristics, destiny, and environmental effects of hydrocarbon waste. To devise effective mitigation strategies, it is essential to have a comprehensive understanding of the sources of hydrocarbon waste. Persistence and toxicity in the environment may be understood by analyzing the many characteristics of hydrocarbon waste, such as its chemical makeup and physical attributes. The impact on ecosystems, animals, and human health is highlighted in the paper, underscoring the necessity of thorough risk assessments and regulatory frameworks. Assessing the environmental impact encompasses its role in climate change, upsets of natural equilibrium, and contamination of soil and water resources. The present chapter is also spotlighted on environmental management and sustainable practices, mitigating the unseen threat to our biosphere.

Keywords Hydrocarbon waste · PAH · Water treatment · Environmental management

1.1 Introduction

Hydrocarbons are organic compounds made up of carbon and hydrogen atoms. Today they make up the backbone of industrial energy requirements primarily in the form of fossil fuels. They are also used to manufacture various synthetic materials used in day-to-day life. However, the extensive use of hydrocarbons has led to the production of copious amounts of hydrocarbon waste. This research paper aims to delve into

---

D. Paramanik · M. Bal (✉)
Department of Chemical Engineering, National Institute of Technology Durgapur, Durgapur, West Bengal 713209, India
e-mail: mbal.che@nitdgp.ac.in

I. D. Behera and A. P. Das (eds.), *Environmental Hydrocarbon Pollution and Zero Waste Approach Towards a Sustainable Waste Management*, Environmental Science and Engineering, https://doi.org/10.1007/978-3-031-81615-4_1

the definition and types of hydrocarbons, emphasizing the critical importance of studying hydrocarbon waste and its environmental implications.

Hydrocarbons are of two types, namely—aliphatic and aromatic. Aliphatic hydrocarbons are compounds which can be linear, branched or cyclic. They may be saturated or unsaturated. They are generally used as fuels or serve as the raw materials for various petrochemicals (Grabato et al. 2024; Behera et al. 2022b). Aromatic hydrocarbons feature conjugated structures with alternating single and double bonds. They are essential components in the pharmaceutical industry and are used in the manufacture of chemical components like benzene, toluene and xylene. They are also used in the synthesis of drugs, dyes and plastics (Speight 2017).

Studying hydrocarbons and the diverse traits of hydrocarbon wastes is important because of the threat they pose to all forms of life (Srivastava et al. 2019). Petroleum hydrocarbons (PHCs) are among the most common environmental poisons, posing a serious threat to all forms of life, from plants to humans (Bala, Anshul et al. 2024; Bala, Dogra et al. 2024; Das, Dey, & Das 2024). PHCs degrade the soil's physicochemical properties, making it unsuitable for agricultural growth (Ali et al. 2023; Lahiri et al. 2021; Behera et al. 2020). A class of extremely resistant molecules known as long-chain aliphatic hydrocarbons decreases the amount of nutrients available in soils, greatly increases their water content, and immobilizes hydrophobic complexes, all of which have an indirect impact on the ecological function of soils (Zuzolo et al. 2021; Behera et al. 2022b).

Industrial and anthropogenic activities have often led to elevated hydrocarbon levels in water. The inherent capacity of water to break down hydrocarbons is often surpassed, resulting in various environmental consequences like interfering with fishes' ability to breathe, destroying algae, plankton and microorganisms (Reeves, n.d.). Crude oil and petroleum derivatives create an impermeable layer on the water's surface, hindering the exchange of oxygen between the environment and the water (Srivastava et al. 2019).

Polycyclic aromatic hydrocarbons (PAHs) are produced through the incomplete combustion of organic substances. Most of them do not have a recognized purpose. Some are utilized in medications and for creating dyes, plastics, and pesticides (Das, Dey, Bala et al. 2024; Dey, Bissoyi et al. 2024; Dey, Dash et al. 2024). Naphthalene, commonly known as mothballs, is employed in the production of dyes, explosives, plastics, lubricants, and as a moth deterrent. Wood preservatives, pesticides, and dyes all include anthracene. PAH compounds like benzopyrene, acenaphthylene, anthracene, and fluoranthene are acknowledged as substances with toxicity, mutagenicity, and carcinogenicity. PAHs represent a significant contributing factor to the occurrence of lung cancer (Manisalidis et al. 2020) and are ranked ninth among chemical compounds threatening to humans (Ifegwu and Anyakora 2015).

A crucial first step in guaranteeing environmental preservation and adherence to water quality regulations is the extraction of hydrocarbon waste from wastewater.

Numerous approaches and technologies are utilized for this specific objective. Conventional wastewater treatment methods are ineffective in completely removing oil compounds and high pollutant concentrations, which can negatively impact the operations and productivity of refineries. Removal procedures should not generate additional harmful pollutants beyond the allowable discharge limits. Sewage filtration through coagulation and dissolved air flotation (DAF) flocculation can be followed by further purification in a biological reactor. Advanced coagulation methods like electrocoagulation and flocculation offer more advanced techniques compared to traditional physical and chemical methods, but they produce significant amounts of unmanageable and often rejected sludge. Physical separation techniques have the capability to isolate significant amounts of petroleum compounds, allowing for the recycling of some of these compounds through diverse processes. However, these methods are associated with high energy requirements and frequently result in blockages and clogging of equipment. Third-party refinement can help to achieve water reuse objectives using nano-filtration, reverse osmosis, and advanced oxidation techniques. Adsorption, which utilizes low-cost materials and adsorbents, is an interim technology that can be applied effectively. By integrating adsorption with one of the advanced techniques, not only can sludge production be reduced, but also the overall process cost (Mohammadi et al. 2020). In methods like Dissolved Air Floatation (DAF), aluminium sulfate and cationic polyelectrolyte are used as flocculants. Aggregates created by flocculation with the cationic polyelectrolyte were found to remove up to 40% of dissolved hydrocarbons whereas alum flocs demonstrated removal of soluble materials, primarily phenols. Overall DAF proved effective in reducing the general hydrocarbon content by 50 to 90% (Galil and Wolf 2001). Different adsorbents like biochar, specifically BCP (biochar from pine wood), ABCP (acid-activated biochar from pine wood), and BBCP (base-activated biochar from pine wood and commercial activated carbon, FS400, are used for the removal of hydrocarbons from aqueous dispersions, simulating bilge waters (Cataldo et al. 2022).

The choice of a particular approach relies on factors such as the composition and number of hydrocarbons in the wastewater, as well as the regulations in place. Combining various methods can enhance the efficiency of removing hydrocarbon waste. It is crucial to emphasize that appropriately disposing or treating the gathered hydrocarbon waste is necessary to avoid additional harm to the environment.

This study endeavours to explore the origins of hydrocarbon waste, the distinguishing features of hydrocarbon waste, the trajectory of hydrocarbon waste, the impact of hydrocarbon waste toxicity on the biosphere, the regulatory frameworks and surveillance methods governing hydrocarbon waste, strategies for alleviating and remediating hydrocarbon waste, advancements in hydrocarbon waste management practices, and finally, the anticipated challenges and future research imperatives in this domain.

1.2 Sources of Hydrocarbon Waste

Hydrocarbon waste can be generated from both anthropogenic activities as well as natural sources. Anthropogenic sources include waste from the oil and gas industry, petrochemical industry, transportation sector and domestic sources. Produced water which is the wastewater separated from oil and gas during production often contains hydrocarbons, dissolved salts, and other contaminants (Fakhru'l-Razi et al. 2009). Produced water, which is generated during the extraction and production of oil and gas, is typically a diverse blend of inorganic and organic substances. It is formed underground and then brought to the surface as part of the oil and gas extraction process (Hoek et al. 2021). Natural compounds such as benzene, toluene, ethylbenzene, and xylene are present in crude oil and gasoline. These petroleum hydrocarbon compounds can be frequently detected in both surface and underground water due to industrial operations, particularly those involving petrochemicals, accidental leakage from reservoirs, or improper disposal of waste. Contaminated water streams from various processes, often containing dissolved hydrocarbons like benzene, toluene, ethylbenzene, and xylene (BTEX) (Mohammadi et al. 2020). The group of reclaimed petroleum hydrocarbons consists of intricate waste streams obtained from the refining of petroleum crude oils and are commonly known as slop oils. Due to their origins from various stages of the refinery process, these slop oils comprise an extensive range of petroleum hydrocarbons and water. The constituents of this category typically comprise liquid substances that are anticipated to exhibit either minimal or substantial vapor pressure and minimal to moderate water solubility (SCREENING-LEVEL HAZARD CHARACTERIZATION Reclaimed Petroleum Hydrocarbons Category: Residual Hydrocarbon Wastes from Petroleum Refining 2011).

Activities related to oil and gas exploration and production, including drilling and extraction operations, can result in the release of hydrocarbons into the environment. Drilling fluids used to lubricate drill bits and stabilize boreholes, often containing hydrocarbons, heavy metals, and other chemicals. Improper disposal or spills can contaminate soil and water (Onwuka et al. 2018). Refining crude oil generates numerous waste streams, including oily sludges, spent catalysts, wastewater containing hydrocarbons and other pollutants, and emissions to air. Accidental releases of oil and gas during pipelines, storage tanks, and refinery operations can lead to significant hydrocarbon waste contamination. Deactivated catalysts used in reactions, often containing heavy metals and hydrocarbon residues (Chemical Risks in Petrochemical Industry—PREVOR, n.d.)

A major source of hydrocarbon waste comes from the manufacturing of various petrochemical products like paints, fertilizers, plastics and detergents (Behera and Das 2023). A frequent byproduct produced during different phases of petrochemical manufacturing are oily sludges, which includes residues from tank bottoms, pipeline and equipment cleaning, as well as reaction mixtures. These sludges often are comprised of complex combinations of hydrocarbons, water, as well as solids (minerals) (Lazar et al. 1999). Yet, research suggests that the composition of petroleum hydrocarbons in oily sludge and contaminated soils differs across various regions. Highly toxic species such as BTEX and PAHs were typically absent, with the hydrocarbons in both the sludge and soil predominantly consisting of long-chain petroleum hydrocarbons. (Al-Mebayedh et al. 2022; Dey et al. 2023). Water streams produced during various stages of the procedure, such as cooling water, condensate streams, and discarded water, become polluted. These wastewater streams may hold dissolved hydrocarbons like BTEX (benzene, toluene, ethylbenzene, xylene) as

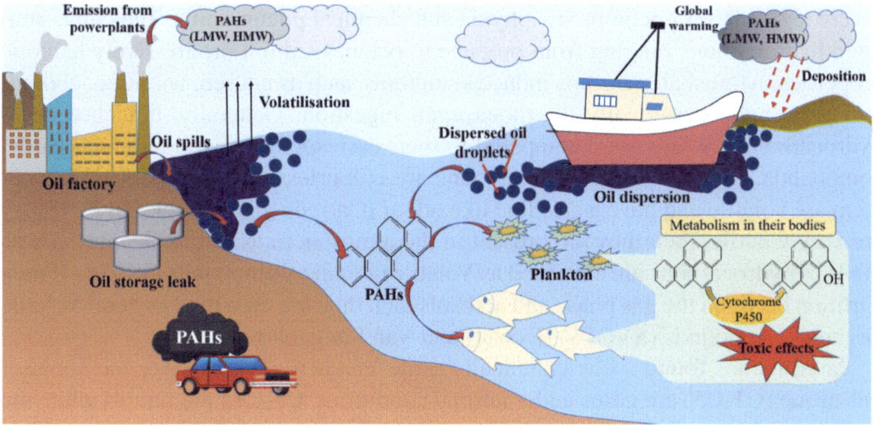

Fig. 1.1 Different sources of hydrocarbon waste

well as organic solvents and other chemicals, depending on the manufactured items (Mohammadi et al. 2020).

Polycyclic aromatic hydrocarbons (PAHs) originate from the incomplete burning of organic substances. (for example meat) as well as fossil fuels, wood stoves and diesel engines (Air Quality and Health n.d.). About 30% of the HC produced into the atmosphere is from the transportation sector (Dey and Mehta 2020).

Although most of the sources of hydrocarbon waste are from anthropogenic sources, they can come from natural sources as well (Fig. 1.1). Seepage of natural gas leads to the emission of substantial quantities of methane and other gases, such as ethane and propane, which contribute to the alteration of the Earth's climate on a global scale (Haque et al. 2022). Some natural sources of hydrocarbon emission include the weathering of minerals, erosions and volcanic activities, forest fires and release of particles from vegetation (Srivastava et al. 2019). The yearly production of XPAHs in waste incinerators ranged from 25.1 to 881 g (Wang et al. 2018).

1.3 Characteristics of Hydrocarbon Waste

Understanding the characteristics of hydrocarbon waste is fundamental to ensuring their safe and environmentally responsible management. This paper delves into the complex world of hydrocarbon waste characteristics, exploring their chemical composition, physical properties, and how these aspects differentiate between crude oil, refined products, and synthetic hydrocarbons.

Hydrocarbons are of two types, aliphatic and aromatic hydrocarbons. Aliphatic hydrocarbons may be cyclic or acyclic compounds that lack an aromatic ring. They act as asphyxiants and depressants of the Central Nervous System (CNS). Their severe toxic effects encompass asphyxia and chemical pneumonitis. Typically, saturated hydrocarbons ranging from propane to octanes exhibit progressively narcotic properties. Almost all paraffins induce symptoms such as nausea, vomiting, abdominal pain, and occasionally diarrhoea upon ingestion. Generally, branched-chain hydrocarbons are less toxic compared to their corresponding straight-chain parent compounds. Most aliphatic hydrocarbons are colourless and flammable. Many of them are volatile and have a gasoline like odour (Carreón and Herrick 2012). If they are very volatile, then they are present in the atmosphere as free molecules. These kinds of hydrocarbons are classified as Volatile Organic Compounds (VOCs). If they partition between the gas phase and aerosols then they are classified as Semi Volatile Organic Compounds (SVOCs) (Dewulf and Van Langenhove 2003).

Paraffins are found to be chemically quite unreactive. The lower members of this group (C1-C5) are gases under normal conditions, C5-C15 are liquids while the higher members are solid. The solubility of paraffins in water increases from methane to butane and subsequently decreases. While paraffins do have narcotic properties they are of a very low nature to be significant for industrial exposure. Olefins like paraffins are gases from C1-C5, liquids from C5-C18 and the heavier members are solids. Olefinic hydrocarbons are more toxic compared to their corresponding saturated compounds. Diolefins increase the narcotic action (Oettingen 1942).

Aromatic compounds are more soluble than aliphatic compounds of a similar size in ionic liquids (Hanke et al. 2003).

The presence of organic materials in soil, namely aliphatic hydrocarbons, has shown to have a positive relation on its hydrophobicity. It is seen that aliphatic hydrocarbons change the structure of hydrophobic soil while compounds like cutin and suberin are resistant to degradation (El Fels et al. 2016). However aliphatic hydrocarbons are generally more susceptible to biodegradation compared to aromatic hydrocarbons as the higher the number of rings, more difficult the degradation (Nzila 2018). From the experimental results of (Chevalier et al. 1990), it is observed that the densities and kinematic viscosities of both aliphatic and aromatic hydrocarbons increase with increasing molecular weight and branching.

Various subgroups of functionalized hydrocarbons polychlorinated biphenyls (PCBs) are categorized as persistence organic solution (Dewulf and Van Langenhove 2003).

1.4 Fate of Hydrocarbon Waste and Toxicity to Biosphere

The fate of hydrocarbon waste refers to what happens to it after it enters the environment. It involves processes like the transport of hydrocarbons (movement of hydrocarbons through different environmental areas such as air, water and soil), the biodegradation and microbial action on hydrocarbons, the bioaccumulation and biomagnification of hydrocarbons and its long-term persistence. Oil pollutants present a significant hazard to both human health and the environment owing to their toxicity, mutagenicity, and carcinogenicity (Mohammadi et al. 2020; Kanungo et al.

2023). Exposure to hydrocarbons can transpire through inhalation, ingestion with or without aspiration, or dermal contact. The atmosphere plays a significant role in facilitating the dispersion of volatile organic compounds (VOCs) throughout the global environment. In the air above seawater coated with crude oil slicks, hexane, cyclohexane, and heptane were identified as the most prevalent volatile organic compounds (VOCs) detected (Kuppusamy et al. 2019). PAHs in the air can travel significant distances, adhering to the surfaces of small solid particles. This airborne transportation can lead to contamination of remote ecosystems, even in areas far from the original source of pollution (Dey et al. 2023; John & Das 2024; Mishra et al. 2024; Mohanty et al. 2024; Paul et al. 2024). PAHs can return to the earth's surface through rainfall or settling onto particles. Alternatively, they can degrade into more persistent products by interacting with sunlight and other chemicals in the atmosphere over several days to weeks. As previously mentioned, PAHs undergo transformation into compounds such as nitro-PAH and hydroxy-PAH, which are more carcinogenic than the original compounds (Mohanraj and Azeez 2003).

Oil spills occur when liquid petroleum hydrocarbons are released into the environment. These oils undergo various natural processes, resulting in the indirect cleansing of polluted areas. When there is an oil spill on water, it tends to float and spread, increasing the surface area for evaporation and dissolution. Eventually, wave action breaks up the floating layers. When hydrocarbons containing 15 or less carbon atoms are spilled on the surface, they tend to evaporate rapidly. Another possible outcome is the production of photochemical smog when volatile hydrocarbons react with nitrogen oxides in the presence of light. Some hydrocarbons, including benzene, toluene, ethylbenzene, and xylenes (collectively known as BTEX), which are constituents of total petroleum hydrocarbons (TPHs), have the capability to easily migrate from floating or sub-surface spills into the water because of their solubility. This presents a notable concern underground, as the soluble fractions have the potential to contaminate wells and surface water. However, a significant portion of oils does not mix with water and may instead disperse (Kuppusamy et al. 2019; Mishra et al. 2021).

Soil, being porous and highly reactive, serves as a potent adsorbent. When an oil spill or leakage occurs on soil, a range of physical, chemical, and biological processes are initiated, which subsequently modify the composition, toxicity, and distribution of petroleum hydrocarbons (PHs) within the soil ecosystem. Physical effects involve the aggregation of soil particles into clumps, reduction of porosity, increased resistance to penetration, and heightened hydrophobicity. Chemical effects encompass the enrichment of soil organic matter resulting from the creation of anaerobic conditions that inhibit aerobic microorganisms and reduce cation exchange capacity (Wang et al. 2021). Soil contamination with petroleum hydrocarbons (PHs) is a significant global concern, as more than 90% of the environmental PHs load is typically retained in soil. This underscores soil's pivotal role as the primary repository for PHs in terrestrial environments, making it a crucial medium for various processes including volatilization, deposition, and degradation of PHs (Kuppusamy et al. 2019). Spills of oil and fuel in soil are among the most expansive and environmentally harmful pollution events, presenting significant risks to both human health and ecosystems, especially in colder regions. The biochemical and physicochemical characteristics of soil suffer deterioration due to the presence of refinery products, which also hinder the growth and development of plants. Soil enzymes, vital biotic components responsible for various biochemical reactions in soil, are adversely affected by petroleum hydrocarbons, further exacerbating the situation. Soil contaminated with petroleum hydrocarbons results in contamination of underground water by organic pollutants, restricting its utilization and causing economic losses and environmental concerns. Additionally, it diminishes the soil's agricultural productivity. The toxicity of petroleum hydrocarbons poses significant risks to microorganisms, plants, animals, and humans, making them vulnerable to adverse effects (Priyadarsini & Das 2024; 2025; Ray et al. 2024; Tripathy et al. 2022; Mohammadi et al. 2020).

When oil spills occur in aquatic environments, the oil is primarily transported between different environmental components via sediments. In water enriched with sediment, spilled oil tends to interact with the sediment particles. This interaction can result in the creation of aggregates known as oil particulate aggregates (OPA), which occur when oil interacts directly with sediments or suspended particulate materials (SPM). At a microscale level, this interaction between oil and fine sediments aids in the extraction of leaked oil from coastlines. The interaction between oil droplets and sediment particles promotes efficient dispersion into the water column, where biodegradation rates are anticipated to be notably high. Consequently, the destiny and movement of oil are greatly impacted by these interactions between oil and sediment. (Kuppusamy et al. 2019).

Cleaning up polluted environments using traditional methods such as physicochemical, thermal, and electromagnetic technologies present a formidable challenge due to its labour-intensive nature and high cost. It's crucial to shift away from costly conventional physicochemical treatment methods towards more environmentally friendly biological treatment technologies. These alternatives not only reduce energy consumption and chemical usage but also lower implementation costs. Moreover, they facilitate the implementation of sustainable, hazard-focused approaches to environmental restoration. Bioremediation offers an eco-friendly, sustainable, and cost-effective approach to restoring and purifying soil polluted with substances like petroleum hydrocarbons in polluted areas. This approach entails the natural degradation of petroleum hydrocarbon contaminants by specialized microorganisms, including bacteria, fungi, yeasts, and algae or their consortia (Behera et al. 2022a; Das et al. 2024). In microbial remediation, microorganisms adhere to petroleum hydrocarbons and break them down by utilizing them as a food source. Enhancing the bioremediation process can be achieved by introducing microbial cultures (biodegraders) and fertilizers. Microbes naturally have the capability to degrade petroleum wastes in the soil, albeit at a very slow pace. While bioremediation is considered an effective strategy for addressing many challenges in successfully remedying petroleum hydrocarbon contamination, it also has limitations. One such limitation is its inability to completely remediate and remove petroleum hydrocarbon wastes from contaminated soils (Ossai et al. 2022; Kuppusamy et al. 2019). Bibliographic analysis of hydrocarbon waste sources, environmental risk, and remediation strategies through VOSviewer is presented in Figs. 1.2 and 1.3.

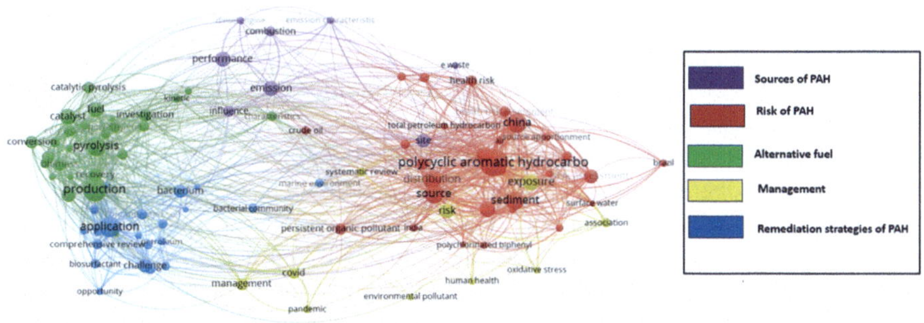

Fig. 1.2 Bibliographic analysis of hydrocarbon waste sources, environmental risk, remediation strategies through VOSviewer

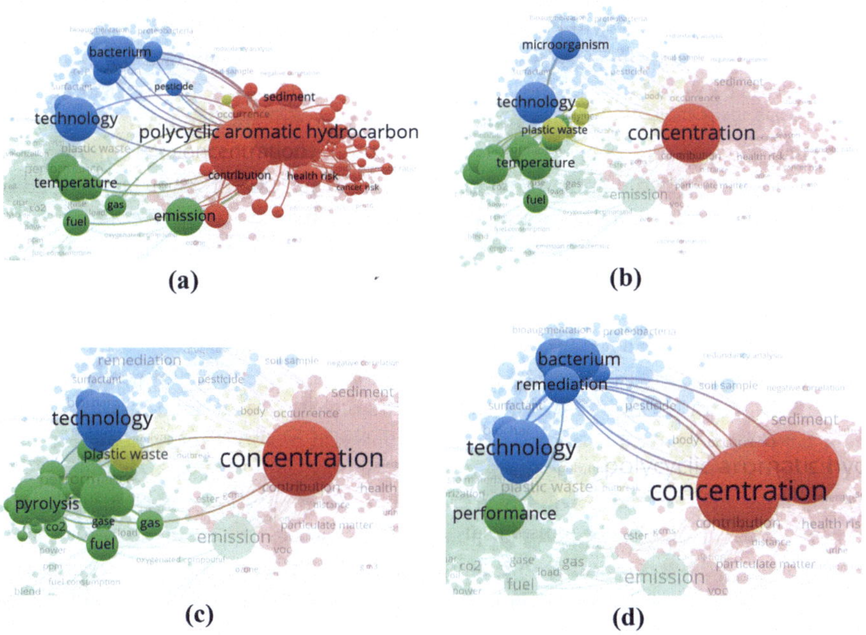

Fig. 1.3 Different clusters of bibliographic analysis on hydrocarbon waste through VOS viewer

1.5 Remediation Strategies

The breakdown of petroleum hydrocarbons is an intricate process influenced by the characteristics, makeup, and concentration of hydrocarbons within the affected environment. It involves a sequence of stages wherein chemical alteration and mineralization of pollutants occur, facilitated by metabolic and enzymatic processes. These processes convert the contaminants into less detrimental and non-perilious substances that become integrated into the biogeochemical cycles. Enzyme-specific biodegradation mechanisms primarily facilitate the degradation of petroleum hydrocarbons (Kuppusamy et al. 2019). Light has emerged as a significant factor in the degradation of petroleum, a departure from earlier estimations where its impact was considered minimal. Recent studies indicate that light-induced degradation can be as influential as biological degradation, particularly in tropical regions (Van Der Heul 2009). Electromagnetic and ultrasound energy can trigger various operations, including volatilization, seepage, desorption, airflow, and microbial activities (Ossai et al. 2022). The chemical processes involved in the degradation of petroleum hydrocarbons include both biotic and abiotic transformations, ultimately resulting in mineralization (Aeppli et al. 2014). Abiotic chemical transformations encompass reactions not triggered by microbes, which reduce pollutant content by breaking down chemicals into simpler or different products through processes such as hydrolysis and oxidation–reduction. Biotic chemical transformation comprises biodegradation and biotransformation. Chemical processes are influenced by both chemical transformation and biodegradation working in tandem. The chosen remediation strategy for a polluted area is tailored to the site's unique characteristics, taking into account factors such as the type and concentration of contaminants, the environmental conditions, and the existing microbial ecosystem. Remediation in a water environment involves addressing contamination in both surface water and groundwater caused by pollutants. Soil remediation, on the other hand, includes addressing contamination in topsoil, subsoil, and sediment. The feasibility of a biological treatment method is heavily influenced by various factors, including the limiting factors and the location of the contaminants. Furthermore, the necessity for remediation depends on whether the contaminated soil, sediments, surface water, and groundwater remain undisturbed in the environment or require removal, excavation, and transport to an offsite facility for treatment. If treatment occurs on-site, it is termed in-situ; if treatment happens off-site, it is termed ex-situ. Biological methods for remediation have shown significant effectiveness for both in-situ and ex-situ processes (Mohanta et al. 2024). They possess the capability to remediate or break down petroleum hydrocarbons and various organic contaminants, transforming them into simpler, non-toxic compounds, while mitigating any potential long-term adverse effects on the affected environments. Various remediation methods with its efficiency and energy consumption of hydrocarbon waste has been shown in Table 1.1.

1.6 Conclusion

The present paper has shed light on the significant but often overlooked threat posed by hydrocarbon waste to our biosphere. Through an exploration of its sources, characteristics, fate, and environmental impact, it has become evident that hydrocarbon waste presents a multifaceted challenge with far-reaching consequences. From its origins in industrial activities, transportation, and natural processes to its diverse

Table 1.1 Efficiency, energy consumption of remediation methods of hydrocarbon waste

Method	Reaction condition	Efficiency	Waste generation	Energy consumption	References
Thermal incineration Fenton	A fixed-bed reactor was employed, utilizing a catalyst composed of MnO_x supported on γ-Al_2O_3. The operating temperature ranged from 443 to 873 K. In contrast, a rotary kiln incinerator operated at approximately 1273 K	> 99% (40 min) 100%	CO, NO_x	Moderate	Huang et al. (2014) and Luengas et al. (2015)
Biological degradation	–	100%	Acetone, acetaldehyde, propanol	Low	Lu et al. (2010)
Adsorption process	A hybrid system combining activated carbon with photocatalytic oxidation was utilized for treating contaminated material	> 90%	Spent adsorbent	Moderate	Jo and Yang (2009)
Fenton oxidation	The wash water from the ion-exchange resin exhibited a pH range of 3–4, containing CH_2O_2 at a concentration of 1200 mg/L and CFe^{2+} at 300 mg/L. Meanwhile, the groundwater contaminated with TCE had a pH of 5.4 and an ORP of 465 mV. Treatment involved the addition of 10 g/L basic oxygen furnace along with 1.0 g/L of H_2O_2	81% (1 h)	CO_2; Cl^-; Fe^{3+}. CE; DCE	High	Tsai et al. (2010) and Luengas et al. (2015)
Ozonation process Plasma	The scenario involved treating contaminated soil using a flushing solvent, specifically acetic acid. The ozone concentration was maintained at 17 ± 2 mg/L, while the temperature remained at 20 ± 2 °C	100% (2 h)	–	High	Alcántara-Garduño et al. (2008)
Condensation	–	Moderate		High	Luengas et al. (2015)

(continued)

Table 1.1 (continued)

Method	Reaction condition	Efficiency	Waste generation	Energy consumption	References
Plasma catalysis	A corona reactor with hybrid pulsed power, featuring adaptable energy density	74–81%	NO_x, formic acid, O_3, carboxylic acids	High	Hoeben et al. (2012)
Photocatalytic oxidation	The experimental setup included gas conditions with a flow rate of 200 mL/min sustained over a period of 30 min. It featured an infrared cell, UV irradiation using a 150 W Xenon lamp, and a catalyst comprising N-doped TiO_2. These gas conditions were established within an O_2 stream under low-pressure circumstances	100% (5 min)	Strong oxidant $OH^{\bullet}$ radicals	Moderate	Shiraishi and Ishimatsu (2009) and Yokosuka et al. (2009)
Membrane separation	–	–	Clogged membranes	High	Huang et al. (2014) and Luengas et al. (2015)
Catalytic ozonation process (COP) Membrane	Ozone production was initiated through the use of a commercial ozone generator advertised as an air purification device. Particle measurements were documented prior to, during, and following the emission of terpenes from a cleaning product formulated with pine oil	100% (2 h)	Secondary organic aerosols	High	Hubbard et al. (2005)
Coagulation/ flocculation with sedimentation	–	98.8%	–	Moderate	Dąbrowski et al. (2005) and Derwent et al. (1998)
Flocculation with DAF Coagulation	An experimental pot study was conducted to examine the impact of biochar derived from green waste through pyrolysis on both the yield of radish (*Raphanus sativus* var. Long Scarlet) and the soil quality of an Alfisol	90%	–	Moderate	Chan et al. (2007)

(continued)

Table 1.1 (continued)

Method	Reaction condition	Efficiency	Waste generation	Energy consumption	References
Coagulation with foam separation	–	95%	–	High	Feron et al. (1992)
Coagulation/ flocculation with MFC	Evaluating aerobic, heterotrophic marine bacteria to identify their potential for producing volatile organic compounds	94%	–	High	Nemecek-Marshall et al. (1995)
Demulsification with centrifugation	The catalytic combustion of acetaldehyde was explored using different metal catalysts supported on oxides, which were prepared via the impregnation method	–	–	–	Mitsui et al. (2008)
Centrifugation	The ramifications of enhancing a slow pyrolysis-centered bioenergy system to prioritize both Biochar and energy production over solely focusing on energy production are significant for greenhouse gas emissions	68.83%	–	–	Gaunt and Lehmann (2008)
Coalescence on granular bed	–	71–99%	–	–	Agnihotri et al. (2007), Ding et al. (2016), and Lehmann (2007)
Membrane separation	Activated carbon as adsorbate	97–99%	–	–	Li et al. (2011), Sun et al. (2011), and Tang et al. (2007)
Electrocoagulation	Adsorbed the phenol and its derivative	99%	–	–	Dąbrowski et al. (2005) and Patnaik (2006)

traits ranging from toxicity to persistence, hydrocarbon waste poses a serious risk to ecosystems and human health alike.

The fate of hydrocarbon waste, whether it be through atmospheric dispersion, soil contamination, or water pollution, underscores the urgency of addressing this issue. Its toxic effects on the biosphere, including disruptions to ecological balance, loss of biodiversity, and adverse health impacts on flora and fauna, emphasize the critical need for mitigation strategies.

However, amidst these challenges, there is hope. Remediation strategies, ranging from biological, chemical, to physical methods, offer promising avenues for mitigating the impacts of hydrocarbon waste. These strategies, when tailored to specific environmental contexts and coupled with rigorous monitoring and enforcement, have shown remarkable efficacy in restoring contaminated sites and safeguarding the integrity of our biosphere.

Addressing the threat of hydrocarbon waste requires concerted efforts from governments, industries, researchers, and communities worldwide. By understanding its origins, traits, fate, and environmental impact, and by implementing effective remediation strategies, we can work towards a sustainable future where the unseen threat of hydrocarbon waste no longer looms over our biosphere.

Acknowledgements This work was supported by the Department of Chemical Engineering, National Institute of Technology Durgapur, West Bengal, India.

References

Aeppli C, Nelson RK, Radović JR, Carmichael CA, Valentine DL, Reddy CM (2014) Recalcitrance and degradation of petroleum biomarkers upon abiotic and biotic natural weathering of deepwater horizon oil. Environ Sci Technol 48(12):6726–6734. https://doi.org/10.1021/ES500825Q/SUPPL_FILE/ES500825Q_SI_001.PDF

Agnihotri S, Zheng Y, Mota JPB, Ivanov I, Kim P (2007) Practical modeling of heterogeneous bundles of single-walled carbon nanotubes for adsorption applications: estimating the fraction of open-ended nanotubes in samples. J Phys Chem C 111(37):13747–13755. https://doi.org/10.1021/JP074183O

Al-Mebayedh H, Niu A, Lin C (2022) Petroleum hydrocarbon composition of oily sludge and contaminated soils in a decommissioned oilfield waste pit under desert conditions. Appl Sci 12(3):1355. https://doi.org/10.3390/APP12031355

Alcántara-Garduño ME, Okuda T, Nishijima W, Okada M (2008) Ozonation of trichloroethylene in acetic acid solution with soluble and solid humic acid. J Hazard Mater 160(2–3):662–667. https://doi.org/10.1016/J.JHAZMAT.2008.03.106

Ali MH, Khan MI, Naveed M, Tanvir MA (2023) Microbe-assisted rhizoremediation of hydrocarbons and growth promotion of chickpea plants in petroleum hydrocarbons-contaminated soil. Sustainability (Switzerland) 15(7). https://doi.org/10.3390/su15076081

Bala K, Anshul, Verma S, Dogra A, Das AP, Kumar R, Nandini (2024) Sustainable management for urban plastic waste generation. In: Sustainable Management of Urban Plastic Waste Through Circular Economic Approaches, pp 211–236. https://doi.org/10.1201/9781003465447-13

Bala K, Dogra A, Lata P, Das AP (2024) Discussing sources and biological-chemical remediation approaches for microplastic pollution. In: Sustainable Microbial Technology for Synthetic and

Cellulosic Microfiber Bioremediation, pp 37–59. https://doi.org/10.1007/978-3-031-62600-5_3

Behera ID, Basak G, Kumar RR, Sen R, Meikap BC (2020) Treatment of petroleum refinery sludge by petroleum degrading bacterium Stenotrophomonas pavanii IRB19 as an efficient novel technology. J Environ Sci Health Part A Toxic/Hazard Substances Environ Eng 56(2). https://doi.org/10.1080/10934529.2020.1866924

Behera ID, Meikap BC, Sen R (2022) Performance evaluation of bacterial consortium for biodegradation of total petroleum hydrocarbon: a comparative strategic biostimulation study. Research Square

Behera ID, Nayak M, Mishra A, Meikap BC, Sen R (2022b) Strategic implementation of integrated bioaugmentation and biostimulation for efficient mitigation of petroleum hydrocarbon pollutants from terrestrial and aquatic environment. Mar Pollut Bull 177:113492. https://doi.org/10.1016/j.marpolbul.2022.113492

Behera ID, Das AP (2023) Impact of petroleum waste on environmental pollution and its sustainable management through circular economy. Springer Nature Switzerland, Imprint: Springer. https://doi.org/10.1007/978-3-031-48220-5

Carreón T, Herrick RL (2012) Aliphatic hydrocarbons. Patty's toxicology, pp 1–102. https://doi.org/10.1002/0471435139.TOX049.PUB2

Cataldo S, Muratore N, Giannici F, Bongiorno D, Chiodo V, Maisano S, Pettignano A (2022) Hydrocarbons removal from synthetic bilge water by adsorption onto biochars of dead *Posidonia oceanica*. Environ Sci Pollut Res 29(60):90231–90247. https://doi.org/10.1007/s11356-022-21998-x

Chan KY, Van Zwieten L, Meszaros I, Downie A, Joseph S (2007) Agronomic values of greenwaste biochar as a soil amendment. Soil Res 45(8):629–634. https://doi.org/10.1071/SR07109

Chemical risks in petrochemical industry—PREVOR (n.d.)

Chevalier JLE, Petrino PJ, Gaston-Bonhomme YH (1990) Viscosity and density of some aliphatic, cyclic, and aromatic hydrocarbons binary liquid mixtures. J Chem Eng Data 35(2):206–212. https://doi.org/10.1021/JE00060A034/ASSET/JE00060A034.FP.PNG_V03

Dąbrowski A, Podkościelny P, Hubicki Z, Barczak M (2005) Adsorption of phenolic compounds by activated carbon—a critical review. Chemosphere 58(8):1049–1070. https://doi.org/10.1016/J.CHEMOSPHERE.2004.09.067

Das AP, Nayak B, Behera ID (2024) Applications of microorganisms in advanced geomicrobiological and biotechnological exploration. Geomicrobiol J 41(4):309–311. https://doi.org/10.1080/01490451.2024.2327059

Das A, Dey S, Bala K, Gnanasekaran R, Das AP (2024) Microbial bioremediation technology for sustainable treatment and management of synthetic microfiber waste. In: Sustainable Microbial Technology for Synthetic and Cellulosic Microfiber Bioremediation, pp 221–238. https://doi.org/10.1007/978-3-031-62600-5_12

Das A, Dey S, Das AP (2024) Ecotoxicological impacts of synthetic microfiber pollutants and development of sustainable mitigation strategies. Environ Chem Ecotoxicol. https://doi.org/10.1016/J.ENCECO.2024.12.003

Derwent RG, Jenkin ME, Saunders SM, Pilling MJ (1998) Photochemical ozone creation potentials for organic compounds in northwest Europe calculated with a master chemical mechanism. Atmos Environ 32(14–15):2429–2441. https://doi.org/10.1016/S1352-2310(98)00053-3

Dewulf J, Van Langenhove H (2003) Hydrocarbons in the atmosphere. Encyclopedia of Life Support Systems (EOLLS): environmental and ecological chemistry

Dey S, Bissoyi A, Koteswara Reddy G, Das AP (2024) Synthetic microfibers: an urban emerging pollutant demanding for sustainable environmental solutions. In: Sustainable Microbial Technology for Synthetic and Cellulosic Microfiber Bioremediation, pp 107–121. https://doi.org/10.1007/978-3-031-62600-5_7

Dey S, Dash N, Ray SK, Priyadarsini S., Chhotaray, C., & Das, A. P. (2024). Occurrence and Quantification of Tire Wear Plastics Pollutants from Highway City Dust of Bhubaneswar

City of Odisha. In: Sustainable Microbial Technology for Synthetic and Cellulosic Microfiber Bioremediation, pp 77–89. https://doi.org/10.1007/978-3-031-62600-5_5

Dey S, Das A, Mallick K, Sahu A, Das AP (2023) Environmental petroleum waste: pollution, toxicity, sustainable remediation. In: Behera ID, Das, AP (eds) Impact of petroleum waste on environmental pollution and its sustainable management through circular economy. Environmental science and engineering. Springer, Cham. https://doi.org/10.1007/978-3-031-48220-5_7

Dey S, Mehta NS (2020) Automobile pollution control using catalysis. Resour Environ Sustain 2. https://doi.org/10.1016/J.RESENV.2020.100006

Dey S, Tripathy B, Kumar MS, Das AP (2023) Ecotoxicological consequences of manganese mining pollutants and their biological remediation. Environ Chem Ecotoxicol. https://doi.org/10.1016/J.ENCECO.2023.01.001

Ding Z, Wan Y, Hu X, Wang S, Zimmerman AR, Gao B (2016) Sorption of lead and methylene blue onto hickory biochars from different pyrolysis temperatures: importance of physicochemical properties. J Ind Eng Chem 37:261–267. https://doi.org/10.1016/J.JIEC.2016.03.035

El Fels L, Lemee L, Ambles A, Hafidi M (2016) Identification and biotransformation of aliphatic hydrocarbons during co-composting of sewage sludge-Date Palm waste using Pyrolysis-GC/MS technique. Environ Sci Pollut Res Int 23(16):16857–16864. https://doi.org/10.1007/S11356-016-6670-9

Fakhru'l-Razi A, Pendashteh A, Abdullah LC, Biak DRA, Madaeni SS, Abidin ZZ (2009) Review of technologies for oil and gas produced water treatment. J Hazard Mater 170(2–3):530–551. https://doi.org/10.1016/j.jhazmat.2009.05.044

Farhan Ul Haque M, Hernández M, Crombie AT, Murrell JC (2022) Identification of active gaseous-alkane degraders at natural gas seeps. ISME J 16(7):1705–1716. https://doi.org/10.1038/s41396-022-01211-0

Feron VJ, Til HP, de Vrijer F, van Bladeren PJ (1992) Toxicology of volatile organic compounds in indoor air and strategy for further research. Indoor Environ 1(2):69–81. https://doi.org/10.1159/000463415

Galil NI, Wolf D (2001) Removal of hydrocarbons from petrochemical wastewater by dissolved air flotation. Water Sci Technol 43(8):107–113. https://doi.org/10.2166/wst.2001.0476

Gaunt JL, Lehmann J (2008) Energy balance and emissions associated with biochar sequestration and pyrolysis bioenergy production. Environ Sci Technol 42(11):4152–4158. https://doi.org/10.1021/ES071361I/SUPPL_FILE/ES071361I-FILE003.PDF

Grabato JRH, Federico SAP, Hizon-Fradejas AB, Mojica E-RE (2024) Pentane. Encyclopedia of toxicology, pp 325–329. https://doi.org/10.1016/B978-0-12-824315-2.00995-7

Hanke CG, Johansson A, Harper JB, Lynden-Bell RM (2003) Why are aromatic compounds more soluble than aliphatic compounds in dimethylimidazolium ionic liquids? A simulation study. Chem Phys Lett 374(1–2):85–90. https://doi.org/10.1016/S0009-2614(03)00703-6

Hoeben WFLM, Beckers FJCM, Pemen AJM, Van Heesch EJM, Kling WL (2012) Oxidative degradation of toluene and limonene in air by pulsed corona technology. J Phys D Appl Phys 45(5):055202. https://doi.org/10.1088/0022-3727/45/5/055202

Hoek EMV, Wang J, Hancock TD, Edalat A, Bhattacharjee S, Jassby D (2021) Oil & gas produced water management. https://doi.org/10.1007/978-3-031-79504-6

Huang B, Lei C, Wei C, Zeng G (2014) Chlorinated volatile organic compounds (Cl-VOCs) in environment—sources, potential human health impacts, and current remediation technologies. Environ Int 71:118–138. https://doi.org/10.1016/J.ENVINT.2014.06.013

Hubbard HF, Coleman BK, Sarwar G, Corsi RL (2005) Effects of an ozone-generating air purifier on indoor secondary particles in three residential dwellings. Indoor Air 15(6):432–444. https://doi.org/10.1111/J.1600-0668.2005.00388.X

Ifegwu OC, Anyakora C (2015) Polycyclic aromatic hydrocarbons: Part I. Exposure. Adv Clin Chem 72:277–304. https://doi.org/10.1016/BS.ACC.2015.08.001

John A, Das AP (2024) Government strategies for achieving circular economy for plastic waste management in Southeast Asia. In: Sustainable Management of Urban Plastic Waste

Through Circular Economic Approaches, 72–82. https://doi.org/10.1201/9781003465447-6/GOVERNMENT-STRATEGIES-ACHIEVING-CIRCULAR-ECONOMY-PLASTICWASTE-MANAGEMENT-SOUTHEAST-ASIA-AKBAR-JOHN-ALOK-PRASAD-DAS

Jo WK, Yang CH (2009) Granular-activated carbon adsorption followed by annular-type photocatalytic system for control of indoor aromatic compounds. Sep Purif Technol 66(3):438–442. https://doi.org/10.1016/J.SEPPUR.2009.02.014

Kanungo J, Sahoo T, Swain LP, & Behera ID (2023) Toxicity of persistent hydrocarbon pollutants, sources and sustainable remediation process. In: Behera ID, Das AP (eds) Impact of petroleum waste on environmental pollution and its sustainable management through circular economy. Springer, pp 39–65. https://doi.org/10.1007/978-3-031-48220-5_2

Kuppusamy S, Maddela NR, Megharaj M, Venkateswarlu K (2019) Total petroleum hydrocarbons: environmental fate, toxicity, and remediation. In: Total petroleum hydrocarbons: environmental fate, toxicity, and remediation. Springer. https://doi.org/10.1007/978-3-030-24035-6/COVER

Lahiri D, Nag M, Dey A, Sarkar T, Joshi S, Pandit S, Das AP, Pati S, Pattanaik S, Tilak VK, Ray RR (2021) Biofilm mediated degradation of petroleum products. Geomicrobiol J 1–10. https://doi.org/10.1080/01490451.2021.1968979

Lazar I, Dobrota S, Voicu A, Stefanescu M, Sandulescu L, Petrisor IG (1999) Microbial degradation of waste hydrocarbons in oily sludge from some Romanian oil fields. J Petrol Sci Eng 22(1–3):151–160. https://doi.org/10.1016/S0920-4105(98)00064-3

Lehmann J (2007) A handful of carbon. Nature 447(7141):143–144. https://doi.org/10.1038/447143a

Li L, Liu S, Liu J (2011) Surface modification of coconut shell based activated carbon for the improvement of hydrophobic VOC removal. J Hazard Mater 192(2):683–690. https://doi.org/10.1016/J.JHAZMAT.2011.05.069

Lu Y, Liu J, Lu B, Jiang A, Wan C (2010) Study on the removal of indoor VOCs using biotechnology. J Hazard Mater 182(1–3):204–209. https://doi.org/10.1016/J.JHAZMAT.2010.06.016

Luengas A, Barona A, Hort C, Gallastegui G, Platel V, Elias A (2015) A review of indoor air treatment technologies. Rev Environ Sci Biotechnol 14(3):499–522. https://doi.org/10.1007/S11157-015-9363-9/METRICS

Manisalidis I, Stavropoulou E, Stavropoulos A, Bezirtzoglou E (2020) Environmental and health impacts of air pollution: a review. Front Public Health 8. https://doi.org/10.3389/fpubh.2020.00014

Mishra S, Dash D, Das AP (2024) Microbial degradation of polyester microfibers using indigenously isolated bacterial strain Exiguobacterium Sp. CLEAN – Soil, Air, Water, 52(12):e202300343. https://doi.org/10.1002/CLEN.202300343

Mishra A, Siddiqi H, Kumari U, Behera ID, Mukherjee S, Meikap BC (2021) Pyrolysis of waste lubricating oil/waste motor oil to generate high-grade fuel oil: a comprehensive review. Renew Sustain Energy Rev 150:111446. https://doi.org/10.1016/j.rser.2021.111446

Mitsui T, Tsutsui K, Matsui T, Kikuchi R, Eguchi K (2008) Catalytic abatement of acetaldehyde over oxide-supported precious metal catalysts. Appl Catal B 78(1–2):158–165. https://doi.org/10.1016/J.APCATB.2007.09.017

Mohammadi L, Rahdar A, Bazrafshan E, Dahmardeh H, Susan MABH, Kyzas GZ (2020) Petroleum hydrocarbon removal fromwastewaters: a review. Processes 8(4). https://doi.org/10.3390/PR8040447

Mohanraj R, Azeez PA (2003) Polycyclic aromatic hydrocarbons in air and their toxic potency. Resonance 8(9):20–27. https://doi.org/10.1007/BF02837918

Mohanta S, Pradhan B, Behera ID (2024) Impact and remediation of petroleum hydrocarbon pollutants on agricultural land: a review. Geomicrobiol J 41(4):345–359. https://doi.org/10.1080/01490451.2023.2243925

Mohanty M, Tripathy B, Koteswara Reddy G, Das AP (2024) Greener technologies for establishing restraint over microplastic pollution. In: Sustainable Microbial Technology for Synthetic and Cellulosic Microfiber Bioremediation, pp 239–258. https://doi.org/10.1007/978-3-031-62600-5_13

Nemecek-Marshall M, Wojciechowski C, Kuzma J, Silver GM, Fall R (1995) Marine Vibrio species produce the volatile organic compound acetone. Appl Environ Microbiol 61(1):44–47. https://doi.org/10.1128/AEM.61.1.44-47.1995

Nzila A (2018) Current status of the degradation of aliphatic and aromatic petroleum hydrocarbons by thermophilic microbes and future perspectives. Int J Environ Res Public Health 15(12). https://doi.org/10.3390/IJERPH15122782

Onwuka OS, Igwe O, Ikenna S, Chinenyem I, Uwom S, Ifediegwu SI, Uwom CS (2018) An assessment of the effectiveness of drilling waste treatment process in X-gas field, Niger Delta, Nigeria. Geol Ecol Landsc 2(4):288–302. https://doi.org/10.1080/24749508.2018.1473751

Ossai IC, Hamid FS, Hassan A, Ossai IC, Hamid FS, Hassan A (2022) Biological treatments for petroleum hydrocarbon pollutions: the eco-friendly technologies. Hazard Waste Manag. https://doi.org/10.5772/INTECHOPEN.102053

Patnaik P (2006) A comprehensive guide to the hazardous properties of chemical substances: third edition, pp 1–1059. https://doi.org/10.1002/9780470134955

Paul A, Priyadarsini S, Karakuş S, Behera JK, Gnanasekaran R, Das AP (2024) Fate and accumulation of synthetic microfibers in the environment. In: Sustainable Microbial Technology for Synthetic and Cellulosic Microfiber Bioremediation, pp 61–76. https://doi.org/10.1007/978-3-031-62600-5_4

Priyadarsini S, Das AP (2024) Analytical and structural characterization of waste lithium-ion batteries for their effective recycling strategy. Environ Chem Ecotoxicol. https://doi.org/10.1016/j.enceco.2024.12.004

Priyadarsini S, Das AP (2025) Lithium bioleaching: A review on microbial-assisted sustainable technology for lithium biocircularity. J Water Process Eng, 69:106744. https://doi.org/10.1016/j.jwpe.2024.106744

Ray SK, Dey S, Das AP, John A (2024) Microplastic waste management through greener technologies: A review. In: Sustainable Management of Urban Plastic Waste Through Circular Economic Approaches, pp 237–261. https://doi.org/10.1201/9781003465447-14

Reeves G (n.d.) Understanding and monitoring hydrocarbons in water

SCREENING-LEVEL HAZARD CHARACTERIZATION Reclaimed Petroleum Hydrocarbons Category: Residual Hydrocarbon Wastes from Petroleum Refining (2011)

Shiraishi F, Ishimatsu T (2009) Toluene removal from indoor air using a miniaturized photocatalytic air purifier including a preceding adsorption/desorption unit. Chem Eng Sci 64(10):2466–2472. https://doi.org/10.1016/J.CES.2009.02.024

Speight JG (2017) Organic chemistry. Environmental organic chemistry for engineers, pp 43–86. https://doi.org/10.1016/B978-0-12-804492-6.00002-2

Srivastava M, Srivastava A, Yadav A, Rawat V (2019) Source and control of hydrocarbon pollution. In: Ince M, Ince OK (eds) Hydrocarbon pollution and its effect on the environment. IntechOpen. https://doi.org/10.5772/intechopen.86487

Sun K, Keiluweit M, Kleber M, Pan Z, Xing B (2011) Sorption of fluorinated herbicides to plant biomass-derived biochars as a function of molecular structure. Biores Technol 102(21):9897–9903. https://doi.org/10.1016/J.BIORTECH.2011.08.036

Tang D, Zheng Z, Lin K, Luan J, Zhang J (2007) Adsorption of p-nitrophenol from aqueous solutions onto activated carbon fiber. J Hazard Mater 143(1–2):49–56. https://doi.org/10.1016/J.JHAZMAT.2006.08.066

Tripathy B, Dash A, Das AP (2022) Detection of environmental microfiber pollutants through vibrational spectroscopic techniques: recent advances of environmental monitoring and future prospects. Critical Rev Anal Chem, 1–11. https://doi.org/10.1080/10408347.2022.2144994

Tsai TT, Sah J, Kao CM (2010) Application of iron electrode corrosion enhanced electrokinetic-Fenton oxidation to remediate diesel contaminated soils: a laboratory feasibility study. J Hydrol 380(1–2):4–13. https://doi.org/10.1016/J.JHYDROL.2009.09.010

Van Der Heul RM (2009) Environmental Degradation of petroleum hydrocarbons.

von Oettingen WF (1942) The toxicity and potential dangers of aliphatic and aromatic hydrocarbons. Yale J Biol Med 15(2):167

Wang L, Cheng Y, Naidu R, Bowman M (2021) The key factors for the fate and transport of petroleum hydrocarbons in soil with related in/ex situ measurement methods: an overview. Front Environ Sci 9:756404. https://doi.org/10.3389/FENVS.2021.756404/BIBTEX

Wang Q, Miyake Y, Tokumura M, Amagai T, Horii Y, Nojiri K, Ohtsuka N (2018) Effects of characteristics of waste incinerator on emission rate of halogenated polycyclic aromatic hydrocarbon into environments. Sci Total Environ 625:633–639. https://doi.org/10.1016/J.SCITOTENV.2017.12.323

Yokosuka Y, Oki K, Nishikiori H, Tatsumi Y, Tanaka N, Fujii T (2009) Photocatalytic degradation of trichloroethylene using N-doped TiO_2 prepared by a simple sol-gel process. Res Chem Intermed 35(1):43–53. https://doi.org/10.1007/S11164-008-0019-Z/METRICS

Zuzolo D, Guarino C, Tartaglia M, Sciarrillo R (2021) Plant-soil-microbiota combination for the removal of total petroleum hydrocarbons (TPH): an in-field experiment. Front Microbiol 11. https://doi.org/10.3389/fmicb.2020.621581

Chapter 2
Environmental Hydrocarbon Pollutants: Sources, Transport, and Effect on Human Health

Madhusmita Mohanty, Jyotirmoyee Mohanty, Sailaja Priyadarsini, Joan Manuel Rodríguez-Díaz, Shalini Yadav, Marina M. S. Cabral-Pinto, Harjeet Nath, Deo Karan Ram, G. Koteswara Reddy, and Alok Prasad Das

Abstract Expansive organic compounds containing hydrogen and carbon atoms liberated into the environment as a result of anthropogenic activity or genuine processes such as industrial processes, vehicular emissions, oil spills, and natural seepage are called environmental hydrocarbon pollutants. Following release, hydrocarbon pollutants undergo a sequence of biological, physical, and chemical alterations. These processes intricately shape their persistence and distribution across soil, water, and air environments. Factors such as advection, oxidation, and microbial degradation contribute to their dynamic behaviour, influencing their long-term

M. Mohanty · J. Mohanty · S. Priyadarsini · A. P. Das (✉)
Department of Life Sciences, Rama Devi Women's University, Bhubaneswar, Odisha, India
e-mail: alok1503@gmail.com

J. M. Rodríguez-Díaz
Técnica de Manabí, S/N, Avenida Urbina y Che Guevara, Portoviejo 130104, Ecuador
e-mail: joan.rodriguez@utm.edu.ec

S. Yadav
Department of Civil Engineering, Rabindranath Tagore University, Raisen 464993, India

M. M. S. Cabral-Pinto
Department of Geoscience, Geobiotec Research Centre, University of Aveiro, 3810-193 Aveiro, Portugal
e-mail: marinacp@ua.pt

H. Nath
Department of Chemical and Polymer Engineering, Tripura University (A Central University), Agartala 799022, India

D. K. Ram
Department of Chemical, Petroleum & Hydrogen Technology, Nims Institute of Engineering & Technology (NIET), Nims University, Jaipur, Rajasthan 303121, India

G. Koteswara Reddy
Department of Biotechnology, Koneru Lakshmaiah Education Foundation (Deemed to be University), Green Fields, Vaddeswaram 522502, India

I. D. Behera and A. P. Das (eds.), *Environmental Hydrocarbon Pollution and Zero Waste Approach Towards a Sustainable Waste Management*, Environmental Science and Engineering, https://doi.org/10.1007/978-3-031-81615-4_2

presence in ecosystems. Hydrocarbon pollutants apply a significant hazard to environmental quality and human wellbeing because of their widespread occurrence and persistence in various ecosystems. Human health is at threat by hydrocarbon pollution in the environment when contaminated air, water, or food is consumed or directly exposed to it causing, can cause cancer, heart disease, respiratory issues, neurological diseases, liver damage, and gastrointestinal ailments. Human exposure to these pollutants can also be increased by consuming polluted seafood and agricultural goods due to the bioaccumulation of hydrocarbons in food chains. Moreover, long-term exposure to certain hydrocarbons, such as polycyclic aromatic hydrocarbons (PAHs) and volatile organic compounds (VOCs), have been linked to carcinogenicity and other chronic health effects. This review seeks to clarify the various origins and methods of transportation for hydrocarbon pollutants, including both natural occurrences and human activities. Furthermore, it provides detailed perception towards the consequences of different hydrocarbons on physiology of human.

Keywords Hydrocarbon · Human health · Environmental pollutants · Toxicity

2.1 Introduction

Environmental pollutants are substances like microplastics (Mallick et al. 2023; Tripathy et al. 2022, 2024), heavy metals (Dey et al. 2023; Pradhan et al. 2023a; Raj and Das 2023), electronic waste (Mishra et al. 2023; Prabhakar et al. 2019) and biomedical waste (Khatun et al. 2023) (Sadasivuni et al. 2019) are introduced into the environment which cause adversative effects on living organisms and ecosystems (Mishra et al. 2019; Priyadarsini and Das 2025). Microplastics and heavy metals significantly harm ecosystems and human health by accumulating in soil, water, and living organisms, thereby interfering with natural processes and food chains (Paul et al. 2023; Priyadarsini and Das 2024; Sahoo et al. 2022). Microplastics, which are generated by the decomposition of bigger plastic components, accumulate in soil and water, entering the food chain and threatening aquatic species and human health (Mishra et al. 2021; Mishra et al. 2022b; Sahu and Das 2024; Tripathy et al. 2024). Heavy metals produced by mining, industrial operations, and poor waste management persist in the environment, polluting water and soil and producing hazardous effects on living species (Das and Ghosh 2023; Das and Mishra 2010; Das et al. 2012; Ghosh and Das 2018; Mishra et al. 2022). Petroleum pollution, which is commonly caused by oil spills and leaks, has a tremendous impact on marine and terrestrial ecosystems, destroying habitats and lowering biodiversity. Addressing these interconnected concerns necessitates comprehensive policies that include stronger laws, sophisticated cleanup technology, and the adoption of sustainable alternatives to reduce environmental and health risks (Das AP et al. 2024b; Mohanty et al. 2024a; Raj and Das 2023).The main cause of the worldwide distribution of hydrocarbons is long-term human pollution (Behera et al. 2020; Kanungo et al. 2024; Lahiri et al. 2021). They are resistant to environmental change and extremely persistent because of their innate features, including hydrophobicity, thermostability, and heterocyclic aromatic ring structures. It has been set up that hydrocarbon pollutants are extremely toxic, immune toxicogenic, teratogenic, carcinogenic, and mutagenic to an extensive number of biological entities (Behera et al. 2021; Das AP et al. 2024a; Patel et al. 2020). Expanding anthropogenic activities triggered by faster urbanization and industrialization release many kinds of environmental pollutants, including PAHs (Das and Mishra 2008; Mojiri et al. 2019). A wide category of varied organic compounds is known as PAHs who have two or more than two cyclic rings. Complex mixtures including hundreds of compounds are typically where PAH are found. Following recent studies, organic pollutants like PAHs are found in a number of aquatic systems, also taking into consideration of the surface waters, groundwater, sewage treatment plant effluent and influent, and seawater (Grandclément et al. 2017). Almost more than 400 different types of PAHs and their effects have been recognized (Pan et al. 2006).

Petroleum exposition influences animals in several direct and indirect ways. During the refining and processing of petroleum, certain by-products arise that are then used to make other highly toxic products (Bal et al. 2020). Unintentional releases of dangerous compounds into the environment occur routinely. If the effects of these releases are related to the worldwide impacts of unintentional crude oil spills, then the combined sources of unrestricted hydrocarbons are the main cause of pollution in the environment. Petroleum products contain a vast number of hydrocarbons, but just a tiny percentage of these compounds have been thoroughly studied for toxicity (Bala, Anshul et al. 2024; Bala, Dogra et al. 2024; Das, Dey, & Das 2024; Das, Dey, Bala et al. 2024). The impact on the environment of petroleum hydrocarbon molecules varies depending on their molecular weight and boiling point, which has a broad distribution.

The pollutants known as PAHs are universal and can be found in aquatic and terrestrial ecosystems, also including to the air (Adeniji et al. 2019). The increased hydrophobicity and poor solubility of PAHs have been observed to hasten their deposition in soil and sediment. Once firmly attached to earth elements, the earth ecosystem serves as the terminal repository for PAHs (Kuppusamy et al. 2017). Both directly and indirectly, PAH pollution has major impacts on health of human and welfare along with the health of other living things on the planet (García-Sánchez et al. 2018; Mohanta et al. 2024). In 1983, the United States Environmental Protection Agency (USEPA) selected sixteen PAH compounds as priority pollutants due to their toxicity, higher concentrations, and elevated exposure (Zheng et al. 2018).

In water, soil, and plants, atmospheric PAHs (aerosols in the gaseous phase) get stuck as particulate matter through dry and wet deposition processes. There are mostly two types of sources of hydrocarbon pollution: sources of natural emissions and sources of anthropogenic emissions. Wild fires, volcanic eruptions, and moorland fires caused by lighting are a few instances of insignificant or less significant natural emission sources (Abdel-Shafy and Mansour 2016).

Manmade sources are the primary generators of PAH pollution, distributed into four main groups: manufacturing, mobile, residential, and farmland emissions. Industrial emissions include electric arc furnaces, coal gasification, oxygen furnaces, and the engines of heavy machinery, such as diesel and gasoline-powered engines. Emissions from agriculture are caused by practices such as burning biomass outdoors and incinerating agricultural waste under improper conditions (Ravindra et al. 2008).

Partially ignited waste, aluminium, steel and iron, coal-tar pitch, cement, colorants, asphalt, production of insecticides and fungicides, rubber tire manufacturing, exhaust from refineries, and generation of electricity are among the many processes of industries that release PAHs into the atmosphere (Dey, Bissoyi et al. 2024; Dey, Dash et al. 2024; Dey et al. 2023; John & Das 2024; Mishra et al. 2024). The main reason these emissions occur is incomplete combustion. Mobile emissions can come from several sources, including the exhaust from ships, aircraft railroads, and off-road heavy-duty and light-duty vehicles (Srogi 2007).

Among the home emission sources are kerosene/wood stoves, cooking on oil or gas burners, burning wood, burning trash, and other home heating techniques (Johnsen and Karlson 2007). In comparison with urban areas, where sources include industrial, mobile, and housing sources, rural areas with higher levels of PAH pollution are mainly driven by domestic and agricultural sources. When food undergoes processing at home or in an industrial setting, such as by grilling, frying, toasting, drying, baking, or barbecuing, PAHs are produced (Rose et al. 2015). Based on a report by the German Environment Agency, PAHs exist in products that people use daily, including numerous sporting goods, bathing shoes, mouse pads, bicycle handles, and toys (Patel et al. 2020).

The variables such as the method, length, and dose of exposure play a key role when deciding the threat PAHs pose to humans. Zheng et al. (2018) discovered that the risk levels of PAHs varied depending on the source. The researchers additionally computed the incremental lifetime cancer risk (ILCR) for individuals exposed to soil-bound PAHs through three distinct routes of exposure; eating posed the highest risk of cancer (98.2–99.5%), followed by skin contact (0.67–1.85%) and inhalation (0.03–0.04%) (Zheng et al. 2018). Two factors that could influence the harmfulness of PAHs are age and pre-health state. The following are examples of acute health effects: dizziness, nausea, vomiting, diarrhoea, rashes, and skin irritation. Naphthene, benzo(a)pyrene and anthracene, and are direct skin irritants and skin sensitizers for humans and animals alike. High levels of naphthalene ingestion or inhaling might cause red blood cell breakdown (Rengarajan et al. 2015). Long-term impacts on health involve lung malfunctions, diminished immune function, problems with breathing, eye cataracts, liver and kidney damage, and asthma-like symptoms. In mammalian systems, PAHs detoxification primarily occurs in the liver through the synthesis of water-soluble epoxide glutathione conjugates through the interactions between multiple oxidase enzymes and cytochrome P450 during catalysis (Yu 2002). The main causes of PAH pollution are anthropogenic sources, which can be divided into four major categories: emissions from transportation, industry, homes, and farming. Coal gasification, oxygen and electric arc furnaces, and heavy machinery engines including diesel and gasoline-powered engines are examples of industrial sources (Behera et al. 2020; Behera et al. 2022b; Mohanta et al. 2016). Agricultural emissions stem from activities like biomass burning and the incineration of farm waste in unsuitable conditions (Rajpara et al. 2017). There is "conceivable," evidence that taking in a lot of red meat increases the risk of rectal and colon cancer. Increased cooking time and temperature have been linked to an increased risk of colorectal cancer and adenoma in some epidemiologic studies (Knekt et al. 1994). Data from 13 prospective studies suggested that a 100 g/d increase in red meat taking was related to enhanced risks of 12% to 17% (Alexander et al. 2015). The generation of PAHs occurs when flesh is cooked on the grill or barbecue, in addition to smoked and cured meats (Phillips 1999). The remediation of pollutants focuses on employing methods and techniques to eliminate, minimize, or neutralize harmful substances in the environment, aiming to restore ecosystems and safeguard human health and biodiversity (Das et al. 2023a; Mishra et al. 2021; Satapathy et al. 2024; Singh et al. 2020). Bioremediation is a sustainable method that utilizes microorganisms, plants, or enzymes to break down, neutralize, or eliminate pollutants from the environment, helping to rehabilitate. (Bhattacharjee et al. 2021; Das et al. 2023c; Mohanty et al. 2024; Paul et al. 2024; Pradhan et al. 2023b). It is an affordable, eco-friendly method that uses natural processes to break down pollutants without harmful chemicals (Das et al. 2015; Das A et al. 2024c). It targets specific contaminants, minimizes damage to non-target species, and supports sustainability through microorganisms or plants. This approach is effective for various pollutants like petroleum, heavy metals, and pesticides (Bala et al. 2024; Dey et al. 2024b; Mishra and Das 2023; Mishra et al. 2020).

N-hexane, a volatile hydrocarbon, was identified to be the origin of poisoning following an in-depth examination of patient exposure histories, workplace field test outcomes indicating n-hexane levels 1.54–3.3 times higher than the national occupational health limits, and clinical presentation demonstrating peripheral nerve injury. Conduction velocities of motor and sensory nerves were slowed, distal latency was prolonged, and wave amplitude decreased. Both motor and sensory nerves were damaged, with the sensory nerve damage being more severe, according to the findings.

In the upper limbs, percutaneous absorption takes place. The human body converted n-hexane into 2,5-hexadione (HD), which led to significant damage to peripheral neurons. At the Ranvier nodes, neurofilament accumulation, recruitment, gathering, and winding are important pathogenic alterations. Myelin degeneration occurs after significant axonal swelling. According to an alternative explanation, autoxidation-prone pyrrole adducts are produced when HD interacts with NFs. Neurofilament-filled axonal swelling and neurofilament covalent crosslinking are the products of the reaction between the nucleophilic protein substrates and oxidized pyrrole. The harm to the nerves worsened even after the exposure stopped.

2.2 Sources

Pollution originates from a variety of sources, both human-induced and natural. Industrial processes like manufacturing, mining, and chemical production release damaging pollutants into the air, water, and soil (Das and Mishra 2009; Das et al.

2023b; Mishra et al. 2022a). Transportation methods, including cars and ships, emit greenhouse gases and particulate matter, exacerbating air pollution (Dey et al. 2024a; Mishra and Das 2021). Hydrocarbons' stable chemical structure and toxicity make them a serious environmental hazard. Hydrocarbons are major pollutants of soil, water, and air (Behera et al. 2022a). Since they are elements of crude oil and products made from it, total petroleum hydrocarbons, or TPH, can be located in numerous locations. Refineries, the places where they are used as feedstocks (in the manufacturing of polymers, for example), producing facilities for produced gas, locations retail service stations, where hydrocarbons are used as fuel or lubricants are some examples of these locations. Additionally, spills and leaks that happened during transit may be the reason they exist (Pinedo et al. 2012). There are some natural sources of TPHs, even though nearly all human activity such as managed spills, accidents, or unforeseen by products of business, industry, or individual endeavors are significant contributors. This category includes organic matter decomposition and seeps from oil deposits. Hydrocarbons can also be synthesized by some higher plants, though these could be only produced in trace amounts and aren't projected to cause substantial contamination (Pinedo et al. 2013). Coastal refineries, other coastal effluents, sediment, surface and groundwater, marine environments, fallout and atmospheric deposition, riverine discharges, wastewater removal, urban runoff and discharges, tanker accidents at sea, tanker operational discharges, offshore production and transport losses, non-tanker shipping losses, and the combustion or pyrolysis of fossil fuels from sources like cars, power plants, processes of industries, and refuse burning can all contribute to the presence of TPHs in these environments (Aj 2016; Mishra et al. 2021; Mohanty et al., 2024; Paul et al. 2024; Priyadarsini & Das 2024, 2025; Ray et al. 2024; Tripathy et al., 2022).

Urban road dust contains high concentrations of hydrocarbons considering they have strong particle associations (Ellis et al. n.d.). Multivariate data analysis techniques have been used in previous research studies to determine the pollutants sources and their contributions to pollutant mixtures (Hussain et al. 2015).

Carlsson et al. (1999) proved that, depending on the type of engine and lubricant used, the hydrocarbon range C17–C40 is mostly derived from non-combusted engine lubricating oil. According to published research studies, diesel fuel may be associated with C12–C16 hydrocarbons (Draper et al. 1996). Non-combusted diesel fuel > non-combusted lubrication oil > tire and asphalt wear is the relative importance order for the different sources in terms of their contributions to the total hydrocarbon load in urban road dust. Sites where crude oil is treated and transferred, such as oil refineries and tanker terminals, may also become polluted. Additional sources of TPH contamination in the soil include coal gasworks sites and shale oil retorting plants, especially those where "benzole recovery" was employed. Chemicals used at home or at work, as well as certain pesticides that use solvents containing TPH components, are additional potential sources (Langley et al. 2003).

2.3 Transport

The transport of pollutants refers to the movement of harmful substances from their source to various parts of the environment via air, water, or soil (Priyadarshini et al. 2024; Ray et al. 2024). PAHs, enter the body through the inhalation of car exhaust fumes, cigarette smoke, contaminated food, contaminated drinking water, and contaminated air from workplaces. One of the main ways that individuals become susceptible to PAHs is through their diet, which is one of these pathways (Diggs et al. 2011). Consumption of contaminated vegetation and pasture is the cause of the build-up of PAHs in foods of animal origin, particularly livestock. Cooking conditions such as type, temperature, length of cooking time, and fat and oil content can all impact the production of PAHs (Perez et al. 2002). The penetration of contaminants to groundwater, vegetation, and food is triggered by the accretion of PAHs in soil and sediment. For example, PAHs are taken up by plant roots from contaminated soil and moved to different areas of the plant. In the current scenarios, exposure to PAHs cannot be avoided (Burchiel and Luster 2001). Many people are first exposed at work; examples include those who work in the food processing and coke manufacturing industries, or traffic officers who inhale road dust that contains PAHs and exhaust from cars (Lee and Vu n.d.). The methods by which that PAHs interact with other pollutants in the atmosphere are influenced by complex physicochemical reactions, photochemical transformations, dry and wet deposition, and interactions with other pollutants (Delgado-Saborit et al. 2010). Depending on the weather, PAHs in the ambient air can either dissolve into vapor or adsorb into airborne particulate matter (relative humidity, ambient temperature, etc.), the aerosol nature (i.e., origin and qualities), and the characteristics of individual PAH (Lima et al. 2005). While in the Arctic, particulate-phase PAHs are more common in the winter or in general, gas-phase PAH concentrations rise in summer or genuinely in tropical regions (Lai et al. 2011). Fruits and vegetables that grow in soil contaminated with PAHs and air deposition may become contaminated (Zelinkova and Wenzl 2015). The use of industrial drying and roasting techniques, heating techniques during preparation, and atmospheric deposition on fresh plants, the two most popular beverages in the world, coffee and tea, are also polluted with PAHs. Duedahl-Olesen et al. (2015) examined 13 brands of coffee and 18 brands of tea for the presence of PAH4 (chrysene, benzo(a)anthracene, benzofluoranthene, and benzo(a)pyrene). Black had the highest PAH4 levels found. Both in instant coffee (2.3–5.2 mg/kg) and tea (25–115 mg/kg) (Duedahl-Olesen et al. 2015) (Fig. 2.1).

2.4 Effect of Hydrocarbons on Human Health

Several PAHs are immunotoxin, teratogenic, mutagenic, and carcinogenic to living things, involving humans, animals, and microorganisms (Bolden et al. 2017). The length of exposure and the type and quantity of TPHs are just two of the many things

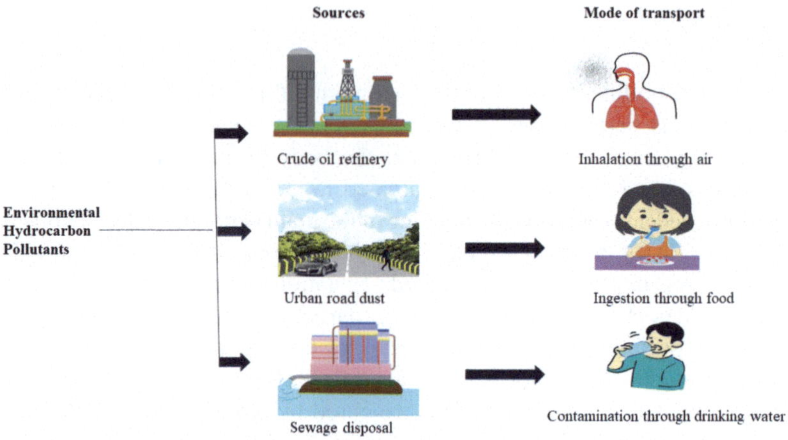

Fig. 2.1 Describe the sources and mode of transport of environmental hydrocarbon pollutants

that affect the health results. Additional variables that need to be taken into consideration include diet, age, sex, family history, way of life, co-exposure to other chemicals, and overall health (Kuppusamy et al. 2019). Table 2.1 describe the different type of hydrocarbon and their impact on the human body.

2.4.1 Effect on the Digestive System

Oral mucositis, gastritis, esophagitis, and disruption of epithelium are among the digestive tract disorders severely harmed by the intake of Petroleum hydrocarbons (PH). There has been evidence from China and Iran (Fagundes et al. 2006) relating the intake of PAHs to oesophageal squamous cell carcinoma, a type of oesophageal cancer. The increased consumption of PAHs through dietary fat results in colorectal cancer (CRC) (Sinha et al. 2005).

2.4.2 Effect on the Respiratory System

The respiratory tract's mucosal membranes can become irritated by gasoline. Still, severe gasoline exposure can cause extensive respiratory tract damage, involving edema, intrapulmonary haemorrhage, pulmonary congestion, and acute exudative tracheobronchitis. Consuming gasoline can cause pulmonary aspiration, which can result in pneumonia. Gasoline and other PH vapours can make the myocardium more sensitive to circulating epinephrine, which could lead to ventricular fibrillation, a potentially lethal state. Two years following the Prestige oil leak, clean-up

Table 2.1 Describe the different type of hydrocarbon and their impact on the human body

Sl. No.	Hydrocarbon	Affected human body part	Impact on the body parts	Reference
1	C9 aromatic hydrocarbon and naphthalene	Lungs	Upper respiratory tract irritation, alveolar/bronchiolar adenoma	Robbins (1951)
2	Diesel hydrocarbon	Digestive tract	Sickness, vomiting, diarrhoea, vexation of the aero-digestive and Gastrointestinal tracts	Tsiaoussis et al. (2019)
3	Diesel hydrocarbon	Eye	Transient pain and/or hyperaemia	Heuss and Glasson (1968)
4	Hydrocarbons like benzene, toluene, ethylbenzene	Nervous system	Neurological issues such as neurasthenia, sleepiness, and slowed sensorimotor speed, chronic toxic encephalopathy (CTE)	Ridgway et al. (2003)
5	Diesel hydrocarbon	Nephron	Acute renal tubular necrosis-related acute renal failure, diarrhea, cramping in the abdomen, nausea, and oliguria that is turning into anuria	Carpenter et al. (1975f, 1975g, 1977b)
6	Hydrocarbon solvent	Skin	Dermal irritation, dryness and defatting of skin	Amoruso et al. (2008)
7	Hydrocarbon solvent	Liver	Hepatocellular hypertrophy	Ennulat et al. (2010)
8	Hydrocarbon	Haematopoietic system	Depletion in haemoglobin amount, bound cell volumes, and/or red blood cell (RBC) counts	Carrillo et al. (2013)
9	1,3,5-trimethylbenzene	Reproductive system	Maternal toxicity and developmental delays of fetus	Saillenfait et al. (2005)
10	Isopropylbenzene (cumene)	Endocrine system	β-hCG the short term and level of progesterone in the long term	Valderrama et al. (2022)

workers had chronic respiratory symptoms and airway damage in their breath condensate (Bosch 2003). The Tasman Spirit oil spill exposed people to a greater number of health effects, such as eye irritation/redness, cough, headache, general illness, nausea, sore throat, runny nose, etc. (Janjua et al. 2006).spill sites are cleaned as part of the clean-up process by "in situ burning" trapped oil on the water column's surface. However, this produces oil sail particulate matter (OSPM), a potentially dangerous compound. OSPM generated superoxide radicles and reactive oxygen species in addition to cytotoxicity in an experimental mouse that was dose- and time-dependent. Reduced systemic oxidative stress, airway inflammation, and body weight gain were observed in mice injected with OSPM. Likewise, in an allergic asthma model of mouse, OSPM was found to be an important factor in increased airway mucus production, peribronchiolar inflammation, and an increase in T helper 2 cells (Th2). These results definitively showed that OSPM has a significant potential to induce inflammation in lungs and modify innate and adaptive immune responses in animals taken for the experiment, endangering the respiratory system during in situ burning oil spill clean-up (Jaligama et al. 2015).

VOCs (xylene, benzene, styrene, and toluene) profiles in patients with, colon, breast, lung and prostate cancer have been demonstrated by Peng et al. The significant results of VOC markers of various kinds of cancer need to be verified in larger clinical studies. Numerous studies investigated whether VOCs could differentiate between other pulmonary diseases and lung cancer (Peng et al. 2010). Malignant pleural mesothelioma (MPM) is an unusual tumour that is mainly caused by asbestos contact. The MPM in a cohort of participants with consistently skilled asbestos presentation could be analysed using VOC profiles. Additionally, cyclohexane was refined by Altomare et al. as an interesting MPM marker (Altomare et al. 2013). When compared to solid controls, Syhre et al. found that patients with pulmonary tuberculosis (TB) had larger magnitudes of breathed-out methyl nicotinate. VOC designs could appropriately differentiate differentiating between TB patients and healthy controls without dynamic TB in a group of patients who were uncertain that they had tuberculosis (Syhre and Chambers 2008).

DEPs have been indicated to boost the making of cytokines of inflammation, involving IL-1, IL-8, and the granulocyte-macrophage colony-stimulating factor and others-stimulating factor derived from bronchial epithelial cells' cyclo-oxygenase stimulation. DEPs have been reported to trigger mast cell degranulation directly, which results in the release of histamine. Exposure to DEPs can cause histamine release, which may cause in allergic rhinosinusitis, conjunctivitis, laryngitis, pharyngitis, and persistent cough.

Exposure to raised concentrations of DEPs greatly compromises macrophages, the primary immune defence system in the lung. This raises the risk of both viral and bacterial bronchitis as well as pneumonia. Many hydrocarbon molecules, which also involve PAH, released by diesel engines are extremely hazardous to the lungs. California declared diesel exhaust to be known to be cancer-causing in 1990. Many lung studies in animals indicate that diesel exhaust particles directly damage DNA and thereby causing carcinogenesis. Studies on the occupational features of truck drivers, there is a significantly higher-than-normal frequency of lung cancer-related deaths among heavy equipment operators and railroad workers. More recent case-control studies on exposure to diesel exhaust at work in Quebec, Montreal, Canada, indicated a minimal connection between lung cancer and exposure both in nonsmokers and smokers. Gustavsson et al. observed that oesophageal cancer was more common in workers who were exposed to combustion products.

2.4.3 Effect on Nervous System

Behaviour and memory may be impacted by low-level petroleum hydrocarbon exposure over a long duration of time almost any hydrocarbon solvent vapor exposure at a relatively high concentration may end up in a broad CNS depression that, at high concentrations, can be fatal. Jet fuel xylene, toluene, and white spirit were inhaled by healthy volunteers for a regulated short period of time (up to 1–2 weeks), during which time their concentration and coordination were found to be impaired at exposure levels over 250, 150, 300, and 200 ppm, respectively. After exposure, these effects are entirely and rapidly reversible (Jones 1990). Peripheral nerves, especially those in the hands and feet, are damaged when melancholy n-hexane and methyl n-butyl ketone are delivered to the brain and spinal cord via petroleum hydrocarbon (PH), which has unsettling effects on sensation and weakens the muscles (Das A et al. 2024b).

If the injury is critical, there may be a loss of motion; healing is usually not severe but it does occur at times. Such injury to the nerves has been seen in shoemakers who use a paste broken up in n-hexane and work in confined areas for extended periods (Wang et al. 2014). Volatile hydrocarbons, like PAH, stick to DEPs and penetrate the lungs rapidly, getting into the central nervous system. There may be a connection between long-term exposure to DEPs and Parkinson's disease because DEPs have been shown to reduce the number of dopaminergic neurons in mouse brain tissue (Block et al. 2004).

A different investigation proved that DEP-induced brain inflammation developed histopathologic shifts that were equivalent to those noticed in Alzheimer's disease patients (Calderón-Garcidueñas et al. 2004).

Defects were also found in tests of perceptual-motor speed, recall memory, problem-solving, and impairments to the visual field, delayed blink reflex latency, and balance (Kilburn 2000).

2.4.4 Effect on Cardiac System

In a recent investigation, Mills and coworkers investigated men who had previously experienced myocardial infarction and who were open to diesel exhaust during moderate activity. Significant ST segment depression and decreased endothelium tissue plasminogen activator release were observed (Mills et al. 2007). In 1816 post-menopausal women exposed to air pollution over an extended period, the most recent research indicated an immediate correlation between the degree of exposure and the risk of disease in circulatory system and death, with diesel exhaust accounting for a significant amount of the exposure (Miller et al. 2007).

According to research conducted in a major European metropolis, traffic exposure and the high concentration of diesel exhaust was linked to the beginning of myocardial infarction within one hour. Peters et al. (2004) found a consistent correlation between increased risk of myocardial infarction to the time spent in traffic. The results in rats by Sakakibara et al. showed that Catalase, superoxide dismutase, atropine, verapamil, diltiazem, diphenhydramine, indomethacin, or propranolol could not stop diesel exhaust particle (DEP)-induced cardiotoxicity. In animal models, diesel exhaust results in variability in heart rate, a notable decrease in left ventricular systolic pressure, an increased left ventricular end-diastolic pressure, and ventricular arrhythmia (Wold et al. 2006). It is predicted that DEP releases superoxide radicals, which harm the heart indefinitely and result in cardiac arrest (Okayama et al. 2006).

Numerous investigations appear to confirm the notion that exposure to diesel exhaust and hypertension are related (Brook et al. 2002). Brief instances of excessive pollution have been linked to transient hypertension, which may be related to how DEPs impact cardiovascular autonomic control Breathing in diluted diesel exhaust may disturb the controlling of vascular tone and endogenous fibrinolysis at levels found in urban environments (Mills et al. 2005).

2.4.5 Effect on Reproductive System

Diesel fume particles (DEPs) have also been shown to reduce a man's ability to conceive. Studies shown that previously DEP exposure aggravated spermatogenesis, leading to daily reductions in sperm motility and production, an increase in sperm morphological deviations from the norm, and modifications to the ultrastructure of mice Leydig cells. After being exposed to DEP, male rodents' testicular capacity was modified, resulting in an increase in serum testosterone levels, a decrease in luteinizing hormone (LH), and a decrease in the production of sperm (Alzahrani and Rajendran 2019). According to Smith et al. (2007), the male reproductive system encountered infertility due to benzo(a)pyrene (Smith et al. 2007). A few reports mention decreased spermatozoa motility and reasonability in addition to spermatozoa's lower acrosin activity discharge that facilitate zona pellucida infiltration; there are also reports of decreased activity of γ-glutamyl transferase and lactate dehydrogenase C4 (LDH-C4) and elevated fructose level as a result of open to xylene. According to one study, the antiandrogenic and estrogenic properties of 4-nitrophenol (PNP) resulted in sterility in vivo. Epidemiologic investigations continue to be done, but one possibility contributing to the rising occurrence of sterility in animals and humans could be the concentration of PNP in the air, water, and soil (Krivoshto et al. 2008). Many international research investigations have connected exposure to diesel exhaust to babies with low birth weight, preterm births, disorders from birth, and an enhanced child death rate (Glinianaia et al. 2004). Low IQ, asthma in children, and a host of behavioral problems in early life have all been related to high prenatal PAH exposure (Perera et al. 2014). Reports of embryotoxicity have been made with lab animals exposed to, benzoanthracene, benzopyrene and naphthene (Rengarajan et al. 2015).

PAHs have the potential to influence essential reproductive hormones regulators like gonadotrophin-releasing hormone, aromatase enzyme, luteinizing hormone, and follicle-stimulating hormone (Bolden et al. 2017).

2.4.6 Effect on the Renal System

Multiple tests showed elevated renal tubule sharpness, reduced urine creatinine, oliguria, tubular necrosis, interstitial edema, and hematuria due to exposure to petroleum hydrocarbon xylene.

The histopathological evaluation indicated a minimal chronic kidney disease. Pee results were common, though, and the main negative effects included higher hyaline bead adjustment in male rats and nephrogenic dysfunction in female rats, which causes cell toxicity (Kum et al. 2007).

2.4.7 Effect on Haematopoietic System

It is also known that TPHs have a serious effect on the hematologic functioning in humans. Oil exposure increases the risk of chronic lymphocytic leukemia, nonlymphocytic leukemia, chronic myelocytic leukemia, and acute lymphocytic leukemia compared to non-exposed individuals. Blood indicators such as blood urea nitrogen (BUN), platelet and WBCs count, creatinine, hemoglobin, hematocrit, and urine phenol were found to be lower in victims of the DWH oil disaster (D'Andrea and Reddy 2014). The spill-affected individuals exhibited elevated levels of alkaline phosphatase, aspartate aminotransferase, and alanine aminotransferase in serum. Two heavy metals—nickel and aluminum—were found in elevated concentrations in blood samples from those who had been exposed to the spill of oil at the Prestige (Das and Mishra 2010; Mishra et al. 2022). In the Ecuadorian Amazon, there is an relation between childhood leukemia incidence and children's proximity to oil fields (Hurtig and San Sebastián 2004).There were 91 cancer cases listed in the 0–14 age range, with extensively higher rates in the 0–4 age group and a relative risk of leukemia ranging from 2.57 to 3.50%. The most dangerous volatile organic compounds (VOCs) found in PHs are known to cause hematologic neoplasms, including leukemia. However, research done in China and the USA also showed that there were significant rates of blood cancer cases among workers of oil field (Sathiakumar et al. 1995). Red blood cell deterioration may result from naphthalene exposure, either by inhalation or absorption (Bisht et al. 2010).

2.4.8 Effect on Immune System

Considering the fact that the exact mechanism of PAH immunotoxicity is unknown, it is known that PAHs cause immunotoxicity in experimental animals such as rodents. A lot of studies show that immunotoxic effects are systemic and are not impacted by the exposure route (Davila et al. 1995). PAHs have been linked to side effects related to immune system such as impaired cytokine synthesis by macrophages and monocytes, disruption of myelopoiesis, death of lymphoid tissues, suppression of B and T cells, myeloid cell and pre-B, pre-T development. By attaching themselves to certain aryl hydrocarbon receptors (AhRs) on lymphocytes and immune system auxiliary cells, PAHs can bind. Along with an increase in cytochrome P450, an AhR-regulated enzyme of metabolism, this interaction results in the production of potentially hazardous oxidative and electrophilic compounds. In mice, very high concentrations of PAHs (15–55 mg/kg of benzo(a)pyrene) are required to suppress humoral and cell-mediated immunity (Burchiel and Gao 2014).

Benzo(a)pyrene, anthracene, and napthene are examples that cause direct skin irritation. Anthracene and benzo(a)pyrene are seen to cause skin sensitivity, which means they may irritate the skin of animals as well as humans (Alegbeleye et al. 2017). In response to PAH exposure, immune suppression is the primary response that can be witnessed. Additional immune system disorders linked to PAH exposure include tumor development, immune potentiation, hypersensitivity expression, autoimmunity, and others. Individuals exposed to oil started showing signs of damaged immune and endocrine systems seven years after the Prestige accident. During three months, workers who manually washed up Prestige oil revealed altered immune systems, including of CD3+ and CD4+ T lymphocytes lower percentage and higher portion of CD8+ T lymphocytes along with plasma concentrations of IL10, IL2, IL4 and IFNγ. likewise, there was a reduction in the proportion of natural killer (NK) cells and an increase in the hydrocortisone concentration. It is essential to keep in mind that there will be adverse consequences from any drop in the quantity of NK cells. A reduced amount of NK cells enhances the risk of cancer and infections by microbes. This is due to NK cells have a crucial role in monitoring the spread of various tumor types and infections by the microbes, preventing tissue damage and subsequent health problems (16). Based on the immunotoxicity index, toluene is described as a "weak" PH, and benzene and PAHs as "strong" PHs (Laffon et al. 2013).

Changes in bone marrow including structural and functional changes triggered by exposure to PAHs have serious implications for health given that red bone marrow is the main organ of formation of blood cells and an essential location for producing cells of immune system (Hrudková et al. 2004).

DEPs are also known to trigger the production of immunoglobulin E and release histamine (Mamessier et al. 2006). Histamine increases P-selectin on the surfaces of endothelial cells and is a strong acute myocardial vasoconstrictor, platelet activator, and Plasma factor IIa activator (Nemmar et al. 2004).

2.4.9 *Effect on the Endocrine System*

People exposed to the Prestige oil disaster saw a considerable reduction in their endocrine hormones, such as lactogenic hormones and cortisol, which are indicators of psychophysiological stress. Researchers evaluated the disruption potential on steroidogenesis using H295R cells, a human adenocarcinoma cell line, in sediment samples near the oil spill named Hebei Spirit, which contained polycyclic aromatic hydrocarbons (PAHs). However, the disruption potentials of these sediment samples in steroid forming pathways declined with hours as a result of oil weathering, which also diminished the remaining oil residues' ability to bind hydrocarbon receptors. However, in vitro experiments employing cell cultures showed that PAHs and their analogs could significantly have impact on steroid forming pathways, leading to a variation in sex-steroid hormones. Key PAHs present in fossil fuels, such as fluorine, naphthalene, phenanthrene, chrysene, and dibenzothiophene, along with their alkylated derivatives, are implicated in promoting the synthesis of steroid hormones (Liu et al. 2018).

Increased levels of E2 (estrogen) were strongly associated with higher PAH concentrations. The rise in E2 levels was primarily caused by polar compounds and aromatic hydrocarbons.

2.5 Conclusion

Human health and pristine environments have been seriously damaged by the biotoxic PAHs pollution from numerous agents. Hydrocarbons are simple yet crucial molecules consisting like natural gas and petroleum. Animals are impacted by petroleum exposure in a variety of direct and indirect ways. Certain by products that are produced during the processing and refining of petroleum are subsequently employed to create other extremely dangerous compounds. Hydrocarbons are released from various sources such as the manufacturing iron and steel cement, aluminium, incineration, production of coal-tar pitch, colorants, rubber tires, and asphalt manufacturing including automobile exhaust. Maximal exposure to such toxic compounds reflects various complicated health issues such as inflammation, dizziness, skin irritation, diarrhea, and vomiting. This review delves into the major sources of hydrocarbons, their medium of transport in the environment and lastly, the effects are mentioned herein. Hydrocarbon sources are now prevalent in the environment with a current scenario of mushrooming of automobile and manufacturing industries. The lethal compounds containing hydrocarbons are inhaled in coke manufacturing and food processing industries. Health complications related to hydrocarbons primarily arise from their exposure through inhalation, ingestion, or skin contact. These complications can change relying on the specific hydrocarbon compound, concentration, interval of exposure, and individual susceptibility. Consequently, all the clinical effects lead to the greatest mortality and morbidity. However, stakeholders can work towards managing hydrocarbons sustainably and would help in minimizing the environmental impact, promoting efficiency, and ensuring long-term viability by incorporating various strategies.

References

Abdel-Shafy HI, Mansour MSM (2016) A review on polycyclic aromatic hydrocarbons: source, environmental impact, effect on human health and remediation. Egypt J Pet 25(1):107–123. https://doi.org/10.1016/J.EJPE.2015.03.011

Adeniji AO, Okoh OO, Okoh AI (2019) Levels of polycyclic aromatic hydrocarbons in the water and sediment of Buffalo River Estuary, South Africa and their health risk assessment. Arch Environ Contam Toxicol 76(4):657–669. https://doi.org/10.1007/s00244-019-00617-w

Aj T (2016) Distribution and sources of aliphatic hydrocarbons in surface sediments of Al-Arbaeen Lagoon, Jeddah, Saudi Arabia. J Fish Livestock Prod 4(2). https://doi.org/10.4172/2332-2608.1000173

Alegbeleye OO, Opeolu BO, Jackson V (2017) Bioremediation of polycyclic aromatic hydrocarbon (PAH) compounds: (acenaphthene and fluorene) in water using indigenous bacterial species isolated from the Diep and Plankenburg rivers, Western Cape, South Africa. Braz J Microbiol [Publ Braz Soc Microbiol] 48(2):314–325. https://doi.org/10.1016/J.BJM.2016.07.027

Alexander DD, Weed DL, Miller PE, Mohamed MA (2015) Red meat and colorectal cancer: a quantitative update on the state of the epidemiologic science. J Am Coll Nutr 34(6):521–543. https://doi.org/10.1080/07315724.2014.992553

Altomare DF, Di Lena M, Porcelli F, Trizio L, Travaglio E, Tutino M, Dragonieri S, Memeo V, De Gennaro G (2013) Exhaled volatile organic compounds identify patients with colorectal cancer. Br J Surg 100(1):144–150. https://doi.org/10.1002/bjs.8942

Alzahrani AM, Rajendran P (2019) Petroleum hydrocarbon and living organisms. In: Hydrocarbon pollution and its effect on the environment. IntechOpen. https://doi.org/10.5772/intechopen.86948

Bala K, Anshul, Verma S, Dogra A, Das AP, Kumar R, Nandini (2024) Sustainable management for urban plastic waste generation. In: Sustainable Management of Urban Plastic Waste Through Circular Economic Approaches, pp 211–236. https://doi.org/10.1201/9781003465447-13

Bal M, Behera ID, Kumari U, Biswas S, Meikap BC (2020) Hydrodynamic study and particulate matter removal in a self priming venturi scrubber. Environ Technol Innov 20. https://doi.org/10.1016/j.eti.2020.101167

Bala K, Dogra A, Lata P, Das AP (2024) Discussing sources and biological-chemical remediation approaches for microplastic pollution. Sustain Microb Technol Synth Cellulosic Microfiber Bioremediat 37–59. https://doi.org/10.1007/978-3-031-62600-5_3

Behera ID, Basak G, Kumar RR, Sen R, Meikap BC (2020) Treatment of petroleum refinery sludge by petroleum degrading bacterium *Stenotrophomonas pavanii* IRB19 as an efficient novel technology. J Environ Sci Health Part A Toxic/Hazard Subst Environ Eng 56(2). https://doi.org/10.1080/10934529.2020.1866924

Behera ID, Nayak M, Biswas S, Meikap BC, Sen R (2021) Enhanced biodegradation of total petroleum hydrocarbons by implementing a novel two-step bioaugmentation strategy using indigenous bacterial consortium. J Environ Manag 292:112746. https://doi.org/10.1016/j.jenvman.2021.112746

Behera ID, Meikap BC, Sen R (2022a) Performance evaluation of bacterial consortium for biodegradation of total petroleum hydrocarbon: a comparative strategic biostimulation study. Res Square

Behera ID, Nayak M, Mishra A, Meikap BC, Sen R (2022b) Strategic implementation of integrated bioaugmentation and biostimulation for efficient mitigation of petroleum hydrocarbon pollutants from terrestrial and aquatic environment. Mar Pollut Bull 177:113492. https://doi.org/10.1016/j.marpolbul.2022.113492

Bhattacharjee J, Mishra S, Das AP (2021) Recent advances in sensor-based detection of toxic dyes for bioremediation application: a review. Appl Biochem Biotechnol 2021:1–20. https://doi.org/10.1007/S12010-021-03767-7

Biological effects of diesel exhaust particles (DEP) on isolated cardiac muscle of guinea pigs. PubMed (n.d.). Retrieved 2 May 2024, from https://pubmed.ncbi.nlm.nih.gov/7531585/

Bisht S, Pandey P, Sood A, Sharma S, Bisht NS (2010) Biodegradation of naphthalene and anthracene by chemo-tactically active Rhizobacteria of *Populus deltoides*. Braz J Microbiol 41:922–930

Block ML, Wu X, Pei Z, Li G, Wang T, Qin L, Wilson B, Yang J, Hong JS, Veronesi B (2004) Nanometer size diesel exhaust particles are selectively toxic to dopaminergic neurons: the role of microglia, phagocytosis, and NADPH oxidase. FASEB J Off Publ Fed Am Soc Exp Biol 18(13):1618–1620. https://doi.org/10.1096/FJ.04-1945FJE

Bolden AL, Rochester JR, Schultz K, Kwiatkowski CF (2017) Polycyclic aromatic hydrocarbons and female reproductive health: a scoping review. Reprod Toxicol 73:61–74. https://doi.org/10.1016/J.REPROTOX.2017.07.012

Bosch X (2003) Exposure to oil spill has detrimental effect on clean-up workers' health. Lancet (Lond, Engl) 361(9352):147. https://doi.org/10.1016/S0140-6736(03)12249-0

Brook RD, Brook JR, Urch B, Vincent R, Rajagopalan S, Silverman F (2002) Inhalation of fine particulate air pollution and ozone causes acute arterial vasoconstriction in healthy adults. Circulation 105(13):1534–1536. https://doi.org/10.1161/01.CIR.0000013838.94747.64

Burchiel SW, Gao J (2014) Polycyclic aromatic hydrocarbons and the immune system. Encycl Immunotoxicol 1–7. https://doi.org/10.1007/978-3-642-27786-3_1192-4

Burchiel SW, Luster MI (2001) Signaling by environmental polycyclic aromatic hydrocarbons in human lymphocytes. Clin Immunol 98(1):2–10. https://doi.org/10.1006/CLIM.2000.4934

Calderón-Garcidueñas L, Reed W, Maronpot RR, Henríquez-Roldán C, Delgado-Chavez R, Calderón-Garcidueñas A, Dragustinovis I, Franco-Lira M, Aragón-Flores M, Solt AC, Altenburg M, Torres-Jardón R, Swenberg JA (2004) Brain inflammation and Alzheimer's-like pathology in individuals exposed to severe air pollution. Toxicol Pathol 32(6):650–658. https://doi.org/10.1080/01926230490520232

Carlsson H, Westerholm R, Östman C (1999) Particulate and semivolatile associated aliphatic hydrocarbon exhaust emission from heavy duty diesel vehicles. Toxicol Environ Chem 68(3–4):413–428. https://doi.org/10.1080/02772249909358674

D'Andrea MA, Reddy GK (2014) Crude oil spill exposure and human health risks. J Occup Environ Med 56(10):1029–1041. https://doi.org/10.1097/JOM.0000000000000217

Das A, Dey S, Bala K, Gnanasekaran R, Das AP (2024) Microbial bioremediation technology for sustainable treatment and management of synthetic microfiber waste. In: Sustainable Microbial Technology for Synthetic and Cellulosic Microfiber Bioremediation, pp 221–238. https://doi.org/10.1007/978-3-031-62600-5_12

Das A, Dey S, Das AP (2024) Ecotoxicological impacts of synthetic microfiber pollutants and development of sustainable mitigation strategies. Environ Chem Ecotoxicol. https://doi.org/10.1016/J.ENCECO.2024.12.003

Das AP, Ghosh S (2023) Mechanism of manganese bioleaching. Manganese Min Microorganisms 87–98. https://doi.org/10.1016/B978-0-12-822147-1.00014-5

Das AP, Mishra S (2008) Hexavalent chromium (VI): health hazards and environmental pollutant. J Environ Res Dev 2(3):386–392

Das AP, Mishra S (2009) Hexavalent chromium [Cr (VI)]: yellow water pollution and its remediation. Sarovar Saurabh ENVIS Newslett Wetland Ecosyst 5(2)

Das A, Mishra S (2010) Biodegradation of the metallic carcinogen hexavalent chromium Cr(VI) by an indigenously isolated bacterial strain. J Carcinog 9. https://doi.org/10.4103/1477-3163.63584

Das AP, Sukla LB, Pradhan N (2012) Microbial recovery of manganese using *Staphylococcus epidermidis*. Int J Nonferrous Metall 2012(02):9–12. https://doi.org/10.4236/IJNM.2012.12002

Das AP, Ghosh S, Mohanty S, Sukla LB (2015) Advances in manganese pollution and its bioremediation. In: Sukla LB, Pradhan N, Panda S, Mishra BK (eds) Environmental microbe biotechnology, 1st edn. Springer, Cham, pp 313–328. https://doi.org/10.1007/978-3-319-19018-1_16

Das AP, Mallavarapu M, Ghosh S (2023a) Ecotoxicity of mining pollutants on the environment and their remediation. Environ Chem Ecotoxicol 5:165–167. https://doi.org/10.1016/J.ENCECO.2023.08.002

Das AP, Dutta K, Khatun R, Behera ID, Singh S, Mishra S (2023b) Microfiber pollution and its microbial mitigation: a review on current trends and future prospects. J Taiwan Inst Chem Eng 105104. https://doi.org/10.1016/J.JTICE.2023.105104

Das AP, Mohanty AK, van Hullebusch ED, Figueiredo G, Das AP, Mohanty AK, van Hullebusch ED, Figueiredo G (2023c) Bioremediation as an emerging technology for the removal of synthetic microplastic pollutants from marine ecosystem. MarPB 194:115297. https://doi.org/10.1016/J.MARPOLBUL.2023.115297

Das AP, Behera ID, Bhanja D (2024a) Sustainable microbial technology for synthetic and cellulosic microfiber bioremediation. https://doi.org/10.1007/978-3-031-62600-5.

Das AP, Behera ID, Das NP (2024b) Renewable energy generation and value addition from environmental microfiber pollution through advanced greener solution, p 278

Das A, Dey S, Das AP (2024a) Ecotoxicological impacts of synthetic microfiber pollutants and development of sustainable mitigation strategies. Environ Chem Ecotoxicol. https://doi.org/10.1016/J.ENCECO.2024.12.003

Das A, Dey S, Bala K, Gnanasekaran R, Das AP (2024b) Microbial bioremediation technology for sustainable treatment and management of synthetic microfiber waste. Sustain Microb Technol Synth Cellulosic Microfiber Bioremediat 221–238. https://doi.org/10.1007/978-3-031-62600-5_12

Davila DR, Davis DP, Campbell K, Cambier JC, Zigmond LA, Burchiel SW (1995) Role of alterations in Ca(2+)-associated signaling pathways in the immunotoxicity of polycyclic aromatic hydrocarbons. J Toxicol Environ Health 45(2):101–126. https://doi.org/10.1080/15287399509531985

Delgado-Saborit JM, Aquilina N, Baker S, Harrad S, Meddings C, Harrison RM (2010) Determination of atmospheric particulate-phase polycyclic aromatic hydrocarbons from low volume air samples. Anal Methods 2(3):231–242. https://doi.org/10.1039/B9AY00157C

Determination of naphthalene in air. PubMed (n.d.). Retrieved 25 Mar 2024, from https://pubmed.ncbi.nlm.nih.gov/14846426/

Dey S, Bissoyi A, Koteswara Reddy G, Das AP (2024) Synthetic microfibers: An urban emerging pollutant demanding for sustainable environmental solutions. In: Sustainable Microbial Technology for Synthetic and Cellulosic Microfiber Bioremediation, pp 107–121. https://doi.org/10.1007/978-3-031-62600-5_7

Dey S, Bissoyi A, Koteswara Reddy G, Das AP (2024a) Synthetic microfibers: an urban emerging pollutant demanding for sustainable environmental solutions. In: Sustainable microbial technology for synthetic and cellulosic microfiber bioremediation, 107–121. https://doi.org/10.1007/978-3-031-62600-5_7

Dey S, Tripathy B, Kumar MS, Das AP (2023) Ecotoxicological consequences of manganese mining pollutants and their biological remediation. Environ Chem Ecotoxicol. https://doi.org/10.1016/J.ENCECO.2023.01.001

Dey S, Dash N, Ray SK, Priyadarsini S, Chhotaray C, Das AP (2024) Occurrence and quantification of tire wear plastics pollutants from highway city dust of Bhubaneswar City of Odisha. In: Sustainable Microbial Technology for Synthetic and Cellulosic Microfiber Bioremediation, pp 77–89. https://doi.org/10.1007/978-3-031-62600-5_5

Dey S, Dash N, Ray SK, Priyadarsini S, Chhotaray C, Das AP (2024b) Occurrence and quantification of tire wear plastics pollutants from highway city dust of Bhubaneswar city of Odisha. Sustain Microb Technol Synth Cellulosic Microfiber Bioremediat 77–89. https://doi.org/10.1007/978-3-031-62600-5_5

Dey S, Tripathy B, Kumar MS, Das AP (2023) Ecotoxicological consequences of manganese mining pollutants and their biological remediation. Environ Chem Ecotoxicol. https://doi.org/10.1016/J.ENCECO.2023.01.001

Diggs DL, Huderson AC, Harris KL, Myers JN, Banks LD, Rekhadevi PV, Niaz MS, Ramesh A (2011) Polycyclic aromatic hydrocarbons and digestive tract cancers: a perspective. J Environ Sci Health Part C Environ Carcinog Ecotoxicol Rev 29(4):324–357. https://doi.org/10.1080/10590501.2011.629974

Draper WM, Dhaliwal JS, Kusum Perera S, Baumann FJ (1996) Determination of diesel fuel and motor oil in water and wastes by a modified diesel-range organics total petroleum hydrocarbon method. J AOAC Int 79(2). https://academic.oup.com/jaoac/article/79/2/508/5684525

Duedahl-Olesen L, Navaratnam MA, Jewula J, Jensen AH (2015) PAH in some brands of tea and coffee. Polycyclic Aromat Compd 35(1):74–90. https://doi.org/10.1080/10406638.2014.918554

Ellis JB, Revitt DM, Lleweilyn N (n.d.) Transport and the environment: effects of organic pollutants on water quality

Fagundes RB, Abnet CC, Strickland PT, Kamangar F, Roth MJ, Taylor PR, Dawsey SM (2006) Higher urine 1-hydroxy pyrene glucuronide (1-OHPG) is associated with tobacco smoke exposure and drinking maté in healthy subjects from Rio Grande do Sul, Brazil. BMC Cancer 6. https://doi.org/10.1186/1471-2407-6-139

García-Sánchez M, Košnář Z, Mercl F, Aranda E, Tlustoš P (2018) A comparative study to evaluate natural attenuation, mycoaugmentation, phytoremediation, and microbial-assisted phytoremediation strategies for the bioremediation of an aged PAH-polluted soil. Ecotoxicol Environ Saf 147:165–174. https://doi.org/10.1016/j.ecoenv.2017.08.012

Ghosh S, Das AP (2018). Metagenomic insights into the microbial diversity in manganese-contaminated mine tailings and their role in biogeochemical cycling of manganese. Sci Rep 8(1):1–12. https://doi.org/10.1038/s41598-018-26311-w

Glinianaia SV, Rankin J, Bell R, Pless-Mulloli T, Howel D (2004) Does particulate air pollution contribute to infant death? A systematic review. Environ Health Perspect 112(14):1365. https://doi.org/10.1289/EHP.6857

Grandclément C, Seyssiecq I, Piram A, Wong-Wah-Chung P, Vanot G, Tiliacos N, Roche N, Doumenq P (2017) From the conventional biological wastewater treatment to hybrid processes, the evaluation of organic micropollutant removal: a review. Water Res 111:297–317. https://doi.org/10.1016/J.WATRES.2017.01.005

Heuss JM, Glasson WA (1968) Hydrocarbon reactivity and eye irritation. Environ Sci Technol 2(12):1109–1116. https://doi.org/10.1021/ES60023A003/ASSET/ES60023A003.FP.PNG_V03

Hrudková M, Fiala Z, Borska L, Novosad J, Smolej L (2004) The effect of polycyclic aromatic hydrocarbons to bone marrow. Acta Med (Hradec Kralove) Suppl

Hurtig AK, San Sebastián M (2004) Incidence of childhood leukemia and oil exploitation in the Amazon basin of Ecuador. Int J Occup Environ Health 10(3):245–250. https://doi.org/10.1179/OEH.2004.10.3.245

Hussain K, Rahman M, Prakash A, Hoque RR (2015) Street dust bound PAHs, carbon and heavy metals in Guwahati city—seasonality, toxicity and sources. Sustain Cities Soc 19:17–25. https://doi.org/10.1016/J.SCS.2015.07.010

Jaligama S, Chen Z, Saravia J, Yadav N, Lomnicki SM, Dugas TR, Cormier SA (2015) Exposure to deepwater horizon crude oil burnoff particulate matter induces pulmonary inflammation and alters adaptive immune response. Environ Sci Technol 49(14):8769. https://doi.org/10.1021/ACS.EST.5B01439

Janjua NZ, Kasi PM, Nawaz H, Farooqui SZ, Khuwaja UB, Najam-ul-Hassan, Jafri SN, Lutfi SA, Kadir MM, Sathiakumar N (2006) Acute health effects of the Tasman Spirit oil spill on residents of Karachi, Pakistan. BMC Public Health 6. https://doi.org/10.1186/1471-2458-6-84

John A, Das AP (2024) Government strategies for achieving circular economy for plastic waste management in Southeast Asia. Sustainable Management of Urban Plastic Waste Through

Circular Economic Approaches, 72–82. https://doi.org/10.1201/9781003465447-6/GOVERNMENT-STRATEGIES-ACHIEVING-CIRCULAR-ECONOMY-PLASTICWASTE-MANAGEMENT-SOUTHEAST-ASIA-AKBAR-JOHN-ALOK-PRASAD-DAS

Johnsen AR, Karlson U (2007) Diffuse PAH contamination of surface soils: environmental occurrence, bioavailability, and microbial degradation. Appl Microbiol Biotechnol 76(3):533–543. https://doi.org/10.1007/S00253-007-1045-2

Jones RD (1990) Xylene/amitraz: a pharmacologic review and profile. Vet Hum Toxicol 32(5):446–448. https://europepmc.org/article/med/2238443

Kanungo J, Sahoo T, Bal M, Behera ID (2024) Performance of bioremediation strategy in waste lubricating oil pollutants: a review. Geomicrobiol J 41(4):360–373. https://doi.org/10.1080/01490451.2023.2245395

Khatun R, Singh S, Dubey NK, Prasad Das A (2023) A review on marine-based phytochemicals and their application in biomedical research. In: Recent frontiers of phytochemicals: applications in food, pharmacy, cosmetics, and biotechnology. Elsevier, Amsterdam, pp 383–395. https://doi.org/10.1016/B978-0-443-19143-5.00013-X

Kilburn KH (2000) Effects of diesel exhaust on neurobehavioral and pulmonary functions. Arch Environ Health 55(1):11–17. https://doi.org/10.1080/00039890009603379

Knekt P, Steineck G, Järvinen R, Hakulinen T, Aromaa A (1994) Intake of fried meat and risk of cancer: a follow-up study in Finland. Int J Cancer 59(6):756–760. https://doi.org/10.1002/IJC.2910590608

Krivoshto IN, Richards JR, Albertson TE, Derlet RW (2008) The toxicity of diesel exhaust: implications for primary care. J Am Board Fam Med 21(1):55–62. https://doi.org/10.3122/jabfm.2008.01.070139

Kum C, Sekkin S, Kiral F, Akar F (2007) Effects of xylene and formaldehyde inhalations on renal oxidative stress and some serum biochemical parameters in rats. Toxicol Ind Health 23(2):115–120. https://doi.org/10.1177/0748233707078218

Kuppusamy S, Thavamani P, Venkateswarlu K, Lee YB, Naidu R, Megharaj M (2017) Remediation approaches for polycyclic aromatic hydrocarbons (PAHs) contaminated soils: technological constraints, emerging trends and future directions. Chemosphere 168:944–968. https://doi.org/10.1016/J.CHEMOSPHERE.2016.10.115

Kuppusamy S, Maddela NR, Megharaj M, Venkateswarlu K (2019) Total petroleum hydrocarbons: environmental fate, toxicity, and remediation. In: Total petroleum hydrocarbons: environmental fate, toxicity, and remediation. Springer International Publishing, Berlin. https://doi.org/10.1007/978-3-030-24035-6

Laffon B, Aguilera F, Ríos-Vázquez J, García-Lestón J, Fuchs D, Valdiglesias V, Pásaro E (2013) Endocrine and immunological parameters in individuals involved in Prestige spill cleanup tasks seven years after the exposure. Environ Int 59:103–111. https://doi.org/10.1016/J.ENVINT.2013.05.014

Lahiri D, Nag M, Dey A, Sarkar T, Joshi S, Pandit S, Das AP, Pati S, Pattanaik S, Tilak VK, Ray RR (2021) Biofilm mediated degradation of petroleum products. Geomicrobiol J 1–10. https://doi.org/10.1080/01490451.2021.1968979

Lai IC, Lee CL, Zeng KY, Huang HC (2011) Seasonal variation of atmospheric polycyclic aromatic hydrocarbons along the Kaohsiung coast. J Environ Manag 92(8):2029–2037. https://doi.org/10.1016/J.JENVMAN.2011.03.026

Langley A (Andrew), Kennedy B, Gilbey M, National Environment Protection Council (Australia), Environment Protection and Heritage Council, National Workshop on the Assessment of Site Contamination (5th: 2002: Adelaide, S. Aust.) (2003) Proceedings of the fifth national workshop on the assessment of site contamination. NEPC Service Corporation

Lee B-K, Vu VT (n.d.) Sources, distribution and toxicity of polycyclic aromatic hydrocarbons (PAHs) in particulate matter 99 X sources. www.intechopen.com

Lima ALC, Farrington JW, Reddy CM (2005) Combustion-derived polycyclic aromatic hydrocarbons in the environment—a review. Environ Forensics 6(2):109–131. https://doi.org/10.1080/15275920590952739

Liu X, Jung D, Zhou K, Lee S, Noh K, Khim JS, Giesy JP, Yim UH, Shim WJ, Choi K (2018) Characterization of endocrine disruption potentials of coastal sediments of Taean, Korea employing H295R and MVLN assays-reconnaissance at 5 years after Hebei Spirit oil spill. Mar Pollut Bull 127:264–272. https://doi.org/10.1016/J.MARPOLBUL.2017.11.055

Mallick K, Sahu A, Dubey NK, Das AP (2023) Harvesting marine plastic pollutants-derived renewable energy: a comprehensive review on applied energy and sustainable approach. J Environ Manag 348:119371. https://doi.org/10.1016/J.JENVMAN.2023.119371

Mamessier E, Nieves A, Vervloet D, Magnan A (2006) Diesel exhaust particles enhance T-cell activation in severe asthmatics. Allergy 61(5):581–588. https://doi.org/10.1111/J.1398-9995.2006.01056.X

Miller KA, Siscovick DS, Sheppard L, Shepherd K, Sullivan JH, Anderson GL, Kaufman JD (2007) Long-term exposure to air pollution and incidence of cardiovascular events in women. N Engl J Med 356(5):447–458. https://doi.org/10.1056/NEJMOA054409

Mills NL, Törnqvist H, Robinson SD, Gonzalez M, Darnley K, MacNee W, Boon NA, Donaldson K, Blomberg A, Sandstrom T, Newby DE (2005) Diesel exhaust inhalation causes vascular dysfunction and impaired endogenous fibrinolysis. Circulation 112(25):3930–3936. https://doi.org/10.1161/CIRCULATIONAHA.105.588962

Mills NL, Törnqvist H, Gonzalez MC, Vink E, Robinson SD, Söderberg S, Boon NA, Donaldson K, Sandström T, Blomberg A, Newby DE (2007) Ischemic and thrombotic effects of dilute diesel-exhaust inhalation in men with coronary heart disease. N Engl J Med 357(11):1075–1082. https://doi.org/10.1056/NEJMOA066314

Mishra S, Dash D, Das AP (2024) Microbial degradation of polyester microfibers using indigenously isolated bacterial strain Exiguobacterium Sp. CLEAN – Soil, Air, Water, 52(12), e202300343. https://doi.org/10.1002/CLEN.202300343

Mishra S, Rath CC, Das AP (2019) Marine microfiber pollution: A review on present status and future challenges. Marine Pollut Bull 140:188–197.https://doi.org/10.1016/J.MARPOLBUL.2019.01.039

Mishra S, Singh RP, Rath CC, Das AP (2020) Synthetic microfibers: source, transport and their remediation. J Water Process Eng 38:101612. https://doi.org/10.1016/J.JWPE.2020.101612

Mishra S, Das AP (2021) Current treatment technologies for removal of microplastic and microfiber pollutants from wastewater. In: Wastewater treatment: cutting-edge molecular tools, techniques and applied aspects, 237–251. https://doi.org/10.1016/B978-0-12-821881-5.00011-8

Mishra S, Rout PK, Das AP (2021) Emerging microfiber pollution and its remediation. In: Prasad R (ed) Environmental and microbial biotechnology. Springer, Singapore, pp 247–266. https://doi.org/10.1007/978-981-15-5499-5_9

Mishra A, Siddiqi H, Kumari U, Behera ID, Mukherjee S, Meikap BC (2021) Pyrolysis of waste lubricating oil/waste motor oil to generate high-grade fuel oil: a comprehensive review. Renew Sustain Energy Rev 150:111446. https://doi.org/10.1016/j.rser.2021.111446

Mishra S, Swain S, Sahoo M, Mishra S, Das AP (2021) Microbial colonization and degradation of microplastics in aquatic ecosystem: a review. Geomicrobiol J. https://doi.org/10.1080/01490451.2021.1983670

Mishra S, Dash D, Das AP (2022a) Detection, characterization and possible biofragmentation of synthetic microfibers released from domestic laundering wastewater as an emerging source of marine pollution. Mar Pollut Bull 185:114254. https://doi.org/10.1016/J.MARPOLBUL.2022.114254

Mishra S, Dash D, Al-Tawaha ARMS, Das AP (2022b) A review on heavy metal ion adsorption on synthetic microfiber surface in aquatic environments. Appl Biochem Biotechnol 194(10):4639–4654. https://doi.org/10.1007/S12010-022-04029-W/METRICS

Mishra S, Singh RP, Rout PK, Das AP (2022c) Membrane bioreactor (MBR) as an advanced wastewater treatment technology for removal of synthetic microplastics. Dev Wastewater Treat Res Processes 45–60. https://doi.org/10.1016/B978-0-323-85583-9.00022-3

Mishra S, Das AP (2023) Microbial remediation of synthetic microfiber contaminated wastewater. In: Shah MP (ed) Microbial technologies in industrial wastewater treatment. Springer, Singapore, pp 337–350. https://doi.org/10.1007/978-981-99-2435-6_17

Mishra S, Ghosh S, van Hullebusch ED, Singh S, Das AP (2023) A critical review on the recovery of base and critical elements from electronic waste-contaminated streams using microbial biotechnology. Appl Biochem Biotechnol 2023:1–30. https://doi.org/10.1007/S12010-023-04440-X

Mohanta S, Sahoo B, Behera ID, Pradhan S (2016) Effect of crushing on near-gravity material distribution in different size fractions of an Indian non-coking coal. J South Afr Inst Min Metall 116(2):209–213. https://doi.org/10.17159/2411-9717/2016/v116n2a12

Mohanta S, Pradhan B, Behera ID (2024) Impact and remediation of petroleum hydrocarbon pollutants on agricultural land: a review. Geomicrobiol J 41(4):345–359. https://doi.org/10.1080/01490451.2023.2243925

Mohanty J, Dey S, Bala K, Das AP (2024a) Recent advances in microplastic pollution for its sustainable management. Sustain Manag Urban Plast Waste Circular Econ Approaches 27–48. https://doi.org/10.1201/9781003465447-3

Mohanty M, Tripathy B, Koteswara Reddy G, Das AP (2024) Greener technologies for establishing restraint over microplastic pollution. In: Sustainable Microbial Technology for Synthetic and Cellulosic Microfiber Bioremediation, pp 239–258. https://doi.org/10.1007/978-3-031-62600-5_13

Mohanty M, Tripathy B, Koteswara Reddy G, Das AP (2024b) Greener technologies for establishing restraint over microplastic pollution. Sustain Microb Technol Synth Cellulosic Microfiber Bioremediat 239–258. https://doi.org/10.1007/978-3-031-62600-5_13

Mojiri A, Zhou JL, Ohashi A, Ozaki N, Kindaichi T (2019) Comprehensive review of polycyclic aromatic hydrocarbons in water sources, their effects and treatments. Sci Total Environ 696. https://doi.org/10.1016/j.scitotenv.2019.133971

Nemmar A, Hoet PHM, Vermylen J, Nemery B, Hoylaerts MF (2004) Pharmacological stabilization of mast cells abrogates late thrombotic events induced by diesel exhaust particles in hamsters. Circulation 110(12):1670–1677. https://doi.org/10.1161/01.CIR.0000142053.13921.21

Okayama Y, Kuwahara M, Suzuki A, Tsubone H (2006) Role of reactive oxygen species on diesel exhaust particle-induced cytotoxicity in rat cardiac myocytes. J Toxicol Environ Health A 69(18):1699–1710. https://doi.org/10.1080/15287390600631078

Pan LQ, Ren J, Liu J (2006) Responses of antioxidant systems and LPO level to benzo(a)pyrene and benzo(k)fluoranthene in the haemolymph of the scallop *Chlamys ferrari*. Environ Pollut 141(3):443–451. https://doi.org/10.1016/J.ENVPOL.2005.08.069

Patel AB, Shaikh S, Jain KR, Desai C, Madamwar D (2020) Polycyclic aromatic hydrocarbons: sources, toxicity, and remediation approaches. Front Microbiol 11. https://doi.org/10.3389/fmicb.2020.562813

Paul A, Dey S, Ram DK, Das AP (2023) Hexavalent chromium pollution and its sustainable management through bioremediation. Geomicrobiol J 1–11. https://doi.org/10.1080/01490451.2023.2218377

Paul A, Priyadarsini S, Karakuş S, Behera JK, Gnanasekaran R, Das AP (2024) Fate and accumulation of synthetic microfibers in the environment. Sustain Microb Technol Synth Cellulosic Microfiber Bioremediat 61–76. https://doi.org/10.1007/978-3-031-62600-5_4

Peng G, Hakim M, Broza YY, Billan S, Abdah-Bortnyak R, Kuten A, Tisch U, Haick H (2010) Detection of lung, breast, colorectal, and prostate cancers from exhaled breath using a single array of nanosensors. Br J Cancer 103(4):542–551. https://doi.org/10.1038/sj.bjc.6605810

Perera FP, Chang HW, Tang D, Roen EL, Herbstman J, Margolis A, Huang TJ, Miller RL, Wang S, Rauh V (2014) Early-life exposure to polycyclic aromatic hydrocarbons and ADHD behavior problems. PLoS ONE 9(11):e111670. https://doi.org/10.1371/JOURNAL.PONE.0111670

Perez C, Lopez De Cerain A, Bello J (2002) Modulation of mutagenic activity in meat samples after deep-frying in vegetable oils. Mutagenesis 17(1)

Peters A, von Klot S, Heier M, Trentinaglia I, Hörmann A, Wichmann HE, Löwel H (2004) Exposure to traffic and the onset of myocardial infarction. N Engl J Med 351(17):1721–1730. https://doi.org/10.1056/NEJMOA040203

Phillips DH (1999) Polycyclic aromatic hydrocarbons in the diet. Mutat Res Genet Toxicol Environ Mutagen 443(1–2):139–147. https://doi.org/10.1016/S1383-5742(99)00016-2

Pinedo J, Ibáñez R, Irabien Á (2012) Risk assessment of total petroleum hydrocarbons (TPHs) fractions 28. www.aidic.it/cet

Pinedo J, Ibáñez R, Lijzen JPA, Irabien Á (2013) Assessment of soil pollution based on total petroleum hydrocarbons and individual oil substances. J Environ Manag 130:72–79. https://doi.org/10.1016/J.JENVMAN.2013.08.048

Prabhakar A, Mishra S, Das AP (2019) Isolation and identification of lead (Pb) solubilizing bacteria from automobile waste and its potential for recovery of lead from end of life waste batteries. Geomicrobiol J 36(10):894–903. https://doi.org/10.1080/01490451.2019.1654044

Pradhan B, Behera ID, Sahoo K, Mohanta S (2023a) Applicability of the teetered bed separator for beneficiating Indian iron ore fines: an experimental study. J Mines Metals Fuels 71(11):1817–1825. https://doi.org/10.18311/jmmf/2023/31100

Pradhan G, Tripathy B, Ram DK, Digal AK, Das AP (2023b) Bauxite mining waste pollution and its sustainable management through bioremediation. Geomicrobiol J 1–10. https://doi.org/10.1080/01490451.2023.2235353

Priyadarshini B, Santhosh Kumar M, Das AP (2024) Microbial degradation of synthetic plastics released from COVID-19 generated single use disposable face masks. Environ Sci Eng Part F2827:1–16. https://doi.org/10.1007/978-3-031-50840-0_1

Priyadarsini S, Das AP (2024) Analytical and structural characterization of waste lithium-ion batteries for their effective recycling strategy. Environ Chem Ecotoxicol. https://doi.org/10.1016/j.enceco.2024.12.004

Priyadarsini S, Das AP (2025) Lithium bioleaching: a review on microbial-assisted sustainable technology for lithium bio-circularity. J Water Process Eng 69:106744. https://doi.org/10.1016/j.jwpe.2024.106744

Raj K, Das AP (2023) Lead pollution: impact on environment and human health and approach for a sustainable solution. Environ Chem Ecotoxicol 5:79–85. https://doi.org/10.1016/J.ENCECO.2023.02.001

Rajpara RK, Dudhagara DR, Bhatt JK, Gosai HB, Dave BP (2017) Polycyclic aromatic hydrocarbons (PAHs) at the Gulf of Kutch, Gujarat, India: occurrence, source apportionment, and toxicity of PAHs as an emerging issue. Mar Pollut Bull 119(2):231–238. https://doi.org/10.1016/J.MARPOLBUL.2017.04.039

Ravindra K, Sokhi R, Van Grieken R (2008) Atmospheric polycyclic aromatic hydrocarbons: source attribution, emission factors and regulation. Atmos Environ 42(13):2895–2921. https://doi.org/10.1016/J.ATMOSENV.2007.12.010

Ray SK, Dey S, Das AP, John A (2024) Microplastic waste management through greener technologies: a review. Sustain Manag Urban Waste Circular Econ Approaches 237–261. https://doi.org/10.1201/9781003465447-14

Rengarajan T, Rajendran P, Nandakumar N, Lokeshkumar B, Rajendran P, Nishigaki I (2015) Exposure to polycyclic aromatic hydrocarbons with special focus on cancer. Asian Pac J Trop Biomed 5(3):182–189. https://doi.org/10.1016/S2221-1691(15)30003-4

Rose M, Holland J, Dowding A, Petch SRG, White S, Fernandes A, Mortimer D (2015) Investigation into the formation of PAHs in foods prepared in the home to determine the effects of frying, grilling, barbecuing, toasting and roasting. Food Chem Toxicol 78:1–9. https://doi.org/10.1016/J.FCT.2014.12.018

Sadasivuni KK, Rattan S, Waseem S, Brahme SK, Kondawar SB, Ghosh S, Das AP, Chakraborty PK, Adhikari J, Saha P, Mazumdar P (2019) Silver nanoparticles and its polymer nanocomposites—synthesis, optimization, biomedical usage, and its various applications. Lect Notes Bioeng 331–373. https://doi.org/10.1007/978-3-030-04741-2_11

Sahoo PP, Singh S, Rout PK, Mishra S, Das AP (2022) Microbial remediation of plastic pollutants generated from discarded and abandoned marine fishing nets. Biotechnol Gnetic Eng Rev. https://doi.org/10.1080/02648725.2022.2152629

Sahu S, Das AP (2024) Microplastics degradation mechanism by aquatic plastiomes: a review. 213–226. https://doi.org/10.1007/978-3-031-50840-0_12

Satapathy S, Kar D, Priyadarsini S, Das AP, Kuanar A (2024) Application of microorganisms in emerging waste management strategies and remediation of synthetic and cellulosic microfibers. Sustain Microb Technol Synth Cellulosic Microfiber Bioremediat 183–206. https://doi.org/10.1007/978-3-031-62600-5_10

Sathiakumar N, Delzell E, Cole P, Brill I, Frisch J, Spivey G (1995) A case-control study of leukemia among petroleum workers. J Occup Environ Med 37(11):1269–1277. https://doi.org/10.1097/00043764-199511000-00005

Singh RP, Mishra S, Das AP (2020) Synthetic microfibers: pollution toxicity and remediation. Chemosphere 257:127199. https://doi.org/10.1016/j.chemosphere.2020.127199

Sinha R, Peters U, Cross AJ, Kulldorff M, Weissfeld JL, Pinsky PF, Rothman N, Hayes RB (2005) Meat, meat cooking methods and preservation, and risk for colorectal adenoma. Can Res 65(17):8034–8041. https://doi.org/10.1158/0008-5472.CAN-04-3429

Smith TL, Merry ST, Harris DL, Joe Ford J, Ike J, Archibong AE, Ramesh A (2007) Species-specific testicular and hepatic microsomal metabolism of benzo(a)pyrene, an ubiquitous toxicant and endocrine disruptor. Toxicol Vitro 21(4):753–758. https://doi.org/10.1016/J.TIV.2007.01.005

Srogi K (2007) Monitoring of environmental exposure to polycyclic aromatic hydrocarbons: a review. Environ Chem Lett 5(4):169–195. https://doi.org/10.1007/S10311-007-0095-0

Syhre M, Chambers ST (2008) The scent of *Mycobacterium tuberculosis*. Tuberculosis 88(4):317–323. https://doi.org/10.1016/j.tube.2008.01.002

Tripathy B, Dash A, Das AP (2022) Detection of environmental microfiber pollutants through vibrational spectroscopic techniques: recent advances of environmental monitoring and future prospects. Crit Rev Anal Chem 1–11. https://doi.org/10.1080/10408347.2022.2144994

Tripathy B, Sahoo PP, Sundaray H, Das AP (2024) In-vitro biodegradation of discarded marine microplastics across the eastern coast of the Bay of Bengal, India using *Exiguobacterium* sp. Environ Chem Ecotoxicol 6:236–247. https://doi.org/10.1016/j.enceco.2024.07.003

Tsiaoussis J, Antoniou MN, Koliarakis I, Mesnage R, Vardavas CI, Izotov BN, Psaroulaki A, Tsatsakis A (2019) Effects of single and combined toxic exposures on the gut microbiome: current knowledge and future directions. Toxicol Lett 312:72–97. https://doi.org/10.1016/J.TOXLET.2019.04.014

Valderrama JFN, Gil VC, Alzate BV, Tavera EA, Noreña E, Porras J, Quintana-Castillo JC, García LJJ, Molina PFJ, Ramos-Contreras C, Sanchez JB (2022) Effects of polycyclic aromatic hydrocarbons on gestational hormone production in a placental cell line: application of passive dosing to in vitro tests. Ecotoxicol Environ Saf 245:114090. https://doi.org/10.1016/J.ECOENV.2022.114090

Wang C, Chen S, Wang Z (2014) Electrophysiological follow-up of patients with chronic peripheral neuropathy induced by occupational intoxication with n-hexane. Cell Biochem Biophys 70(1):579–585. https://doi.org/10.1007/s12013-014-9959-7

Wold LE, Simkhovich BZ, Kleinman MT, Nordlie MA, Dow JS, Sioutas C, Kloner RA (2006) In vivo and in vitro models to test the hypothesis of particle-induced effects on cardiac function and arrhythmias. Cardiovasc Toxicol 6(1):69–78. https://doi.org/10.1385/CT:6:1:69

Yu H (2002) Environmental carcinogenic polycyclic aromatic hydrocarbons: photochemistry and phototoxicity. J Environ Sci Health Part C Environ Carcinog Ecotoxicol Rev 20(2):149–183. https://doi.org/10.1081/GNC-120016203

Zelinkova Z, Wenzl T (2015) The occurrence of 16 EPA PAHs in food—a review. Polycyclic Aromat Compd 35(2–4):248–284. https://doi.org/10.1080/10406638.2014.918550

Zheng H, Xing X, Hu T, Zhang Y, Zhang J, Zhu G, Li Y, Qi S (2018) Biomass burning contributed most to the human cancer risk exposed to the soil-bound PAHs from Chengdu Economic Region, western China. Ecotoxicol Environ Saf 159:63–70. https://doi.org/10.1016/J.ECOENV.2018.04.065

Chapter 3
Sources of Micro and Nano Plastic Hydrocarbon Pollutants in Marine Bodies and Their Treatment

Suprada Swain, M. Purnima Sudha, Sudeshna Dey, Yugantika Badu, Subhadra Rani Bhoi, Maha Awjan Alreshidi, Maytham T. Qasim, Harjeet Nath, Deo Karan Ram, and Alok Prasad Das

Abstract Through recent studies, micro and nano plastics (MNPs) have proven to be a subtle environmental problem. Evidence has been discovered, establishing MNPs as a noteworthy concern in the marine environment and posing a risk to human well-being. The MNP contamination, formerly believed to exist independently in land, sea, and air, is interconnected. The global rise in environmental contamination caused by micro and nanoplastics has become a pressing issue that requires immediate response. The current strong demand for plastic, driven by its low cost, inertness, flexibility, resistance to oxidation, and durability, has significantly increased environmental pollution. Urban areas, which are the primary contributors to these pollutants, must enhance community awareness and education to reduce the presence of pollution-causing substances. These pollutants not only degrade the environment but also, according to current studies, have an equally detrimental impact on human health. Hence its high time to take proper steps to reduce the amount of MNP pollutants and make the treatment processes more efficient further on. This systematic review aims at the summarization of effects of MNPs as pollutants in marine ecology, their impact on the marine world and human health along with the effective treatment

S. Swain · M. Purnima Sudha · S. Dey · Y. Badu · S. R. Bhoi · A. P. Das (✉)
Department of Life Sciences, Rama Devi Women's University, Bhubaneswar, Odisha, India
e-mail: alok1503@gmail.com

M. A. Alreshidi
Department of Chemistry, University of Ha'il, Ha'il 81441, Saudi Arabia

M. T. Qasim
College of Health and Medical Technology, Al-Ayen University, Thi-Qar 64001, Iraq

H. Nath
Department of Chemical and Polymer Engineering, Tripura University (A Central University), Agartala 799022, India

D. K. Ram
Department of Chemical, Petroleum & Hydrogen Technology, Nims Institute of Engineering & Technology (NIET), Nims University, Jaipur, Rajasthan 303121, India

I. D. Behera and A. P. Das (eds.), *Environmental Hydrocarbon Pollution and Zero Waste Approach Towards a Sustainable Waste Management*, Environmental Science and Engineering, https://doi.org/10.1007/978-3-031-81615-4_3

processes to reduce their amount as pollutants and to help future researchers to take the required steps to enhance the treatment technologies, accordingly.

Keywords Microplastics (MPs) · Nano plastics (NPs) · Toxicity · Sustainable development · Wastewater treatment

3.1 Introduction

In the current period, plastic has become an indispensable part of human life, with many uses in everyday tasks. Recent findings show that even in the most isolated locations on Earth, there are significant levels of micro and nano pollution (Allen et al. 2022). At first, it was believed that micro and nano plastics (MNPs) pollution was predominantly a marine problem having a significant effect on the marine biota (Gall and Thompson 2015), and originating primarily from land-based sources (Ashrafy et al. 2023) and is estimated to be 443 million tons by 2025. According to recent studies, approximately 10% of plastics are recycled, 11% of them are destroyed by burning, 9% are enshrouded into land and 70% are simply released to the environment (Enfrin et al. 2019). Pollution from micro and nano plastics (MNPs), which mainly originate on land, was formerly believed to be a marine problem (Allen et al. 2022), having a significant effect on marine biota (Gall and Thompson 2015). The oceans were thought to be the ultimate resting place for pollution due to the release of various types of pollutants (Das et al. 2023a, b; Mallick et al. 2024; Das and Ghosh 2023d, e). These pollutants include petroleum waste, plastic debris, heavy metals, and agricultural runoff (Kanungo et al. 2023a, b; Das et al. 2023a, b; Behera et al. 2021, 2022b; Biswas et al. 2020). However, as the research into MNP progressed, it became clear that the pollution was prevalent in various ecosystems, including freshwater, groundwater, soil, sediment, air, and even re-suspension from the ocean through ocean spray (Wagner et al. 2014).

As per (Ashrafy et al. 2023), microplastics (MPs) are synthetic materials that come from primary and secondary sources and have a diameter of less than 5 mm. Nano plastics (NPs) are those having a size range between 1 nm and 1 mm. Primary and secondary microplastics are the two groups into which microplastics are divided according to the method of release (Moharana et al. 2024; Pradhan et al. 2024; Sahu and Das 2024). Primary microplastics are purposely manufactured as small particles and released directly through a variety of non-biological sources, in contrast to secondary microplastics, which are produced as a result of the abiotic and biotic breakdown of relatively bigger plastic materials after they have been introduced into the environment (Das et al. 2023a; Das and Ghosh 2023b; Mohanty et al. 2024). Secondary microplastics can come from a variety of sources, including industrial manufacturing, tourism destinations, fishing grounds, textile businesses, and debris from abandoned car parts (De Falco et al. 2019). Microplastics are more likely to infiltrate the marine ecosystem through runoff as a result of the degradation of common plastic into smaller particles, which is facilitated by ultraviolet radiation and lower temperatures. Primary microplastics (MPs) are the predominant type of plastic that enters marine environments such as seas (Tiwari et al. 2019), as well as freshwater bodies and rivers (Eriksen et al. 2013). This occurrence of pollutants is due to household sewage, wastewater or household products. Moreover, it has been found that even the personal care product like facial cleanser, detergent, toothpaste, shaving foam contains nano plastic particles which are being released into the aquatic mediums through municipal sewage (Mishra and Das 2021, 2023a). Approximately 67 million microplastic particles (MPs) are being discharged into the marine environment through sewage treatment systems (Prata 2018). Plastic pollution, including plastic fragments and microplastics like single-use plastics such as plastic bottles and polythene bags, is easily noticeable and captures the attention of everyone (Allen et al. 2022). Nevertheless, MNPs are currently encountering a significant issue as they lack a direct human impact and are not visible to the naked eye. However, MNPs are facing a distal problem due to a lack of direct personal impact and their invisibility to the naked eye.

MNPs not only harm the environment but also pose a concern to human health, human faeces contain several environmental, indicating consumption from multiple sources. Micro-nanoparticles (MNPs) contaminate the environment and may be harmful to human health (Das et al. 2015; Dey et al. 2023; Raj and Das 2023). Human faeces contain several MNPs, suggesting they may be consumed from various sources. MNPs were found in colectomy specimens, indicating that they can move into the human colon (Liu et al. 2022) found MNPs in pregnant women's placentas, raising concerns. MNPs have been found in human blood (Ali et al. 2023), suggesting they can circulate and be absorbed in organs. MNPs can be ingested, inhaled, or come into contact with the skin (Revel et al. 2018; Prata et al., 2018). The MNPs not only pollute the environment but can be a potential threat to human health, too. Numerous MNPs have been detected in human faeces, suggesting the ingestion of MNPs from various sources. Micro nanoparticles (MNPs) not only contaminate the environment but also pose a possible risk to human health. A multitude of MNPs have been identified in human faeces, indicating the potential consumption of MNPs from diverse origins. Recent investigations have demonstrated the presence of MNPs in human blood as well, suggesting that MNPs are capable of circulating throughout the body and becoming absorbed in organs. Humans are primarily exposed to MNPs through three main routes: ingestion, inhalation, and skin contact. Recent research has shown that oral consumption is the main way people are exposed to a significant number of MNPs present in packaged food goods, drinking water bottles, and other sources (Mishra et al. 2023; Singh et al. 2020). Nanoplastics can permeate cell membranes and overcome the intestinal barrier, entering the bloodstream (Yang et al. 2022) and potentially reaching other organs in the body. These MNPs are not restricted to the aforementioned; furthermore, their capacity to cross the blood–brain barrier and accumulate in the brain, thereby inducing neurotoxicity, renders them potentially hazardous. MNPs may also reach the human body via respiration; upon entering the respiratory zone, they become deposited and have the potential to migrate to secondary organs. The MNPs can also enter the human body through respiration, which on entering the respiratory zone gets deposited and can translocate to secondary organs. Currently, the cutaneous exposure to nano plastics is being effectively managed due to the implementation of bans on the use of microbeads in personal care products and detergents by several nations. The growing amount of MNPs are found to have toxic effects on the marine species including inhibition in growth, reproductive disorders and immune stress (Araújo et al. 2021). A study of in vitro mammalian research shows the potential adverse effects include cytotoxicity, oxidative stress, inflammatory response, genotoxicity, embryotoxicity, neurotoxicity, and even carcinogenicity.

The increasing presence of micro and nano plastics in the environment poses a threat to both land and marine ecosystems. According to (Boyle and Örmeci 2020), the introduction of MNPs into the food chain may pose dangers to both human and environmental health. It's time to address the issue of MNPs. Wastewater treatment plants, the primary recipients of terrestrial MNPs, act as a barrier, preventing their entry into the marine environment. Wastewater treatment plants, the primary recipients of terrestrial MNPs, act as a barrier, preventing their entry into the marine environment. Although wastewater treatment plants are not intended for this purpose but still can manage in efficiently remove the MNPs to some extent. According to a report by (Rout et al. 2022), the content of MNPs in the wastewater treatment plant in the UK decreases by roughly 8%, 70%, 93%, and 97% after the preliminary, primary, secondary, and tertiary treatment procedures, respectively. Therefore, it is crucial to acquire knowledge about the elimination mechanisms of MNPs and incorporate them into the current treatment strategies to enhance the prevention of the discharge of these pollutants into the environment. Advanced oxidation process and electrocoagulation are among the advancing treatment methods used to remove MNPs. This review paper provides a concise overview of the advent of micro and nano plastics (MNPs), their impact on the marine environment and human health, and proposes measures to improve the efficient treatment of these MNPs.

3.2 Sources of Micro and Nano Plastics

Plastic particles smaller than 0.5 mm in diameter are referred to as microplastics; nanoplastics are even smaller than microplastics. Nanoplastics are smaller than 100 nm or (1/10000 mm). Micro and nano plastics are so tiny that we can't see them with the naked eye even under an optical microscope. These tiny, micro, and nano plastics may be generated by the degradation of larger plastics or may be manufactured for commercial use. Microplastics are classified as either primary or secondary. Figure 3.1 shows different sources of hydrocarbon pollutants. Microfibers are examples of primary microplastics created for commercial applications. Primary microplastics; are manufactured for commercial use such as microfibers. Secondary microplastics: generated from the degradation of large plastics. Sources of micro and nano plastics may be land-based or ocean-based (Mishra et al. 2022a; Mishra et al. 2020). Various wastes are accumulated in the ocean, of all the waste plastics comprise

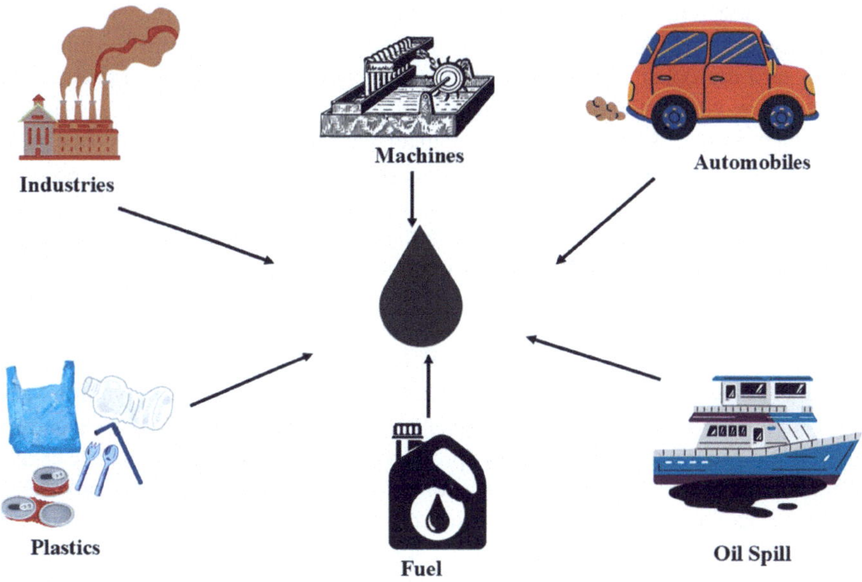

Fig. 3.1 Sources of hydrocarbon pollutants

approximately 76%. Microplastics may enter into the body of living organisms by ingestion. These aquatic organisms are already affected by microplastics and subsequently deliver microplastics into the intestines of humans or other aquatic organisms consuming living organisms (Mishra et al. 2019; Mishra et al. 2021a).

How Do Micro and Nano-Plastics Get into the Environment?

Today we can't live without using a single plastic material. Plastics have become our daily need starting from our toothbrushes, hair comb, cosmetics and personal care, plastic bottles, carry bags to many daily use accessories. Due to human and environmental activity, the larger plastics degraded or weathered to form smaller fragments. These smaller pieces can flow via the wind and can reach anywhere, mostly these smaller size particles end up in the ocean producing marine waste and causing various diseases among aquatic organisms. Besides our daily use, industrial generation of primary microplastics such as microfiber is also released into the ecosystem (Mishra and Das 2022). While producing plastic utilities the raw material required to manufacture such materials may also be released in some amounts before incorporating into manufacturing material. Other main channels by which primary microplastics enter the ecosystem are; road runoff, and wind transfer. The packaged foods we eat are provided in plastic wrap. When we tear it apart, the small piece of that wrap gets transferred by the wind from one place to various other places and finally gets into the hydrosphere ecosystem. Road runoff is equally responsible, the vehicles moving on the road carry and transfer microplastics along with them. Marine coatings and marine activities such as shipping, fishing, and tourism all contribute to the spread of microplastics and hydrocarbon pollutants in the oceans (Mallick et al. 2023). Approximately 90% of microplastics generated each year get swept by rainwater directly into the ocean. Now, the ocean is the biggest reservoir of microplastics. Further, the accumulation of petroleum hydrocarbon waste in the ocean is a complex environmental issue with significant ecological and economic implications (Bal et al. 2020; Behera et al. 2022a, b, 2020; Mohanta et al. 2024). Accidental releases from oil tankers, drilling rigs, pipelines, urban runoff, industrial discharges, and improper waste disposal are major sources of petroleum pollution that make the ocean difficult to clean up (Mishra et al. 2021a, b; Srivastava et al. 2019).

Microfibers

Microfibers are synthetic fabrics consisting of ultra-fine fibres. Microfibre is made up of various polyesters; polyamides; and a combination of polyester, polyamide, and polypropylene. The diameter of the microfibre is less than 10 μm. It is a secondary microplastic generated from artificially produced textiles. It has no second use except polluting the environment and posing a threat to the ecosystem and various living organisms. Due to their slow degradation plastics have now become omnipresent from topsoil to ponds, sediments of rivers, and sludge. The majority of microfibers detected in oceans, as per are the consequence of interior and outdoor laundering, domestic drainage, and the direct dumping of waste garments into rivers. The rate of microfibers released into the marine environment increased by approximately 14% over two years. Microfibers are discharged into the ocean annually at a rate of approximately 2 million tons. When we wash our clothes using a washing machine the effluent generated is released to be treated in wastewater treatment plants (WWTPs). The wastewater treatment plant can regulate the microfibers present in the effluent by trapping it but some amounts are still released into the environment by surface carriage (Mishra et al. 2022b). As most of the microfibers end up in the ocean it is uptake by marine animals, hampering their normal digestion and respiration process. In the food chain, organisms of higher trophic levels when feeding upon marine animals are also affected by microplastic pollution. Microfibers are also present in table salt. Yoga pants, furniture, paint, mattresses, fishing nets, and plastic bags are the primary sources of microfiber pollution, and their fragments ultimately wind up in the ocean. The demand for synthetic garments will result in a continued increase in global microfiber pollution (Mishra and Das 2022). Global microfiber pollution will continue increasing in demand for synthetic clothing. Synthetic microfiber can be replaced by natural microfibers. In rural areas, people wash their clothes in the local river or water stream where the direct release of microfibre from synthetic clothes enters into the river and subsequently into the ocean.

3.2.1 Cosmetics and Personal Care

Personal care and cosmetic products contain plastic microparticles that are discharged directly into sewers and are unable to be recycled or filtered due to their diminutive size. Then, they eventually accumulate in the water bodies, resulting in an ecological risk. The small size of these plastic microparticles in personal care and cosmetic products impedes their ability to be filtered or recycled, resulting in their direct discharge into sewers. Eventually, they accumulate in the water bodies, which poses an ecological risk. When we wash our face with a face wash containing micro beads which are made up of microfibers are washed along the water and get into the soil or the local water bodies. Various cosmetic and skin care products we buy are packed in plastic bottle; after using the product we usually don't throw the empty bottles in the right way which again causes plastic pollution (Jagatee et al. 2024). When we close a plastic bottle may be any kind of, we twist the cap around the bottle, because of that approximately 1000 microplastics are get into the bottle or released outside the bottle. The personal care products that contain micro beads, these micro beads may penetrate our body and enter into the body.

3.2.2 Ocean-Based Sources of Microplastics

Marine vehicles (ships, boats), commercial fisheries, and tourism are among the ocean-based sources from which approximately 10–20% of MNPs are discharged into the aquatic ecosystem. When tourists come to enjoy the beauty of the aquatic ecosystem; they enjoy it though the packages of the food products they brought along them, it finally gets into the ocean and makes it polluted. The issue is further exacerbated by the microplastic pollution that is emitted from naval vessels and shipping. Additionally, a substantial amount of plastic waste is discharged into the marine environment from offshore industries, including the petrochemical, pharmaceutical, and textile sectors, as well as adjacent landfills. Polymers, epoxy coatings, vinyl, lacquers, and metals are among the numerous varieties of marine coatings that are applied to the hulls of marine vessels. These coatings contribute to the microplastic load in the environment when they are weathered, scrapped, sanded, disposed of, or divided during the application process (Sajwan et al. 2024). In aquaculture, fishing nets may get lost or damaged, and such activities can easily introduce microplastics into the water source. When transportation via waterway is carried out from one country to another loss of one or few containers carrying the product may also cause microplastic pollution in aquatic ecosystems.

3.2.3 MNPs in Agricultural Use

Nutrient pills are polymer-coated nutrient mixtures that facilitate the diffusion of nutrients into the adjacent soil, thereby increasing yields and reducing the necessity for continuous fertilizer applications. Mulches for temperature and moisture regulation, silage and fumigation films, and anti-bird and vegetation protection are among the numerous applications of other microplastics that contain polymer-based products. With the growing population, the demand for agricultural products such as staple food rice, wheat, pulses, spices, cotton, fruits, and vegetables has increased (Benya et al. 2023). To yield good quality products and produce more in less time requires the application of synthetic fertilizer which causes soil microplastic pollution. The primary sources of microplastics in agricultural soil are generally recognized as the use of plastic greenhouses, plastic films, domestic effluent, irrigation, and the land use of waste in agricultural production. It has been shown that the administration of organic fertilizer, an emerging input source, is a significant method for microplastics to infiltrate the soil in recent years.

3.2.4 Road Markings and Automobile Tires

During their wear away, hot melt paints that are employed in road markings contain 15–25% polymer binders, which contribute to microplastic contamination. Tyre-wear is an unavoidable source of microplastic pollution. If we take a look into the composition of the tire, it is made up of natural rubber (19%), synthetic polymer (24%), antioxidants, anti-contents, curing system (14%), fillers (26%), textiles (4%), steel (12%). Synthetic polymer is the main highlight. Brake pads also generate microplastic pollution. Because of people's less awareness and laziness, they don't follow the instructions of the government. The government has given various facilities such as to dispose of wet waste in a green color dustbin and dry waste in a yellow color dustbin, some people throw the plastic waste anywhere in the local place due to this the plastics get carried out by air to other places or it enters into the soil causing soil microplastic pollution. One-time used products such as plastic glass, straws, and plates are randomly thrown into the environment which takes a very long period to degrade. Due to soil microplastic pollution microorganisms living in the soil found difficulties in absorbing their food and living slowly causing their death.

3.2.5 Microplastics from Urban Areas

Plastic particulates are utilized in the polymeric fills of turf to produce a shock-absorbing effect. Microscopic fibers are present in both interior and exterior building coatings, which enhance the elasticity and resilience of the paint by thickening it. If cured paint is removed, microplastics may be released into the environment. In a process, plastic granules are also employed as an abrasive material and are frequently found in blasting media and scrubbers for machinery and other metallic surfaces. Various construction works are daily carried out in urban areas the cement used as the raw material is packed in usually packed in plastic bags. Upon the finishing of construction, these plastic bags are not disposed of in the correct way which slowly degrades and remains in the environment for a long period. Various non-biodegradable and easily disposable menstrual products, like sanitary pads, and tampons harm the environment and produce microplastics that are released into the environment. When we wrap gifts for birthdays and other functions these also shed microplastics. In a wedding the use of plastic bottles, spoons, and plates is common and after their use incorrect disposal is also common so a huge amount of microplastics is generated each year from urban as well as from rural areas. Various transportations are carried out daily from one place to another, for commercial as well as personal purposes, during transportation some plastics may be unknowingly released from the carrier vehicle. When landfills are not managed properly the plastics deposited as waste get into the soil causing soil microplastic pollution. Various medical waste, pharmaceutical products, or used syringe which has no further use also contribute to micro and nano plastic pollution.

3.2.6 Plastic Utilities and Plastic Pellets

From our daily used products approximately 80–85% are directly plastic or partially hold plastic materials. Starting from toothbrush, comb, bottles, writing pen, polythene, personal care products and cosmetics, plastic containers, bags, constructing materials, laundry detergents, washing powder, plastic organizers, sandals, synthetic clothes, other accessories such as table, chair, toys of children etc. Single-use products consisting of polymeric plastic, such as straws, beverage bottles, cutlery, and coffee cups, have been identified as a significant contributor to plastic pollution in the environment. An additional factor contributing to the contamination of MNPs is the sudden rise in the use of single-use face masks during the COVID-19 pandemic. A wide range of plastic products are manufactured using plastic granules or nurdles as feedstock. When it gets lost accidentally during processing it adds micro plastic pollution. Pellets may be relatively larger in size when it gets into the environment as compared to other micro plastics, it slowly degrades to form micro plastics.

3.3 Impact of Microplastics & Nano-Plastics on Environment & Human Health

As we are aware plastic has become a matter of global concern. At the same time, microplastic pollution has emerged as one of the major environmental challenges in the twenty-first century. Microplastics are extremely intricated sources having a mixed bag of occurrences with precarious effects posing a greater challenge in disclosing their toxicity and risks to human health and environment. Assessment of microplastics and nanoplastics qualitatively and quantitatively has sealed the field of research on their origin and their noxious impact on the ecosystem (Tripathy et al. 2024a, b; Sahoo et al. 2022). The toxicity of microplastics and nano plastics is increasing rapidly day by day. Various works on microplastics and nanoplastics revealed their varied toxic effects which include oxidative stress, neurotoxicity, immune disorders, and reproductive and developmental damages. Moreover, there is no technical advancement in the consumption, relocation, modification, and cumulation of different environmental pollutants (Das et al. 2011; Mishra et al. 2019).

3.3.1 Impact of Microplastics & Nano Plastics on the Environment

Microplastics and nano plastics interact with the terrestrial ecosystem thereby bringing about a change in the geochemistry and biophysical environment due to their toxic effects. Microplastics and nanoplastics are brought to the agricultural fields by irrigation, surface water run-off from various sites, and sewage sludge which get integrated into the fertile soil altering the biochemistry and structural characteristics of soil biota and reducing soil fertility. As a result, they slow down the water and nutrient contents taken by the soil fauna and their transformation along the food chain disturbs the ecological balance (Das and Singh 2011). The germinating seeds upon exposure to microplastics show stunted growth and reduced shoot. Moreover, microplastics slow down the water absorption capacity of soil and disrupt the functional link between terrestrial microorganisms and water-soluble soil masses. A recent example of microplastic pollution is the extensive use of face masks during COVID-19—19 which has impaired the terrestrial and aquatic ecosystems showing that the pandemic has not lessened the risks of plastic pollution in the environment (Pradhan et al. 2024; Sahu et al. 2024).

Microplastics and nano plastics have also been detected in the atmosphere. A survey on microplastic for the first time investigated microplastic fibers in indoor and outdoor air. The indoor microplastic concentration ranges between 1.0 and 60.0 fibers/m^3 and the outdoor concentration ranges between 0.3 and 1.5 fibers/m^3. Some natural events such as hurricanes, convection, and turmoil in water bodies is responsible for microplastic transportation and accumulation. These phenomena result in the transportation of microplastic and nanoplastic particles to ocean water surfaces and even remote areas affecting their ecosystem significantly. Microplastics and nano-plastics reach water bodies through wind and ocean currents. These pollutants also alter the link between primary producers and higher trophic-level organisms of the water ecosystem. Microplastic and nanoplastic are sources of heavy metals like cadmium, zinc, and lead which have detrimental effects on aquatic organisms causing heavy metal toxicity and heavy metal contamination in water bodies (Bansal et al. 2023).

3.3.2 *Impact of Microplastics & Nanoplastics on Human Health*

Not only the environment but also human health is affected by the widespread distribution and presence of micro and nano plastics. Humans get exposed to such particles through multiple ways such as ingestion, inhalation, and skin contact (Priyadarsini & Das 2024, 2025; Ray et al. 2024; Das, Dey, & Das 2024). Even human stool contains microplastics which are ingested through their diet. Microplastics and nano plastics are also found in food packets, water bottles, and plastic packaging which pose a significant threat to the human digestive tract. Another way of entering micro and nano plastics into the human body is inhalation which damages the respiratory system by blocking air passage and causing adverse health effects. Those minute particles also enter the human body through dermal contact either from personal care products or detergents causing inflammation and irritation on the skin (Das et al. 2024a; Dey et al. 2023b; Dey et al. 2024a, b).

When microplastics and nano plastics come in contact with a biological system, they interact with biomacromolecules such as proteins and lipids which leads to the formation of soft and hard nanoplastic coronas. These nano-plastic coronas bring about a change in the physiological properties paving the way for health complications. Macrophages like lipids and proteins interact with coronas to control the entry and modification of nanoplastics inside cells. One intriguing finding was that, in contrast to the mice's robust physical condition, the corona surrounding silver nanoparticles exacerbated inflammatory responses in those with metabolic disorders. Nanoplastics can be transported by two ways-Passive transport and Active transport. Passive transport is generated due to potential differences in the concentration of nanoplastics inside and outside of the plasma membrane. Conversely, active transport needs ATP to function and is carried against the concentration gradient. To transport and assimilate nanoplastics from their surroundings, both passive and active movement take place simultaneously. Under normal physiological conditions, nanoplastics only passively pass across the plasma membrane if they can easily insert themselves into the surface pores. A clarifying example can be observed in cancerous cells, where the increased surface pores of the membrane allow nanoplastics to enter the cell.

The chemicals and different additions to microplastics and nano plastics enhance their qualities but they are harmful to the well-being of living beings. They even absorb chemicals from their surroundings. These substances can leak out of the polymer and into their surroundings. When humans are exposed to polycyclic aromatic hydrocarbons (PAHs), microplastics absorb them and cause negative effects. Certain hazardous chemical additions phthalates, triclosan, bisphenol A (BPA), organotins, bisphenol A (BPA), and brominated flame retardants are specifically harmful to humans. It is unsure whether these additives enter the biological system directly or indirectly. Leached BPA, an additive used for the production of polycarbonate plastic and epoxy resin present in food and beverage cans leads to endocrine abnormality and worsen human health. Significant research shows that

BPA damages growing fetuses, alters brain systems, and impairs liver function and insulin resistance. BPA functions as an agonist for estrogen receptors, inhibiting transcription mediated by thyroid hormone and altering the activity of pancreatic beta cells. These additives increase the risk of obesity and cardiovascular ailment, other reproductive and developmental problems. Phthalates on exposure to humans are another harmful agent impairing sexual development and congenital anomalies. In addition to, Butyl—benzyl phthalate and di-2-ethylhexyl phthalate (DHEP) are likely carcinogenic to humans. BPA too interrupts the hormonal functions of human beings by obstructing hormone production and action. It also has a detrimental effect on brain development in the growing embryo causing damage to it. Furthermore, research on phthalate esters, including butyl benzyl and di-2-ethylhexyl phthalate, has shown that these substances can cause tumors in humans and may even be carcinogenic. Human internal organs are exposed to micro and nanoplastics through a variety of channels, which has detrimental effects on cells. It is still unclear how harmful micro and nanoplastics are to human health. Our lives are difficult because we don't know how much of these particles to take within. Knowing the effective internal dose is essential for diagnosing human health issues. In addition, nothing is known about the toxicity of these tiny particles, which have added to the complexity of the entire planet, and there is scant information on human exposure to microplastics.

3.4 Present Treatment and Technological Strategies for Micro and Nano Plastic

Regarding the dangers of micro and nano plastic as well as additional pollutants, scientists have developed different methods for removing the contamination of MNPs, and the treatment of these pollutants is shown in Fig. 3.2.

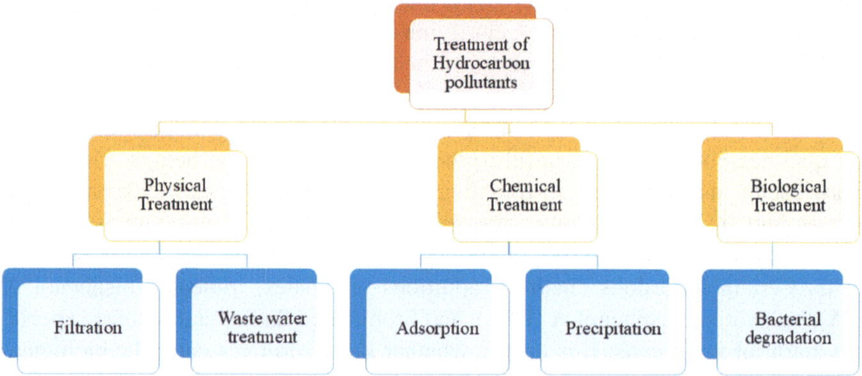

Fig. 3.2 Various treatment method of hydrocarbon pollutants

3.4.1 *Removal of Micro and Nano Plastics via Bacterial Enzyme*

Plastics can disperse micro and nano plastics via biodegradable and non-biodegradable processes (Mandal et al. 2024). The non-biodegradable methods can increase surface area, change mechanical characteristics, and break down the building blocks, which increases physicochemical reactions and interactions with microbes (Mandal et al. 2024). For example, the plastic surface is affected at the molecular level by photo-or oxidative degradation processes, which produce new functional groups like esters, ketone, and acetones that, when they react with OH radicals or N and O oxides, lessen the plastic's hydrophobic properties. Plastic polymer structures contain heteroatoms that serve as locations for hydrolytic enzymatic reactions. Extracellular hydrolytic enzymes are released by eukaryotic cells or microorganisms, aiding in the process of biodegradation. Biofilm production elevates hetero aggregate formation due to which MNPs stick along because of The biofilm matrix's extracellular hydrolytic enzymes. According to reports, the ability of these bacteria to accelerate the breakdown process is due to their special extracellular digestion and evolutionary development of multifunctional enzymes. In instance, contaminants are among the numerous complex chemical compounds that bacteria can naturally alter by using enzymes for metabolic purpose. The process "Enzymatic degradation" uses Enzymes act as organic compounds' natural biocatalysts, helping to break down complicated molecules like polymers of plastic into smaller units. The resulting goods are thought to be less harmful to the environment. More than 90 bacteria have been found to contain enzymes that break down plastic. Taking into account the abundance of MPs and NPs in the biosphere and the striking genetic and metabolic diversity of microbial populations, some microorganisms have evolved the capacity to utilize plastic. To use enzymatic degradation, there is involvement of culture-based approach. In this process, the desired microorganisms are grown and screened to view specific activity.

Two main types of enzymes used in the degradation of plastic are Extracellular enzymes and Intracellular enzymes (Gu 2003; Tang et al. 2022). The extracellular enzymes show many processes, including hydrolytic and oxidative activities hence are widely used for MNP biodegradation. These enzymes help in plastic degradation by depolymerization which breaks down the long chains of carbon into short fragments. These smaller fragments are capable of passing through semi-permeable outer membranes and provide microbial cells with a carbon source (Ray et al. 2024). Extracellular hydrolases, Enzymes that cut long carbon chains into smaller pieces include lipases, proteases, and cellulases. For example, it was found that the degradation of nylon was mediated by three hydrolases found in strains of Flavobacterium. Similarly, Bacillus subtilis MZA-75, a bacterium isolated from soil by Wilkes and Aristilde (2017), degrades polyurethane. Enzymes that break down plastic can have their catalytic activity increased through protein engineering. Glycoside hydrolases are also useful in degradation of MPs. The unique bacteria Ideonella sakaiensis 201-F6 is capable of using poly (ethylene terephthalate) as a source of carbon and energy.

Poly (ethylene terephthalate) is hydrolyzed by two enzymes to produce environmentally safe monomers such terephthalic acid and ethylene glycol. Currently, microbial enzymatic degradation of MNPs is not as effective because this is a slow process and cannot degrade the MNPs completely (Das and Ghosh 2023e; Mishra and Das 2023b).

3.4.2 By Filtration

Filtration is the oldest method of removing MNPs. Depending on the size of MNPs, the filtration process is separated into three stages: nano-filtration (2 nm), ultrafiltration (2–100 m), and microfiltration (0.1–1 μm). Yogarathinam et al. (2022) developed an inexpensive ceramic hollow fiber microporous membrane based on silica from guinea corn husk ash in order to eradicate MPs. Fu et al. (2023) discovered that the MP removal effectiveness may surpass 60% when they used an ultra-light, sustainable sponge to filter soup. According to Wang et al. (2023), MPs with small particle sizes can also be filtered by employing low-cost biochar in filter systems that use a bifunctional silver nanowire membrane to filter NPs with good retention (86.7% for 50 nm and nearly 95.0% for 100–1000 nm). Bifunctional silver nanowire membrane used to filter nano plastic had good retention and this idea was proposed by Yang and the efficiency was around 87% for 50 nm and 97% for 100–1000 nm. PMMA (polymethyl methacrylate) and PS (polystyrene) NPs were filtered using electrospun polyurethane nanofiber membranes, and the retention rate was higher than 92%.

3.4.3 By Adsorption

Adsorption is a method that is used to remove a variety of environmental pollutants, including pesticides, heavy metals, PAHs, and antibiotics (Behera 2022; Das and Ghosh 2023j). This process makes use of mechanisms like ion exchange, hydrogen bond interactions, hydrophobic contacts, and π π interactions. These techniques differ from more conventional adsorption techniques like activated carbon and $K(SO4)_2$. Coordination bonds hold the organic molecules and metal ligands in the crystalline porous substance known as the metal–organic framework (MOF) together. Wang et al. (2023) created Zr-MOF-based foam materials, which were effectively used to remove MNPs with a removal efficiency of 95.5% and good recycling capability. Cárdenas-Alcaide et al. (2022) created a three-dimensional graphene-like carbon-built layered double-oxide material to eliminate MNPs over a wide pH range. The elimination of MNP was 80% effective when the pH range was 3–11. The effectiveness was $\geq$ 60% at pH values ranging from 1 to 13. Magnet extraction is an additional efficient way to extract MNPs. According to Wang et al. (2021), effective Mg/Zn-modified magnetic biochar adsorbents were suggested for the removal of MNPs. MNPs and modified magnetic biochar had two different removal efficiencies of 98 and 99%, respectively.

The chemical bonding and electrostatic interactions between MNPs and biochar are responsible for these removal efficiencies. Shi et al. (2022) employed nano Fe_3O_4 to adsorb MPs in diverse situations.

3.4.4 By Precipitation

MNPs and other colloidal and particle materials are eliminated using this technique. PMMA and PVA (polyvinylacetate) can be effectively removed from water using the calcium carbonate ($CaCO_3$) co-precipitation method, according to research by Wang et al. (2023) following a half-hour of continuous application of low-concentration Na_2CO_3 and $CaCl_2$, PMMA and PVA nanoparticles were completely eliminated. By adjusting the pH and stirring speed, coagulation-flocculation and sedimentation were utilized to remove MPs, as described by the conditions that led to the highest PS removal effectiveness (98.9%) were pH 5, 3.4 mg Al/L coagulant, and 7 min of flocculation followed by 30 min of sedimentation. Fe-based coagulants had more efficiency to remove PE MPs in drinking water rather than conventional coagulation methods studied by (Ma et al. 2019).

3.4.5 Chemical Degradation Methods

Chemical strategies are also used to eliminate the MNPs which mostly include degradation of MNPs. Biodegradation and abiotic degradation are the degradation processes used for removal of MNPs. The molecular mass of MNPs does not change when we degrade it mechanically thus, degrading it chemically actually changes its molecular mass. Of all the abiotic degradation mechanisms, photocatalytic degradation is thought to be the most efficient and environmentally beneficial. It entails an interaction between organic substrates and ROS produced on semiconductor surfaces, rupturing the chemical bonds that hold organic pollutants together (Lahiri et al. 2021; Mishra et al. 2021b; Mohanty and Das 2021). Electrons that transition from valence bands to conduction bands produce photogenerated holes. According to Chen et al. (2022), these pores can directly oxidize organic molecules, producing CO_2 and H_2O. TiO_2 is the least toxic and least priced photocatalyst available effective in degrading MNPs and resistant to both acids and alkali. The rate of 400 nm PS degradation in the presence of TiO_2 nanoparticle films made by Triton X-100 within 12 h of UV light irradiation is 98.40%, according to Nabi et al. (2020).

3.4.6 Elimination of MNPs from Wastewater Treatment Plants

WWTPs are the primary source dumping MNPs into the water (Rout et al. 2022). Approximately 50 types of MNPs have been found in PES, PAM, and PET WWTPs. These are produced by the textile sector as well as from the food packaging, cosmetics, and personal care product industries. The physicochemical properties of MNPs results varying in the removal rates. First phases of WWTPs are screening, grit chamber and sedimentation. MNPs are eliminated by screens intercepting MNPs, skimming aggregated surface MNPs, and separating heavy MNPs by gravity during sedimentation. Air bubbles are produced during the aeration process, and these bubbles cling to the light-floating MNPs before rising to the surface and being removed by skimming. Dissolved air flotation (DAF) removes about 95% of MNPs, according to Talvitie et al. (2017). Even coagulation is essential for getting rid of MNPs. During the first treatment, coagulants such as aluminum and ferric salts are added, which cause the surface charge to become unstable and combine with the MNPs in the wastewater to create flocs. These are then removed by skimming or settling. The process of removal of MNPs by coagulation is around 90% effective (Bui et al. 2020). Hydrophobic and electrostatic contacts are the two types of interactions that aid in MNP extraction. Lipids are adhered to MNPs of the main sludge by hydrophobic mediated interaction, and MNPs are electrostatically mediated onto the surface of sludge particles.

Plastic biodegradation is a process in which the characteristics of plastics undergo alteration through modifications in molecular weight, shape, chemical structure, color, and tensile strength of flexible polymers due to microbial breakdown. Microbes degrade various environmental pollutants through different bioremediation methods (Das and Ghosh 2023b, c, f, g; Das et al. 2024). The enzymatic and non-enzymatic degradation by microorganisms, primarily fungi and bacteria and consortium, is a key aspect of this process. Biodegradation progresses in alignment with the original polymer, its biological composition, and favourable environmental degradation conditions. The mechanical attributes of biodegradable substances are impacted by their chemical makeup, production, processing, storage, application environments, and ageing. Various microorganisms, including bacteria, fungi, and algae, possess the capability to degrade petroleum hydrocarbons, converting them into less toxic substances. Key strategies comprise bioaugmentation, which introduces specific hydrocarbon-degrading microbes to contaminated sites, and biostimulation, which enhances the natural microbial degradation process through nutrient addition (Behera et al. 2022c, 2022b, 2020; Mohanta et al. 2024). Plastics generally decompose aerobically in natural settings and landfills, anaerobically in compost and soil, and, to some extent, aerobically in sediments (Nag et al. 2023). Microorganisms derive energy by facilitating energy-generating chemical reactions, involving the disruption of chemical bonds and extraction of electrons from pollutants. This form of biological response is known as an oxidoreduction reaction, wherein the organic contaminant is oxidized, leading to the reduction of the electron-receiving molecule

(Das and Ghosh 2023h, g; Ghosh et al. 2023). The pollutants function as electron acceptors, while the electron donors transfer electrons. The energy harnessed from these electron transfers is utilized for the production of fresh cells, incorporating a minor portion of carbon from the contaminants (Das et al. 2023b; Das and Ghosh 2023a; Kanungo et al. 2023b).

3.5 Conclusion

As we all are becoming aware of different microplastic sources, we must prioritize our self-awareness towards substituting the sources of microplastics with natural biodegradable products. A tiny chocolate wrap can degrade into micro plastics and can stay for long in the environment with its non-biodegradable nature. People's awareness is the key to safeguard the environment. All the MNPs that are already in the environment, there is less chances of removal them from the ecosystem specially from the ocean. Although micro and nano plastics cause damage to both environment and humans, it cannot be completely eradicated from the whole world. The only way to get rid of this adverse situation is to make effective use of it. Humans can employ simplicity which is a manifestation of doing more with less. It is to make purposeful choices and purchasing what truly needed because over consumption is the major cause of natural resources exploitation. Energy scavenging from rejected plastic pollutants can be innovative and best way offering a sustainable solution to the prevailing environmental pollution through closed loop economy, eco—friendly technologies, waste management are some remedies to deduce micro and nano plastics pollution from the environment. Usage of disposed plastic products for renewable resource production can be a better alternative to the present-day energy generation and at the same time can rescue the existence of millions of organisms deceasing each year due to microplastic consumption. Government should initiate microplastic waste related policies and sensitize people by educating through group discussions and organizing community activities. Advanced technologies of this modern world are exceptionally necessary for energy rejuvenation and providing an aspiration opt for power harvesting plant location and rehabilitation. These energy harvesting processes can be methodical and acceptable for microplastics recycling to recuperate energy. Moreover, proper operational procedure for explosive increase of microplastics waste can be quantum jump to avoid acute environmental and socio-economic obstacles and strengthening economically ideal society. The manufacturing of plastic has enormously increased due to high demand and usage by mankind. Plastic being non-biodegradable cannot be naturally degraded. Because of that scientists have to find alternative ways to degrade plastic and lower their negative impact on our environment. The plastics when degrade via biodegradable or non-biodegradable processes gives micro and nano plastics. Currently MNPs are extensively distributed in the atmosphere, land, oceans which is very difficult to extract and recycle. In this review paper we have discussed about different enzyme-based biodegradation of

MNPs as a green analytical tool for their sustainable mitigation. Moreover, advancement in genetic engineering and development of novel nanoenzymes have a great opportunity in the future for effective degradation of micro and nano plastics. In this review paper we have discussed about several methods for extraction of MNPs.

References

Ali N, Katsouli J, Marczylo EL, Gant TW, Wright S, Bernardino De La Serna J (2023) The potential impacts of micro-and-nano plastics on various organ systems in humans. http://creativecommons.org/licenses/by/4.0/

Allen S, Allen D, Karbalaei S, Maselli V, Walker TR (2022) Micro(nano)plastics sources, fate, and effects: what we know after ten years of research. J Hazard Mater Adv 6:100057. https://doi.org/10.1016/j.hazadv.2022.100057

da Costa Araújo AP, Rocha TL, e Silva DM, Malafaia G (2021) Micro (nano) plastics as an emerging risk factor to the health of amphibian: a scientometric and systematic review. Chemosphere 283:131090. https://doi.org/10.1016/J.CHEMOSPHERE.2021.131090

Ashrafy A, Liza AA, Islam MN, Billah MM, Arafat ST, Rahman MM, Rahman SM (2023) Microplastics pollution: a brief review of its source and abundance in different aquatic ecosystems. J Hazard Mater Adv 9. https://doi.org/10.1016/j.hazadv.2022.100215

Bal M, Behera ID, Kumari U, Biswas S, Meikap BC (2020) Hydrodynamic study and particulate matter removal in a self priming venturi scrubber. Environ Technol Innov 20:101167. https://doi.org/10.1016/j.eti.2020.101167

Bansal S, Singh Kushwah S, Garg A, Sharma K (2023) Utilization of plastic waste in construction industry in India—a review. Mater Today Proc. https://doi.org/10.1016/J.MATPR.2023.03.693

Behera ID (2022) Biodegradation of petroleum sludge by indigenously developed bacterial consortium IIT Kharagpur

Behera ID, Basak G, Kumar RR, Sen R, Meikap BC (2020) Treatment of petroleum refinery sludge by petroleum degrading bacterium *Stenotrophomonas pavanii* IRB19 as an efficient novel technology. J Environ Sci Health Part A 56(2):226–239. https://doi.org/10.1080/10934529.2020.1866924

Behera ID, Meikap BC, Sen R (2022a) Performance evaluation of bacterial consortium for biodegradation of total petroleum hydrocarbon: a comparative strategic biostimulation study. Research square

Behera ID, Nayak M, Biswas S, Meikap BC, Sen R (2021) Enhanced biodegradation of total petroleum hydrocarbons by implementing a novel two-step bioaugmentation strategy using indigenous bacterial consortium. J Environ Manag 292:112746. https://doi.org/10.1016/j.jenvman.2021.112746

Behera ID, Nayak M, Mishra A, Meikap BC, Sen R (2022b) Strategic implementation of integrated bioaugmentation and biostimulation for efficient mitigation of petroleum hydrocarbon pollutants from terrestrial and aquatic environment. Mar Pollut Bull 177:113492. https://doi.org/10.1016/j.marpolbul.2022.113492

Biswas S, Diwakar RK, Behera ID, Meikap BC, Sen TK, Khiadani M (2020) Aqueous phase phenol removal from synthetic and real steel plant effluents through a batch and Semifluidized bed column operation: experimental and model analysis. J Environ Chem Eng 8(5):104441. https://doi.org/10.1016/j.jece.2020.104441

Benya A, Mohanty S, Hota S, Das AP, Rath CC, Achary KG, Singh S (2023) Endangered *Curcuma caesia* Roxb.: qualitative and quantitative analysis for identification of industrially important elite genotypes. Ind Crop Prod 195:116363. https://doi.org/10.1016/J.INDCROP.2023.116363

Boyle K, Örmeci B (2020) Microplastics and nanoplastics in the freshwater and terrestrial environment: a review. Water (Switz) 12(9). MDPI AG. https://doi.org/10.3390/w12092633

Bui XT, Vo TDH, Nguyen PT, Nguyen VT, Dao TS, Nguyen PD (2020) Microplastics pollution in wastewater: characteristics, occurrence and removal technologies. Environ Technol Innov 19:101013. https://doi.org/10.1016/J.ETI.2020.101013

Das AP, Behera ID, Bhanja D (2024) Sustainable microbial technology for synthetic and cellulosic microfiber bioremediation. Springer Nature. https://doi.org/10.1007/978-3-031-62600-5

Das A, Dey S, Das AP (2024) Ecotoxicological impacts of synthetic microfiber pollutants and development of sustainable mitigation strategies. Environ Chem Ecotoxicol. https://doi.org/10.1016/J.ENCECO.2024.12.003

Das A, Dey S, Das AP (2024a) Ecotoxicological impacts of synthetic microfiber pollutants and development of sustainable mitigation strategies. Environ Chem Ecotoxicol. https://doi.org/10.1016/J.ENCECO.2024.12.003

Das AP, Dutta K, Khatun R, Behera ID, Singh S, Mishra S (2023a) Microfiber pollution and its microbial mitigation: a review on current trends and future prospects. J Taiwan Inst Chem Eng 105104. https://doi.org/10.1016/J.JTICE.2023.105104

Das AP, Ghosh S (2023a) Cost-effectiveness of the biomining process. Manganese Min Microorg 143–151. https://doi.org/10.1016/B978-0-12-822147-1.00008-X

Das AP, Ghosh S (2023b) Evolution of biomining technology. Manganese Min Microorg 5–19. https://doi.org/10.1016/B978-0-12-822147-1.00012-1

Das AP, Ghosh S (2023c) Factors affecting bioleaching. Manganese Min Microorg 77–86. https://doi.org/10.1016/B978-0-12-822147-1.00004-2

Das AP, Ghosh S (2023d) Introduction. Manganese Min Microorg 1–4. https://doi.org/10.1016/B978-0-12-822147-1.00005-4

Das AP, Ghosh S (2023e) Manganese and its application. Manganese Min Microorg 29–40. https://doi.org/10.1016/B978-0-12-822147-1.00007-8

Das AP, Ghosh S (2023f) Mechanism of manganese bioleaching. Manganese Min Microorg 87–98. https://doi.org/10.1016/B978-0-12-822147-1.00014-5

Das AP, Ghosh S (2023g) Metagenomic insights into unculturable microbial diversity. Manganese Min Microorg 63–76. https://doi.org/10.1016/B978-0-12-822147-1.00002-9

Das AP, Ghosh S (2023h) Microorganisms in manganese biomining processes. Manganese Min Microorg 55–61. https://doi.org/10.1016/B978-0-12-822147-1.00013-3

Das AP, Ghosh S (2023i) Mixed culture bioleaching: an insight into manganese biomining process and efficacy. Manganese Min Microorg 129–141. https://doi.org/10.1016/B978-0-12-822147-1.00001-7

Das AP, Ghosh S (2023j) Role of metabolites and metabolomics for understanding of manganese bioleaching. Manganese Min Microorg 117–127. https://doi.org/10.1016/B978-0-12-822147-1.00011-X

Das AP, Ghosh S, Mohanty S, Sukla LB (2015) Consequences of manganese compounds: a review. Toxicol Environ Chem 96(7):981–997. https://doi.org/10.1080/02772248.2015.1005428

Das AP, Mohanty AK, van Hullebusch ED, Figueiredo G, Das AP, Mohanty AK, van Hullebusch ED, Figueiredo G (2023b) Bioremediation as an emerging technology for the removal of synthetic microplastic pollutants from marine ecosystem. MarPB 194:115297. https://doi.org/10.1016/J.MARPOLBUL.2023.115297

Das A, Ray SK, Mohanty M, Mohanty J, Dey S, Das AP (2024b) Ecotoxicity of microplastic wastes and their sustainable management: A review. Environ Chem Ecotoxicol 6:144–152

Das AP, Singh S (2011) Occupational health assessment of chromite toxicity among Indian miners. Indian J Occup Environ Med 15(1):6. https://doi.org/10.4103/0019-5278.82998

Das AP, Sukla LB, Pradhan N, Nayak S (2011) Manganese biomining: a review. Biores Technol 102(16):7381–7387. https://doi.org/10.1016/J.BIORTECH.2011.05.018

De Falco F, Di Pace E, Cocca M, Avella M (2019) The contribution of washing processes of synthetic clothes to microplastic pollution. Sci Rep 9(1). https://doi.org/10.1038/s41598-019-43023-x

Dey S, Das A, Mallick K, Sahu A, Das AP (2023b) Environmental petroleum waste: Pollution, toxicity, sustainable remediation, pp 159–175. https://doi.org/10.1007/978-3-031-48220-5_7

Dey S, Dash A, Das AP (2024a) Bioremediation as a sustainable and circular solution for mining waste management and bioenergy generation. In: Sustainable management of mining waste and tailings: a circular economy approach, pp 107–119. https://doi.org/10.1201/9781003442455-6/BIOREMEDIATION-SUSTAINABLE-CIRCULAR-SOLUTION-MINING-WASTEMANAGEMENT-BIOENERGY-GENERATION-SUDESHNA-DEY-AKANSHYA-DASH-ALOK-PRASAD-DAS

Dey S, Dash N, Ray SK, Priyadarsini S, Chhotaray C, Das AP (2024b) Occurrence and quantification of tire wear plastics pollutants from highway city dust of Bhubaneswar City of Odisha. In: Sustainable microbial technology for synthetic and cellulosic microfiber bioremediation, pp 77–89. https://doi.org/10.1007/978-3-031-62600-5_5

Dey S, Tripathy B, Kumar MS, Das AP (2023) Ecotoxicological consequences of manganese mining pollutants and their biological remediation. Environ Chem Ecotoxicol. https://doi.org/10.1016/J.ENCECO.2023.01.001

Enfrin M, Dumée LF, Lee J (2019) Nano/microplastics in water and wastewater treatment processes—origin, impact and potential solutions. Water Res 161:621–638. https://doi.org/10.1016/J.WATRES.2019.06.049

Eriksen M, Mason S, Wilson S, Box C, Zellers A, Edwards W, Farley H, Amato S (2013) Microplastic pollution in the surface waters of the Laurentian Great Lakes. Mar Pollut Bull 77(1–2):177–182. https://doi.org/10.1016/J.MARPOLBUL.2013.10.007

Gall SC, Thompson RC (2015) The impact of debris on marine life. Mar Pollut Bull 92(1–2):170–179. https://doi.org/10.1016/J.MARPOLBUL.2014.12.041

Ghosh S, Tripathy B, Dey S, Das AP (2023) Comparative investigation of fungal and bacterial manganese biomining mechanisms. Manganese Min Microorg 99–115. https://doi.org/10.1016/b978-0-12-822147-1.00016-9

Gu JD (2003) Microbiological deterioration and degradation of synthetic polymeric materials: recent research advances. Int Biodeterior Biodegradation 52(2):69–91. https://doi.org/10.1016/S0964-8305(02)00177-4

Jagatee S, Priyadarshini S, Rath CC, Das AP (2024) Synthetic microfiber: an enduring environmental problem linked to sustainable development. Environ Sci Eng Part F2333 93–112. https://doi.org/10.1007/978-3-031-51792-1_6

Kanungo J, Sahoo T, Bal M, Behera ID (2023a) Performance of bioremediation strategy in waste lubricating oil pollutants: a review. Geomicrobiol J. https://doi.org/10.1080/01490451.2023.2245395

Kanungo J, Sahoo T, Swain LP, Behera ID (2023b) Toxicity of persistent hydrocarbon pollutants, sources and sustainable remediation process. In: Behera ID, Das AP (eds) Impact of petroleum waste on environmental pollution and its sustainable management through circular economy. Springer Nature Switzerland. pp 39–65. https://doi.org/10.1007/978-3-031-48220-5_2

Lahiri D, Nag M, Dey A, Sarkar T, Joshi S, Pandit S, Das AP, Pati S, Pattanaik S, Tilak VK, Ray RR (2021) Biofilm mediated degradation of petroleum products. Geomicrobiol J 1–10. https://doi.org/10.1080/01490451.2021.1968979

Liu Z, Zhuan Q, Zhang L, Meng L, Fu X, Hou Y (2022) Polystyrene microplastics induced female reproductive toxicity in mice. J Hazard Mater 424:127629. https://doi.org/10.1016/J.JHAZMAT.2021.127629

Ma B, Xue W, Ding Y, Hu C, Liu H, Qu J (2019) Removal characteristics of microplastics by Fe-based coagulants during drinking water treatment. J Environ Sci 78:267–275. https://doi.org/10.1016/J.JES.2018.10.006

Mallick K, Sahu A, Dubey NK, Das AP (2023) Harvesting marine plastic pollutants-derived renewable energy: a comprehensive review on applied energy and sustainable approach. J Environ Manage 348:119371. https://doi.org/10.1016/J.JENVMAN.2023.119371

Mallick K, Sahu S, Sahu A, Dey S, Das AP (2024) A critical review of marine microfiber pollution routes, toxicity, and its sustainable remediation. Environ Sci Eng Part F2333 189–211. https://doi.org/10.1007/978-3-031-51792-1_10

Mandal M, Roy A, Popek R, Sarkar A (2024) Micro-and Nano-plastic degradation by bacterial enzymes: a solution to "White Pollution." Microbe 100072. https://doi.org/10.1016/j.microb.2024.100072

Mishra S, Das AP (2021). Current treatment technologies for removal of microplastic and microfiber pollutants from wastewater. Wastewater Treat: Cut-Edge Mol Tools, Tech Appl Asp 237–251. https://doi.org/10.1016/B978-0-12-821881-5.00011-8

Mishra S, Das AP (2022) Treatment of the wastewater polluted with synthetic microfiber released from washing machine. In: Das BB, Hettiarachchi H, Sahu PK, Nanda S (eds) Lecture notes in civil engineering, vol 207. Springer Science and Business Media Deutschland GmbH, pp 109–117. https://doi.org/10.1007/978-981-16-7509-6_9/COVER

Mishra S, Das AP (2023a) Microbial remediation of synthetic microfiber contaminated wastewater. In: Shah MP (ed) Microbial technologies in industrial wastewater treatment. Springer, Singapore, pp 337–350. https://doi.org/10.1007/978-981-99-2435-6_17

Mishra S, Das AP (2023b) Microbial remediation of synthetic microfiber contaminated wastewater. Microb Technol Ind Wastewater Treat 337–350. https://doi.org/10.1007/978-981-99-2435-6_17/COVER

Mishra S, Dash D, Das AP (2022a) Detection, characterization and possible biofragmentation of synthetic microfibers released from domestic laundering wastewater as an emerging source of marine pollution. Mar Pollut Bull 185:114254. https://doi.org/10.1016/J.MARPOLBUL.2022.114254

Mishra S, Dash D, Das AP (2023). Aquatic microbial diversity on plastisphere: colonization and potential role in microplastic biodegradation. Geomicrobiol J 1–12. https://doi.org/10.1080/01490451.2023.2209750

Mishra S, Charan Rath C, Das AP (2019) Marine microfiber pollution: a review on present status and future challenges. Mar Pollut Bull 140:188–197. https://doi.org/10.1016/J.MARPOLBUL.2019.01.039

Mishra S, Rout PK, Das AP (2021a) Emerging microfiber pollution and its remediation. In: Prasad R (ed) Environmental and microbial biotechnology. Springer, Singapore, pp 247–266. https://doi.org/10.1007/978-981-15-5499-5_9

Mishra S, Singh RP, Rath CC, Das AP (2020) Synthetic microfibers: source, transport and their remediation. J Water Process Eng 38:101612. https://doi.org/10.1016/J.JWPE.2020.101612

Mishra S, Singh RP, Rout PK, Das AP (2022b) Membrane bioreactor (MBR) as an advanced wastewater treatment technology for removal of synthetic microplastics. Dev Wastewater Treat Res Process 45–60. https://doi.org/10.1016/B978-0-323-85583-9.00022-3

Mishra S, Swain S, Sahoo M, Mishra S, Das AP (2021b) Microbial colonization and degradation of microplastics in aquatic ecosystem: a review. Geomicrobiol J. https://doi.org/10.1080/01490451.2021.1983670

Mohanta S, Pradhan B, Behera ID (2024) Impact and remediation of petroleum hydrocarbon pollutants on agricultural land: a review. Geomicrobiol J 41(4):345–359. https://doi.org/10.1080/01490451.2023.2243925

Mohanty M, Mohanty J, Dey S, Dutta K, Shah MP, Das AP (2024) The face mask: a tale from protection to pollution and demanding sustainable solution. Emerg Contam 10(2):100298. https://doi.org/10.1016/J.EMCON.2023.100298

Mohanty SS, Das AP (2021) A systematic study on the microbial degradation of glyphosate: a review. Geomicrobiol J. https://doi.org/10.1080/01490451.2021.1998255

Moharana S, Dey S, Priyadarsini S, Kumar MS, Das AP (2024) Identification and characterization of microplastic pollutants from the marine sediments of Paradeep coast of Bay of Bengal, India for their sustainable management. Environ Sci Eng Part F2333 57–73. https://doi.org/10.1007/978-3-031-51792-1_4

Nag M, Lahiri D, Ghosh S, Sarkar T, Pati S, Das AP, Ram DK, Bhattacharya D, Ray RR (2023) Application of microorganisms in biotransformation and bioremediation of environmental contaminant: a review. Geomicrobiol J. https://doi.org/10.1080/01490451.2023.2261443

Pradhan G, Dey S, Priyadarsini S, Kumar MS, Das AP (2024) Characterization and quantification of microplastics pollutants in sediment samples from daya river of odisha state in india for their appropriate management. Environ Sci Eng Part F2333 75–92. https://doi.org/10.1007/978-3-031-51792-1_5

Prata JC (2018) Airborne microplastics: consequences to human health? Environ Pollut 234:115–126. https://doi.org/10.1016/J.ENVPOL.2017.11.043

Priyadarsini S, Das AP (2024) Analytical and structural characterization of waste lithium-ion batteries for their effective recycling strategy. Environ Chem Ecotoxicol. https://doi.org/10.1016/j.enceco.2024.12.004

Priyadarsini S, Das AP (2025) Lithium bioleaching: A review on microbial-assisted sustainable technology for lithium biocircularity. J Water Process Eng 69:106744. https://doi.org/10.1016/j.jwpe.2024.106744

Raj K, Das AP (2023) Lead pollution: impact on environment and human health and approach for a sustainable solution. Environ Chem Ecotoxicol 5:79–85. https://doi.org/10.1016/J.ENCECO.2023.02.001

Ray SK, Dey S, Das AP, John A (2024) Microplastic waste management through greener technologies: a review. In: Sustainable management of urban plastic waste through circular economic approaches, pp 237–261. https://doi.org/10.1201/9781003465447-14

Revel M, Châtel A, Mouneyrac C (2018) Micro(nano)plastics: a threat to human health? Curr Opin Environ Sci Health 1:17–23. https://doi.org/10.1016/J.COESH.2017.10.003

Rout PR, Mohanty A, Sharma A, Miglani M, Liu D, Varjani S (2022) Micro-and nanoplastics removal mechanisms in wastewater treatment plants: a review. J Hazard Mater Adv 6. https://doi.org/10.1016/j.hazadv.2022.100070

Sahoo PP, Singh S, Rout PK, Mishra S, Das AP (2022) Microbial remediation of plastic pollutants generated from discarded and abandoned marine fishing nets. Biotechnol Gnetic Eng Rev. https://doi.org/10.1080/02648725.2022.2152629

Sahu A, Mallick K, Das AP (2024) Environmental impact of single-use synthetic face mask and its recycling: a sustainable approach. In: Das AP, Mishra S (eds) Impact of COVID-19 waste on environmental pollution and its sustainable management. Environmental science and engineering. Springer, Cham. https://doi.org/10.1007/978-3-031-50840-0_11

Sahu S, Das AP (2024) Microplastics degradation mechanism by aquatic plastiomes: a review, pp 213–226. https://doi.org/10.1007/978-3-031-50840-0_12

Sajwan D, Sharma A, Sharma M, Krishnan V (2024) Upcycling of plastic waste using photo-, electro-, and photoelectrocatalytic approaches: a way toward circular economy. ACS Catal 4865–4926. https://doi.org/10.1021/ACSCATAL.4C00290/ASSET/IMAGES/MEDIUM/CS4C00290_0028.GIF

Singh RP, Mishra S, Das AP (2020) Synthetic microfibers: pollution toxicity and remediation. Chemosphere 257:127199. https://doi.org/10.1016/j.chemosphere.2020.127199

Srivastava M, Srivastava A, Yadav A, Rawat V (2019) Source and control of hydrocarbon pollution. IntechOpen. https://doi.org/10.5772/intechopen.86487

Tang KHD, Lock SSM, Yap PS, Cheah KW, Chan YH, Yiin CL, Ku AZE, Loy ACM, Chin BLF, Chai YH (2022) Immobilized enzyme/microorganism complexes for degradation of microplastics: A review of recent advances, feasibility and future prospects. Sci Total Environ 832:154868. https://doi.org/10.1016/J.SCITOTENV.2022.154868

Tiwari M, Rathod TD, Ajmal PY, Bhangare RC, Sahu SK (2019) Distribution and characterization of microplastics in beach sand from three different Indian coastal environments. Mar Pollut Bull 140:262–273. https://doi.org/10.1016/J.MARPOLBUL.2019.01.055

Tripathy B, Dey S, Dubey NK, Lo W-C, Das AP (2024a) An omics-based approach for managing the risk of COVID-19-generated wastes, pp 51–66. https://doi.org/10.1007/978-3-031-50840-0_4

Tripathy B, Sahoo PP, Sundaray H, Das AP (2024b) In-vitro biodegradation of discarded marine microplastics across the eastern coast of the Bay of Bengal, India using *Exiguobacterium* sp. Environ Chem Ecotoxicol 6:236–247. https://doi.org/10.1016/j.enceco.2024.07.003

Wagner M, Scherer C, Alvarez-Muñoz D, Brennholt N, Bourrain X, Buchinger S, Fries E, Grosbois C, Klasmeier J, Marti T, Rodriguez-Mozaz S, Urbatzka R, Vethaak AD, Winther-Nielsen M, Reifferscheid G (2014) Microplastics in freshwater ecosystems: what we know and what we need to know. Environ Sci Eur 26(1):1–9. https://doi.org/10.1186/s12302-014-0012-7
Yang X, Man YB, Wong MH, Owen RB, Chow KL (2022) Environmental health impacts of microplastics exposure on structural organization levels in the human body. Sci Total Environ 825:154025. https://doi.org/10.1016/J.SCITOTENV.2022.154025

Chapter 4
Advanced Waste Management Technique for Liquid Hydrocarbon Waste

B. Thirumalaiyammal and P. F. Steffi

Abstract Liquid hydrocarbon waste is a matter of grave concern when analysing environmental risks, given its inherent toxicity and the potential for contaminating precious water sources. Conventional waste management methods like containment ponds and physical separation strategies fall short in their capability to adequately address the treatment and disposal needs associated with such hazardous waste materials. It is against this backdrop that a novel waste management approach emerges; custom-tailored to cater specifically to the intricate nature of liquid hydrocarbon waste. The crux of this innovative technique lies in the strategic amalgamation of cutting-edge advanced oxidation processes (AOPs) and membrane separation technology, directing its focus towards the dual goals of efficient treatment and the concurrent recovery of valuable hydrocarbons from the polluted waste stream. By harnessing the potential of these advanced technologies in tandem, this method showcases a promising pathway towards more sustainable and effective solutions in the realm of liquid hydrocarbon waste management. The synergy between AOPs and membrane separation technology paves the way for a comprehensive and intuitive process flow that not only addresses the existing gaps but also sets a benchmark for future advancements in the field. It is through such targeted and forward-thinking initiatives that the daunting challenges posed by liquid hydrocarbon waste can be met head-on, promising a more ecologically sound and resource-efficient approach to waste management on a global scale.

Keywords Liquid hydrocarbon waste · Advanced oxidation processes · Membrane separation · Waste management · Environmental remediation

---

B. Thirumalaiyammal (✉)
Research Scholar, PG and Research, Department of Microbiology, Cauvery College for Women (Autonomous, Affiliated to Bharathidasan University), Trichy, Tamil Nadu, India
e-mail: puvi.divs@gmail.com

P. F. Steffi
Assistant Professor, PG and Research, Department of Microbiology, Cauvery College for Women (Autonomous, Affiliated to Bharathidasan University), Trichy, Tamil Nadu, India

I. D. Behera and A. P. Das (eds.), *Environmental Hydrocarbon Pollution and Zero Waste Approach Towards a Sustainable Waste Management*, Environmental Science and Engineering, https://doi.org/10.1007/978-3-031-81615-4_4

4.1 Introduction

Liquid hydrocarbon waste is a byproduct of various industrial processes, including oil refining, petrochemical production, and transportation activities. Due to its complex composition and hazardous nature, effective management of liquid hydrocarbon waste is essential to prevent environmental contamination and protect human health. Traditional disposal methods, such as incineration and landfilling, may not be suitable for liquid hydrocarbon waste due to its persistence and potential to leach harmful substances into the environment. Therefore, the development and implementation of advanced waste management techniques are necessary to address the challenges associated with liquid hydrocarbon waste. This chapter explores innovative approaches including bioremediation, chemical oxidation, membrane filtration, and advanced separation technologies for the treatment and remediation of liquid hydrocarbon waste. Table 4.1. Shows various advancement techniques for degradation of waste. Liquid hydrocarbon waste management poses significant challenges due to its potential environmental hazards. Conventional methods often struggle to adequately address the complexities of hydrocarbon contamination. However, emerging technologies offer promising solutions to enhance the efficiency and sustainability of waste management practices. One such advanced technique gaining attention is the utilization of graphene-based materials for the treatment of liquid hydrocarbon waste (Bala, Anshul et al. 2024; Bala, Dogra et al. 2024; Das, Dey, & Das 2024; Das, Dey, Bala et al. 2024). Graphene's unique properties, including its high surface area, exceptional electrical conductivity, and chemical inertness, make it an ideal candidate for various environmental applications, including waste remediation. The management of liquid hydrocarbon waste presents a critical environmental challenge, particularly in industries such as petroleum refining, chemical manufacturing, and transportation. Traditional waste management techniques often fall short in effectively mitigating the ecological impact of hydrocarbon contamination. However, advancements in technology offer promising solutions to address this issue.

4.2 Nanomaterials in Liquid Hydrocarbon Waste Management

Nanomaterials have garnered significant attention due to their unique properties, including but not limited to their high surface area-to-volume ratios, surface reactivity, and the ability to fine-tune their surface chemistry, positioning them as prime choices for a wide array of environmental applications. In the specific realm of managing liquid hydrocarbon waste, nanomaterials present a plethora of benefits that cannot be overlooked: First and foremost, their enhanced surface area-to-volume ratios allow for more efficient interaction and adsorption of contaminants present in the liquid hydrocarbon waste. This means that a relatively small amount of nanomaterial can effectively treat a larger volume of waste, thereby maximizing efficiency and

Table 4.1 Summarizing advanced waste management techniques for liquid hydrocarbon waste

Technique	Description
Bioremediation	Use of microorganisms to break down hydrocarbon pollutants into harmless byproducts through metabolism
Chemical oxidation	Employing oxidizing agents like hydrogen peroxide or ozone to degrade hydrocarbon molecules into less harmful substances
Thermal treatment	Combustion or thermal degradation of hydrocarbon wastes at high temperatures to destroy organic contaminants and reduce waste volume
Membrane separation	Utilizing membrane technologies to selectively separate hydrocarbons from liquid streams, allowing recovery or concentration
Supercritical fluid extraction	Using supercritical fluids like carbon dioxide to extract hydrocarbons from contaminated materials, offering high selectivity and efficiency
Adsorption	Application of adsorbent materials such as activated carbon or zeolites to capture hydrocarbon molecules from liquid wastes
Electrocoagulation	Applying an electrical current to destabilize and aggregate contaminants in water, facilitating their removal through precipitation
Hydrothermal liquefaction	Conversion of wet biomass or organic waste, including hydrocarbons, into bio-oil under high temperature and pressure conditions

minimizing the resources needed for the process. Moreover, the innate surface reactivity exhibited by nanomaterials plays a pivotal role in catalysing reactions that can lead to the degradation or conversion of harmful hydrocarbons into less toxic byproducts. This reactivity can be tailored and optimized to target specific contaminants within the waste stream, enhancing the selectivity and overall efficacy of the treatment process. Furthermore, the tunable surface chemistry of nanomaterials enables precise customization of their properties to suit the requirements of diverse waste management scenarios. By strategically modifying the surface functional groups or coatings of nanomaterials, researchers and engineers can tailor their interactions with different types of hydrocarbons, allowing for enhanced adsorption, separation, or even transformation mechanisms. Additionally, the nanoscale dimensions of these materials confer them with unique transport characteristics, facilitating their dispersion in liquid waste streams and ensuring thorough coverage and contact with the contaminants. This attribute can be particularly advantageous in situations where uniform distribution and efficient contact between the treatment agent and the waste constituents are critical for successful remediation efforts. In conclusion, the multifaceted advantages offered by nanomaterials in liquid hydrocarbon waste management underscore their immense potential to revolutionize environmental remediation practices. From their exceptional surface properties to their tailored reactivity and customizable surface chemistry, nanomaterials present a versatile toolkit for addressing the complex challenges posed by hydrocarbon contamination, paving the way for more sustainable and effective waste treatment solutions.

4.2.1 Adsorption-Based Remediation

Several research investigations have unequivocally illustrated the remarkable efficacy exhibited by nanomaterials in adsorbing hydrocarbon contaminants present in aqueous environments. Noteworthy examples of such nanomaterials encompass graphene-based materials, carbon nanotubes, and metal–organic frameworks, all of which have displayed substantial adsorption capabilities owing to their expansive surface areas and modifiable surface chemistry properties. Through these unique features, these nanomaterials have proven to be exceedingly efficient in the elimination of various hydrocarbons such as crude oil, gasoline, and diesel from water sources that have been contaminated with these substances (Dey, Bissoyi et al. 2024; Dey, Dash et al. 2024; Dey et al. 2023; John & Das 2024; Mishra et al. 2024; Mohanty et al. 2024). The versatility and adaptability of these nanomaterials enable them to effectively bind to the hydrocarbon molecules, thereby facilitating their removal from aqueous solutions and contributing significantly to the remediation of polluted water bodies. Furthermore, the structural characteristics of these nanomaterials play a crucial role in enhancing their adsorption performance, as the intricate arrangements of atoms and molecules on their surfaces create an ideal environment for the efficient capture and retention of hydrocarbon contaminants. The continuous advancements in nanomaterial synthesis and design have further expanded the range of applications for these materials in environmental remediation efforts, underscoring their importance in tackling pollution challenges associated with hydrocarbon contamination in water systems. Through ongoing research and development initiatives, scientists and engineers are continuing to explore novel ways to enhance the adsorption capacities and overall efficiency of nanomaterials for the removal of hydrocarbons from aqueous environments, thereby fostering sustainable solutions for addressing environmental pollution concerns.

4.2.2 Photocatalytic Degradation

Photocatalytic degradation, a process enhanced by the inclusion of nanomaterials like titanium dioxide (TiO_2) nanoparticles and zinc oxide (ZnO) nanoparticles, represents a highly promising avenue for breaking down hydrocarbon pollutants when exposed to UV radiation. Through this method, the inherent photocatalytic properties of these materials spur the creation of reactive oxygen species which play a pivotal role in transforming hydrocarbons into less harmful byproducts, ultimately expediting the overall remediation process. This innovative approach harnesses the powerful effects of nanotechnology to address environmental challenges by utilizing the unique properties of TiO_2 and ZnO nanoparticles to drive the photocatalytic breakdown of complex hydrocarbons under UV light. The resultant chemical reactions triggered by the introduction of these nanoparticles lead to the generation of oxygen radicals that decisively alter the structure of hydrocarbons, guiding them towards a pathway of reduced toxicity. Such a mechanism not only holds the potential to greatly increase the efficiency of pollution remediation efforts but also significantly minimizes the environmental impact of hydrocarbon pollutants, thereby outlining a sustainable solution to a pressing global issue. Ultimately, the controlled integration of TiO_2 and ZnO nanoparticles into the photocatalytic process showcases the innovative ways in which nanomaterials can be leveraged to address complex environmental challenges, underscoring the importance of advanced materials and technology in paving the way towards a cleaner, healthier future.

4.2.3 Chemical Transformation

Certain nanomaterials, with their unique properties at the nanoscale, demonstrate remarkable catalytic abilities that facilitate the crucial chemical conversion of hydrocarbon contaminants into less harmful forms. For example, zero-valent iron nanoparticles have garnered significant attention in the scientific community due to their exceptional capability to break down chlorinated hydrocarbons via reductive dechlorination reactions. In addition to iron nanoparticles, bimetallic nanoparticles comprising different metal combinations have exhibited even greater catalytic efficiency in dealing with the degradation of intricate hydrocarbon mixtures, showcasing a promising avenue for more effective environmental remediation strategies. The synergistic effects observed in these bimetallic systems underscore the potential for nanomaterial-based solutions in tackling diverse and persistent environmental pollutants, highlighting the burgeoning field of nanocatalysis as a key player in sustainable remediation technologies. Furthermore, ongoing research efforts are focused on fine-tuning the properties of nanomaterials to enhance their catalytic performance, aiming to address the evolving challenges posed by various hydrocarbon contaminants in our environment. Through continued exploration and innovation, nanomaterials hold immense promise in advancing the field of environmental remediation and fostering a cleaner, healthier planet for generations to come.

4.3 Environmental Applications

Nanomaterial-based adsorbents and catalysts represent cutting-edge tools utilized across a diverse array of environmental settings to address the pressing issue of liquid hydrocarbon waste remediation. These advanced materials have found extensive applications in critical areas such as groundwater remediation, wastewater treatment, and soil remediation. In groundwater remediation, nanomaterial-based adsorbents and catalysts have demonstrated remarkable effectiveness in removing harmful hydrocarbons and restoring water quality to acceptable levels. Similarly, in wastewater treatment processes, the intrinsic properties of nanomaterials enable them to efficiently adsorb and degrade hydrocarbon pollutants, leading to cleaner effluents being discharged back into the environment (Paul et al. 2024; Priyadarsini & Das 2024, 2025; Ray et al. 2024; Tripathy et al. 2022). Moreover, in soil remediation efforts, nanomaterials have proven to be invaluable in facilitating the cleanup of contaminated sites by sequestering, degrading, or transforming liquid hydrocarbon waste pollutants. The adaptability and versatility of nanomaterials play a pivotal role in tailoring bespoke solutions to specific contamination scenarios, thereby enhancing the overall efficacy and success of remediation efforts. By harnessing the unique properties of nanomaterials, environmental experts can design innovative strategies that precisely target and address liquid hydrocarbon waste contamination, contributing to a more sustainable and cleaner environment for present and future generations.

4.4 Bioremediation of Liquid Hydrocarbon Waste

Bioremediation, characterized by its sustainable, environmentally friendly, and economically viable nature, is a methodology that harnesses the impressive metabolic capabilities of microorganisms to effectively degrade and neutralize harmful contaminants in the ecosystem (Behera et al. 2022b; Kanungo et al. 2024). Within the domain of managing liquid hydrocarbon waste specifically, the process of bioremediation predominantly relies on the extraordinary metabolic prowess of a wide array of organisms, including bacteria, fungi, and various microscopic life forms, to metabolize hydrocarbons, thereby converting them into benign byproducts such as carbon dioxide and water. The success of bioremediation is intricately tied to numerous critical factors, encompassing the level of microbial activity, prevailing environmental conditions, nutrient availability, and oxygen levels (Behera et al. 2022). Through the exploration of several compelling case studies, the implementation of bioremediation strategies in various environmental settings has been exemplified, underscoring its effectiveness in combatting soil and water contamination challenges arising from the widespread presence of hydrocarbon pollutants.

Key Mechanisms of Bioremediation:

1. **Aerobic Biodegradation**: In aerobic conditions (with oxygen present), aerobic bacteria and fungi utilize hydrocarbons as carbon and energy sources through enzymatic reactions, ultimately mineralizing them into non-toxic end products.
2. **Anaerobic Biodegradation**: In anaerobic conditions (without oxygen), anaerobic bacteria perform hydrocarbon degradation through fermentation or other anaerobic metabolic pathways, producing methane, carbon dioxide, and other byproducts.
3. **Bioaugmentation**: This involves the introduction of specialized microbial cultures or genetically engineered organisms to enhance the biodegradation capabilities of indigenous microbial communities, especially in environments with low microbial activity.
4. **Biostimulation**: Biostimulation aims to enhance the activity of indigenous microorganisms by providing nutrients (such as nitrogen and phosphorus) or other growth-promoting factors to stimulate their metabolic activity and optimize hydrocarbon degradation rates.

Applications of Bioremediation in Liquid Hydrocarbon Waste Management: Bioremediation has been successfully applied in various scenarios involving liquid hydrocarbon waste, including:

- Treatment of oil spills in marine environments, where indigenous hydrocarbon-degrading bacteria help mitigate the environmental impact of oil spills by breaking down the oil components.
- Remediation of contaminated groundwater and soil at industrial sites, such as petroleum refineries or gas stations, where hydrocarbon leaks or spills have occurred.
- Management of produced water from oil and gas extraction operations, where microbial processes can help remove hydrocarbons and other contaminants before discharge or reuse.

Case Studies and Success Stories:

i. Exxon Valdez Oil Spill: Following the Exxon Valdez oil spill in 1989, bioremediation techniques, including the application of fertilizers to stimulate microbial activity, were employed to facilitate the natural degradation of oil in the affected areas.
ii. Gulf of Mexico Deepwater Horizon Oil Spill: After the Deepwater Horizon oil spill in 2010, microbial communities in the Gulf of Mexico played a crucial role in breaking down the released oil, highlighting the potential of bioremediation as a response strategy for large-scale oil spills.

4.5 Chemical Oxidation of Liquid Hydrocarbon Waste

Chemical oxidation has emerged as a highly promising and efficient technique for addressing the treatment of liquid hydrocarbon waste. This method involves utilizing a variety of powerful oxidizing agents to expedite the breakdown of complex hydrocarbon molecules into compounds that are substantially less harmful to the environment. Among the commonly employed oxidants in these processes are hydrogen peroxide, ozone, and persulfate, each playing a crucial role in facilitating the degradation and mineralization of hydrocarbon contaminants. The reaction between these oxidizing agents and the hydrocarbon pollutants results in a transformative process that significantly reduces the environmental impact of these harmful substances. The versatility of chemical oxidation allows for its application in both in situ and ex situ treatments, enabling flexibility in addressing diverse site-specific conditions and remediation objectives. Through careful consideration of the unique circumstances at each contaminated site, environmental practitioners can effectively determine the most appropriate approach to harness the power of chemical oxidation. This adaptability is key to optimizing the outcomes of remediation efforts and ensuring the successful treatment of liquid hydrocarbon waste. Over the years, numerous case studies have underscored the remarkable effectiveness of chemical oxidation in remedying various forms of liquid hydrocarbon waste. From petroleum hydrocarbons to chlorinated solvents and even emerging contaminants, this technique has consistently demonstrated its capacity to tackle a wide range of environmental challenges (Behera et al. 2020; Lahiri et al. 2021). These real-world examples serve as compelling evidence of the significant impact that chemical oxidation can have in transforming hazardous substances into harmless byproducts, thereby contributing to the restoration and preservation of our ecosystems.

Key Mechanisms of Chemical Oxidation:

1. **Direct Oxidation**: Oxidizing agents directly react with hydrocarbon molecules, breaking chemical bonds and transforming them into smaller, less complex compounds. For example, hydrogen peroxide can oxidize hydrocarbons to form alcohols, ketones, or carboxylic acids.
2. **Free Radical Reactions**: Some oxidizing agents, such as persulfate or ozone, generate free radicals upon activation. These free radicals are highly reactive species that can attack hydrocarbon molecules, initiating chain reactions that lead to their degradation.
3. **Generation of Reactive Oxygen Species (ROS)**: Oxidants like hydrogen peroxide can produce reactive oxygen species such as hydroxyl radicals ($\bullet OH \bullet OH$), which are extremely reactive and capable of oxidizing a wide range of organic contaminants, including hydrocarbons.

Applications of Chemical Oxidation in Liquid Hydrocarbon Waste Management: Chemical oxidation has been successfully applied in various scenarios involving liquid hydrocarbon waste, including:

- Treatment of contaminated groundwater and soil at industrial sites, where hydrocarbon spills or leaks have occurred due to industrial activities such as petroleum refining or chemical manufacturing.
- Remediation of oil and gas production wastewater, where organic contaminants like hydrocarbons need to be removed before discharge or reuse.
- Cleanup of oil spills in marine environments, where chemical oxidation can be used to mitigate the environmental impact of oil contamination by breaking down **hydrocarbons into less toxic compounds.**

Case Studies and Success Stories:

i. **Superfund Sites Cleanup**: Chemical oxidation has been used at Superfund sites contaminated with liquid hydrocarbon waste to remediate groundwater and soil. In some cases, advanced oxidation processes combining hydrogen peroxide with catalysts like iron or UV light have been employed to enhance contaminant degradation.
ii. **Oil Spill Remediation**: During the cleanup of the Prestige oil spill off the coast of Spain in 2002, chemical oxidation techniques, including the application of hydrogen peroxide and ozone, were used to treat contaminated seawater and shoreline areas, accelerating the degradation of spilled oil.

4.6 Membrane Filtration for Liquid Hydrocarbon Waste Treatment

Membrane filtration technologies play a crucial role in efficiently and selectively separating hydrocarbon contaminants from liquid waste streams by capitalizing on differences in size, charge, and molecular weight. These innovative processes, including reverse osmosis, ultrafiltration, and nanofiltration, are widely embraced in the treatment of liquid hydrocarbon waste due to their ability to attain high removal efficiencies and generate clean water suitable for reuse or safe discharge back into the environment. However, it is important to note that challenges like membrane fouling and concentration polarization are common occurrences within membrane filtration operations, necessitating regular maintenance and cleaning to safeguard process performance. The practical application of membrane filtration technologies in handling hydrocarbon-contaminated wastewater from various industrial and remediation sites is effectively illustrated through compelling case studies, which serve to underscore the flexibility and efficiency of these methods in addressing complex environmental challenges with precision and effectiveness.

Key Mechanisms of Membrane Filtration:

1. **Size Exclusion**: Membranes with specific pore sizes selectively allow molecules smaller than the pore size to pass through while retaining larger molecules. This mechanism is particularly relevant for microfiltration and ultrafiltration

membranes, which can effectively separate hydrocarbons from water based on size differences.

2. **Selective Permeability**: Membranes can exhibit selective permeability based on the chemical properties of the molecules being separated. For example, nanofiltration and reverse osmosis membranes utilize differences in solute concentration and electrostatic interactions to separate hydrocarbons from water.
3. **Pressure-Driven Separation**: Membrane filtration processes often involve the application of pressure gradients across the membrane to drive the separation process. This pressure-driven mechanism helps overcome resistance to flow and facilitates the passage of hydrocarbons through the membrane while retaining water and other contaminants.

Applications of Membrane Filtration in Liquid Hydrocarbon Waste Management: Membrane filtration has been successfully applied in various scenarios involving liquid hydrocarbon waste, including:

- Treatment of produced water from oil and gas extraction operations, where membranes are used to remove hydrocarbons, suspended solids, and other contaminants before discharge or reuse.
- Separation and recovery of hydrocarbons from oily wastewater generated by industrial processes such as petroleum refining, chemical manufacturing, and food processing.
- Concentration of hydrocarbons from contaminated water sources, enabling their subsequent treatment or disposal through other methods such as bioremediation or chemical oxidation.

Case Studies and Success Stories:

i. **Oilfield Produced Water Treatment**: Membrane filtration systems have been deployed at oilfield sites to treat produced water and recover valuable hydrocarbons for reuse or resale. These systems offer a cost-effective and environmentally friendly solution for managing produced water while minimizing freshwater consumption.

ii. **Industrial Wastewater Reclamation**: Membrane filtration technologies have been implemented in various industrial sectors to reclaim wastewater and recover valuable resources, including hydrocarbons. By separating hydrocarbons from water, membranes enable the reuse of treated water and the recovery of valuable products from wastewater streams.

4.7 Advanced Separation Technologies for Liquid Hydrocarbon Waste

Advanced separation technologies, including distillation, extraction, and adsorption, offer additional options for the treatment and recovery of valuable components from liquid hydrocarbon waste streams. Distillation techniques such as fractional distillation and vacuum distillation enable the separation of hydrocarbon mixtures based on their boiling points, allowing for the recovery of pure solvents and fuels. Extraction methods such as solvent extraction and supercritical fluid extraction facilitate the selective removal of target compounds from complex matrices, enhancing the efficiency of waste treatment processes. Adsorption techniques utilizing activated carbon, zeolites, and other adsorbents can effectively remove hydrocarbon pollutants from aqueous solutions through physical adsorption mechanisms. Case studies demonstrate the application of advanced separation technologies in various industries, including oil and gas, chemical manufacturing, and environmental remediation.

4.7.1 Membrane Distillation (MD)

Membrane distillation is an innovative separation process that utilizes hydrophobic membranes to separate water from hydrocarbon-containing liquid streams through vapor-phase transport. This technique relies on the vapor pressure difference across the membrane to drive water vapor through the membrane pores, leaving behind concentrated hydrocarbons. MD offers several advantages, including high selectivity, low energy consumption, and compatibility with high salinity or oily wastewater streams. It is particularly suitable for desalination, concentration of brine streams, and treatment of produced water from oil and gas operations (Alkhudhiri et al. 2012).

4.7.2 Liquid–Liquid Extraction (LLE)

Liquid–liquid extraction, also known as solvent extraction, involves the transfer of target compounds between two immiscible liquid phases based on their solubility differences. In the context of liquid hydrocarbon waste treatment, LLE can be employed to selectively extract hydrocarbons from aqueous solutions or organic solvents using appropriate extraction solvents. Advanced LLE techniques, such as reactive extraction or facilitated transport, enhance selectivity and efficiency by incorporating chemical reactions or carrier molecules to improve mass transfer rates. LLE is widely used in the petrochemical industry for hydrocarbon recovery, purification of specialty chemicals, and removal of contaminants from process streams (Treybal 1980).

4.7.3 Carbon Molecular Sieve (CMS) Membranes

Carbon molecular sieve membranes are a novel class of membranes with sub-nanometer-sized pores that selectively adsorb gas molecules based on their size and shape. These membranes exhibit high permeability to small molecules like hydrogen and hydrocarbons while rejecting larger molecules such as nitrogen and carbon dioxide. CMS membranes are well-suited for gas separation applications, including hydrocarbon recovery from refinery off-gases, natural gas upgrading, and hydrogen purification. Their robustness, thermal stability, and chemical resistance make them attractive for harsh operating conditions encountered in hydrocarbon processing (Koros and Mahajan 2000).

4.8 Conclusion

Advanced waste management techniques, such as bioremediation, chemical oxidation, membrane filtration, and advanced separation technologies, offer innovative solutions for addressing the treatment and remediation of liquid hydrocarbon waste. These techniques are crucial in managing the challenges associated with the complex composition and hazardous nature of liquid hydrocarbon waste. Bioremediation, leveraging the natural capabilities of microorganisms to break down contaminants, represents a sustainable and cost-effective method for dealing with liquid hydrocarbon waste. Ongoing research and technological advancements in bioremediation continuously enhance its efficacy as a versatile tool for environmental restoration and pollution control, ultimately contributing to the advancement of greener and more sustainable waste management practices. Chemical oxidation, another effective approach, plays a vital role in the treatment of liquid hydrocarbon waste by offering rapid and targeted degradation of contaminants. Through continuous research and technological innovations, chemical oxidation remains a key player in environmental remediation efforts, promoting the development of sustainable waste management practices that are crucial in safeguarding the environment and human health. In the realm of membrane filtration, this technology serves as a versatile and efficient method for liquid hydrocarbon waste treatment, providing selective separation and concentration of hydrocarbons from contaminated liquid streams. Ongoing research and technological advancements in membrane filtration further solidify its position as a fundamental technology in environmental remediation efforts, contributing significantly to the progress of sustainable waste management practices. Moreover, advanced separation technologies significantly contribute to the resolution of challenges associated with liquid hydrocarbon waste management by providing efficient and sustainable solutions for separation, purification, and recovery processes. Cutting-edge techniques like membrane distillation, liquid–liquid extraction, and carbon molecular sieve membranes offer advanced approaches that allow for the selective separation of hydrocarbons from intricate liquid streams. Continual research and development efforts in these technologies drive innovation and progress in the field of liquid hydrocarbon waste management, underscoring the importance of interdisciplinary collaboration and stakeholder engagement in addressing the multifaceted challenges of managing liquid hydrocarbon waste.

References

Alkhudhiri A, Darwish N, Hilal N (2012) Membrane distillation: a comprehensive review. Desalination 287:2–18

Alvarez PJJ, Illman WA (2015) In: Bioremediation and natural attenuation: process fundamentals and mathematical models. Wiley and Sons

Andrade LH, Alvarez PJ (2011) Advanced oxidation technologies: sustainable solutions for environmental remediation. Green Chem 13(10):2700–2710

Atlas RM, Hazen TC (2011) Oil biodegradation and bioremediation: a tale of the two worst spills in U.S. history. Environ Sci Technol 45(16):6709–6715

Bala K, Dogra A, Lata P, Das AP (2024) discussing sources and biological-chemical remediation approaches for microplastic pollution. In: Sustainable Microbial Technology for Synthetic and Cellulosic Microfiber Bioremediation, pp 37–59. https://doi.org/10.1007/978-3-031-62600-5_3

Baker RW (2004) Membrane technology and applications. Wiley & Sons

Bala K, Anshul, Verma S, Dogra A, Das AP, Kumar R, Nandini (2024) Sustainable management for urban plastic waste generation. In: Sustainable Management of Urban Plastic Waste Through Circular Economic Approaches, pp 211–236. https://doi.org/10.1201/9781003465447-13

Behera ID, Basak G, Kumar RR, Sen R, Meikap BC (2020) Treatment of petroleum refinery sludge by petroleum degrading bacterium Stenotrophomonas pavanii IRB19 as an efficient novel technology. J Environ Sci Health—Part A Toxic/Hazardous Substances and Environ Eng 56(2). https://doi.org/10.1080/10934529.2020.1866924

Behera ID, Meikap BC, Sen R (2022) Performance evaluation of bacterial consortium for biodegradation of total petroleum hydrocarbon: a comparative strategic biostimulation study. Research Square

Behera ID, Nayak M, Mishra A, Meikap BC, Sen R (2022b) Strategic implementation of integrated bioaugmentation and biostimulation for efficient mitigation of petroleum hydrocarbon pollutants from terrestrial and aquatic environment. Marine Pollut Bullet 177:113492. https://doi.org/10.1016/j.marpolbul.2022.113492

Chen J et al (2009) Long-term environmental impacts of nanomaterials: current knowledge and research gaps. J Hazard Mater 42(2):180–195

Chen L et al (2017) ZnO nanoparticles for the photocatalytic degradation of liquid hydrocarbon waste: recent advances and future perspectives. J Hazard Mater 63(5):192–201

Crittenden BD, Trussell RR, Hand DW, Howe KJ, Tchobanoglous G (2012) MWH's water treatment: principles and design. 3rd ed. Wiley and Sons

Das A, Dey S, Das AP (2024) Ecotoxicological impacts of synthetic microfiber pollutants and development of sustainable mitigation strategies. Environ Chem Ecotoxicol. https://doi.org/10.1016/J.ENCECO.2024.12.003

Dey S, Bissoyi A, Koteswara Reddy G, Das AP (2024) Synthetic microfibers: An urban emerging pollutant demanding for sustainable environmental solutions. In: Sustainable Microbial Technology for Synthetic and Cellulosic Microfiber Bioremediation, pp 107–121. https://doi.org/10.1007/978-3-031-62600-5_7

Das A, Dey S, Bala K, Gnanasekaran, R. Das AP (2024) Microbial bioremediation technology for sustainable treatment and management of synthetic microfiber waste. In: Sustainable Microbial Technology for Synthetic and Cellulosic Microfiber Bioremediation, pp 221–238. https://doi.org/10.1007/978-3-031-62600-5_12

Dey S, Dash N, Ray SK, Priyadarsini S, Chhotaray C, Das AP (2024) Occurrence and quantification of tire wear plastics pollutants from highway city dust of Bhubaneswar City of Odisha. In: Sustainable Microbial Technology for Synthetic and Cellulosic Microfiber Bioremediation, pp 77–89. https://doi.org/10.1007/978-3-031-62600-5_5

Dey S, Tripathy B, Kumar MS, Das AP (2023) Ecotoxicological consequences of manganese mining pollutants and their biological remediation. Environ Chem Ecotoxicol. https://doi.org/10.1016/J.ENCECO.2023.01.001

Dionysiou DD, Suidan MT (eds) (2015) In: Chemical oxidation technologies for the nineties: an overview. vol 26. Springer Science and Business Media

Drews A (2010) Membrane fouling in membrane bioreactors—characterisation, contradictions, cause and cures. J Membr Sci 363(1–2):1–28

Fenton HJH (1894) Oxidation of tartaric acid in presence of iron. J Chem Soc Trans 65:899–910

John A, Das AP (2024) Government strategies for achieving circular economy for plastic waste management in Southeast Asia. In: Sustainable Management of Urban Plastic Waste Through Circular Economic Approaches, pp 72–82. https://doi.org/10.1201/9781003465447-6/GOVERNMENT-STRATEGIES-ACHIEVING-CIRCULAR-ECONOMY-PLASTICWASTE-MANAGEMENT-SOUTHEAST-ASIA-AKBAR-JOHN-ALOK-PRASAD-DAS

Kanungo J, Sahoo T, Swain LP, Behera ID (2023) Toxicity of persistent hydrocarbon pollutants, sources and sustainable remediation process. In: Behera ID, Das AP (eds) Impact of petroleum waste on environmental pollution and its sustainable management through circular economy. Springer Nature Switzerland, pp 39–65. https://doi.org/10.1007/978-3-031-48220-5_2

Koros WJ, Mahajan R (2000) Pushing the limits on possibilities for large scale gas separation: which strategies? J Membr Sci 175(2):181–196

Lahiri D, Nag M, Dey A, Sarkar T, Joshi S, Pandit S, Das AP, Pati S, Pattanaik S, Tilak VK, Ray RR (2021) Biofilm mediated degradation of petroleum products. Geomicrobiol J 1–10.https://doi.org/10.1080/01490451.2021.1968979

Leahy JG, Colwell RR (1990) Microbial degradation of hydrocarbons in the environment. Microbiol Rev 54(3):305–315

Li Z et al (2011) Toxicity of nanomaterials in environmental applications: current understanding and future perspectives. Environ Sci Technol 48(3):315–328

Li J et al (2018) Photocatalytic degradation of hydrocarbon pollutants using TiO_2 nanoparticles: mechanisms and applications. Appl Catal B 90(4):457–468

Liu X et al (2012) Nanomaterial-based adsorbents and catalysts for soil remediation: a comprehensive review. J Hazard Mater 65(4):210–225

Liu X et al (2019) Metal-organic frameworks for the adsorption and catalytic degradation of liquid hydrocarbon waste: a comprehensive review. Water Res 78(1):102–115

Mishra S, Dash D, Das AP (2024) Microbial degradation of polyester microfibers using indigenously isolated bacterial strain Exiguobacterium Sp. CLEAN – Soil, Air, Water, 52(12):e202300343. https://doi.org/10.1002/CLEN.202300343

Mohanty M, Tripathy B, Koteswara Reddy G, Das AP (2024) Greener technologies for establishing restraint over microplastic pollution. In: Sustainable Microbial Technology for Synthetic and Cellulosic Microfiber Bioremediation, pp 239–258. https://doi.org/10.1007/978-3-031-62600-5_13

Mulder M (1996) Basic principles of membrane technology. Kluwer Academic Publishers

Paul A, Priyadarsini S, Karakuş S, Behera JK, Gnanasekaran R, Das AP (2024) Fate and accumulation of synthetic microfibers in the environment. In: Sustainable Microbial Technology for Synthetic and Cellulosic Microfiber Bioremediation, pp 61–76. https://doi.org/10.1007/978-3-031-62600-5_4

Priyadarsini S, Das AP (2024) Analytical and structural characterization of waste lithium-ion batteries for their effective recycling strategy. Environ Chem Ecotoxicol. https://doi.org/10.1016/j.enceco.2024.12.004

Priyadarsini S, Das AP (2025) Lithium bioleaching: A review on microbial-assisted sustainable technology for lithium bio-circularity. J Water Process Eng, 69:106744. https://doi.org/10.1016/j.jwpe.2024.106744

Ray SK, Dey S, Das AP, John A (2024) Microplastic waste management through greener technologies : A review. In: Sustainable Management of Urban Plastic Waste Through Circular Economic Approaches, pp 237–261. https://doi.org/10.1201/9781003465447-14

Ron EZ, Rosenberg E (2014) Enhanced bioremediation of oil spills in the sea. Curr Opin Biotechnol 27:191–194

Smith A et al (2016) Zero-valent iron nanoparticles for the remediation of chlorinated hydrocarbon waste: a review. Environ Sci Technol 49(2):602–615

Treybal RE (1980) Liquid extraction. McGraw-Hill

Tripathy B, Dash A, Das AP (2022) Detection of environmental microfiber pollutants through vibrational spectroscopic techniques: Recent advances of environmental monitoring and future prospects. Critical Reviews of Analytical Chemistry, 1–11. https://doi.org/10.1080/10408347.2022.2144994

Wang H et al (2010) Scalability of nanomaterial production for environmental applications: challenges and opportunities. Environ Sci Technol 49(5):602–615

Wang Y et al (2013) Nanomaterials in wastewater treatment: current status and future perspectives. Water Res 52(6):180–195

Wang Y et al (2015) Bimetallic nanoparticles for the catalytic degradation of complex hydrocarbon mixtures: current status and future prospects. J Hazard Mater 72(3):210–225

Wang J, Zhang X, Li F (2016) Applications of advanced oxidation processes for wastewater treatment: a review. Water Sci Technol 73(1):113–122

Wang H et al (2020) Carbon nanotubes for the remediation of liquid hydrocarbon waste: mechanisms and applications. J Hazard Mater 55(2):120–135

Wang Y, Li Z, Hu Y, Li J (2022) Graphene-based materials for liquid hydrocarbon waste treatment: current status and future perspectives. Environ Sci Technol 36(4):457–468

Wickramasinghe SR, Gupta BS (2016) Membrane separation principles and applications: from material selection to mechanisms and industrial uses. Elsevier

Zhang L et al (2014) Nanomaterial-based adsorbents for groundwater remediation: challenges and opportunities. Environ Sci Technol 38(1):315–328

Zhang Y et al (2021) Graphene-based nanomaterials for the adsorption of hydrocarbon pollutants: a review. Environ Sci Technol 45(3):210–225

Zhang H, Lv X, Wang F, Li Y, Wu Z (2023) Enhanced electrochemical degradation of liquid hydrocarbon waste using graphene-derived electrodes. J Hazard Mater 285:192–201

Chapter 5
Management Techniques, Existing and Recent Advancements for Treatment of Hydrocarbons from Liquid Wastes

Geetanjali Basak and Ipsita Dipamitra Behara

Abstract Aqueous solutions containing hydrocarbons are a major concern of various oil and gas industries. The worldwide demand for the dispense of hydrocarbon-containing aqueous samples, simultaneously with the strict regulations regarding industrial spent off, necessitates the stimulations for the treatment of oily wastewater. Advanced technologies have served to have overtaken the existing traditional technology as promising approaches. The chapter aims to provide insight into the existing advanced technologies for the management of oily aqueous solutions in order to maintain environmental sustainability. This chapter not only deals with the present advanced technologies but also emphasise on recent modern technologies along with nanotechnological methods for treatment of hydrocarbons present in the oily wastewater. The chapter also deals with the future scopes regarding research requirements for management of hydrocarbons containing liquid samples.

Keywords Oily liquid waste · Hydrocarbons · Organic content · Petroleum

5.1 Introduction

Oily liquid wastes are elucidated as an aqueous sample containing combinations of grease, lipids, and fats along with highly concentrated inorganic and organic dissolution (Wei et al. 2020). These kinds of samples are released from different commercial enterprises that relate to dairy, poultry, oil, and petroleum and even from restaurants as well (Sanghamitra et al. 2021). Owing to its volatile and steady properties, these unsinkable compounds are tough to mitigate. Therefore, the generation and release

G. Basak (✉)
Raniganj Girls' College, Affiliated to Kazi Nazrul University, Raniganj, West Bengal 713358, India
e-mail: geetanjalibasak@gmail.com; tgeetanjali_microbiology@raniganjgirlscollege.org

I. D. Behara
Indira Gandhi Institute of Technology, Sarang, Dhenkanal 75b9146, India

I. D. Behera and A. P. Das (eds.), *Environmental Hydrocarbon Pollution and Zero Waste Approach Towards a Sustainable Waste Management*, Environmental Science and Engineering, https://doi.org/10.1007/978-3-031-81615-4_5

of this kind of oily aqueous discharge impose a hindrance on the rapid development and industrialisation of urban areas (Kuyukina et al. 2020). To meet the specific rules and regulations for wastewater dispense and simultaneously combine recycling and re-usage policies, several parameters such as aqueous suspension constituents, capital, and enhanced product recovery are considered for the reclamation of oily wastewater (Rajasulochana and Preethy 2016). Consequences of petroleum hydrocarbons towards the ecological systems may interfere with the well-being of microbes and their physiological activities (Bala, Anshul et al. 2024; Bala, Dogra et al. 2024; Das, Dey, & Das 2024; Das, Dey, Bala et al. 2024). Mineralisation and mobility of hydrocarbons and their bi-products in the environment is one of its fates, while the other one involves the evaporation of these volatile compounds that are being stimulated by external environmental parameters (Lahiri et al. 2021)). Once the hydrocarbon is spilled into the water bodies, most of it escapes as light aliphatic chains through vapourization (Mango 1997; Behera et al. 2020; Das et al. 2024b; Mohanta et al. 2016). Oil-containing liquid samples are considered as dangerous United States Environmental Protection Agency (USEPA) (USEPA 2004) if not adequately mitigated as they might pose a serious threat not only to the aquatic and terrestrial ecosystems but also to the air (Ibrahim et al. 2017). Environmental management is the major concern for scientific researchers to check whether industries are maintaining environmental sustainability (Basak and Basak 2023; Mallick et al. 2023; Dey et al. 2024; Biswas et al. 2020). Intake of excessive dissolved oxygen by microbes in aquatic environment and harmful dispense of ammonia with hydrogen sulphide from the hydrocarbon containing wastewater makes them intolerable in the aquatic ecosystem due to unavailability of oxygen in it (Seveso et al. 2021; Das et al. 2023; Behera et al. 2022a; Sahu et al. 2024; Dey, Bissoyi et al. 2024; Dey, Dash et al. 2024; Dey et al. 2023; John & Das 2024; Mishra et al. 2024). Thus, oil containing effluents releasing nitrogen and phosphorus may lead to algal blooms (Lurling and Mucci 2020). Indirectly it might affect the health of humans by carrying carcinogenic phenolic compounds (Mearns et al. 2020). Moreover, this hydrocarbon mixed water cannot be utilised for farming as it may reduce the quality and hamper the agricultural property of the soil, thereby affecting plant growth and germination (Sutton et al. 2013). Nowadays, researchers are working on its remediation and reuse as well. Reviews are being given on its reuse and recycle (Basak 2024). Therefore, the present chapter aims to give an overall view of the existing and advanced methods for oily liquid waste management. The chapter also discusses on the constraints and advantages of each approach showing future strategies to troubleshoot all the difficulties faced during the operation of these approaches.

5.2 Produced Water

Other than oil containing wastewater, produced water is one such huge liquid source that carries dispenses from oily industries showcasing nearly 80% of the aqueous suspension (Igunnu and Chen 2014). About 60% of produced water is released into the sea with each tonnes of hydrocarbon generation (Jimenez et al. 2018). Produced water is a suspension of formation water, salt containing water from aquifer and injected water to enhance recovery of oil, chemicals and other products obtained through segregation methods. On the whole presence of mixed or suspended oil, dissolved solids, or gases along with aliphatic and aromatic hydrocarbons constitute the produced water (Ossai et al. 2020). Certain minerals like radioactive substances, toxic heavy metals, and positive or negatively charged ions are present, along with the existence of additional materials containing alcohol, dispersed solids, and gases in the produced water (Surkatti et al. 2020; Das et al. 2011). The constitution of produced water varies based on its source of existence, types of the oil industries their processing and formation of products as well.

5.3 Traditional Methods

In order to meet the policies while dispensing the produced water into the water bodies, it must be managed by physico-chemical and biological methods as it may contain various oily contaminants that may harmful for the ecosystem and remains there for long period.

5.3.1 Hydrocyclone

Total suspended and distributed materials are set aside depending on the divergence in denseness of the particle. However, the materials that are being mixed with the liquid sample cannot be separated. This method is simple to operate and requires less energy. It is used for managing produced water at commercial level as nearly more than 70% of oil, 45% of aromatic hydrocarbons and 30% of phenol can be extracted from produced water by this process. Although the method is approachable but its erroneous property creating heat and stress during operation imparts a disadvantage for this approach.

5.3.2 Coagulation/flocculation

The organic content of this kind of oily liquid waste can be treated by methods called Coagulation/flocculation. A series of chemical solutions are added to reduce the oxygen demand or organic content of the oily wastewater or produced water. Two types of combining speed is applied such as (i) high-to make the particles unstable and (ii) low-to clog the materials present inside the oily aqueous sample and remains at the bottom due to its heavy weight. Again, the method is not suitable for removing impurities that are liquified within the sample. Thus, this is the initial measure taken before other methods. Examples of coagulants are calcium hydroxide and ferric chloride (Altas and Buyukgungor 2008), poly-zinc silicate (Zeng et al. 2007) that removed nearly 100% of chemical oxygen demand and oil from oil-containing aqueous samples. However, the requirement of coagulants at high amounts leads to hazardous consequences on the economic and ecological implementation of this method. Moreover, it may lead to secondary pollution that requires further management later on.

This method is coupled with the generation of positively and negatively charged ions in the presence of electricity. Small globules made by the hydrogen gas clings to the material and conjugates them that are taken out from the top of the oily liquid sample. Approximately 100% of phenol is removed from petroleum-aqueous samples by this method (Abdelwahab et al. 2008). Thus, this technology needs large energy input in terms of electric current and cannot be used at the industrial level.

5.3.3 Adsorption

A process that involves the usage of a solid agent known as adsorbent and sequesters the contaminants superficially from water samples is a fast, easy to operate and low-cost process. However, the process is influenced by a series of parameters. A number of waste materials have been used, viz. sugarcane bagasse, shells of egg, peels of banana, activated carbon, and rice husk for the removal of aliphatic and aromatic hydrocarbons from the oily effluent (Hasan et al. 2019; Kanungo et al. 2023). Certain benchmarks are required to be satisfied in order to consider adsorbing agents as suitable. They are as follows: should be capable of adhering high amounts of contaminants; capability of adhering quickly; should be able to reuse and recycle and easily available.

5.3.4 Membrane Technology

A consolidated, highly potent method that necessitates the incorporation of fewer chemicals. This mainly involves filters of different diameters starting from the bigger to lower size of the pores, i.e. microfilters, ultrafilters, nanofilters, and reverse osmosis (Gisi et al. 2016). Dispersed solid material is taken off by the first filtration step followed by the scooping out of the oil from the produced water or oily wastewater sample. The third step involves the bypassing of toxic metals or ions, and mixed solids are withdrawn by the final step. There are certain membranes that are steady even at high temperatures. Polymeric membranes and ceramic membranes are of that category. Ceramic membranes are more beneficial than polymeric membranes as they are more potent in liquid recovery from oil. But its operational cost still remains to be as one of the disadvantages. Maintaining membrane life is one of the important issue as several particles get attached to the holes of the filter and blocks the pores.

A numbers of reports have been given on the alteration of membrane superficially by coating with nanomaterials of titanium, iron, graphene in oxide form (Yalcinkaya et al. 2020).

5.3.5 *Oxidation*

Oxidation means utilising an agent that, along with an activator, will oxidise to form OH^- ions in order to remove organic and inorganic content from the oil containing aqueous suspensions. The agent mineralises the contaminant to a simpler form. Examples of these agents include ozone, permanganate, chlorine dioxide, hydrogen peroxide, and UV radiation (Liang et al. 2018). Although this method is quite easy to carry on and efficient in water removal, still its operation cost and generation of harmful residual products are disadvantages. The use of hydrogen peroxide with different oxidants like iron and ozone is common for the separation of total organic content from various oily liquid samples that might be potent with a significant removal of more than 75%. Metal oxides in the form of nanomaterials are also efficient in removing organic content from oily wastewater samples. This method is suitable for industrial applications as it is ecologically benign and low cost.

5.3.6 *Thermal Approach*

It is a powerfully built and sturdy process that involves two steps for obtaining water from oilfield wastewater with better attributes. Initially, the water evaporates and then cools and concentrates on a desired quality product (Liang et al. 2018). Temperature and pressure play an important role in this process. Therefore, these factors decide the type of process to be followed. The differences between these three types are given in Table 5.1.

5.4 Modern Approaches

Removal of contaminants through microorganisms by a phenomenon called remediation in existence or in-existence of oxygen, where these small microscopic worlds use these contaminants as carbon or nitrogen substrate with simultaneous generation of important metabolite are in demand these days (Adetunji and Olaniran 2021a, b). Although this method is highly appreciable in terms of cost, the requirement of capital and continuous treatment of waste deposition has come with constraints (Mohanty et al. 2024; Paul et al. 2024; Priyadarsini & Das 2024, 2025; Ray et al. 2024; Tripathy et al. 2022). The main target of the scientists while treating petroleum hydrocarbons is to break down the total hydrocarbon content present in the petroleum sample of

Table 5.1 Difference between multistage flash, multi-effect and vapour compression distillation unit

S. No.	MSD	MED	VSD
Full form	Multistage flash distillation	Multi-effect distillation	Vapour compression distillation
	Oil water sample is heated up under low pressure and changed to steam	Water sample is vaporized and then cooled. Requires less temperature	It requires temperature for vapourization and compression. It can work even at low temperatures in which MED and MSF cannot work
Cost	Low cost	Less	Minimum
Stability	Almost two decades	Yes	Yes
Disadvantage	It leads to formation of dirt that must be cleaned at regular intervals with acids	Formation of scales stops the method to be used in industries	Can be used at commercial level along with the aim to remove other pollutants from the aqueous sample of oil industries
References	Ettouney et al. 2002	Jimenez et al. 2019	Ettouney et al. 2002

sludge or water by biodegradation process (Behara et al. 2020; Mohanta et al. 2024; Behera and Das 2023; Mishra et al. 2021).

5.4.1 *Plant Mediated Remediation*

This method is also known as Phytoremediation. A process where pollutants are broken down by the action between the indigenous microbes present in the soil around the plant root. Various biomolecules secreted by the radicles of plants induce microorganisms in the biodegradation process. The utilisation of water hyacinths and biscuit grasses is common for the separation of oil from aqueous solutions (Sansui et al. 2012). The efficiency of this process is based on several parameters, viz., variety of plants, the constitution of microbes and soil, microbial composition, soil properties, the rectification of soil and quality of the soil. Using biowaste material, including the skin of bananas and cereals from the brewing industry, removes hydrocarbons. Several disadvantages of this approach include huge land footprints, environmental factors, the necessity of various minerals and a lot of time to execute this operation.

5.4.2 *Microbiological Methods*

Microbes play a crucial role in degrading inorganic and organic pollutants present in aqueous oilfield solutions (Kanungo et al. 2024; Das et al. 2024a; Tripathy et al. 2024). As a matter of fact, they utilise this organic molecule as their source of carbon and nitrogen nutrition, which helps them grow and mineralise as well. After growth, the biomass is segregated through proper gathering methods. Oxygen is one of the most important factors that decide the mode of biomineralization. Microbes use biological aerated filters (BAFs), activated sludge process, and aerobic treatment mediated by algae in presence of oxygen to remove floated and mixed particles from the oil industry's effluents. This is an aerobic approach. Among the three types, biological aerated filters are considered to be the easiest in terms of operation as they can remove approximately 80% of organic content, monocyclic aromatic hydrocarbons and nearly 95% of oil from the oily effluents (Wei et al. 2020; Zhao et al. 2006). However, the continuous dispersal of secondary waste that requires post-management is one of its hindrances.

Although another process where sludge is being activated in the existence of oxygen, microbes present there could efficiently remove nearly 100% of hydrocarbons along with retrieval of approximately 90% of oil from the produced water sample (Li et al. 2017). Again, the formation of a huge mass of solid waste and the compulsory necessity of oxygen, along with post-processing requirements, inhibits the process from being followed at commercial levels. The mode of action for the reclamation of the oily liquid samples by algae is different from the aforementioned processes. Here it requires carbon dioxide to form oxygen. Even though micro or macro algae utilise hydrocarbons as the sole source of carbon, an excessive amount of it may be harmful to the biological agent. Limited quantities of nitrogen and phosphorus in wastewater commenced from oil industries inhibit the cultivation of algae. Several species of microalgae are common for treating oily aqueous suspensions. Example: *Chlorella* and *Neochloris.* A consortium of bacteria and algae, with their positive, beneficial interaction, might increase the efficiency of the treatment process. Reports are there that state the breaking down of alkyl-containing aromatic hydrocarbons (Jones et al. 2008).

Drawbacks of aerobic process in terms of generation of huge waste can be overcome by a process which involves degradation without implementation of oxygen. This process is called an anaerobic process that, along with less formation of waste in terms of sludge, also produces several other biological products by breaking organic contaminants (Behera 2022). Anaerobic digestion (AD) is one such type of anaerobic process that leads to the formation of biologically generated gas and volatile fatty acids (Kong et al. 2019). Based on the substrate, the formation of biologically desired products is produced, such as in the presence of hydrogen and carbon dioxide. A lot of external factors are needed to be controlled. Therefore, the approach is cumbersome and costly, along with the generation of solid waste and gas.

5.5 Large-Scale Technologies

In order to increase the efficacy and continuity of the process with endurance towards noncompliant contaminants, reactors mediated through the microbial world are taken into account. A series of reactors are used to manage oily wastewater. In a fixed

Table 5.2 List of works showcasing immobilisation of microbes and bioreactors involved in the removal of organic content of oily liquid samples

Technique	Species	Organic content removal (%)	Reference
Free and immobilized cells	B350M-immobilized B350-free	78 64	Zhao et al. (2006)
Polyvinyl alcohol	*Bacillus* sp. M-12	90	Li et al. (2005)
Chitosan	*Sphingobium* sp.	~ 73	Khondee et al. (2012)
Polyurethane foam	*Gordonia* sp.	50	Chanthamalee et al. (2013)
Polyethylene plastic pellet	*Pseudoxanthomanas* sp. RN 402	83	Nopcharoenkul et al. (2013)

bed reactor in the presence of oxygen under subsurface conditions, nearly 90% of organic content and mixed solids could be removed within 250 days of the operation (Vendramel et al. 2015). Similarly, other bioreactors like fixed film bioreactors and sequencing batch bioreactors are also used.

Microbes incorporated within a shell called as immobilized cells are also used in bioreactors for effective treatment of oily liquid samples. This technique not only enhances the potency of the biological agent but also opens a chance to re-utilize for several cycles as it is protected through the shell from external unfavourable calamities, thereby lowering the operational charges (Adetunji and Olaniran 2018). A list of works showcasing the immobilisation of microbes and bioreactors involved in the removal of the organic content of oily liquid samples is given in Table 5.2. When the continuous operation is carried out in the absence of air or under anaerobic conditions, it is advantageous in terms of energy utilisation, generation of methane gas, less expenditure and release of low amount of waste. However, the deposition of oil and gas on the surface of the waste and body hinders the process from being carried out successfully (Shende and Pophali 2020). For example, an aqueous sample packed in biochar, when run through a column reactor, could remove 95% of organic content from the sample with the formation of biomethane and biodiesel (Lopez et al. 2014).

5.5.1 Membrane Bioreactors (MBRs)

A novel technique that involves the usage of a small sheet made up of organic or inorganic layers to allow certain materials to pass through them. This mechanism is involved in the segregation of oil from oily liquid suspension (Makisha 2020). Biological reactors with these membranes, pores containing material, mainly target aliphatic and aromatic hydrocarbons. Along with 90% of total organic content with simultaneous 90% recovery of oil and gas is being separated by this approach for synthetic and industrial oily aqueous samples as well (Pendashteh et al. 2012; Mallick et al. 2024). Although the process does not need any extra material and energy and is easy going, it still requires pressure to be carried out. Four types of filtration processes are based on the size of the porous material, i.e. ultra, micro, nano and reverse osmosis (Pendergast and Hoek 2011; Bal et al. 2020).

Three types of material are there for membranes, viz. (i) polymeric, (ii) ceramic and/or (iii) nanomaterials.

Polymeric membranes are low-cost, tiny, less energy-consuming, highly efficient membranes that can scoop out the oil spread on the surface of oily wastewater. Examples of some porous materials are polyethersulfone and polyamide, etc. One of the major disadvantages is its low attraction to water and tendency to produce foul smells during the operation. The second membrane is capable of withstanding harmful external factors with very tiny pores and excessive fouling. The materials are made of inorganic components like boric acid, sodium metasilicate, zirconia, etc. The third one has increase in surface area and flux rate containing nanomaterials of different metals embedded on the membrane to promote additional recovery of oil and its separation from oily waste liquid.

5.5.2 Electrochemical Technology

Using electricity for removing hydrocarbon contaminants from oily liquid samples is an ecologically benign, skilled, low-cost cost and energy-profitable technology (Treviño-Roséndez et al. 2021; Behera et al. 2021; Pradhan et al. 2023). Several factors are responsible for the operation, such as process timing, saline incorporation, the material of the cathode/anode, the varied concentration of the phenol, pH and temperature (de Almeida et al. 2014). Two types of these methods are (i) electro-coagulation and (ii) electro floatation (Chen 2004). The first technology includes the dispensing of coagulating agents through electrolysis by releasing H_2 and OH^- ions (Tetteh and Rathilal 2020). It does not need any chemical intervention and has the additional advantages of withstanding diversified contaminants and liquid samples, being easy to operate and being ecologically benign. Certain drawbacks include operational costs and high energy consumption (Uludag Demirer et al. 2020; Behera et al. 2022a, b). For petroleum, petroleum-containing industrial plant wastewater was processed to remove nearly 87% and 90% of organic content and dispensed solids, respectively (Ibrahim et al. 2013).

The second aforementioned technique permits harmful materials to float on the upper part of the liquid sample through electrolysis and release small globules of H_2 gas and O_2 gases from the anode and cathodes. Then, these globules conjugate with the harmful contaminants and come on top of the liquid sample. Finally, they are being removed by scooping (Hosny 1992). Stainless steel showed better removal of oil than aluminium electrodes, where approximately 100% of oil separation was dominating over approximately 91% of oil segregated through stainless steel and aluminium electrodes, respectively (Adetunji and Olaniran 2021a, b).

5.5.3 Oxidation Process by Advanced Methods

These are non-conventional processes where a method is used applying electricity while releasing an oxidising agent in order to remove electrons from the contaminants residing in oily liquid sources (Abou-Taleb et al. 2020). This phenomenon is called the electrooxidation process. The addition of oxygen takes place either on the upper layer of the electrode induced by the recent weight and speed at which the contaminants are getting mixed off or indirectly with the help of some material released electrochemically, being dependent on the same factors (Chen 2004). There are several instances given to prove that the technology is quite applicable as follows:

(i) In an experiment, lead oxide-strengthened tantalum, and boron-conjugated diamond electrodes were examined for removal of organic content from the oily aqueous sample. Results of boron conjugated diamond showed improvement over the previous one with separation of 85.0% and 96.0%, respectively (Gargouri et al. 2014).

Ruthenium oxide conjugated with titanium electrodes showed a removal of 90.0% of polycyclic aromatic hydrocarbons within 1 h and 30 min (Tran et al. 2009).

Another innovative method induced by Fenton oxidation process where hydrogen peroxide is released from negatively charged electrodes and Fe^{2+} ions are dispersed from acidic solutions of iron. These hydrogen peroxide are strong mineralisers that carry out this process, either straightforwardly or mediated through OH ions and materials that help in the aggregation of organic matter by flocculation. (Kakavandi and Ahmadi 2019). Although this process is ecologically sustainable, universal, and system driven its inadequate generation of hydrogen peroxide imparts a disadvantage to it.

The assemblage of the above two methods can also be applied to treat hydrocarbons containing wastewater or produce water samples. Using different types of electrodes (aluminium and iron) in this process may promote hydrocarbon removal from 60.0 to 99.0%.

5.5.4 Photocatalytic Oxidation

This technique is one of the most promising, cheap, highly stable under different ecological stresses. The technology is driven by any light source (natural or artificial) that leads to the generation of OH^- and O_2^- on the body of agents that favours the reaction in the presence of a light source (Basak et al. 2020). A number of environmental parameters influence the treatment process, such as initial concentration of hydrocarbons, the strength of light, pH, temperature, and types of nanomaterial being used in the experiment (Sundar and Kanmani 2020). The process is vigorous when it is conjugated with nanoparticles. For example, UV irradiation in the presence of titanium nanoparticles could remove approximately 86% of oil when compared only in the presence of UV irradiation; less removal of 70% oil was noted (Kang et al. 2011). Similarly

Polymer conjugated with nanomaterials, e.g. polyaziridine, along with titanium dioxide formed by layering over itself in the presence of UV could remove 98% of chemical oxygen from petroleum-containing industrial wastewater (Mansouri et al. 2014).

5.5.5 *Adsorption Process*

A physicochemical phenomenon that utilizes the mechanism of the electrostatic force of attraction between the contaminants and adsorbing agents. Pollutant that is present in the aqueous sample is the desired component to be removed is known as adsorbate. The solidified, aqueous or gaseous state where the sorbate gets sequestered on the adsorbent. When the adsorbent is made of biological substances, it is known as a biosorbent. This approach is used for the treatment of inorganic as well as organic materials like oil using biosorbents obtained from biological waste material (Jun et al. 2020). Several biosorbents like outer skeleton of crab, straws of burley, polymers conjugated with chemical as well as biological surfactants could remove more than 70.0% of organic content from the oily aqueous sample. Although granular activated carbon alone is quite efficient in exhibiting an adsorption process, still when it is implemented along with surface-modified CTAB (Cetrimonium bromide), it could remove about 90.0% of oil from the oily liquid sample (Jahi et al. 2015).

The process is again affected by different external parameters, just like other processes. Initially, the process is optimised by varying the different external parameters one by one to obtain the maximum separation after standardising the procedure. The mechanism of the above two processes can be well understood under the increased magnitude of the light source so that the electrons get elicited betwixt conduction and valence band in case of hydrocarbons with more than one benzene ring (PHC) (Shanker et al. 2017). The demonstration is highlighted in Fig. 5.1.

$$\text{PHC} + \text{irradiation} \rightarrow (\text{PHC})^*$$
$$\text{nanomaterials} + \text{irradiation} \rightarrow (\text{PHC})\ e^-\text{CB} + e^+\text{VB}$$
$$e^- + O_2 \longrightarrow \cdot O_2^-$$
$$O_2^- + \text{PHC} \longrightarrow \text{PHC} + \text{OO}$$
$$O_2^- + H_2O + H^+ \longrightarrow H_2O_2 + O_2$$
$$H^+ + H_2O \longrightarrow H^+ + \cdot OH^-$$
$$H^+ + OH \longrightarrow \cdot OH$$
$$(\text{PHC})^* + \cdot OH \longrightarrow \text{PHC (mineralized)} \longrightarrow \text{Broken components } (CO_2, NO_3^{2-}, NH_4^+, SO_4^{3-})$$

Fig. 5.1 Mechanism of hydrocarbon degradation by photocatalytic oxidation process

5.5.6 Conjunctive Management Concept

Since the aqueous waste obtained from petroleum industries is a mixture of different components, the use of individual techniques may not be feasible for the reclamation of those samples. Thus, the use of more than one technology is in demand in the form of a combination to extract or segregate the oil spills from the liquid samples. Recently, scientists have been interested in formulating conjugated technologies to fulfil their aims of remediation of contaminant-containing oily wastewater (Han et al. 2019). Mixed implementation of electrocoagulation and biological reactor with free cells in the presence of oxygen could remove over 80.0% of total petroleum hydrocarbon content from oily liquid solutions. The potentiality was enhanced when the microbes were encapsulated within a sheath, the total hydrocarbon content was increased to 98. 0% (Perez et al. 2016). Similarly, the combination of various techniques like a photocatalytic reactor with ultrafiltration in the presence of titanium dioxide nanoparticles, electrocoagulation with ultrafiltration and nanofiltration methods in the presence of nanoconjugates are also mentioned by the report given (Adetunji and Olaniran 2021a, b). An overall illustration of various technologies is given in Fig. 5.2.

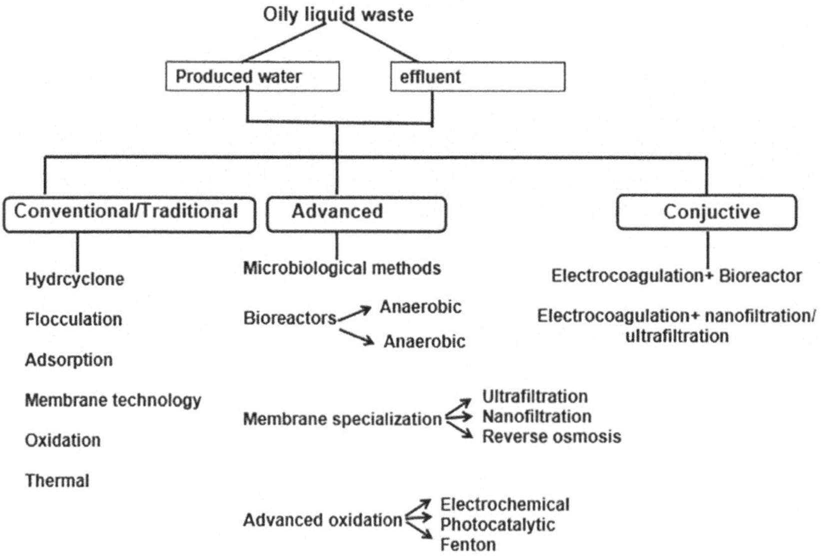

Fig. 5.2 Overall illustration of liquid waste treatment approaches

5.6 Conclusion

Impurities generated through aqueous waste from various petroleum oil-containing industries serve to be a major issue because of their lethal consequences over the ecological balance. Therefore, many traditional approaches have been practised to reduce this problem. However, certain disadvantages dominated viz.

(i). They are incapable of complete eradication of these contaminants as these materials can, in turn, lead to the generation of solid waste that gets clogged on its surface and may hinder the process.
(ii). A single chemical process such as flocculation/coagulation requires a secondary method like the microbe-mediated reactors to completely remove the oily contaminants from the environment.

Thus, using advanced technology like adsorption through biological mediators provides an additional advantage during the process. Conjunctive methods are more efficient and capture the recent interest of the researchers to eliminate these wastes from the environment. The choice of exact method is based on the implementation charges, versatility of components present in the liquid sample, policies of treatment and final application of the treated water.

5.7 Future Opportunities

- Balancing the implementation rate and maintaining charges while handling the biological reactor is a challenge that can be combated through the fabrication of low-cost reactors with maximum potency. The built-up should also try to maintain the energy utilisation rate, fouling smell release of various gaseous materials.
- The smell generated from the membrane technique by altering the composition with extra improvement so that the material becomes tolerant towards the odour and increased attraction towards water.
- Fabrication of low-cost electrodes with increased steadiness is required.
- The liquid sample can be treated priorly with chemical or physical agents to remove oil and gas naturally in photocatalytic processes. Moreover, surface alterations of materials that favour reaction can be done so as to obtain a versatile capability of the agents with increased potentiality. Further care must be taken while processing so that it can be reused.
- Attention must be paid to the implementation of biological-based waste products while undergoing the adsorption process so that the method will be of low cost, ecologically benign, and more efficient.

Acknowledgements The authors express their gratitude to the Department of Chemical Engineering, Indira Gandhi Institute of Technology, Sarang. The IDB expresses gratitude to the Odisha State Higher Education Council for their support in the form of the Seed Fund Research Project (Project ID: 22SF/CE/022 Chemical Engineering).

References

Abdelwahab O, Amin NK, El Ashtoukhy ESZ (2008) Electrochemical removal of phenol from oil refinery wastewater. J Hazard Mater 163:711–716

Abou-Taleb EM, Hellal MS, Kamal KH (2020) Electro-oxidation of phenol in petroleum wastewater using a novel pilot-scale electrochemical cell with graphite and stainless-steel electrodes. Water Environ J 35:259–268

Adetunji AI, Olaniran AO (2018) Treatment of lipid-rich wastewater using a mixture of free or immobilized bioemulsifer and hydrolytic enzymes from indigenous bacterial isolates. Desalin Water Treat 132:274–280

Adetunji AI, Olaniran AO (2021a) Production and potential biotechnological applications of microbial surfactants: an overview. Saudi J Biol Sci 28(1):669–679

Adetunji AI, Olaniran AO (2021b) Treatment of industrial oily wastewater by advanced technologies: a review. Appl Wat Sci. 11:98

Altas L, Buyukgungor H (2008) Sulfide removal in petroleum refinery wastewater by chemical precipitation. J Hazard Mater 153:462–9

Bal M, Behera ID, Kumari U, Biswas S, Meikap BC (2020) Hydrodynamic study and particulate matter removal in a self priming venturi scrubber. Environ Technol Innov 20:101167. https://doi.org/10.1016/j.eti.2020.101167

Bala K, Anshul, Verma S, Dogra A, Das AP, Kumar R, Nandini (2024) Sustainable management for urban plastic waste generation. In: Sustainable Management of Urban Plastic Waste Through Circular Economic Approaches, pp 211–236. https://doi.org/10.1201/9781003465447-13

Bala K, Dogra A, Lata P, Das AP (2024) Discussing sources and biological-chemical remediation approaches for microplastic pollution. In: Sustainable Microbial Technology for Synthetic and Cellulosic Microfiber Bioremediation, pp 37–59. https://doi.org/10.1007/978-3-031-62600-5_3

Basak G, Hazra C, Sen R (2020) Biofunctionalized nanomaterials for in situ clean-up of hydrocarbon contamination: a quantum jump in global bioremediation research. J Env Manag 256(15):109913

Basak G (2024) Recycling, re-using, regeneration and recovering of value added products petroleum hydrocarbons through circular economic based approaches. In: Impact of petroleum waste on environmental pollution and its sustainable management through circular economy. Springer nature

Basak S, Basak G (2023) Green HRM and IPR: a way towards environmentalism. Himal J Econo Busi Manag 4

Behara ID, Basak G, Kumar RR, Sen R, Meikap BC (2020) Treatment of petroleum refinery sludge by petroleum degrading bacterium *Stenotrophomonas pavanii* IRB19 as an efficient novel technology. J Env Sci Heal Part A 56(2):226–239

Behera ID (2022) Biodegradation of petroleum sludge by indigenously developed bacterial consortium. IIT, Kharagpur

Behera ID, Das AP (2023) Impact of petroleum waste on environmental pollution and its sustainable management through circular economy. Springer

Behera ID, Basak G, Kumar RR, Sen R, Meikap BC (2020) Treatment of petroleum refinery sludge by petroleum degrading bacterium *Stenotrophomonas pavanii* IRB19 as an efficient novel technology. J Environ Sci Health-Part Toxic/Hazard Subst Environ Eng 56(2). https://doi.org/10.1080/10934529.2020.1866924

Behera ID, Nayak M, Biswas S, Meikap BC, Sen R (2021) Enhanced biodegradation of total petroleum hydrocarbons by implementing a novel two-step bioaugmentation strategy using indigenous bacterial consortium. J Environ Manag 292:112746. https://doi.org/10.1016/j.jenvman.2021.112746

Behera ID, Meikap BC, Sen R (2022a) Performance evaluation of bacterial consortium for biodegradation of total petroleum hydrocarbon: a comparative strategic biostimulation study. Res Sq 1–24

Behera ID, Nayak M, Mishra A, Meikap BC, Sen R (2022b) Strategic implementation of integrated bioaugmentation and biostimulation for efficient mitigation of petroleum hydrocarbon pollutants from terrestrial and aquatic environment. Mar Pollut Bull 177:113492. https://doi.org/10.1016/j.marpolbul.2022.113492

Biswas S, Diwakar RK, Behera ID, Meikap BC, Sen TK, Khiadani M (2020) Aqueous phase phenol removal from synthetic and real steel plant effluents through a batch and semifluidized bed column operation: experimental and model analysis. J Environ Chem Eng 8(5):104441. https://doi.org/10.1016/j.jece.2020.104441

Chanthamalee J, Wongchitphimon T, Luepromchai E (2013) Treatment of oily bilge water from small fshing vessels by PUF-immobilized Gordonia sp JC11. Water Air Soil Pollut 224:1601

Chen G (2004) Electrochemical technologies in wastewater treatment. Separ Purif Technol 38:11–41

Das AP, Sukla LB, Pradhan N, Nayak S (2011) Manganese biomining: a review. Biores Technol 102(16):7381–7387. https://doi.org/10.1016/J.BIORTECH.2011.05.018

Das AP, Nayak B, Behera ID (2024b) Applications of microorganisms in advanced geomicrobiological and biotechnological exploration. Geomicrobiol J 41(4):309–311. https://doi.org/10.1080/01490451.2024.2327059

Das AP, Dutta K, Khatun R, Behera ID, Singh S, Mishra S (2023) Microfiber pollution and its microbial mitigation: a review on current trends and future prospects. J Taiwan Inst Chem Eng 105104. https://doi.org/10.1016/j.jtice.2023.105104

Das AP, Behera ID, Das NP (2024) Renewable energy generation and value addition from environmental microfiber pollution through advanced greener solution. Springer Nature

Dey S, Bissoyi A, Koteswara Reddy G, Das AP (2024) Synthetic microfibers: an urban emerging pollutant demanding for sustainable environmental solutions. In: Sustainable Microbial Technology for Synthetic and Cellulosic Microfiber Bioremediation, pp 107–121. https://doi.org/10.1007/978-3-031-62600-5_7

Das A, Dey S, Bala K, Gnanasekaran R, Das AP (2024) Microbial bioremediation technology for sustainable treatment and management of synthetic microfiber waste. In: Sustainable Microbial Technology for Synthetic and Cellulosic Microfiber Bioremediation, pp 221–238. https://doi.org/10.1007/978-3-031-62600-5_12

Das A, Dey S, Das AP (2024) Ecotoxicological impacts of synthetic microfiber pollutants and development of sustainable mitigation strategies. Environ Chem Ecotoxicol. https://doi.org/10.1016/J.ENCECO.2024.12.003

Dey S, Dash N, Ray SK, Priyadarsini S, Chhotaray C, Das AP (2024) Occurrence and quantification of tire wear plastics pollutants from highway city dust of Bhubaneswar City of Odisha. In: Sustainable Microbial Technology for Synthetic and Cellulosic Microfiber Bioremediation, pp 77–89. https://doi.org/10.1007/978-3-031-62600-5_5

Dey S, Tripathy B, Kumar MS, Das AP (2023) Ecotoxicological consequences of manganese mining pollutants and their biological remediation. Environ Chem Ecotoxicol. https://doi.org/10.1016/J.ENCECO.2023.01.001

de Almeida CC, da Costa PRF, de Macedo Melo MJ, dos Santos EV, Martínez-Huitle CA (2014) Application of electrochemical technology for water treatment of Brazilian industry efuents. J Mex Chem Soc 58:276–286

De Gisi S, Lofrano G, Grassi M, Notarnicola M (2016) Characteristics and adsorption capacities of low-cost sorbents for wastewater treatment: a review. Sustainable Mater Technol 9:10–14

Ettouney HM, El-Dessouky HT, Faibish RS, Gowin PJ. Evaluating the economics of desalination. Chem Eng Prog 98:32–39

Gargouri B, Gargouri OG, Gargouri B, Trabelsi SK, Abdelhedi R, Bouaziz M (2014) Application of electrochemical technology for removing petroleum hydrocarbons from produced water using lead dioxide and boron-doped diamond electrodes. Chemosphere 117:309–315

Han M, Zhang J, Chu W, Chen J, Zhou G (2019) Research progress and prospects of marine oily wastewater treatment: a review. Water 11:2517

Hasan R, Chong CC, Setiabudi H (2019) Synthesis of KCC-1 using rice husk ash for Pb removal from aqueous solution and petrochemical wastewater. Bull Chem React Eng Catal 14:196–204

Hosny AY (1992) Separation of oil from oilwater emulsions using an electroflotation cell with insoluble electrodes. Filtr Sep 29:419–423

Ibrahim DS, Lathalakshmi M, Muthukrishnaraj A, Balasubramanian N (2013) An alternative treatment process for upgrade of petroleum refnery wastewater using electrocoagulation. Pet Sci 10:421–430

Ibrahim I, Hassan MA, Abd-Aziz S, Shirai Y, Andou Y, Othman MR, Ali AAM, Zakaria MR (2017) Reduction of residual pollutants from biologically treated palm oil mill efuent fnal discharge by steam activated bioadsorbent from oil palm biomass. J Clean Prod 141:122–127

Igunnu ET, Chen GZ (2014) Produced water treatment technologies. Int J Low-Carbon Technol 9(3):157–177

Jahi N, Ling ES, Othaman R, Ramli S (2015) Modifcation of oil palm plantation wastes as oil adsorbent for palm oil mill efuent (POME). Malays J Anal Sc 19:31–40

Jimenez S, Mico MM, Arnaldos M, Medina F, Contreras S (2018) State of the art of produced water treatment. Chemosphere 192:186–208

Jimenez S, Andreozzi M, Mico MM, Alvarez MG, Contreras S (2019) Produced water treatment by advanced oxidation processes. Sci Total Environ 666:12–21

Jones D, Head I, Gray N, Adams J, Rowan A, Aitken C et al (2008) Crude-oil biodegradation via methanogenesis in subsurface petroleum reservoirs. Nature 451:176–180

John A, Das AP (2024) Government strategies for achieving circular economy for plastic waste management in Southeast Asia. In: Sustainable Management of Urban Plastic Waste Through Circular Economic Approaches, pp 72–82. https://doi.org/10.1201/9781003465447-6/GOVERNMENT-STRATEGIES-ACHIEVING-CIRCULAR-ECONOMY-PLASTIC-WASTE-MANAGEMENTSOUTHEAST-ASIA-AKBAR-JOHN-ALOK-PRASAD-DAS

Jun KC, Abdul Raman A, Buthiyappan A (2020) Treatment of oil refnery efuent using bio-adsorbent developed from activated palm kernel shell and zeolite. RSC Adv 40:23510–24158

Kakavandi B, Ahmadi M (2019) Efcient treatment of saline recalcitrant petrochemical wastewater using heterogenous UV-assisted sono-Fenton process. Ultrason Sonochem 56:25–36

Kang J, Lu L, Zhan W, Li B, Li D, Ren Y, Liu D (2011) Photocatalytic pretreatment of oily wastewater from the restaurant by a vacuum ultraviolet/TiO2 system. J Hazard Mater 186:849–854

Kanungo J, Sahoo T, Bal M, Behera ID (2024) Performance of bioremediation strategy in waste lubricating oil pollutants: a review. Geomicrobiol J 41(4):360–373. https://doi.org/10.1080/01490451.2023.2245395

Kanungo J, Sahoo T, Swain LP, Behera ID (2023) Toxicity of persistent hydrocarbon pollutants, sources and sustainable remediation process. In: Behera ID , Das AP (eds). Impact of petroleum waste on environmental pollution and its sustainable management through circular economy. Springer Nature Switzerland, pp 39–65. https://doi.org/10.1007/978-3-031-48220-5_2

Kong Z, Li L, Xue Y, Yang M, Li Y-Y. Challenges and prospects for the anaerobic treatment of chemical-industrial organic wastewater: a review. J Cleaner Prod 231

Kuyukina MS, Krivoruchko AV, Ivshina IB (2020) Advanced bioreactor treatment of hydrocarbon–containing wastewater. Appl Sci 10:831

Lahiri D, Nag M, Dey A, Sarkar T, Joshi S, Pandit S, Das AP, Pati S, Pattanaik S, Tilak VK, Ray RR (2021) Biofilm mediated degradation of petroleum products. Geomicrobiol J 1–10. https://doi.org/10.1080/01490451.2021.1968979

Li J, Sun S, Yan P, Fang L, Yu Y, Xiang Y et al (2017) Microbial communities in the functional areas of a biofilm reactor with anaerobic-aerobic process for oily wastewater treatment. Bioresour Technol 238:7–15

Liang Y, Ning Y, Liao L, Yuan B (2018) Chapter fourteen—special focus on produced water in oil and gas fields: origin, management, and reinjection practice. In: Yuan B, Wood DA (eds) Formation damage during improved oil recovery. Gulf Professional Publishing, pp 515–586

Lopez RJ, Higgins SR, Pagaling E, Yan T, Cooney MJ (2014) High rate anaerobic digestion of wastewater separated from grease trap waste. Renew Energy 62:234–242

Lürling M, Mucci M (2020) Mitigating eutrophication nuisance: in-lake measures are becoming inevitable in eutrophic waters in the Netherlands. Hydrobiologia. https://doi.org/10.1007/s10750-020-04297-9

Makisha N (2020) Research of oily wastewater treatment by means of membrane methods. Int Sci Technol Conf Earth Sci 459:042015

Mallick K, Sahu S, Sahu A, Dey S, Das AP (2024) A critical review of marine microfiber pollution routes, toxicity, and its sustainable remediation. In: Das AP, Behera ID, Das NP (eds) Renewable energy generation and value addition from environmental microfiber pollution through advanced greener solution. Environ Sci Eng. Springer, Cham. https://doi.org/10.1007/978-3-031-51792-1_10

Mango FD (1997) The light hydrocarbons in petroleum: a critical review. Org Geochem 26:417–440

Mansouri AM, Shahrezaei F, Zinatizadeh AAL, Azandaryani AH, Pirsaheb M, Sharafi K (2014) Preparation of poly ethyleneimine (PEI)/nanotitania (TiO2) multilayer flm on quartz tube by layer-by-layer self-assembly and its applictions for petroleum refnery wastewater treatment. J Taiwan Insti Chem Eng 45:2501–2510

Mearns AJ, Morrison AM, Arthur C, Rutherford N, Bissell M (2020) Efects of pollution on marine organisms. Water Environ Res 92:1510–1532

Mishra S, Dash D, Das AP (2024) Microbial degradation of polyester microfibers using indigenously isolated bacterial strain Exiguobacterium Sp. CLEAN – Soil, Air, Water, 52(12):e202300343. https://doi.org/10.1002/CLEN.202300343

Mishra A, Siddiqi H, Kumari U, Behera ID, Mukherjee S, Meikap BC (2021) Pyrolysis of waste lubricating oil/waste motor oil to generate high-grade fuel oil: a comprehensive review. Renew Sustain Energy Rev 150:111446. https://doi.org/10.1016/j.rser.2021.111446

Mohanta S, Pradhan B, Behera ID (2024) Impact and remediation of petroleum hydrocarbon pollutants on agricultural land: a review. Geomicrobiol J 41(4):345–359. https://doi.org/10.1080/01490451.2023.2243925

Mohanta S, Sahoo B, Behera ID, Pradhan S (2016) Effect of crushing on near-gravity material distribution in different size fractions of an Indian non-coking coal. J South Afr Inst Min Metall 116(2):209–213. https://doi.org/10.17159/2411-9717/2016/v116n2a12

Mohanty M, Tripathy B, Koteswara Reddy G, Das AP (2024) Greener technologies for establishing restraint over microplastic pollution. In: Sustainable Microbial Technology for Synthetic and Cellulosic Microfiber Bioremediation, pp 239–258. https://doi.org/10.1007/978-3-031-62600-5_13

Nopcharoenkul W, Netsakulnee P, Pinyakong O (2013) Diesel oil removal by immobilized *Pseudoxanthomonas* sp. RN402. Biodegradation 24:387–393

Ossai IC, Ahmed A, Hassan A, Hamid FS (2020) Remediation of soil and water contaminated with petroleum hydrocarbon: a review. Environ Technol Innovation 17:100526

Paul A, Priyadarsini S, Karakuş S, Behera JK, Gnanasekaran R, Das AP (2024) Fate and accumulation of synthetic microfibers in the environment. In: Sustainable Microbial Technology for Synthetic and Cellulosic Microfiber Bioremediation, pp 61–76. https://doi.org/10.1007/978-3-031-62600-5_4

Priyadarsini S, Das AP (2024) Analytical and structural characterization of waste lithium-ion batteries for their effective recycling strategy. Environ Chem Ecotoxicol. https://doi.org/10.1016/j.enceco.2024.12.004

Priyadarsini S, Das AP (2025) Lithium bioleaching: A review on microbial-assisted sustainable technology for lithium bio-circularity. J Water Process Eng, 69:106744. https://doi.org/10.1016/j.jwpe.2024.106744

Pendashteh AR, Abdullah LC, Fakhru'l-Razi A, Madaeni SS, Abidin ZZ, Biak DRA (2012) Evaluation of membrane bioreactor for hypersaline oily wastewater treatment. Proc Saf Environ Prot 90:45–55

Pendergast MM, Hoek EMV (2011) A review of water treatment membrane nanotechnologies. Energy Environ Sc 4:1946–1971

Pérez LS, Rodriguez OM, Reyna S, Sánchez-Salas JL, Lozada JD, Quiroz MA, Bandala ER (2016) Oil refnery wastewater treatment using coupled electrocoagulation and fxed flm biological processes. Phys Chem Earth Parts a/b/c 91:53–60
Pradhan B, Behera ID, Sahoo K, Mohanta S (2023). Applicability of the teetered bed separator for beneficiating Indian Iron ore fines: an experimental study. J Mines Metals Fuels 71(11):1817–1825. https://doi.org/10.18311/jmmf/2023/31100
Rajasulochana P, Preethy V (2016) Comparison on efficiency of various techniques in treatment of waste and sewage water-a comprehensive review. Resource-Efcient Technol 2:175–184
Ray SK, Dey S, Das AP, John A (2024) Microplastic waste management through greener technologies: A review. In: Sustainable Management of Urban Plastic Waste Through Circular Economic Approaches, pp 237–261. https://doi.org/10.1201/9781003465447-14
Sahu A, Mallick K, Das AP (2024) Environmental impact of single-use synthetic face mask and its recycling: a sustainable approach. In: Das AP, Mishra S (eds) Impact of COVID-19 waste on environmental pollution and its sustainable management. Environmental science and engineering. Springer, Cham. https://doi.org/10.1007/978-3-031-50840-0_11
Sanghamitra P, Mazumder D, Mukherjee S (2021) Treatment of wastewater containing oil and grease by biological method-a review. J Environ Sci Health A. https://doi.org/10.1080/10934529.2021.1884468
Sanusi S, Abdullah S, Idris M (2012) Preliminary test of phytoremediation of hydrocarbon contaminated soil using *paspalum vaginatum* sw. Aust J Basic Appl Sci 6:39–42
Seveso D, Louis YD, Montano S, Galli P, Saliu F (2021) The Mauritius oil spill: what's next? Pollutants 1:18–28
Shende AD, Pophali GR (2020) Anaerobic treatment of slaughterhouse wastewater: a review. Environ Sci Pollut Res. https://doi.org/10.1007/s11356-020-10921-x
Sundar KP, Kanmani S (2020) Progression of photocatalytic reactors and its comparison: a review. Chem Eng Res Des 154:135–150
Sungur S, Özkan A (2017) Characterization of wastewaters obtained from Hatay tanneries.Nat Eng Sci 2:111–118
Surkatti R, El-Naas MH, Van Loosdrecht MCM, Benamor A, Al-Naemi F, Onwusogh U (2020) Biotechnology for gas-to-liquid (GTL) wastewater treatment: a review. Water 12(8):2126
Sutton NB, Maphosa F, Morillo JA, Al-Soud WA, Longenhof AAM, Grotenhuis T, Rijnaarts HHM, Smidt H (2013) Impact of longterm diesel contamination on soil microbial community structure. Appl Environ Microbiol 79:619–630
Tetteh EK, Rathilal S (2020) Evaluation of diferent polymeric coagulants for the treatment of oily refnery wastewater. Cogent Eng 7:1785756
Tran LH, Drogui P, Mercier G, Blais JF (2009) Electrochemical degradation of polycyclic aromatic hydrocarbons in creosote solution using ruthenium oxide on titanium expanded mesh anode. J Hazard Mater 164:1118–1129
Treviño-Roséndez J, Medel A, Meas Y (2021) Electrochemical technologies for treating petroleum industry wastewater. Curr Opin Electrochem. https://doi.org/10.1016/j.coelec.2021.100690
Tripathy B, Dash A, Das AP (2022) Detection of environmental microfiber pollutants through vibrational spectroscopic techniques: Recent advances of environmental monitoring and future prospects. Crit Rev Anal Chem, 1–11. https://doi.org/10.1080/10408347.2022.2144994
Tripathy B, Sahoo PP, Sundaray H, Das AP (2024) In-vitro biodegradation of discarded marine microplastics across the eastern coast of the Bay of Bengal, India using *Exiguobacterium* sp. Environ Chem Ecotoxicol 6:236–247. https://doi.org/10.1016/j.enceco.2024.07.003
Uludag-Demirer S, Olson N, Ives R, Nshimyimana JP, Rusinek CA, Rose JB, Liao W (2020) Techno-economic analysis of electrocoagulation on water reclamation and bacterial/viral indicator reductions of a high-strength organic wastewater-anaerobic digestion efuent. Sustainability 12:2697
USEPA (2004) Technical development document for the final effluent limitations guidelines and standards for the meat and poultry products point source category; final Rule. Fed Reg 69:54476–54555

Vendramel S, Bassin JP, Dezotti M, Jr Sant'Anna GL (2015) Treatment of petroleum refnery wastewater containing heavily polluting substances in an aerobic submerged fxed-bed reactor. Environ Technol 36:2052–2059

Wei Y, Jin Y, Zhang W (2020) Treatment of high-concentration wastewater from an oil and gas feld via a paired sequencing batch and ceramic membrane reactor. Int J Environ Res Public Health 17:1953

Wei X, Kazemi M, Zhang S, Wolfe F (2020) Petrochemical wastewater and produced water: treatment technology and resource recovery. Water Environ Res 92

Yalcinkaya F, Boyraz E, Maryska J (2020) A review on membrane technology and chemical surface modification for the oily wastewater treatment. Materials 13

Zeng Y, Yang C, Zhang J, Pu W (2007) Feasibility investigation of oily wastewater treatment by combination of zinc and PAM in coagulation/flocculation. J Hazard Mater 147(3):991–996

Zhao X, Wang Y, Ye Z, Borthwick AGL, Ni J (2006) Oil field wastewater treatment in biological aerated filter by immobilized microorganisms. Process Biochem 41(7):1475–1483

Chapter 6
Zero Waste Initiatives in Industrial Hydrocarbons and Waste Management

Koteswara Reddy G, Sai Sree Thanay Allam, Amel Gacem, Geya Govind, Praveen Tummalacharla, Gnanasekaran Ramakrishnan, Krishna Kumar Yadav, Harjeet Nath, Deo Karan Ram, and Alok Prasad Das

Abstract Liquid waste poses significant risk globally. It is reported that about 75% of earth is covered with water. Liquid waste that is improperly disposed of can easily spread, soak into other materials, or contaminate the soil. Industrial hydrocarbon waste encompasses the byproducts and residues resulting from the production, processing, and utilization of hydrocarbon-based products across diverse industries. Hydrocarbons, composed chiefly of hydrogen and carbon atoms, are fundamental organic compounds widely present in industrial operations owing to their roles as fuels, solvents, and essential raw materials in manufacturing processes. The industrial sector heavily depends on hydrocarbons for energy generation, transportation, the fabrication of plastics, chemicals, and a myriad of other commodities. Nonetheless,

---

K. R. G (✉) · S. S. T. Allam · G. Govind · P. Tummalacharla · G. Ramakrishnan · A. P. Das
Department of Biotechnology, Koneru Lakshmaiah Education Foundation (Deemed to Be University), Green Fields, Vaddeswaram 522502, India
e-mail: koteswarareddy@kluniversity.in

A. Gacem
Department of Physics, Faculty of Sciences, University 20 Août 1955, Skikda, Algeria

K. K. Yadav
Faculty of Science and Technology, Madhyanchal Professional University Ratibad, Bhopal, MP 462044, India

Environmental and Atmospheric Sciences Research Group, Scientific Research Center, Al-Ayen University, Thi-Qar, Nasiriyah 64001, Iraq

Department of Lifesciences, Ramadevi Women's University, Bhubaneswar, Odisha, India

H. Nath
Department of Chemical and Polymer Engineering, Tripura University (A Central University), Agartala 799022, India

D. K. Ram
Department of Chemical, Petroleum and Hydrogen Technology, Nims Institute of Engineering and Technology (NIET), Nims University, Jaipur, Rajasthan 303121, India

I. D. Behera and A. P. Das (eds.), *Environmental Hydrocarbon Pollution and Zero Waste Approach Towards a Sustainable Waste Management*, Environmental Science and Engineering, https://doi.org/10.1007/978-3-031-81615-4_6

the extensive utilization of hydrocarbons also engenders substantial waste generation, thereby presenting environmental and health hazards if inadequately handled or managed. This liquid waste can be transformed into energy by using waste-to-energy (WTE) technologies to satisfy various needs of people. The WTE technologies like gasification, microbial fuel cell (MFC), Anaerobic digestion (biogas), pyrolysis, Hydrothermal carbonisation (HTC), Circulating fluidized bed (CFB), supercritical water gasification, supercritical water oxidation and Dendro liquid energy (DLE). Out of all these technologies, HTC and DLE are the recent technologies in the waste-to-energy space. The predominant forms of energy released through the implementation of these methods are primarily thermal energy and electrical energy. These methods used to convert liquid waste to energy which promotes sustainable way of living and eco-friendly environment. This developing technology can be used to save our ecosystem and solve the energy problems in the world. Nevertheless, the scale of energy generation using WTE technologies is still in beginning phase at present, but it can be made into great energy solution soon.

Keywords Hydrocarbons · Gasification · Pyrolysis · Hydrothermal carbonisation

6.1 Introduction

In the evolving landscape of global industrial practices, the imperative for sustainable and environmentally conscious operations has gained unprecedented prominence (Das and Ghosh 2022). The industrial practice increases contaminants like heavy metals (Das et al. 2011; Ghosh et al. 2016; Prabhakar et al. 2019; Raj and Das 2023), microplastics (Das Dutta et al. 2023a, b; Das Mohanty et al. 2023a, b; Mallick et al. 2023; Mishra Rath et al. 2019a, b), electronic waste (Mishra et al. 2023) and biomedical waste (Sahu et al. 2024; Khatun et al. 2023a, 2023b; Sadasivuni et al. 2019). At the forefront of this transformative shift lies the formidable challenge of effectively managing industrial hydrocarbon (IHC) waste—an integral aspect of industrial processes requiring innovative and transformative solutions. The overarching goal of realizing a circular economy and diminishing the ecological footprint has culminated in the conceptualization and implementation of Zero Waste Initiatives in Industrial Hydrocarbon Waste Management (Basile et al. 2021). This holistic approach signifies a departure from conventional linear models of production and waste disposal, prompting industries to reassess their roles within the broader ecosystem (Ibn-Mohammed et al. 2021). Within this transformative framework, the emphasis extends beyond mere waste reduction to a complete reimagining of waste as a potential resource (Das and Ghosh 2018; Mishra et al. 2019; Mishra and Das 2023; Mohanty et al. 2018; Behera et al. 2021). Amidst escalating global concerns about climate change, resource scarcity, and environmental degradation, zero waste initiatives emerge as strategic pathways for industries to navigate the intricate intersection of economic growth and environmental responsibility (Wagh et al. 2024; Ghosh and Das 2017; Pradhan et al. 2023a). The context for Zero Waste Initiatives in Industrial Hydrocarbon Waste Management is rooted in the acknowledgment that conventional industrial processes, particularly those involving hydrocarbons, have historically generated substantial waste streams with adverse environmental consequences (Bal et al. 2018; Kanungo et al. 2023; Mishra et al. 2022a, b; Mishra and Das 2021; Suman et al. 2023). These waste streams, often laden with intricate hydrocarbons and hazardous components, present challenges ranging from soil and water contamination to greenhouse gas emissions (Kumar et al. 2023; Mishra et al. 2022a, b; Mishra and Das 2022). Against this backdrop, the concept of zero waste takes shape as a visionary and proactive response, aiming to address not only the symptoms but also the root causes of environmental degradation associated with industrial activities (Dey et al. 2024a, b, c). At its core, zero waste in the industrial hydrocarbon context signifies a commitment to reconfigure the industrial ecosystem into a closed-loop system, systematically eliminating, reducing, or converting waste into valuable resources (Glavič et al. 2023). This paradigm shift aligns with the broader principles of a circular economy, where materials and resources circulate for an extended duration through strategies like recycling, reusing, and repurposing. The journey toward zero waste demands a comprehensive and integrated approach, encompassing technological advancements, regulatory frameworks, stakeholder collaboration, and a cultural shift within industries. Technological innovations play a pivotal role in realizing zero waste goals in industrial hydrocarbon waste management. Advanced processes, including supercritical water gasification, catalytic conversion, and sustainable solvent extraction, gain prominence for their efficacy in converting hydrocarbon waste into valuable products or benign by-products (Muh et al. 2021). Moreover, the integration of artificial intelligence and data analytics in waste monitoring and management optimizes resource utilization, minimizing inefficiencies. On the regulatory front, governments and international bodies increasingly recognize the importance of fostering a conducive environment for zero waste initiatives (Da Costa et al. 2020). Incentive programs, stringent waste disposal regulations, and eco-labelling mechanisms are being implemented to encourage industries to adopt sustainable waste management practices. Compliance with these regulations not only aligns industries with environmental goals but also contributes to building a positive public image and meeting the expectations of environmentally conscious consumers (Aripin et al. 2023). Stakeholder collaboration stands as another cornerstone of successful zero waste initiatives. Engaging with suppliers, consumers, and local communities fosters a collective responsibility for waste reduction and resource conservation (Kanade et al. 2024). Moreover, cultivating a culture of sustainability within organizations, where employees actively participate in waste reduction initiatives, creates a sense of ownership and commitment to the broader environmental cause (Dey et al. 2023). Zero Waste Initiatives in Industrial Hydrocarbon Waste Management embody a visionary and transformative approach to harmonize industrial growth with environmental sustainability (Kirby and El-Kaffass 2022). Industries are aiming not only to reduce the negative impacts of waste but also to fundamentally redefine industrial processes. Through technological innovations, regulatory support, and collaborative efforts, the zero-waste paradigm offers a roadmap for industries to navigate the intricate balance between economic prosperity and ecological well-being, ushering in an era where waste is no longer a burden but an asset for a regenerative and sustainable future.

6.1.1 Sources and Effects of Hydrocarbons

Industrial hydrocarbons emanate from diverse processes, with refining and petrochemical manufacturing playing pivotal roles. In the realm of refining, processes such as distillation and cracking yield a spectrum of hydrocarbons essential to produce fuels and various petrochemical products (Kianfar and Salimi 2020). Concurrently, petrochemical manufacturing involves the synthesis of hydrocarbons in the production of plastics, solvents, and various chemicals (Kulprathipanja et al. 2021; Lahiri et al. 2021). Energy production represents a major contributor to emissions is the burning of fossil fuels like coal, oil, and natural gas, which releases hydrocarbons into the atmosphere, consequently causing air pollution. Das, Dey, & Das (2024) power plants, reliant on hydrocarbon-based fuels, also play a significant role in hydrocarbon emissions. Transportation activities contribute significantly to hydrocarbon emissions through vehicle emissions, where combustion engines release hydrocarbons as by-products (Jahirul et al. 2022; Behera et al. 2020, 2022). Waste disposal methods, such as the incineration of materials like plastics, release hydrocarbons into the atmosphere. Accidental spills and leaks from industrial facilities handling hydrocarbons, termed industrial accidents, also contribute to hydrocarbon release (Bahadori 2020). The effects of industrial hydrocarbons span various environmental facets (Priyadarsini et al. 2024a, b; Mohanty et al. 2024a, b, c). They significantly impact air quality by contributing to smog formation and the release of particulate matter. Water contamination arises from runoff and leaks during industrial processes, while soil quality is compromised through spills and leaks, impacting plant and microbial life (Tripathy et al. 2024). Human health is adversely affected by respiratory issues and potential carcinogenicity associated with hydrocarbon exposure (Kuppusamy et al. 2020a; Kanungo et al. 2024). Moreover, industrial hydrocarbons contribute to climate change through the release of greenhouse gases during combustion. Ecosystems, both aquatic and terrestrial, are disrupted as hydrocarbons contaminate water bodies and disrupt soil and air quality (Kolawole and Iyiola 2023). Ground-level ozone formation, a consequence of hydrocarbon presence, further impacts respiratory health. Hydrocarbons accumulate in organisms, posing risks to higher trophic levels through bioaccumulation, especially in persistent organic pollutants (POPs). The long-lasting effects of certain hydrocarbons contribute to environmental persistence, impacting ecosystems over time (Kuppusamy et al. 2020a). Additionally, hydrocarbons contribute to habitat disruption, resulting in the loss of biodiversity within affected ecosystems. Understanding these sources and effects is imperative for implementing effective mitigation strategies and promoting sustainable industrial practices that minimize the environmental impact of industrial hydrocarbons.

6.2 Current Liquid WTE Technologies

The various liquid waste-to-energy technologies are used by many contraries in the world, and these technologies are advanced catalytic conversion, gasification, pyrolysis, Anaerobic digestion (biogas), pyrolysis, Hydrothermal carbonisation (HTC), Circulating fluidized bed (CFB), supercritical water gasification, supercritical water oxidation and Dendro liquid energy (DLE).

6.2.1 Advanced Catalytic Conversion

Advanced Catalytic Conversion for hydrocarbon treatment entails deploying sophisticated catalytic processes to convert hydrocarbons efficiently and selectively into valuable products (Cruchade et al. 2022). This technology leverages advanced catalysts and precise reaction engineering to enhance the efficiency, selectivity, and sustainability of diverse hydrocarbon conversion processes. Key applications within this framework encompass catalytic cracking, hydroprocessing, dehydrogenation, reforming, metathesis reactions, and selective oxidation, as shown in Fig. 6.1 (Abbas-Abadi et al. 2023). These advanced catalytic processes play a pivotal role in disassembling intricate hydrocarbons, upgrading feedstocks, and generating high-value fuels, chemicals, and intermediates. The focal point is the advancement of catalysts with enhanced activity, stability, and selectivity to optimize conversion yields while minimizing undesirable by-products and environmental impact. Advanced Catalytic Conversion serves as a fundamental element in the pursuit of more sustainable and economically viable solutions for hydrocarbon treatment within the petrochemical and bio-based industries.

6.2.2 Anaerobic Digestion

The process of oxidation of biodegradable waste with the help of microorganisms under anaerobic conditions is known as anaerobic digestion (Zamri et al. 2021). In this the sophisticated organic compounds are converted into simpler organic molecules in anaerobic conditions, resulting in end products such as methane gas, carbon dioxide and many more. This process requires high organic matter, which is found mostly in sewage, sludge, domestic waste, etc. One of its outputs is methane gas, which is also considered biogas; methane is an odourless and colourless gas and is a member of the paraffin series of hydrocarbons (Abubakar et al. 2022). This readily burns in the air, forming water vapour and carbon dioxide. Anaerobic digestion is carried out in anaerobic digesters, which incorporates the decomposition process. In this digester, anaerobic bacteria called acetogens and methanogens are present; these bacteria are classified as anaerobic as they do not utilize oxygen as their primary electron donor

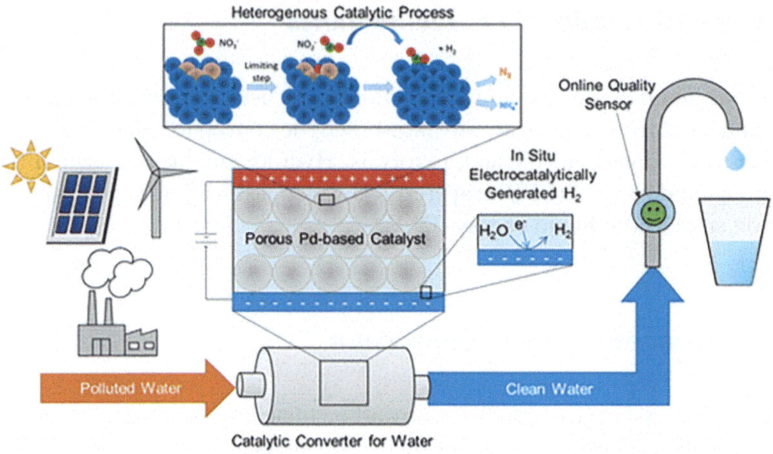

Fig. 6.1 The process of a catalytic converter in the conversion of petrochemical industrial waste into pure water. (Heck et al. 2019)

(Harirchi et al. 2022). Instead, they accept electrons from acetate and methane to produce energy.

Anaerobic waste treatment is used to convert hydrocarbons into useful form of energy source (Shi et al. 2020). In this, the concentrated liquid organic waste mainly consists of food and beverage industry waste, waste from paper industries, chemicals from petrochemical industries, household waste, etc (Priyadarsini et al. 2025; Paul et al. 2023, 2024). These wastes are filled into specialized treatment plants in which the further process is continued under anaerobic conditions like hydrolysis, acidogenesis, acetogenesis, and methanogenesis, representing the sequential reactions occurring within the anaerobic treatment plant. Hydrolysis is the process of breaking the chains of complex molecules by hydrolysing enzymes is known as hydrolysis; this process is also recognized as the liquefaction of complex molecules (Kumar and Verma 2020). In the acidolysis process, the acid breakdown of oligo polymers into simpler molecules takes place. Next comes the acetogenesis process in which the formation of acetic acid formation facilitated by acetogens is noted (Kremp and Müller 2021). The final step is methanogenesis, in which methane hydrogen and carbon dioxide are produced from their intermediates (Kurth et al. 2020; Lee et al. 2023). The biogas is then used to convert into heat or electricity. Many countries like the USA, India, and many more are following this method to produce electricity or heat energy.

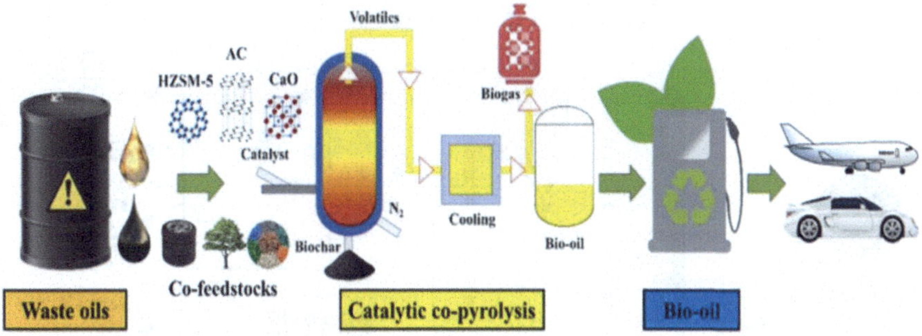

Fig. 6.2 Process of pyrolysis (Su et al. 2022)

6.2.3 Pyrolysis

The process of subjecting substances to high temperatures in an inert atmosphere to ease the thermal decomposition leads to a chemical change in the substance (Tomassetti et al. 2022). The pyrolysis of Hydrocarbon waste will produce various volatile products, which leave out the solid residue that is used as an efficient fuel (Mishra et al. 2021). The output of this process includes a liquid phase bio-oil, a gaseous stream, and a solid product termed biochar. The bio-oil possesses significant economic value, as it serves as one of the limited carbon–neutral alternatives. It can be utilized in various forms of energy production, including electrical energy, heat energy, mechanical energy, and more. Figure 6.2 illustrates the pyrolysis process.

6.2.4 Gasification

Gasification is a process that transforms carbon-containing materials like coal, petrochemical waste, and trash into synthetic gas (Speight 2020). Initially developed in the eighteenth century to produce town gas for illumination, gasification facilities now serve various purposes such as fuel, chemical, fertilizer, and power production. In a closed reactor, gasification converts carbon-based materials directly into synthetic gas or syngas without extensive combustion or oxygen use. This method releases energy from biomass and municipal solid waste, providing an alternative to landfilling or incineration. Gasification operates at temperatures higher than pyrolysis (typically above 700 °C) through partial oxidation, resulting in syngas formation composed mainly of carbon monoxide (CO) and hydrogen (H_2), with smaller amounts of methane (CH_4) and carbon dioxide (CO_2) (Mariyam et al. 2022). The choice of gasifying agent, whether air, steam, or steam-oxygen, defines the process. Various reactor types like fixed beds, fluidized beds, and entrained flow gasifiers are commonly used in gasification. Syngas finds applications in combined heat and power generation, hydrogen production through pressure swing adsorption and

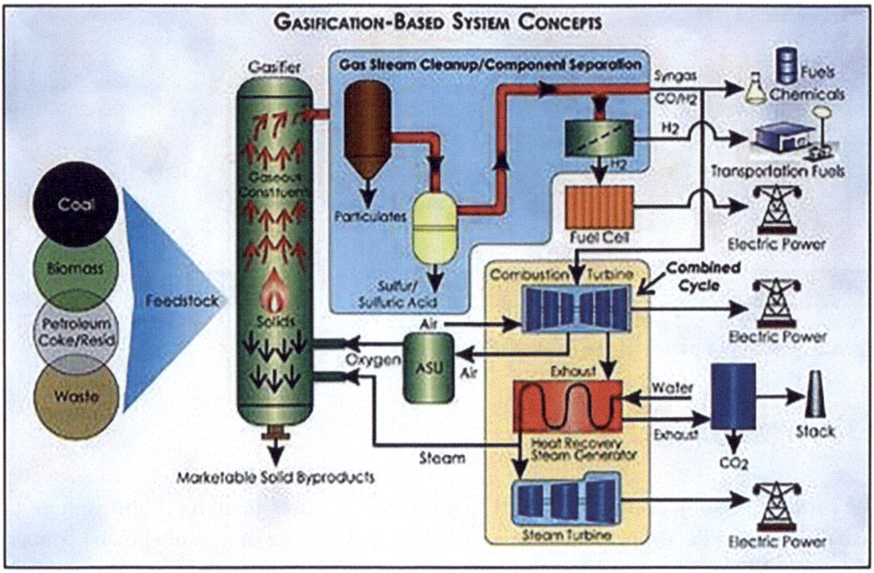

Fig. 6.3 Process of gasification (U.S. Department of Energy 2003)

water–gas shift reactions, and Fischer–Tropsch synthesis for liquid fuels (Martinelli et al. 2020). Managing wastewater sludge gasification composition is crucial for optimizing energy recovery in downstream processes such as combined heat and power recovery and Fischer–Tropsch synthesis for liquid fuel production. A diagrammatic representation of the gasification system is presented in Fig. 6.3.

6.2.5 Hydrothermal Technologies

Hydrothermal technologies refer to chemical and physical transformations that occur under high-temperature (200–600 °C) and high-pressure (5–40 MPa) conditions in liquid or supercritical water (Namratha et al. 2022). One such technology is hydrothermal carbonization, which accelerates the gradual conversion of hydrocarbon waste from petrochemical industries into hydro char using an acid catalyst under high pressure and heat, simulating the generation of fossil fuels. Advanced technical developments in hydrothermal liquefaction (HTL) have led to limited studies exploring its application in treating effluents of petroleum industries (Bal et al. 2020).

6.2.6 Supercritical Water Gasification and Supercritical Water Oxidation (SCWG & SCWO)

Supercritical water exhibits unique and controllable physicochemical properties that provide advantages in Waste management, organic compound gasification, and manufacturing processes (Okolie et al. 2020). Supercritical water oxidation (SCWO) is mainly utilized for neutralization harmful waste by making organic molecules and oxygen completely miscible in water under supercritical conditions (Zhang et al. 2022). This allows oxidation to occur within one homogeneous fluid phase consisting of efficient transport capabilities, enabling many organic molecules to be fully oxidized in less than a minute at the right temperature. SCWO finds applications in various areas, including wastewater and sewage treatment.

Supercritical water gasification (SCWG) breaks down organic molecules through reductive hydrothermal processes without excess oxygen. It leads to a gasification reaction network of complex organic compounds, yielding light fuel gases like H2, CO, and CH4 under elevated temperatures and reduced feedstock concentrations (Leong et al. 2021). However, SCWG can also produce refractory products such as Phenolic compounds, furfurals, and char resulting from polymerization processes or difficulties in breaking aromatic rings, along with the formation of metal oxides and inorganic acids in the presence of heteroatoms or metals. Both SCWG and SCWO are increasingly used for wastewater sludge treatment due to their ability to bypass the energy-intensive drying process required by other methods like hydrothermal liquefaction (HTL) (Wei et al. 2021). SCWG produces syngas like conventional gasification, which can be used as a gaseous fuel or further processed through Fischer–Tropsch (FT) synthesis to create liquid fuels. Companies like General Atomics are developing innovative solutions for industrial waste management, as depicted in Fig. 6.4.

6.2.7 Dendro Liquid Energy

The most recent and promising technology for converting hydrocarbon waste into energy is known as dendro liquid energy (DLE), which generates electricity with minimal emissions (Mallick et al. 2022). DLE plants are significantly more efficient in energy generation compared to other waste-to-energy (WTE) systems, achieving approximately four times higher efficiency due to their operation at moderate temperatures ranging from 150 °C to 250 °C (Da Costa et al. 2020; Kuppusamy et al. 2020b). These plants produce clean fuels such as hydrogen and carbon monoxide, forming syngas by utilizing both wet and dry waste materials. One of the notable advantages of DLE is its lack of combustion, which eliminates the need for expensive emission control equipment, resulting in cost savings and eco-friendliness. With an energy conversion efficiency exceeding 80% and minimal emissions, DLE produces syngas and byproducts free from particles or tar. This technology's low operating costs make

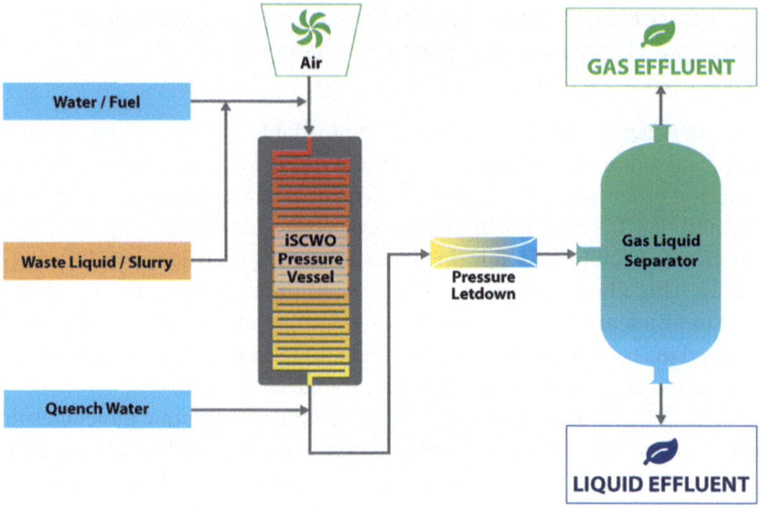

Fig. 6.4 Industrial supercritical water oxidation (iSCWO) system by general atomics (Choi et al. 2020)

it a favourable option for various communities. The waste-to-energy sector anticipates wider adoption of DLE in the future, given its promising market potential and the growing need for energy independence, particularly during challenging times such as recent energy crises (Khan et al. 2022). Enhancing independence in food and energy production is crucial, and waste-to-energy systems like DLE contribute significantly to achieving this goal. Various methods used in wastewater treatment Processes as shown in Table 6.1.

6.2.8 *H_2O_2/UV Advanced Oxidation Process*

The paragraph explores the effectiveness of hydrogen peroxide (H_2O_2) as an oxidizing agent in the wastewater processing, particularly when coupled with ultraviolet (UV) radiation to initiate the formation of HO· radicals. It elucidates the various chemical reactions integral to this process and underscores the necessity of maintaining an optimal concentration of H_2O_2 for efficient pollutant degradation. Special attention is given to the pivotal role played by UV radiation, especially wavelengths exceeding 300 nm, in catalysing the decomposition of H_2O_2 and subsequent generation of HO· radicals. Additionally, it acknowledges the diminished efficacy of hydroperoxyl radicals relative to hydroxyl radicals owing to their lower reduction potential (Rott et al. 2018).

Furthermore, the paragraph delineates the advantages associated with the H_2O_2/UV process, including H_2O_2's aqueous solubility, absence of mass transfer limitations, and its function as a proactive source of HO· radicals. It also delves into the

Table 6.1 Various methods used in wastewater treatment processes

S. No	Process name	Date of initiation	Raw materials used	Applications	References
1	Advanced catalytic conversion	Mid-twentieth century	Heavy hydrocarbons, such as crude oil fractions or vacuum gas oils	Production of valuable fuels, including gasoline and diesel, by breaking down large hydrocarbons	(Cruchade et al. 2022; Abbas-Abadi et al. 2023)
2	Pyrolysis	Late 19th and early twentieth centuries	Plastic waste, rubber, waste oil	Plastic and rubber recycling, bioenergy production	(Tomassetti et al. 2022)
3	Gasification	Nineteenth and twentieth centuries	Coal, Biomass, Petrochemical Industrial Waste	Power generation, chemical production	(Mariyam et al. 2022; Martinelli et al. 2020)
4	Hydrothermal technologies	Mid-twentieth century	Oil and petroleum industries, Biomass, Sewage Sludge, Organic Waste	Supercritical water oxidation	(Namratha et al. 2022)
5	SCWG & SCWO	Twentieth century	Insulating oils, tar, gasoline	Supercritical water oxidation, supercritical water gasification	(Zhang et al. 2022; Leong et al. 2021)
6	Dendro liquid energy	Recently discovered	Diesel fuel, hydraulic fluids, motor oils, lubricants biomass, sewage, organic waste	Power generation, biogas production	(Mallick et al. 2022; Khan et al. 2022)

influence of operating pH on the segregation of radicals and enhancement of degradation rates, with an emphasis on the significance of maintaining a low pH environment (Liu et al. 2021). The discourse extends to the examination of studies assessing the application of H_2O_2 in treating wastewater from petroleum refineries and oil production facilities, where it underscores the challenges in achieving complete degradation of specific pollutants like total petroleum hydrocarbons (TPH) and methyl tert-butyl ether (MTBE).

Moreover, the paragraph provides detailed insights into the outcomes of studies analysing pollutant removal efficiencies, particularly focusing on compounds like MTBE, and elucidates the factors impacting these removal rates, such as H_2O_2 concentration and competitive interactions with HO· radicals. Lastly, it alludes to a comparative examination of diverse studies employing H_2O_2- utilizing advanced oxidation processes for the treatment of various wastewater types, offering a comprehensive perspective on both the potential and constraints of this approach in wastewater treatment research.

6.2.9 Fenton and Photo-Fenton Advanced Oxidation Processes

Henry John Horstman Fenton's discovery in 1894 unveiled a chemical amalgam comprising hydrogen peroxide (H_2O_2) and ferric ions, famously dubbed Fenton's chemical. Recognizing its potent oxidizing capabilities within an acidic milieu, Fenton sparked subsequent exploration into its utility among researchers seeking to mitigate organic contaminants. Nonetheless, considerable discourse persists regarding the precise mechanisms underlying this reaction, including the intricacies of Fenton's reaction (Raji and Mirbagheri 2021). The traditional Fenton method entails the aqueous amalgamation of hydrogen peroxide and ferrous ions within an acidic milieu. This fusion instigates the breakdown of hydrogen peroxide into hydroxyl ions and hydroxyl radicals, concomitant with the oxidation from ferrous ions to ferric ions. The kinetics governing this reaction are intricate, entailing the involvement of diverse intermediary compounds, such as aqueous ferric and ferrous species, which interact with hydrogen peroxide to engender hydroxyl radicals (Ghime and Ghosh 2019). The reaction's efficacy hinges upon various factors, including pH, the hydrogen peroxide-to-ferrous ions ratio, and the presence of organic acids.

Operating at pH 3, deemed optimal for Fenton's reaction, precipitates the predominance of ferric hydroxide compounds. Conversely, at markedly low pH levels (< 2.5), distinct compounds emerge, exhibiting sluggish reactivity with hydrogen peroxide, thereby diminishing hydroxyl radical production and impeding system efficiency (Zouanti et al. 2020). Conversely, basic pH fosters the formation of iron hydroxide compounds, inert to hydrogen peroxide, thus thwarting the Fenton reaction. Consequently, pH adjustment prior to Fenton reagent addition in wastewater treatment is imperative. The hydrogen peroxide-to-ferrous ions ratio similarly impacts reaction efficiency. While a 1:2 ratio may yield heightened degradation rates, a 1:5 ratio is typically advocated due to commensurate outcomes and diminished reagent requirements. However, sole reliance on the Fenton process may prove inadequate for effectual mineralization of organic compounds (Villegas-Guzman et al. 2017). The progressive conversion of ferrous ions to ferric ions during the reaction could disrupt hydroxyl radical generation, thereby impeding organic compound degradation.

In efforts to augment the Fenton process, researchers have explored UV radiation, culminating in the advent of the photo-Fenton method. UV irradiation substantially enhances the degradation rate of organic contaminants by facilitating ferrous ion regeneration through ferric compound photolysis. This iterative process curtails the exigency for supplementary ferrous ions and heightens reaction efficiency. Nonetheless, both Fenton and photo-Fenton processes exhibit limitations necessitating low pH values (< 2) and potentially mandating iron separation post-reaction (Perini et al. 2018). Additionally, certain ions, such as chloride ions, could obfuscate reaction dynamics by scavenging hydroxyl radicals. Optimization of operational parameters, encompassing pH modulation and reagent ratio control, is indispensable for maximizing efficiency whilst minimizing chemical consumption (Shokri 2018). Studies corroborate the efficacy of Fenton and photo-Fenton methods in mitigating diverse wastewater contaminants, spanning hydrocarbons to phenolic compounds. Despite their promise, further research is indispensable to fine-tune efficiency and abate chemical utilization, especially within expansive applications like industrial wastewater treatment.

6.2.10 Ozone Based Advanced Oxidation Processes

Ozone (O_3) is widely utilized for the purification of drinking water and wastewater treatment because of its strong oxidizing properties. Serving as a selective oxidant, ozone rapidly reacts with electron-rich organic compounds, particularly under neutral pH conditions, with high reaction rate constants (KO_3) (Boczkaj et al. 2017). Moreover, the generation of hydroxyl radicals (OH·) from ozone consumption further boosts its oxidative potential, enabling the targeted degradation of various organic compounds with controlled diffusion rates. The formation of radicals, including superoxide radicals (O2 − ·) and hydroxyl radicals (HO·), is catalysed by complex reactions between ozone and water. Ozone degradation in water is accelerated at higher pH levels (Rekhate and Srivastava 2020). The addition of hydrogen peroxide (H_2O_2) to the ozone decomposition cycle enhances hydroxyl radical production, thereby improving degradation rates. Various factors, such as contact time, water alkalinity, and ozone concentrations, influence the effectiveness of the process. Combining hydrogen peroxide with ozone enhances degradation rates, making the H_2O_2/O_3 system the most efficient advanced oxidation process (AOP) for water purification.

Researchers have investigated both homogeneous and heterogeneous catalysts to expedite ozonation. Metal ions and metal oxides exhibit varying degrees of catalytic activity depending on the circumstances (Elmobarak et al. 2021; Pradhan et al. 2023b). Parameters such as ozone concentration and process duration significantly affect degradation rates. The ozone/catalyst system is particularly effective at reducing chemical oxygen demand (COD) and total organic carbon (TOC) levels, especially at high pH levels. Additionally, the direct absorption of ozone through UV radiation generates hydrogen peroxide as an intermediate, which then decomposes into highly reactive hydroxyl radicals. Despite the efficacy of ozone based AOPs in degrading organic pollutants, challenges such as high operational costs, energy intensity, and potential bromate formation persist. Nonetheless, optimization strategies such as adjusting pH, increasing H_2O_2/O_3 ratios, or enhancing hydroxyl radical production help mitigate these challenges (Krishnan et al. 2017). Ozone-based AOPs play a crucial role in industrial water treatment, particularly in managing wastewater from petroleum refineries. Studies have shown that combining ozone-based processes with UV radiation effectively degrades volatile organic compounds (VOCs) and reduces chemical and biochemical oxygen demand (COD/BOD). Improved reactor designs facilitate the formation of radical species, ensuring efficient contaminant degradation without mass transfer limitations.

6.2.11 Combined AOP Processes

In recent years, advanced oxidation processes (AOPs) have emerged as effective techniques for treating petroleum wastewater, providing diverse strategies to improve efficiency (Khatri et al. 2018). The combination of different AOP methods with conventional wastewater treatment approaches has demonstrated enhanced efficacy in addressing contaminated industrial wastewater sources. Factors such as waste stream properties, environmental regulations, and cost considerations guide the selection of these methods. Efforts to improve the degradation and separation of contaminants have led to the integration of AOPs with biological methods and other processes. Biological treatment methods, such as the activated sludge process (ASP), have proven effective in treating petroleum refinery wastewater, achieving COD separation efficiencies ranging from 70 to 96% (Razavi and Miri 2015). Additionally, integrating binary or extra-biological methods, along with combining biological methods with membrane based AOP techniques, holds promise for wastewater treatment (Jagatee et al. 2024; Das et al. 2024a, b, c). However, challenges such as the generation of large quantities of sludge, the need for trained personnel, and frequent maintenance are associated with many of these procedures. Some Alternative methods also necessitate exceedingly long hydraulic retention times (HRT), which pose constraints to Bio-AOP techniques. An integrated advanced oxidation process merges photobioreactors (PBR) with conventional oxic/anoxic methods to treat petrochemical wastewater, resulting in significant COD reduction rates (Liang et al. 2019). This demonstrates the potential of electrocoagulation-based AOP methods to effectively treat high-concentration petrochemical water.

Several studies have investigated the integration of different AOPs, such as photocatalysis combined with ultrasound, ozonation paired with ultrasound, and the combination of Fenton and photo-Fenton processes with ultrasound (An et al. 2017; Giwa et al. 2020). These integrations have shown improved degradation rates compared to standalone AOP processes, with the synergistic impact playing a vital role in evaluating system efficiency. AOP-bioremediation represents a novel integrated approach applied to industrial wastewater for the separation of micropollutants, particularly in industrial wastewater from petroleum processing containing persistent biological contaminants. While this dual-phase procedure holds promise, recent literature suggests challenges in its practical application due to varying characteristics of effluent, operational parameters, and structural designs (Zhai et al. 2018). Membrane-based AOPs, closely linked to photocatalytic processes employing semiconductor catalysts, offer an innovative technique for pollutant degradation in wastewater. By utilizing membranes for filtration, this method disperses nanoparticle catalysts in the reactor suspension, with the membrane serving as a barrier for the catalyst particles. This approach is anticipated to enhance the quality of permeate via the reduction–oxidation process, generating active radicals to effectively degrade pollutants.

6.3 Future Scope

The future presents optimistic opportunities for employing a range of advanced processes to convert petroleum industry effluents into valuable energy resources. Combined Advanced Oxidation Processes (AOP), amalgamating various oxidation methodologies such as ozone-based AOP, Fenton, Photo-Fenton AOP, and H2O2/UV AOP, exhibit improved efficacy in decomposing intricate hydrocarbons into more manageable compounds. Dendro liquid energy technologies demonstrate the potential to extract energy from biomass, offering sustainable alternatives. Supercritical water gasification and oxidation (SCWG & SCWO) hold significant promise in effectively transforming organic matter into usable energy forms, while hydrothermal technologies present innovative avenues for converting diverse organic compounds under elevated temperature and pressure conditions. Additionally, the evolution of gasification, pyrolysis, and anaerobic digestion processes offers feasible pathways for recovering energy from varied organic waste streams. Advanced catalytic conversion techniques play a critical role in augmenting conversion rates and product selectivity, thereby optimizing the energy recovery process from petroleum industry effluents. Collectively, these technologies constitute a diverse toolkit poised to meet the escalating demand for sustainable energy solutions while mitigating the environmental repercussions associated with petroleum waste disposal.

6.4 Conclusion

In conclusion, the integration of Combined Advanced Oxidation Processes (AOP) demonstrates improved effectiveness in decomposing complex hydrocarbons, with Dendro liquid energy technologies offering sustainable alternatives through the extraction of energy from biomass. Furthermore, Supercritical water gasification and oxidation (SCWG & SCWO) alongside hydrothermal technologies present promising avenues for converting organic matter into usable energy forms under extreme conditions. Additionally, the advancement of gasification, pyrolysis, and anaerobic digestion processes offers viable pathways for recovering energy from diverse organic waste streams. The crucial role of advanced catalytic conversion techniques in enhancing conversion rates and product selectivity remains paramount. Collectively, these technologies form a diverse toolkit ready to meet the rising demand for sustainable energy solutions while concurrently mitigating the environmental impacts linked with petroleum waste disposal.

References

Abbas-Abadi MS, Ureel Y, Eschenbacher A, Vermeire FH, Varghese RJ, Oenema J, Van Geem KM (2023) Challenges and opportunities of light olefin production via thermal and catalytic pyrolysis of end-of-life polyolefins: towards full recyclability. Prog Energy Combust Sci 96:101046

Abubakar AM, Silas K, Aji MM (2022) An elaborate breakdown of the essentials of biogas production. J Eng Res Sci 1(4):93–118

An C, Huang G, Yao Y, Zhao S (2017) Emerging usage of electrocoagulation technology for oil removal from wastewater: a review. Sci Total Environ 579:537–556

Aripin Z, Mulyani SR, Haryaman A (2023) Marketing strategy in project sustainability management efforts in extractive industries: building a reciprocity framework for community engagement. Kriez Acad J Dev Commun Serv 1(1):25–38

Bahadori A (2020) Waste management in the chemical and petroleum industries. John Wiley & Sons

Bal B, Ghosh S, Das AP (2018) Microbial recovery and recycling of manganese waste and their future application: a review. Geomicrobiol J 36(1):85–96. https://doi.org/10.1080/01490451.2018.1497731

Bal M, Behera ID, Kumari U, Biswas S, Meikap BC (2020) Hydrodynamic study and particulate matter removal in a self priming venturi scrubber. Environ Technol Innov 20:101167. https://doi.org/10.1016/j.eti.2020.101167

Basile V, Capobianco N, Vona R (2021) The usefulness of sustainable business models: analysis from oil and gas industry. Corp Soc Responsib Environ Manag 28(6):1801–1821

Behera ID, Basak G, Kumar RR, Sen R, Meikap BC (2020) Treatment of petroleum refinery sludge by petroleum degrading bacterium Stenotrophomonas pavanii IRB19 as an efficient novel technology. J Environ Sci Health Part Toxic/hazardous Subst Environ Eng 56(2):226–239. https://doi.org/10.1080/10934529.2020.1866924

Behera ID, Nayak M, Biswas S, Meikap BC, Sen R (2021) Enhanced biodegradation of total petroleum hydrocarbons by implementing a novel two-step bioaugmentation strategy using indigenous bacterial consortium. J Environ Manag 292:112746. https://doi.org/10.1016/j.jenvman.2021.112746

Behera ID, Nayak M, Mishra A, Meikap BC, Sen R (2022) Strategic implementation of integrated bioaugmentation and biostimulation for efficient mitigation of petroleum hydrocarbon pollutants from terrestrial and aquatic environment. Marine Pollut Bull 177:113492. https://doi.org/10.1016/j.marpolbul.2022.113492

Boczkaj G, Fernandes A, Makoś P (2017) Study of different advanced oxidation processes for wastewater treatment from petroleum bitumen production at basic pH. Ind Eng Chem Res 56(31):8806–8814

Choi H, Schleicher RW, Bolin J (2020) Performance analysis of silicon carbide composite clad uranium carbide fuel for a long-life gas-cooled fast reactor under normal operation—part I: design criteria and material data. Nucl Technol 206(7):993–1009

Cruchade H, Medeiros-Costa IC, Nesterenko N, Gilson JP, Pinard L, Beuque A, Mintova S (2022) Catalytic routes for direct methane conversion to hydrocarbons and hydrogen: current state and opportunities. ACS Catal 12(23):14533–14558

Da Costa JP, Mouneyrac C, Costa M, Duarte AC, Rocha-Santos T (2020) The role of legislation, regulatory initiatives and guidelines on the control of plastic pollution. Front Environ Sci 8:104

Das A, Dey S, Das AP (2024) Ecotoxicological impacts of synthetic microfiber pollutants and development of sustainable mitigation strategies. Environ Chem Ecotoxicol. https://doi.org/10.1016/J.ENCECO.2024.12.003

Das AP, Ghosh S (2018) Bioleaching of Manganese from mining waste materials. Mater Today Proceed 5(1):2381–2390. https://doi.org/10.1016/J.MATPR.2017.11.459

Das AP, Sukla LB, Pradhan N, Nayak S (2011) Manganese biomining: a review. Biores Technol 102(16):7381–7387. https://doi.org/10.1016/J.BIORTECH.2011.05.018

Das A, Ghosh S (2022) Role of microorganisms in extenuation of mining and industrial wastes. Geomicrobiol J 39:173–175. https://doi.org/10.1080/01490451.2022.2038953

Das AP, Dutta K, Khatun R, Behera ID, Singh S, Mishra S (2023a) Microfiber pollution and its microbial mitigation: a review on current trends and future prospects. J Taiwan Inst Chem Eng. https://doi.org/10.1016/J.JTICE.2023.105104

Das AP, Mohanty AK, van Hullebusch ED, Figueiredo G, Das AP, Mohanty AK, van Hullebusch ED, Figueiredo G (2023b) Bioremediation as an emerging technology for the removal of synthetic microplastic pollutants from marine ecosystem. MarPB 194:115297. https://doi.org/10.1016/J.MARPOLBUL.2023.115297

Das A, Ray SK, Mohanty M, Mohanty J, Dey S, Das AP (2024a) Ecotoxicity of microplastic wastes and their sustainable management: a review. Environ Chem Ecotoxicol 6:144–152. https://doi.org/10.1016/J.ENCECO.2024.05.003

Das A, Dey S, Bala K, Gnanasekaran R, Das AP (2024b) Microbial bioremediation technology for sustainable treatment and management of synthetic microfiber waste. In: Sustainable microbial technology for synthetic and cellulosic microfiber bioremediation, pp 221–238. https://doi.org/10.1007/978-3-031-62600-5_12

Das A, Dey S, Das AP (2024c) Ecotoxicological impacts of synthetic microfiber pollutants and development of sustainable mitigation strategies. Environ Chem Ecotoxicol. https://doi.org/10.1016/J.ENCECO.2024.12.003

Dey S, Tripathy B, Kumar MS, Das AP (2023) Ecotoxicological consequences of manganese mining pollutants and their biological remediation. Environ Chem Ecotoxicol. https://doi.org/10.1016/J.ENCECO.2023.01.001

Dey S, Bissoyi A, Koteswara Reddy G, Das AP (2024a). Synthetic microfibers: an urban emerging pollutant demanding for sustainable environmental solutions. In: Sustainable microbial technology for synthetic and cellulosic microfiber bioremediation, pp 107–121. https://doi.org/10.1007/978-3-031-62600-5_7

Dey S, Dash A, Das AP (2024b) Bioremediation as a sustainable and circular solution for mining waste management and bioenergy generation. In: Sustainable management of mining waste and tailings, pp 107–119. https://doi.org/10.1201/9781003442455-6/BIOREMEDIATION-SUSTAINABLE-CIRCULAR-SOLUTION-MINING-WASTE-MANAGEMENT-BIOENERGY-GENERATION-SUDESHNA-DEY-AKANSHYA-DASH-ALOK-PRASAD-DAS

Dey S, Dash N, Ray SK, Priyadarsini S, Chhotaray C, Das AP (2024c) Occurrence and quantification of tire wear plastics pollutants from highway city dust of Bhubaneswar city of Odisha. In: Sustainable microbial technology for synthetic and cellulosic microfiber bioremediation, pp 77–89. https://doi.org/10.1007/978-3-031-62600-5_5

Elmobarak WF, Hameed BH, Almomani F, Abdullah AZ (2021) A review on the treatment of petroleum refinery wastewater using advanced oxidation processes. Catalysts 11(7):782

Ghime D, Ghosh P (2019) Removal of organic compounds found in the wastewater through electrochemical advanced oxidation processes: a review. Russ J Electrochem 55:591–620

Ghosh S, Das AP (2017) Bioleaching of manganese from mining waste residues using Acinetobacter sp. Geol Ecol Landscapes 1(2):77–83. https://doi.org/10.1080/24749508.2017.1332847

Ghosh S, Mohanty S, Nayak S, Sukla LB, Das AP (2016) Molecular identification of indigenous manganese solubilising bacterial biodiversity from manganese mining deposits. J Basic Microbiol 56(3):254–262. https://doi.org/10.1002/JOBM.201500477

Giwa A, Yusuf A, Dindi A, Balogun HA (2020) Polygeneration in desalination by photovoltaic thermal systems: a comprehensive review. Renew Sustain Energy Rev 130:109946

Glavič P, Pintarič ZN, Levičnik H, Dragojlović V, Bogataj M (2023) Transitioning towards net-zero emissions in chemical and process industries: a holistic perspective. Processes 11(9):2647

Harirchi S, Wainaina S, Sar T, Nojoumi SA, Parchami M, Parchami M, Taherzadeh MJ (2022) Microbiological insights into anaerobic digestion for biogas, hydrogen or volatile fatty acids (VFAs): a review. Bioengineered 13(3):6521–6557

Heck KN, Garcia-Segura S, Westerhoff P, Wong MS (2019) Catalytic converters for water treatment. Acc Chem Res 52(4):906–915

Ibn-Mohammed T, Mustapha KB, Godsell J, Adamu Z, Babatunde KA, Akintade DD, Koh SCL (2021) A critical analysis of the impacts of COVID-19 on the global economy and ecosystems and opportunities for circular economy strategies. Resour Conserv Recycl 164:105169

Jahirul MI, Rasul MG, Schaller D, Khan MMK, Hasan MM, Hazrat MA (2022) Transport fuel from waste plastics pyrolysis—a review on technologies, challenges and opportunities. Energy Convers Manag 258:115451

Jagatee S, Priyadarshini S, Rath CC, Das AP (2024) Synthetic microfiber: an enduring environmental problem linked to sustainable development. Environ Sci Eng Part F 2333:93–112. https://doi.org/10.1007/978-3-031-51792-1_6

Kanade TM, Joseph J, Ansari S, Varghese MAM, Savale T (2024) Solid waste management for environmental sustainability and human health. J Inform Educ Res, 4(1)

Kanungo J, Sahoo T, Bal M, Behera ID (2024) Performance of bioremediation strategy in waste lubricating oil pollutants: a review. Geomicrobiol J 41(4):360–373. https://doi.org/10.1080/01490451.2023.2245395

Kanungo J, Sahoo T, Swain LP, Behera ID (2023) Toxicity of persistent hydrocarbon pollutants, sources and sustainable remediation process. In ID Behera, AP Das (eds), Impact of petroleum waste on environmental pollution and its sustainable management through circular economy (pp 39–65). Springer Nature Switzerland. https://doi.org/10.1007/978-3-031-48220-5_2

Khan MS, Mubeen I, Caimeng Y, Zhu G, Khalid A, Yan M (2022) Waste to energy incineration technology: recent development under climate change scenarios. Waste Manage Res 40(12):1708–1729

Khatri I, Singh S, Garg A (2018) Performance of electro-Fenton process for phenol removal using Iron electrodes and activated carbon. J Environ Chem Eng 6(6):7368–7376

Khatun R, Singh S, Dubey NK, Prasad Das A (2023a) A review on marine-based phytochemicals and their application in biomedical research. Recent Front Phytochem Appl Food Pharm Cosmet Biotechnol. https://doi.org/10.1016/B978-0-443-19143-5.00013-X

Khatun R, Singh S, Dubey NK, Prasad Das A (2023b) A review on marine-based phytochemicals and their application in biomedical research. In: Recent frontiers of phytochemicals: applications in food, pharmacy, cosmetics, and biotechnology (pp 383–395). Elsevier. https://doi.org/10.1016/B978-0-443-19143-5.00013-X

Kianfar E, Salimi M (2020) A review on the production of light olefins from hydrocarbons cracking and methanol conversion. Adv Chem Res 59:1–81

Kirby DA, El-Kaffass I (2022) The characteristics of a green, innovative and transformational entrepreneur: an example of transformative entrepreneurship in an efficiency-driven economy. Int J Technol Learn Innov Dev 14(1–2):7–22

Kolawole AS, Iyiola AO (2023) Environmental pollution: threats, impact on biodiversity, and protection strategies. In: Sustainable utilization and conservation of Africa's biological resources and environment (pp 377–409). Singapore: Springer Nature Singapore

Kremp F, Müller V (2021) Methanol and methyl group conversion in acetogenic bacteria: biochemistry, physiology and application. FEMS Microbiol Rev 45(2):040

Krishnan S, Rawindran H, Sinnathambi CM, Lim J W (2017) Comparison of various advanced oxidation processes used in remediation of industrial wastewater laden with recalcitrant pollutants. In: IOP Conference series: materials science and engineering. IOP Publishing. 206, 1, 012089

Kulprathipanja S, Rekoske JE, Wei D, Slone RV, Pham T, Liu C (2021) Modern petrochemical technology: methods. John Wiley & Sons, Manufacturing and Applications

Kumar M, Pandit S, Patel V, Khanna N, Nag M, Lahiri D, Ray RR, Das A, Das D (2023) Maximization of energy recovery from starch processing wastewater by thermophilic dark fermentation coupled with microbial fuel cell technology. Geomicrobiol J. https://doi.org/10.1080/01490451.2023.2209555

Kumar B, Verma P (2020) Application of hydrolytic enzymes in biorefinery and its future prospects. Microbial strategies for techno-economic biofuel production, 59–83

Kuppusamy S, Maddela NR, Megharaj M, Venkateswarlu K, Kuppusamy S, Maddela NR, Venkateswarlu K (2020a) Impact of total petroleum hydrocarbons on human health. Total Petrol Hydrocarbon Environ Fate Toxicity Rem, 139–165

Kuppusamy S, Maddela NR, Megharaj M, Venkateswarlu K, Kuppusamy S, Maddela NR, Venkateswarlu K (2020b) Ecological impacts of total petroleum hydrocarbons. Total Petrol Hydrocarbon Environ Fate Tox Rem, 95–138

Kurth JM, Op den Camp HJ, Welte CU (2020) Several ways one goal—methanogenesis from unconventional substrates. Appl Microbiol Biotechnol 104(16):6839–6854

Lahiri D, Nag M, Dey A, Sarkar T, Joshi S, Pandit S, Das AP, Pati S, Pattanaik S, Tilak VK, Ray RR (2021) Biofilm mediated degradation of petroleum products. Geomicrobiol J. https://doi.org/10.1080/01490451.2021.1968979

Lee ES, Park SY, Kim CG (2023) Feasibility test anaerobically enhancing methane yield under the injection of hydrogen and carbon dioxide. Renew Energy 212:761–768

Leong YK, Chen WH, Lee DJ, Chang JS (2021) Supercritical water gasification (SCWG) as a potential tool for the valorization of phycoremediation-derived waste algal biomass for biofuel generation. J Hazard Mater 418:126278

Liang J, Mai W, Tang J, Wei Y (2019) Highly effective treatment of petrochemical wastewater by a super-sized industrial scale plant with expanded granular sludge bed bioreactor and aerobic activated sludge. Chem Eng J 360:15–23

Liu Y, Zhao Y, Wang J (2021) Fenton/Fenton-like processes with in-situ production of hydrogen peroxide/hydroxyl radical for degradation of emerging contaminants: advances and prospects. J Hazard Mater 404:124191

Mallick K, Sahu A, Dubey NK, Das AP (2023) Harvesting marine plastic pollutants-derived renewable energy: a comprehensive review on applied energy and sustainable approach. J Environ Manage 348:119371. https://doi.org/10.1016/J.JENVMAN.2023.119371

Mallick D, Sharma SD, Kushwaha A, Brahma HS, Nath R, Bhowmik R (2022) Emerging commercial opportunities for conversion of waste to energy: aspect of gasification technology. In: Waste-to-energy approaches towards zero waste. Elsevier. pp 105–127

Mariyam S, Shahbaz M, Al-Ansari T, Mackey HR, McKay G (2022) A critical review on co-gasification and co-pyrolysis for gas production. Renew Sustain Energy Rev 161:112349

Martinelli M, Gnanamani MK, LeViness S, Jacobs G, Shafer WD (2020) An overview of Fischer-Tropsch synthesis: XtL processes, catalysts and reactors. Appl Catal A 608:117740

Mishra S, Das AP (2021) Current treatment technologies for removal of microplastic and microfiber pollutants from wastewater. Wastewater Treat Cutting-Edge Mol Tools Tech Appl Aspects. https://doi.org/10.1016/B978-0-12-821881-5.00011-8

Mishra S, Rath CC, Das AP (2019a) Marine microfiber pollution: a review on present status and future challenges. Mar Pollut Bull 140:188–197. https://doi.org/10.1016/J.MARPOLBUL.2019.01.039

Mishra A, Siddiqi H, Kumari U, Behera ID, Mukherjee S, Meikap BC (2021) Pyrolysis of waste lubricating oil/waste motor oil to generate high-grade fuel oil: a comprehensive review. Renew Sustain Energy Rev 150:111446. https://doi.org/10.1016/j.rser.2021.111446

Mishra S, Dash D, Das AP (2022a) Detection, characterization and possible biofragmentation of synthetic microfibers released from domestic laundering wastewater as an emerging source of marine pollution. Mar Pollut Bull 185:114254. https://doi.org/10.1016/J.MARPOLBUL.2022.114254

Mishra S, Singh RP, Rout PK, Das AP (2022b) Membrane bioreactor (MBR) as an advanced wastewater treatment technology for removal of synthetic microplastics. Dev Wastewater Treat Res Process. https://doi.org/10.1016/B978-0-323-85583-9.00022-3
Mishra S, Ghosh S, van Hullebusch ED, Singh S, Das AP (2023) A critical review on the recovery of base and critical elements from electronic waste-contaminated streams using microbial biotechnology. Appl Biochem Biotechnol 2023:1–30. https://doi.org/10.1007/S12010-023-04440-X
Mishra S, Das AP (2022) Treatment of the wastewater polluted with synthetic microfiber released from washing machine. In BB Das, H Hettiarachchi, PK Sahu, S Nanda (eds), Lecture notes in civil engineering. Springer Science and Business Media Deutschland GmbH. Vol 207, pp 109–117. https://doi.org/10.1007/978-981-16-7509-6_9/COVER
Mishra S, Das AP (2023) Microbial remediation of synthetic microfiber contaminated wastewater. In MP Shah (ed), Microbial technologies in industrial wastewater treatment (pp 337–350). Springer, Singapore. https://doi.org/10.1007/978-981-99-2435-6_17
Mishra S, Rout PK, Das AP (2019) Solar photovoltaic panels as next generation waste: a review. B Res Appl Chem, 9(6)
Mohanty S, Ghosh S, Bal B, Das AP (2018) A review of biotechnology processes applied for manganese recovery from wastes. Rev Environ Sci Bio/Technol 17(4):791–811. https://doi.org/10.1007/S11157-018-9482-1
Mohanty J, Dey S, Bala K, Das AP (2024a) Recent advances in microplastic pollution for its sustainable management. In: Sustainable management of urban plastic waste through circular economic approaches, pp 27–48. https://doi.org/10.1201/9781003465447-3
Mohanty M, Mohanty J, Dey S, Dutta K, Shah MP, Das AP (2024b) The face mask: a tale from protection to pollution and demanding sustainable solution. Emerg Contam 10(2):100298. https://doi.org/10.1016/J.EMCON.2023.100298
Mohanty M, Tripathy B, Koteswara Reddy G, Das AP (2024c) Greener technologies for establishing restraint over microplastic pollution. In: Sustainable microbial technology for synthetic and cellulosic microfiber bioremediation, pp 239–258. https://doi.org/10.1007/978-3-031-62600-5_13
Muh E, Tabet F, Amara S (2021) Biomass conversion to fuels and value-added chemicals: a comprehensive review of the thermochemical processes. Curr Altern Energy 4(1):3–25
Namratha K, Nayan MB, Darshan MS, Basavarajappa HT, Madesh P, Byrappa K (2022) Hydrothermal—from geology to technology (Part 1). J Geol Soc India 98(3):353–362
Okolie JA, Nanda S, Dalai AK, Berruti F, Kozinski JA (2020) A review on subcritical and supercritical water gasification of biogenic, polymeric and petroleum wastes to hydrogen-rich synthesis gas. Renew Sustain Energy Rev 119:109546
Paul A, Dey S, Ram DK, Das, AP (2023) Hexavalent chromium pollution and its sustainable management through bioremediation. https://doi.org/10.1080/01490451.2023.2218377
Paul A, Priyadarsini S, Karakuş S, Behera JK, Gnanasekaran R, Das AP (2024) Fate and accumulation of synthetic microfibers in the environment. In: Sustainable microbial technology for synthetic and cellulosic microfiber bioremediation, pp 61–76. https://doi.org/10.1007/978-3-031-62600-5_4
Perini JAL, Tonetti AL, Vidal C, Montagner CC, Nogueira RFP (2018) Simultaneous degradation of ciprofloxacin, amoxicillin, sulfathiazole and sulfamethazine, and disinfection of hospital effluent after biological treatment via photo-Fenton process under ultraviolet germicidal irradiation. Appl Catal B 224:761–771
Prabhakar A, Mishra S, Das AP (2019) Isolation and identification of lead (Pb) solubilizing bacteria from automobile waste and its potential for recovery of lead from end of life waste batteries. Geomicrobiol J 36(10):894–903. https://doi.org/10.1080/01490451.2019.1654044
Pradhan B, Behera ID, Sahoo K, Mohanta S (2023a) Applicability of the teetered bed separator for beneficiating Indian Iron ore fines: an experimental study. J Mines Metals Fuels 71(11):1817–1825. https://doi.org/10.18311/jmmf/2023/31100

Pradhan G, Tripathy B, Ram DK, Digal AK, Das AP (2023b) Bauxite mining waste pollution and its sustainable management through bioremediation. Geomicrobiol J, 1–10. https://doi.org/10.1080/01490451.2023.2235353

Priyadarshini B, Das AP (2024a) The adverse impact of synthetic microplastics released from disposable face masks pollutants on aquatic microbiota, aquatic species and humans: a review. In: Environmental science and engineering, Part F2827, pp 67–81. https://doi.org/10.1007/978-3-031-50840-0_5

Priyadarsini S, Das AP (2024b) Analytical and structural characterization of waste lithium-ion batteries for their effective recycling strategy. Environ Chem Ecotoxicol. https://doi.org/10.1016/j.enceco.2024.12.004

Priyadarsini S, Das AP (2025) Lithium bioleaching: a review on microbial-assisted sustainable technology for lithium bio-circularity. J Water Process Eng 69:106744. https://doi.org/10.1016/j.jwpe.2024.106744

Raj K, Das AP (2023) Lead pollution: Impact on environment and human health and approach for a sustainable solution. Environ Chem Ecotoxicol 5:79–85. https://doi.org/10.1016/J.ENCECO.2023.02.001

Raji M, Mirbagheri SA (2021) A global trend of Fenton-based AOPs focused on wastewater treatment: a bibliometric and visualization analysis. Water Pract Technol 16(1):19–34

Razavi SMR, Miri T (2015) A real petroleum refinery wastewater treatment using hollow fiber membrane bioreactor (HF-MBR). J Water Process Eng 8:136–141

Rekhate CV, Srivastava JK (2020) Recent advances in ozone-based advanced oxidation processes for treatment of wastewater-a review. Chem Eng J Adv 3:100031

Rott E, Kuch B, Lange C, Richter P, Kugele A, Minke R (2018) Removal of emerging contaminants and estrogenic activity from wastewater treatment plant effluent with UV/chlorine and UV/H2O2 advanced oxidation treatment at pilot scale. Int J Environ Res Public Health 15(5):935

Sadasivuni KK, Rattan S, Waseem S, Brahme SK, Kondawar SB, Ghosh S, Das AP, Chakraborty PK, Adhikari J, Saha P, Mazumdar P (2019) Silver nanoparticles and its polymer nanocomposites—synthesis, optimization, biomedical usage, and its various applications. Lect Notes Bioeng. https://doi.org/10.1007/978-3-030-04741-2_11

Sahu A, Mallick K, Das AP (2024) Environmental impact of single-use synthetic face mask and its recycling: a sustainable approach. In: Das AP, Mishra S (eds) impact of COVID-19 waste on environmental pollution and its sustainable management. Environmental Science and Engineering. Springer, Cham. https://doi.org/10.1007/978-3-031-50840-0_11

Shi J, Xu C, Han H (2020) The concept of anaerobic hydrolysis to produce clean and stable energy from the case study of utilization of three types of pollutants in coal pyrolysis wastewater. Biores Technol 298:122570

Shokri A (2018) Application of Sono–photo-Fenton process for degradation of phenol derivatives in petrochemical wastewater using full factorial design of experiment. Int J Ind Chem 9:295–303

Speight JG (2020) Handbook of gasification technology: science, processes, and applications. John Wiley & Sons

Su G, Ong HC, Mofijur M, Mahlia TI, Ok YS (2022) Pyrolysis of waste oils for the production of biofuels: a critical review. J Hazard Mater 424:127396

Suman S, Rajak DK, Kumar G, Kumar B, Jibran JA (2023) A detailed review on the environmental problem and remediation of anthropogenic biomass waste. Anthropogenic Environ Hazards Compensation Mitig, 183–201

Tomassetti S, Aquilanti A, Muciaccia PF, Coccia G, Mankel C, Koenders EA, Di Nicola G (2022) A review on thermophysical properties and thermal stability of sugar alcohols as phase change materials. J Energy Storage 55:105456

Tripathy B, Sahoo PP, Sundaray H, Das AP (2024) In-vitro biodegradation of discarded marine microplastics across the eastern coast of the Bay of Bengal, India using Exiguobacterium sp. Environ Chem Ecotoxicol 6:236–247. https://doi.org/10.1016/j.enceco.2024.07.003

U.S. Department of Energy (2003) Office of fossil fuel energy. http://www.fe.doe.gov/programs/powersystems/gasification/howgasificationworks.html

Villegas-Guzman P, Giannakis S, Rtimi S, Grandjean D, Bensimon M, De Alencastro LF, Pulgarin C (2017) A green solar photo-Fenton process for the elimination of bacteria and micropollutants in municipal wastewater treatment using mineral iron and natural organic acids. Appl Catal B 219:538–549

Wagh MS, Sowjanya S, Nath PC, Chakraborty A, Amrit R, Mishra B, Mohanta YK (2024) Valorisation of agro-industrial wastes: circular bioeconomy and biorefinery process–a sustainable symphony. Process Saf Environ Prot

Wei N, Xu D, Hao B, Guo S, Guo Y, Wang S (2021) Chemical reactions of organic compounds in supercritical water gasification and oxidation. Water Res 190:116634

Zamri MFMA, Hasmady S, Akhiar A, Ideris F, Shamsuddin AH, Mofijur M, Mahlia TMI (2021) A comprehensive review on anaerobic digestion of organic fraction of municipal solid waste. Renew Sustain Energy Rev 137:110637

Zhai J, Ma H, Liao J, Rahaman MH, Yang Z, Chen Z (2018) Comparison of Fenton, ultraviolet–Fenton and ultrasonic–Fenton processes on organics and colour removal from pre-treated natural gas produced water. Int J Environ Sci Technol 15:2411–2422

Zhang F, Li Y, Liang Z, Wu T (2022) Energy conversion and utilization in supercritical water oxidation systems: a review. Biomass Bioenerg 156:106322

Zouanti M, Bezzina M, Dhib R (2020) Experimental study of degradation and biodegradability of oxytetracycline antibiotic in aqueous solution using Fenton process. Environ Eng Res 25(3):316–323

Chapter 7
Recent Advancements in the Remediation of Marine Hydrocarbon and Plastic Pollutants

G. Reddy Koteswara, S. Bhargavi, D. Yamuna, M. Anjani Devi, G. Jeevana Lakshmi Kumar, T. Manas, Srikaram Adarsh, Vazrala Bhavya Sai, Harjeet Nath, Maha Awjan Alreshidi, Hussein Togun, Deo Karan Ram, and Alok Prasad Das

Abstract Marine pollution constitutes a critical environmental challenge, with plastic waste disposal emerging as a primary culprit threatening the health of aquatic ecosystems globally. Improper waste management leads to the accumulation of plastic debris in our oceans, where it degrades into microplastics—particles smaller than five millimeters—posing a significant challenge for removal and remediation efforts. The inadvertent ingestion of these microplastics by marine organisms disrupts digestive processes and frequently leads to mortality. Furthermore, plastic waste disposal exacerbates the phenomenon of eutrophication, a condition characterized

G. Reddy Koteswara (✉) · S. Bhargavi · D. Yamuna · M. Anjani Devi ·
G. Jeevana Lakshmi Kumar · T. Manas · S. Adarsh · V. Bhavya Sai
Department of Biotechnology, Koneru Lakshmaiah Education Foundation (Deemed to Be University), Green Fields, Vaddeswaram 522502, India
e-mail: koteswarareddy@kluniversity.in

H. Nath
Department of Chemical and Polymer Engineering, Tripura University (A Central University), Agartala 799022, India

M. A. Alreshidi
Department of Chemistry, University of Ha'il, Ha'il 81441, Saudi Arabia
e-mail: m.alrashedi@uoh.edu.sa

H. Togun
Department of Mechanical Engineering, College of Engineering, University of Baghdad, Baghdad, Iraq

D. K. Ram
Department of Chemical, Petroleum and Hydrogen Technology, Nims Institute of Engineering and Technology (NIET), Nims University, Jaipur, Rajasthan 303121, India

A. P. Das
Department of Lifesciences, Ramadevi Women's University, Bhubaneswar, Odisha, India

I. D. Behera and A. P. Das (eds.), *Environmental Hydrocarbon Pollution and Zero Waste Approach Towards a Sustainable Waste Management*, Environmental Science and Engineering, https://doi.org/10.1007/978-3-031-81615-4_7

by excessive nutrient enrichment that introduces harmful toxins into marine environments. The situation is alarming, as estimates indicate that approximately eight million metric tons of plastic enter our oceans each year. Ocean currents then concentrate this plastic debris, forming vast plastic gyres, even in the most remote deep-sea regions like the Mariana Trench. The pervasive presence of these pollutants has severe health repercussions for marine life, jeopardizing the delicate balance of our marine ecosystems.

Keywords Plastic pollution · Eutrophication marine pollution · Polymers · Plastics · Marine pollution · Plastic waste disposal · Microplastics · Eutrophication

7.1 Introduction

Amidst the global challenge of marine pollution, researchers and innovators are vigorously pursuing novel strategies to mitigate the contamination of our oceans (Mishra et al. 2019). This study delves into recent advancements in addressing marine hydrocarbon and plastic pollution, presenting innovative solutions for a cleaner environmental future (Das et al. 2024; Mallick et al. 2023; Mishra et al. 2022a, b, c; Sahoo et al. 2022). The profound impact of hydrocarbon spills and plastic waste on marine ecosystems cannot be overstated, from oil slicks on the ocean's surface to microplastics infiltrating the food chain, highlighting the critical need for effective cleanup methods (Mohanta et al. 2024; Hale et al. 2020; Mishra and Das 2021).

Thankfully, scientists and engineers are stepping up to confront this challenge, leveraging biotechnology, nanotechnology, and innovative cleanup techniques to revolutionize our approach to marine pollution (Das et al. 2023. From bioengineered microorganisms capable of digesting hydrocarbons to autonomous robots tasked with removing plastic debris, these breakthroughs hold significant promise for restoring oceanic health (Priyadarsini & Das 2024, 2025). With sustained support and investment in these cutting-edge technologies, we can strive towards a cleaner and healthier marine environment for future generations (Nikoleta et al. 2021). Prioritising the development and deployment of these solutions is imperative in combating the escalating threat of plastic pollution in our oceans (Mishra et al. 2021a, b, 2023a; Mishra and Das 2023; Singh et al. 2020). By harnessing the power of innovation and scientific inquiry, we can meet the urgent need for sustainable interventions to safeguard marine ecosystems (Dan et al. 2019; Khatun et al. 2023). Through collaborative efforts and concerted action, we can drive meaningful change and foster a more sustainable future for our planet (Mohanty et al. 2024; Raj and Das 2023; Pradhan et al. 2023). Approximately 70% of the Earth's surface is covered by oceans, playing a pivotal role in global climate regulation, and supplying half of the atmospheric oxygen (Bigg et al. 2003). The vast expanse of the five oceans constitutes the largest continuous ecosystem on Earth, housing diverse habitats that support myriad organisms.

Marine bacteria and archaea inhabit virtually all marine ecosystems, comprising up to 90% of the total marine biomass (Jörg and Cendrella 2016). These microorganisms play essential roles in oceanic biogeochemical cycles, influencing carbon, oxygen, nitrogen, phosphorous, sulfur, and iron cycling (Das and Ghosh 2024; Nag et al. 2023). Beyond their ecological significance, marine bacteria serve as valuable resources for biotechnological applications (Mohanty et al. 2018; Behera et al. 2021). Adapted to thrive in dynamic marine environments, these microorganisms produce bioactive compounds with potential applications in bioremediation, pharmaceuticals, cosmetics, and food industries (Olugbemi Tope and Adetunji 2021; Bala, Anshul et al. 2024; Bala, Dogra et al. 2024; Das, Dey, & Das 2024; Das, Dey, Bala et al. 2024). Exploiting the biotechnological capabilities of marine bacteria involves the systematic exploration of valuable products, including genes, proteins, and secondary metabolites. Both culture-dependent and culture-independent methods have been employed to explore marine bacterial diversity and biotechnological capabilities (Mishra et al. 2023a, b; Behera et al. 2022a; Kanungo et al. 2023). Culture techniques have led to the isolation of bacterial species and consortia with high degradation capacities, polymer production, and unique enzyme expression. Metagenomic approaches coupled with functional screenings have identified bioactive compounds from unculturable marine bacteria, expanding the repertoire of biotechnological tools (Bighneswar 2020).

Despite significant progress, many marine environments and bacterial communities remain underexplored (Ghosh et al. 2016; Prabhakar et al. 2019). Innovative approaches are being developed to characterize rare marine bacteria and uncover novel bioproducts, holding promise for addressing contemporary environmental challenges such as oil and plastic pollution (Das et al. 2024; Mishra et al. 2021a, b; Mishra and Das 2021). According to the International Union of Pure and Applied Chemistry, plastics are polymeric materials that might include additives aimed at improving performance or cutting costs. Over the past five decades, global plastic waste production has surged, posing significant environmental and health risks. Microplastics and nanoplastics, resulting from the fragmentation of discarded plastic, have emerged as pervasive environmental pollutants, threatening ecosystems and human health (Siva et al. 2020; Mishra et al. 2022a, b, c; Mishra et al. 2020; Mishra et al. 2022a, b, c; Mishra and Das 2022). Microplastics are particles smaller than 5 mm, continually fragmenting in the environment until they reach nanoparticle size (Danlian et al. 2021; Tripathy et al. 2022). Aquatic ecosystems, including wastewater treatment plants, act as significant origins of microplastics and nanoplastics, potentially leading to terrestrial contamination through irrigation and soil improvement processes.

The ubiquity of microplastics and nano-plastics across environmental compartments underscores their environmental persistence and mobility. These particles act as vectors for toxic chemical contaminants and pathogenic microbes, posing threats to biota and ecosystem health (Sahu et al., 2024; Pawan et al. 2023; Mallick et al. 2023; Mishra and Das 2021).

While significant strides have been made in understanding the fate, transport, and toxicity of microplastics and nanoplastics, gaps persist in elucidating their ecological effects. This review seeks to address knowledge gaps by examining the interplay between nanomaterial science, microbial ecology, and the ecological impacts of microplastics and nanoplastics. Furthermore, this review explores the role of biofilms on plastic surfaces, microbe-plastic symbiosis, and technologies for remediation, including membrane systems, biodegradation, biopolymers, and phytoremediation. Establishing management approaches for plastic waste is essential to implementing effective remediation strategies in conjunction with regulatory measures. In conclusion, unravelling the complexities of microbe-plastic interactions and their ecological ramifications is crucial for developing comprehensive strategies to mitigate plastic pollution. By integrating physicochemical and biotechnological principles with regulatory frameworks, we can work towards a more sustainable future and safeguard the health of our planet (Pawan et al. 2023).

7.1.1 The Impact of Marine Pollution

Marine pollution, particularly stemming from hydrocarbons and plastics, presents a significant threat to the intricate balance of marine ecosystems and the diverse wildlife reliant upon them. The toxicity of hydrocarbons can induce severe harm to marine life, manifesting in immediate poisoning and long-term health and reproductive complications (Behera et al. 2020; Lahiri et al. 2021). Incidents such as oil spills, whether accidental or intentional, yield immediate and enduring repercussions on vulnerable species such as seabirds, marine mammals, and fish. In contrast, plastic contamination introduces a distinct array of challenges. The accumulation of plastic waste in oceans has reached alarming levels, with projections suggesting a potential surplus of plastic over fish by 2050 if current trends persist. Marine organisms frequently mistake plastic debris for food, leading to internal injuries, obstructions, and starvation (Dey, Bissoyi et al. 2024; Dey, Dash et al. 2024; Dey et al. 2023). Additionally, microplastics, resulting from the fragmentation of larger plastic items, can be ingested by small organisms, infiltrating the food chain and ultimately posing threats to larger predators, including humans. Beyond the direct harm inflicted upon wildlife, marine pollution precipitates broader consequences. Ecosystem degradation disrupts the delicate balance of marine food webs, triggering cascading effects throughout the ecosystem (Reid et al. 2018). Furthermore, waterborne contaminants jeopardize seafood safety, posing risks to human health. These pressing challenges necessitate immediate action. The adverse effects of marine pollution on both wildlife and human populations can be mitigated through the enactment of stringent waste disposal regulations and the promotion of sustainable practices. Collaboration among governments, businesses, and individuals is imperative to safeguard our oceans and ensure the enduring well-being of all living beings. Stricter legislation governing plastic waste disposal, coupled with the promotion of sustainable practices, can diminish the impact of marine pollution on wildlife and human populations alike

(Sahoo et al. 2022). By proactively safeguarding our oceans, we can foster a healthier environment for future generations. The implementation of effective waste management strategies and the advocacy for eco-friendly alternatives play pivotal roles in attenuating the deleterious effects of marine pollution. All stakeholders must prioritize the preservation of our oceans and endeavor toward long-term solutions to address this pressing issue.

The United Nations Convention on the Law of the Sea defines pollution as the introduction, directly or indirectly by humans, of substances or energy into the marine environment, including estuaries. This introduction results in harmful effects such as damage to living resources and marine life, threats to human health, interference with marine activities like fishing, degradation of seawater quality, and loss of amenities. Williams criticized the traditional division of pollution into categories like air, water, and land, arguing that all pollutants eventually make their way to the ocean. Pollution typically originates from human settlements, resource exploitation, and activities such as infrastructure development, agriculture, industry, urbanization, and tourism. Persistent organic pollutants, nutrients, oils, radionuclides, heavy metals, pathogens, sediments, litter, and debris are among the contaminants of concern (Priyadarsini and Das 2025). Coastal areas worldwide have suffered significant damage from pollution, particularly impacting commercial coastal and marine fisheries (Shahidul and Tanaka 2004). Consequently, controlling aquatic pollution is urgently needed to sustainably manage and conserve these vital fisheries and aquatic resources. However, as described by Williams, the pollution issue is complex, with interconnected problems, uncertainties, conflicts, and constraints, making effective control challenging. Additionally, due to gaps in scientific understanding of marine pollution, there is a lack of knowledge hindering the development of effective management strategies. This paper aims to achieve three objectives: firstly, to review major coastal and marine pollution areas and their ecosystem impacts; secondly, to summarize the current status of polluted coastal and marine fisheries; and thirdly, to propose a conceptual model for improved pollution management to ensure sustainable resource utilization.

7.1.2 Exploration of Current Methods

The ongoing battle against marine hydrocarbon and plastic contamination has persisted for decades, highlighting the formidable nature of the challenge and the limitations inherent in current methodologies. Traditional approaches to oil spill cleanup predominantly involve the deployment of booms, skimmers, and dispersants (Yaw Kwakye et al. 2022). Booms serve as floating barriers to contain spills, while skimmers are employed to extract oil from the water's surface. Dispersants, chemical agents applied to break down oil into smaller droplets, complement natural degradation processes. Despite their efficacy to a certain degree, these methods encounter challenges, particularly in adverse sea conditions and delicate ecosystems. Similarly, the management of plastic waste has heavily relied on manual labor, with global initiatives organizing beach cleanups and ocean clean-up campaigns. However, these endeavors are often labor-intensive and time-consuming, unable to match the pace of plastic waste accumulation in oceans. Moreover, the vast expanse of the ocean and the dispersion of plastic debris present challenges in locating and removing all sources of pollution. To address these issues, researchers and engineers are spearheading the development of innovative technologies such as autonomous drones and robotic systems to assist in detecting and removing plastic waste in the ocean (Das, A. Prasad. & Ghosh, Shreya 2023). These autonomous drones and robotic systems hold promise for enhancing efficiency and effectiveness in marine environment cleanup efforts (Aimee and Justin 2017). Furthermore, international collaborations and partnerships are emerging to tackle the global issue of plastic pollution, recognizing the need for coordinated efforts to combat this pervasive problem. Through collaborative initiatives, nations can share resources and expertise to implement more sustainable waste management practices and mitigate further harm to marine ecosystems. Continuous research and cooperation are essential in the pursuit of a cleaner and healthier ocean for future generations. This collective action is crucial in developing innovative solutions and technologies to address existing plastic waste and prevent future pollution. Governments, organizations, and individuals must unite to protect our oceans and preserve marine life for generations to come.

7.2 Breakthrough Technologies in Marine Pollution Remediation

To address the challenges posed by current methods and combat the magnitude of marine pollution, scientists and engineers have embraced innovative technologies and strategies. These advancements present promising solutions for mitigating hydrocarbon and plastic contamination and rejuvenating the health of our oceans. By investing in these technologies and supporting ongoing research endeavors, we can advance towards a more sustainable future for our marine environments. Through concerted efforts by experts and stakeholders, substantial progress can be made in shielding our oceans from further degradation. Collaboratively, we can scale up the implementation of these innovative solutions and make tangible improvements to the health of our oceans. It is imperative to prioritize funding and resources towards these advancements to ensure the establishment of a cleaner and healthier marine ecosystem for future generations. With the appropriate investments and unwavering commitment, we can reverse the damage inflicted upon our oceans and cultivate a more resilient environment for marine life to flourish. Taking immediate action is paramount in safeguarding the well-being of our oceans, benefiting all living creatures on Earth. By proactively implementing measures now, we can avert irreparable harm to our oceans and secure a sustainable future for marine life. It is incumbent upon us to act swiftly and decisively in safeguarding and preserving the delicate balance of our marine ecosystems.

7.2.1 *Bioremediation: Harnessing Nature's Power*

Bioremediation involves utilizing natural organisms to break down pollutants. Certain bacteria and fungi could digest and metabolize hydrocarbons, converting them into harmless byproducts (Yamini and Rajeswari 2023). Researchers explore the use of bioengineered microorganisms with enhanced hydrocarbon degradation capabilities. These genetically modified organisms can be released into affected areas to accelerate the natural breakdown of oil spills. Bioremediation offers a more efficient and eco-friendlier alternative to traditional cleanup methods by expediting degradation processes (Behera et al. 2022b; Das et al. 2024). Furthermore, bioremediation holds promise in minimizing the long-term impact of oil spills on ecosystems and wildlife. By leveraging nature's inherent abilities, this approach restores polluted areas to their natural state. It also proves cost-effective compared to other cleanup techniques, making it a practical solution for addressing oil spills. Moreover, the use of genetically modified organisms in bioremediation can be customized to target specific contaminants, thereby enhancing their efficacy in cleanup efforts (Narasimhan et al. 2017). Through ongoing research and innovation, bioremediation continues to evolve as a sustainable solution for environmental restoration and conservation.

7.2.2 *Innovative Filtration Systems for Microplastic Remediation*

The development of advanced filtration systems holds promise in addressing the pervasive issue of microplastic contamination in the ocean (Tariq et al. 2023). These systems deploy innovative filtration technologies, such as membranes and sieves, to selectively capture microplastic particles while allowing seawater to permeate. By targeting specific size ranges and employing materials with a high affinity for microplastics, these filtration systems effectively extract these minuscule particles from the water column (Nastaran et al. 2021; Tripathy et al. 2024). Moreover, some filtration systems integrate nanotechnology, utilizing engineered nanomaterials to enhance microplastic capture efficiency. These advancements have the potential to significantly diminish the presence of microplastics in the ocean, thereby mitigating their adverse effects on marine life. Ongoing research endeavours are focused on improving the scalability and cost-effectiveness of these filtration systems to make them more accessible for widespread deployment. Through the synergistic integration of diverse technologies and materials, scientists aim to devise efficient solutions to combat the menace of microplastic pollution in our oceans. Leveraging engineered nanomaterials to augment microplastic capture efficiency represents a promising avenue for reducing the prevalence of microplastics in marine environments. Additionally, ongoing efforts to enhance the scalability and cost-effectiveness of these filtration systems aim to facilitate their broader adoption and implementation, thereby contributing to the preservation of marine ecosystems for future generations. By harnessing the power of innovation and collaboration, we can pave the way for a cleaner and healthier marine environment, free from the scourge of microplastic pollution. It is imperative that we prioritize the development and implementation of these advanced filtration systems to safeguard the integrity of our oceans and ensure the well-being of marine ecosystems and biodiversity. Through concerted efforts and sustained investment in research and technology, we can make significant strides towards mitigating the impact of microplastic pollution and securing a sustainable future for our planet's oceans.

7.2.3 Integration of Robotics and Artificial Intelligence (AI)

The convergence of robotics and artificial intelligence (AI) introduces innovative avenues for marine pollution remediation. Autonomous robots, equipped with sensors, cameras, and manipulators, demonstrate remarkable capabilities in navigating ocean environments, proficiently identifying, collecting, and removing plastic debris (Jennifer et al. 2024). These robots are programmable to target specific high-contamination areas or patrol extensive coastlines autonomously, efficiently retrieving plastic waste. Augmented by AI algorithms, these robots adapt to environmental changes and optimize cleanup strategies, rendering cleanup operations more efficient, cost-effective, and scalable (Morteza et al. 2024). The utilization of these advanced technologies holds promise for significantly mitigating the impact of plastic pollution on marine ecosystems and biodiversity. Furthermore, data collected by these robots provide researchers and policymakers with crucial insights into the origins and distribution patterns of plastic waste in our oceans, facilitating the development of targeted interventions and strategies to curb further pollution (João 2021). In essence, the integration of robotics and AI into cleanup endeavors offers a promising avenue for addressing the global challenge of plastic waste in marine environments. Additionally, robotic technology can access areas inaccessible to humans, such as deep-sea locations or remote coastlines, enhancing the comprehensiveness of cleanup operations. This comprehensive approach holds potential for more effective and efficient cleanup initiatives, thereby fostering a healthier marine ecosystem for future generations.

7.2.4 Nanotechnology: Advancing Cleanup Solutions for Cleaner Oceans

Nanotechnology stands as a cutting-edge discipline, offering innovative solutions to combat marine pollution, specifically the pervasive contamination caused by hydrocarbons and plastics. At the nanoscale, a diverse array of materials, ranging from graphene to carbon nanotubes and nanoscale zero-valent iron (nZVI), exhibit extraordinary properties such as high surface area and unique chemical characteristics (Hao et al. 2021). These properties render them highly effective for adsorbing pollutants from marine environments, presenting a promising avenue for remediation efforts. One of the remarkable features of nanomaterials is their versatility. They can be dispersed in water or integrated into absorbent materials, allowing for efficient capture of pollutants across a wide spectrum of marine settings. This versatility enhances their applicability and effectiveness in addressing marine pollution challenges. Moreover, catalytic nanoparticles, including titanium dioxide (TiO2) and iron oxide nanoparticles, hold significant potential for expediting the degradation of hydrocarbons in marine environments (Mphilisi et al. 2015). Through processes such as photocatalysis or Fenton-like reactions, these nanoparticles catalyze the generation of reactive species, leading to the breakdown of hydrocarbons into less harmful compounds. This catalytic activity contributes to the restoration of marine ecosystems and mitigates the adverse effects of pollution. In addition to standalone nanomaterials, the development of nanocomposites represents a burgeoning field in remediation technology. Nanocomposites involve the dispersion of nanoscale particles within a matrix material, resulting in enhanced properties and performance (Rong et al. 2013). These nanocomposites can be utilized in various remediation applications, including sorbents, membranes, and coatings, effectively improving pollutant removal capabilities and ecological outcomes.

Overall, the integration of nanotechnology into environmental remediation efforts holds immense promise for achieving cleaner oceans and safeguarding marine biodiversity. Continued research and development in this interdisciplinary field are essential for unlocking new strategies and advancing technologies to address the complex challenges posed by marine pollution on a global scale.

7.2.5 Use of Biochar

Biochar, derived from organic sources such as agricultural waste, forestry residues, or biomass, has emerged as a focal point in addressing marine hydrocarbon and plastic pollution due to its remarkable potential for remediation applications (Qiang et al. 2019). Characterized by its porous structure and extensive surface area, biochar serves as an efficient adsorbent capable of capturing various pollutants, including hydrocarbons and plastics, in marine environments contaminated by oil spills or plastic debris. Through both physical and chemical interactions, biochar effectively binds these pollutants to its surface, thereby facilitating their removal from the water column or sediment. In remediation efforts targeting contaminated marine sediments, biochar plays a pivotal role in mitigating the adverse impacts of hydrocarbon and plastic pollutants. Whether applied directly to sediment layers through surface application or incorporated via sediment mixing, biochar functions to sequester and immobilize pollutants, thereby diminishing their bioavailability and potential harm to marine organisms (Yuanyuan et al. 2021). This approach not only aids in the restoration of affected ecosystems but also contributes to the overall health and resilience of marine habitats. Moreover, beyond its adsorption capabilities, biochar exhibits a unique capacity to enhance biodegradation processes within marine environments (Li et al. 2019). Acting as a substrate for microbial activity, biochar fosters the proliferation of microorganisms capable of metabolizing adsorbed pollutants. Through this synergistic interaction between biochar and microbial communities, pollutants are metabolized and degraded into less harmful compounds, thereby promoting the natural attenuation of contaminants in marine ecosystems. Furthermore, the inherent stability of biochar in marine environments underscores its suitability for long-term pollutant remediation initiatives. With the potential to persist for centuries without significant degradation, biochar ensures sustained effectiveness in pollutant sequestration and removal, offering enduring solutions for environmental management and habitat restoration efforts. Additionally, the utilization of biochar derived from renewable biomass sources not only facilitates pollutant remediation but also contributes to carbon sequestration and climate change mitigation endeavors, thereby aligning with broader sustainability objectives (Radheshyam and Wusirika 2023). In summation, the multifaceted properties and applications of biochar position it as an asset in the quest for cleaner and healthier marine environments, offering tangible benefits for both environmental restoration and long-term ecosystem resilience.

7.3 Highlighting Collaborative Efforts

Addressing marine pollution is an urgent global challenge necessitating collaborative efforts and cohesive international initiatives. Global organizations are increasingly pooling their knowledge, resources, and best practices to collectively combat marine contamination. International frameworks, such as the United Nations Environment Programme's Global Partnership on Marine Litter and the Global Ocean Cleanup Initiative, serve as vital platforms for coordination and cooperation. These initiatives facilitate the exchange of innovative ideas, provide support for the development of advanced technologies, and enable the implementation of effective cleanup strategies. By fostering collaboration on a global scale, we can maximize our collective impact in cleansing our oceans. The enormity of the challenge underscores the necessity of unified action, as no single country or organization can address it alone. Through robust international partnerships, significant progress can be achieved in restoring the health and vitality of our marine ecosystems for future generations (Carlos et al. 2020). Together, we can pave the way for a more sustainable future for our oceans by fostering collaboration and sharing knowledge across borders. By uniting efforts and sharing best practices, we can amplify our effectiveness in combating marine pollution and safeguarding the well-being of our oceans. Through collaborative endeavours, we can work towards a cleaner and healthier marine environment that benefits all life on Earth.

7.4 Future Challenges and Prospects

While recent breakthroughs in cleaning up marine hydrocarbon and plastic contamination instil hope for a cleaner future, formidable challenges await resolution. Expanding the reach of these technologies and implementing them globally presents significant logistical and financial obstacles. Moreover, careful evaluation of the long-term environmental impacts of these innovative solutions is imperative to prevent unintended consequences. Additionally, complementary preventive measures, such as reducing hydrocarbon usage and plastic waste production, are vital to augment cleanup endeavours. A sustained commitment to research, innovation, and investment is essential to surmount these challenges and ensure the enduring sustainability of our oceans. Innovative technologies, such as ocean cleanup systems and sustainable fishing practices, hold promise in addressing marine pollution. However, their scalability and global deployment face considerable hurdles, compounded by logistical and financial constraints. A thorough assessment of the environmental ramifications of these solutions is crucial to mitigate potential adverse effects. Collaborative engagement among governmental bodies, private enterprises, and non-profit organizations is indispensable in advancing ocean conservation initiatives (Rebecca et al. 2021). Through synergistic collaboration, the consolidation of resources and expertise can yield comprehensive solutions that safeguard our oceans for generations to come.

7.5 Conclusion

The issue of marine pollution, particularly from hydrocarbons and plastics, requires urgent attention and concerted efforts from all stakeholders. The latest breakthroughs in cleaning up marine contamination offer hope for a cleaner and more sustainable future. By harnessing the power of biotechnology, nanotechnology, and robotics, we can develop innovative solutions that revolutionize the way we tackle marine pollution. Yet, it is vital to acknowledge that solely focusing on cleanup efforts is insufficient. Preventive measures, such as reducing our reliance on hydrocarbons and adopting more sustainable practices, are equally important. Continued research, investment, and collaboration are essential to preserving the health and biodiversity of our oceans for generations to come. Let us join forces and turn the tide on marine contamination, for the sake of our planet and future generations. By integrating advanced technologies like artificial intelligence, nanotechnology, and robotics, we can develop innovative solutions that revolutionize the way we tackle marine pollution. However, it is crucial to recognize that solely focusing on cleanup efforts is not enough. Preventive measures, such as reducing our reliance on hydrocarbons and adopting more sustainable practices, are equally important in protecting our oceans. Continued research, investment, and collaboration are essential to preserving the health and biodiversity of our oceans for generations to come. Let us join forces and turn the tide on marine contamination, for the sake of our planet and future generations. Together, we can work towards a cleaner and healthier marine environment by implementing comprehensive strategies that address both the root causes and effects of pollution. A comprehensive approach to ocean conservation ensures a sustainable future for all creatures that rely on healthy oceans for survival.

References

Aimee VW, Justin D (2017) The dawning of the ethics of environmental robots

Bala K, Anshul, Verma S, Dogra A, Das AP, Kumar R, Nandini (2024) Sustainable management for urban plastic waste generation. In: Sustainable Management of Urban Plastic Waste Through Circular Economic Approaches, pp 211–236. https://doi.org/10.1201/9781003465447-13

Bala K, Dogra A, Lata P, Das AP (2024) Discussing sources and biological-chemical remediation approaches for microplastic pollution. In: Sustainable Microbial Technology for Synthetic and Cellulosic Microfiber Bioremediation, pp 37–59. https://doi.org/10.1007/978-3-031-62600-5_3

Behera ID, Basak G, Kumar RR, Sen R, Meikap BC (2020) Treatment of petroleum refinery sludge by petroleum degrading bacterium Stenotrophomonas pavanii IRB19 as an efficient novel technology. J Environ Sci Health Part A 56(2):226–239. https://doi.org/10.1080/10934529.2020.1866924

Behera ID, Nayak M, Biswas S, Meikap BC, Sen R (2021) Enhanced biodégradation of total petroleum hydrocarbons by implementing a novel two-step bioaugmentation strategy using indigenous bacterial consortium. J Environ Manag 292:112746. https://doi.org/10.1016/j.jenvman.2021.112746

Behera ID, Nayak M, Mishra A, Meikap BC, Sen R (2022b) Strategic implementation of integrated bioaugmentation and biostimulation for efficient mitigation of petroleum hydrocarbon pollutants from terrestrial and aquatic environment. Mar Pollut Bull 177:113492. https://doi.org/10.1016/j.marpolbul.2022.113492

Behera ID, Meikap BC, Sen R (2022a) Performance evaluation of bacterial consortium for biodegradation of total petroleum hydrocarbon: a comparative strategic biostimulation study. Res Square

Bigg GR, Jickells TD, Liss PS, Osborn TJ (2003) The role of the oceans in climate

Bighneswar B (2020) Metagenomics: the approach and techniques for finding new bioactive compounds

Carlos MD, Agusti S, Barbier E, Britten GL, Castilla JC, Gattuso JP, Fulweiler RW, Hughes TP, Knowlton N, Lovelock CE, Lotze HK, Predragovic M, Poloczanska E, Roberts C, Worm B (2020). Rebuilding marine life

Cole M, Lindeque P, Halsband C, Galloway TS (2011) Microplastics as contaminants in the marine environment: a review. Mari Pollut Bull 62(12):2588–2597

Dan L, John M. Baxter, Diva J. Amon, Duncan E.J. Currie, Craig A. Downs, Jason M. Hall-Spencer, Harriet Harden-Davies, Richard Page, Chris P. Reid, Callum M. Roberts, Alex Rogers, Thorsten Thiele, Charles R.C. Sheppard, Rashid U. Sumaila, Lucy C. Woodall (2019). Eight urgent, fundamental and simultaneous steps needed to restore ocean health, and the consequences for humanity and the planet of inaction or delay.

Danlian H, Jiaxi T, Cheng M, Deng R, Chen S, Yin L, Li R (2021) Microplastics and nanoplastics in the environment: macroscopic transport and effects on creatures

Das A, Dey S, Bala K, Gnanasekaran R, Das AP (2024) Microbial bioremediation technology for sustainable treatment and management of synthetic microfiber waste. In: Sustainable Microbial Technology for Synthetic and Cellulosic Microfiber Bioremediation, pp 221–238. https://doi.org/10.1007/978-3-031-62600-5_12

Das A, Dey S, Das AP (2024) Ecotoxicological impacts of synthetic microfiber pollutants and development of sustainable mitigation strategies. Environ Chem Ecotoxicol. https://doi.org/10.1016/J.ENCECO.2024.12.003

Das AP, Dutta K, Khatun R, Behera ID, Singh S, Mishra S (2023) Microfiber pollution and its microbial mitigation: A review on current trends and future prospects. J Taiwan Inst Chem Eng, 105104. https://doi.org/10.1016/J.JTICE.2023.105104

Das A. Prasad, Ghosh, Shreya (2023) Manganese mining microorganisms

Das AP, Nayak B, Behera ID (2024) Applications of microorganisms in advanced geomicrobiological and biotechnological exploration. Geomicrobiol J 41(4):309–311. https://doi.org/10.1080/01490451.2024.2327059

Dey S, Bissoyi A, Koteswara Reddy G, Das AP (2024) Synthetic microfibers: An urban emerging pollutant demanding for sustainable environmental solutions. In: Sustainable Microbial Technology for Synthetic and Cellulosic Microfiber Bioremediation, pp 107–121. https://doi.org/10.1007/978-3-031-62600-5_7

Dey S, Dash N, Ray SK, Priyadarsini S, Chhotaray C, Das AP (2024) Occurrence and quantification of tire wear plastics pollutants from highway city dust of Bhubaneswar City of Odisha. In: Sustainable Microbial Technology for Synthetic and Cellulosic Microfiber Bioremediation, pp 77–89. https://doi.org/10.1007/978-3-031-62600-5_5

Dey S, Tripathy B, Kumar MS, Das AP (2023) Ecotoxicological consequences of manganese mining pollutants and their biological remediation. Environmental Chemistry and Ecotoxicology. https://doi.org/10.1016/J.ENCECO.2023.01.001

Ghosh S, Mohanty S, Nayak S, Sukla LB, Das AP (2016) Molecular identification of indigenous manganese solubilising bacterial biodiversity from manganese mining deposits. J Basic Microbiol 56(3):254–262. https://doi.org/10.1002/JOBM.201500477

Hale RC, Seeley ME, La Guardia MJ, Mai L, Zeng EY (2020) A global perspective on microplastics

Hao T, Wang J, Zhang S, Pang H, Wang X, Chen Z, Li M, Song G, Qiu M, Yu S (2021) Recent advances in nanoscale zero-valent iron-based materials: characteristics, environmental remediation and challenges

Jennifer AC, Samadikhoshkho Z, Rehman AU, Valle-Pérez AU, Ponce de León EH, Hauser CA, Feron EM, Ahmad R (2024) A systematic review of robotic efficacy in coral reef monitoring techniques

João NLC (2021) Artificial intelligence as a strategic tool to contribute to sustainable development

Jörg O, Cendrella L (2016) Marine bacteria and archaea: diversity, adaptations, and culturability

József L, Sadhbh M (2021) Microplastics in the food chain: food safety and environmental aspects

Kanungo J, Sahoo T, Swain LP, Behera ID (2023) Toxicity of persistent hydrocarbon pollutants, sources and sustainable remediation process. In I. D Behera, AP Das (eds), Impact of petroleum waste on environmental pollution and its sustainable management through circular economy. Springer Nature Switzerland. pp 39–65 https://doi.org/10.1007/978-3-031-48220-5_2

Shankar MK, Srivastava SK (2017) Genetically modified microbes for bioremediation of oil spills in marine environment

Khatun R, Singh S, Dubey NK, Prasad Das A (2023) A review on marine-based phytochemicals and their application in biomedical research. Recent Front Phytochem Appl Food Pharm Cosmet Biotechnol. https://doi.org/10.1016/B978-0-443-19143-5.00013-X

Lahiri D, Nag M, Dey A, Sarkar T, Joshi S, Pandit S, Das AP, Pati S, Pattanaik S, Tilak VK, Ray RR (2021) Biofilm mediated degradation of petroleum products. Geomicrobiol J. https://doi.org/10.1080/01490451.2021.1968979

Li W, Tse H, Fok L (2016) Plastic waste in the marine environment: a review of sources, occurrence and effects. Sci Total Environ 566–567:333–349

Li L, Zou D, Xiao Z, Zeng X, Zhang L, Jiang L, Wang A, Ge D, Zhang G, Liu F (2019) Biochar as a sorbent for emerging contaminants enables improvements in waste management and sustainable resource use

Mallick K, Sahu A, Dubey NK, Das AP (2023) Harvesting marine plastic pollutants-derived renewable energy: a comprehensive review on applied energy and sustainable approach. J Environ Manage 348:119371. https://doi.org/10.1016/J.JENVMAN.2023.119371

Mishra S, Das AP (2021) Current treatment technologies for removal of microplastic and microfiber pollutants from wastewater. Wastewater Treat Cutt Edge Mol Tools Tech Appl Aspects. https://doi.org/10.1016/B978-0-12-821881-5.00011-8

Mishra S, Rath CC, Das AP (2019) Marine microfiber pollution: a review on present status and future challenges. Mar Pollut Bull 140:188–197. https://doi.org/10.1016/J.MARPOLBUL.2019.01.039

Mishra S, Singh RP, Rath CC, Das AP (2020) Synthetic microfibers: source, transport and their remediation. J Water Process Eng 38:101612. https://doi.org/10.1016/J.JWPE.2020.101612

Mishra S, Swain S, Sahoo M, Mishra S, Das AP (2021b) Microbial colonization and degradation of microplastics in aquatic ecosystem: a review. Geomicrobiol J. https://doi.org/10.1080/01490451.2021.1983670

Mishra S, Dash D, Al-Tawaha ARMS, Das AP (2022a) A review on heavy metal ion adsorption on synthetic microfiber surface in aquatic environments. Appl Biochem Biotechnol 194(10):4639–4654. https://doi.org/10.1007/S12010-022-04029-W/METRICS

Mishra S, Dash D, Das AP (2022b) Detection, characterization and possible biofragmentation of synthetic microfibers released from domestic laundering wastewater as an emerging source of marine pollution. Mar Pollut Bull 185:114254. https://doi.org/10.1016/J.MARPOLBUL.2022.114254

Mishra S, Singh RP, Rout PK, Das AP (2022c) Membrane bioreactor (MBR) as an advanced wastewater treatment technology for removal of synthetic microplastics. Dev Wastewater Treat Res Process. https://doi.org/10.1016/B978-0-323-85583-9.00022-3

Mishra S, Dash D, Das AP (2023a) Aquatic microbial diversity on plastisphere: colonization and potential role in microplastic biodegradation. Geomicrobiol J. https://doi.org/10.1080/01490451.2023.2209750

Mishra S, Ghosh S, van Hullebusch ED, Singh S, Das AP (2023b) A critical review on the recovery of base and critical elements from electronic waste-contaminated streams using microbial biotechnology. Appl Biochem Biotechnol 2023:1–30. https://doi.org/10.1007/S12010-023-04440-X

Mishra S, Das AP (2022) Treatment of the wastewater polluted with synthetic microfiber released from washing machine. In BB Das, H Hettiarachchi, PK Sahu, S Nanda (eds), Lecture notes in civil engineering. Springer Science and Business Media Deutschland GmbH. Vol 207, pp 109–117. https://doi.org/10.1007/978-981-16-7509-6_9/COVER

Mishra S, Das AP (2023) Microbial remediation of synthetic microfiber contaminated wastewater. In MP Shah (ed), Microbial Technologies in Industrial Wastewater Treatment. Springer, Singapore. pp. 337–350. https://doi.org/10.1007/978-981-99-2435-6_17

Mishra S, Rout PK, Das AP (2021) Emerging microfiber pollution and its remediation. In R Prasad (ed), Environmental and microbial biotechnology. Springer, Singapore. pp. 247–266. https://doi.org/10.1007/978-981-15-5499-5_9

Mitrano DM, Wohlleben W (2020) Microplastic regulation should be more precise to incentivize both innovation and environmental safety

Mohanta S, Pradhan B, Behera ID (2024) Impact and remediation of petroleum hydrocarbon pollutants on agricultural land: a review. Geomicrobiol J 41(4):345–359. https://doi.org/10.1080/01490451.2023.2243925

Mohanty S, Ghosh S, Bal B, Das AP (2018) A review of biotechnology processes applied for manganese recovery from wastes. Reviews in Environmental Science and Bio/technology 17(4):791–811. https://doi.org/10.1007/S11157-018-9482-1

Mohanty M, Mohanty J, Dey S, Dutta K, Shah MP, Das AP (2024) The face mask: a tale from protection to pollution and demanding sustainable solution. Emerg Contam 10(2):100298. https://doi.org/10.1016/J.EMCON.2023.100298

Mohapatra, Mersky RL (n.d) Circular Economy in the Construction Industry, (eds) (1st ed., pp 61–68). CRC Press. https://doi.org/10.1201/9781003217619-9
Morteza S, Why NK, Yadollahi M, Kamyab H, Cheng J, Khorami M (2024) Cultivating a sustainable future in the artificial intelligence era: a comprehensive assessment of greenhouse gas emissions and removals in agriculture
Mphilisi MM, Ngila CJ, Mamba BB (2015) Recent developments in environmental photocatalytic degradation of organic pollutants: the case of titanium dioxide nanoparticles
Nag M, Lahiri D, Ghosh S, Sarkar T, Pati S, Das AP, Ram DK, Bhattacharya D, Ray RR (2023) Application of microorganisms in biotransformation and bioremediation of environmental contaminant: a review. Geomicrobiol J. https://doi.org/10.1080/01490451.2023.2261443
Narasimhan MK, Muthukumaran C, Sharmila G, Gurunathan B (2017) Genetically modified organisms and its impact on the enhancement of bioremediation
Nastaran R, Hamidian AH, Wu C, Zhang Y, Yang M (2021) Microplastic sampling techniques in freshwaters and sediments
Nikoleta B, Gambardella C, Karantzalos K, Monteiro JG, Canning-Clode J, Kemna S, Arrieta-Giron CA, Lemmen C (2021) Global assessment of innovative solutions to tackle marine litter
Olugbemi OT, Adetunji CO (2021) Biological, biochemical, and biodiversity of biomolecules from marine-based beneficial microorganisms: industrial perspective
Ossai I, Ahmed A, Hassan A, Hamid F (2020) Remediation of soil and water contaminated with petroleum hydrocarbon: a review. Environ Technol Innov 17:100526
Pacwa-Płociniczak M, Płaza G, Piotrowska-Seget Z, Cameotra S (2011) Environmental applications of biosurfactants: recent advances. Int J Mol Sci 12:633–654
Pawan KR, Yadav S, Kataria N, Khoo KS (2023) Microplastics and nanoplastics in the terrestrial food chain: Uptake, translocation, trophic transfer, ecotoxicology, and human health risk
Prabhakar A, Mishra S, Das AP (2019) Isolation and identification of lead (Pb) solubilizing bacteria from automobile waste and its potential for recovery of lead from end of life waste batteries. Geomicrobiol J 36(10):894–903. https://doi.org/10.1080/01490451.2019.1654044
Pradhan B, Behera ID, Sahoo K, Mohanta S (2023) Applicability of the teetered bed separator for beneficiating indian iron ore fines: an experimental study. J Mines Metals Fuels 71(11):1817–1825. https://doi.org/10.18311/jmmf/2023/31100
Priyadarsini S, Das AP (2024) Analytical and structural characterization of waste lithium-ion batteries for their effective recycling strategy. Environ Chem Ecotoxicol. https://doi.org/10.1016/j.enceco.2024.12.004
Priyadarsini S, Das AP (2025) Lithium bioleaching: a review on microbial-assisted sustainable technology for lithium bio-circularity. J Water Process Eng, 69, 106744. https://doi.org/10.1016/j.jwpe.2024.106744
Qiang H, Song S, Chen Z, Hu B, Chen J, Wang X (2019) Biochar-based materials and their applications in removal of organic contaminants from wastewater
Radheshyam Y, Wusirika R (2023) Biochar as an environment-friendly alternative for multiple applications
Raj K, Das AP (2023) Lead pollution: impact on environment and human health and approach for a sustainable solution. Environ Chem Ecotoxicol 5:79–85. https://doi.org/10.1016/J.ENCECO.2023.02.001
Rebecca LG, Enrici A, Betsill M, Le Cornu E, Basurto X, Co-Designers R (2021) Opening the black box of conservation philanthropy: a co-produced research agenda on private foundations in marine conservation
Reid NC, Zinnert J, Young DR (2018) Degradation of coastal ecosystems: causes, impacts and mitigation efforts
Rochman C, Hoh E, Hentschel B, Kaye S (2013) Long-term field measurement of sorption of organic contaminants to five types of plastic pellets: implications for plastic marine debris. Environ Sci Technol 47(3):1646–1654
Rong MZ, Zhang MQ, Ruan WH (2013) Surface modification of nanoscale fillers for improving properties of polymer nanocomposites

Sahoo PP, Singh S, Rout PK, Mishra S, Das AP (2022) Microbial remediation of plastic pollutants generated from discarded and abandoned marine fishing nets. Biotechnol Gnetic Eng Rev. https://doi.org/10.1080/02648725.2022.2152629

Sahu A, Mallick, K., Das, A.P. (2024). Environmental impact of single-use synthetic face mask and its recycling: a sustainable approach. In: Das AP, Mishra S (eds) Impact of COVID-19 waste on environmental pollution and its sustainable management. Environmental Science and Engineering. Springer, Cham. https://doi.org/10.1007/978-3-031-50840-0_11

Shahidul IM, Tanaka M (2004) Impacts of pollution on coastal and marine ecosystems including coastal and marine fisheries and approach for management: a review and synthesis

Singh RP, Mishra S, Das AP (2020) Synthetic microfibers: pollution toxicity and remediation. Chemosphere 257:127199. https://doi.org/10.1016/j.chemosphere.2020.127199

Siva SS, Dogiparthi LK, Gangadhar L, Chakravorty A, Abhishek N (2020) Effects of microplastics and nanoplastics on marine environment and human health

Tariq M, Mustafa B, Mackenzie K, Ali W, Sabir RI, Anum W, Gaurav GK, Riaz U, Liu X, Peng L (2023) Recent developments in microplastic contaminated water treatment: Progress and prospects of carbon-based two-dimensional materials for membranes separation

Tripathy B, Dash A, Das AP (2022) Detection of environmental microfiber pollutants through vibrational spectroscopic techniques: recent advances of environmental monitoring and future prospects. Crit Rev Anal Chem. https://doi.org/10.1080/10408347.2022.2144994

Tripathy B, Sahoo PP, Sundaray H, Das AP (2024) In-vitro biodegradation of discarded marine microplastics across the eastern coast of the Bay of Bengal, India using Exiguobacterium sp. Environ Chem Ecotoxicol 6:236–247. https://doi.org/10.1016/j.enceco.2024.07.003

Xu X, Liu W, Tian S, Wang W, Qi Q, Jiang P, Yu H (2018) Petroleum hydrocarbon-degrading bacteria for the remediation of oil pollution under aerobic conditions: a prospective analysis. Front Microbiol, 9

Yamini V, Rajeswari DV (2023) Metabolic capacity to alter polycyclic aromatic hydrocarbons and its microbe-mediated remediation

Yaw KA, Nyankson E (2022) Dispersants as an oil spill clean-up technique in the marine environment

Yuanyuan Y, Ye S, Zhang C, Zeng G, Tan X, Song B, Zhang P, Yang H, Li M, Chen Q (2021) Application of biochar for the remediation of polluted sediments

Zhang H, Pap S, Taggart M, Boyd K, James N, Gibb S (2019) A review of the potential utilisation of plastic waste as adsorbent for removal of hazardous priority contaminants from aqueous environments. Environ Pollut, 113698

Chapter 8
Novel Methods and Environment-Friendly Techniques for the Remediation of Environmental Petroleum Pollutants

Jeffrey Baloyi and Ayanda Mafunda

Abstract The generation of oily sludge and petroleum wastewater is the primary pollutant associated with the petroleum industry. Without adequate treatment, discharged petroleum wastewater effluent and oily sludge can disturb aquatic ecosystems, contaminate the environment, and are considered carcinogenic and detrimental to human health. Over 80% of waste effluents produced worldwide are dumped into aquatic bodies without undergoing any prior treatment, raising concerns about the direct discharge of petroleum effluent into the environment without removing or lowering hazardous contaminants. Industries must treat their effluent before discharging it into the environment. Conventional systems for treating petroleum wastewater and oily sludge have demonstrated significant shortcomings, such as low efficiency, high capital, and operating costs, and they typically produce toxic secondary waste streams, nonbiodegradable substances, and high volumes of sludge. In this investigation, novel methods and environmentally friendly techniques for treating petroleum wastewater while minimizing or eliminating secondary pollution will be explored. Conventional and current physical, physicochemical, chemical, biological, and integrated petroleum wastewater treatment technologies were discussed. In addition, several techniques to tackle petroleum hydrocarbons in oily sludge have been investigated. These technologies were classified into oil recovery, sludge disposal and remediation methods, along with the benefits and limitations as well as the improvements that have been made to make them more environmentally sustainable. This research not only adds value by highlighting sustainable wastewater treatment technologies but also sheds light on ways to valorize the waste streams produced to offset the frequently high treatment costs.

J. Baloyi (✉) · A. Mafunda
Smart Places Cluster, Council for Scientific and Industrial Research (CSIR), Pretoria 0001, South Africa
e-mail: SBaloyi1@csir.co.za

I. D. Behera and A. P. Das (eds.), *Environmental Hydrocarbon Pollution and Zero Waste Approach Towards a Sustainable Waste Management*, Environmental Science and Engineering, https://doi.org/10.1007/978-3-031-81615-4_8

Keywords Petroleum wastewater · Oily sludge · Sustainable wastewater treatment · Physical treatment techniques · Chemical treatment techniques · Biological treatment techniques · Advanced oxidation processes · Valorisation

8.1 Introduction

Rapid population growth and urbanization have resulted in a heightened demand for petroleum-based resources and a surge in global energy requirements. Projections indicate that by the year 2035, the global demand for energy will surge by approximately 37% compared to current levels, with a significant portion of this escalating demand relying on petroleum-derived products (Jafarinejad and Jiang 2019).

Petroleum processing activities such as extraction of oil, refining, transportation, and storage result in the production of oily sludge which is a substantial waste byproduct within the petroleum industry. Due to the high concentration of toxic petroleum hydrocarbons in this waste stream, it is considered hazardous. Its composition typically comprises 30–50% oil, 30–50% water, and 10–12% particulate matter (Lin et al. 2018). The recovery of oily sludge presents tremendous obstacles due to the abundance of refractory heavy petroleum hydrocarbons (Zhang et al. 2013). Moreover, waste sludge treatment and disposal expenses can represent a significant portion, the operational costs of wastewater treatment facilities typically range between 50 and 60%, and improper disposal practices have the potential to result in severe environmental contamination (Lin et al. 2017).

Petroleum industries have a substantial demand for water in their hydrocarbon processing and crude oil refining operations. Consequently, a significant proportion, roughly 80–90%, of the provided water ultimately becomes wastewater (Jafarinejad and Jiang 2019). Typical pollutants found in petroleum industry effluents include oil and grease, petroleum hydrocarbons, heavy metals, sulphides, ammonia, naphthenic acids, and phenols (Aljuboury et al. 2017; Elmobarak et al. 2021; Kanungo et al. 2023). Due to these contaminants' slow breaking down, ecosystems are seriously threatened by environmental risks and toxicological issues (Diya'uddeen et al. 2010).

Untreated petroleum wastewater released into the environment has the potential to harm aquatic ecosystems in a number of ways. Its oily nature can impede sunlight penetration into the water, while also increasing the chemical oxygen demand and biological oxygen demand (COD and BOD), hindering plant growth (Abuhasel et al. 2021). Additionally, the prevalence of hydrocarbons in this wastewater is considered to have the ability to cause cancer or genetic mutations in humans (Abuhasel et al. 2021; Elmobarak et al. 2021).

Due to strict environmental regulations in the petroleum industry, there has been a recent shift towards cleaner technologies and environmentally sustainable manufacturing processes (Kuokkanen et al. 2021). Concerns about pollution and water scarcity have led to a surge in the demand to recover clean water, energy, and useful materials from petroleum effluents (Bala, Anshul et al. 2024; Bala, Dogra et al. 2024; A. Das, Dey, & Das 2024; A. Das, Dey, Bala et al. 2024). In addition to promoting sustainable living, a green economy, and waste reduction, efficient resource recovery can also lower recovery costs and improve environmental safety (Varjani et al. 2020; Mallick et al. 2023). Protecting aquatic ecosystems and people's health is crucial, and extensive research on environmentally acceptable remediation methods and sustainable treatment technology has been initiated (Abuhasel et al. 2021; Ani et al. 2018).

8.2 Petroleum Wastewater Treatment Technology Strategy and Approach

The substantial contribution of wastewater discharges from petroleum refineries to various water bodies has created challenges, especially for water-constrained countries. Wastewater treatment facilities are unable to handle the discharge of untreated effluent from petroleum refineries. The use of in-plant wastewater treatment technology can lessen these issues (Behera et al. 2020; Kanungo et al. 2024).

When designing or renovating petroleum refineries, it is imperative to make reducing water consumption and pollution a top priority. Several key improvements can be implemented to achieve this objective. These include using more efficient catalysts that require less frequent regeneration, replacing barometer condensers with air fan coolers or surface condensers to eliminate a significant source of oil–water emulsion, and employing more recent hydrocracking and hydrotreating methods that generate lower effluent loadings than conventional techniques.

Moreover, in order to use less caustics, acids, water washes, and filter particles that need to be disposed of, refineries can also improve their drying, sweetening, and finishing operations. Sustainability can also be increased by implementing water reuse techniques like recycling stripper bottoms, feeding low-pressure boilers with blowdown from high-pressure boilers, and using treated effluent as makeup water. Implementing water recycling systems, like using steam condensate as boiler feed-water, and adopting effective maintenance practices, such as dry-cleaning methods and proactive leak management, are essential steps. Moreover, optimizing cooling towers for multiple water recycling cycles can significantly reduce the quantity of cooling water used for single-time purposes. By embracing these advancements, refineries can operate more sustainably and environmentally responsibly while maintaining efficiency and productivity.

When designing or renovating petroleum refineries, it is imperative to make reducing water consumption and pollution a top priority. Several key improvements can be implemented to achieve this objective. These include using more effective catalysts that need less frequent regeneration, changing out barometer condensers for air fan coolers or surface condensers to get rid of an important component of oil–water emulsion, and using more recent hydrocracking and hydrotreating techniques that produce fewer effluent loadings than traditional methods.

Additionally, refineries can optimize their drying, sweetening, and finishing processes to reduce the use of water washes, acids, caustics, and filter solids that need disposal. Water reuse strategies, such as repurposing stripper bottoms, utilizing cleaned effluent as makeup water and high-pressure boiler blowdown as feed for low-pressure boilers, can further contribute to sustainability. Implementing water recycling systems, like using condensate water vapour as boiler feedwater, and adopting effective maintenance practices, such as dry-cleaning methods and proactive leak management, are essential steps. Moreover, optimizing cooling towers for multiple water recycling cycles can significantly reduce the quantity of cooling water used for single-time purposes. By embracing these advancements, refineries can operate more sustainably and environmentally responsibly while maintaining efficiency and productivity (Dey, Bissoyi et al. 2024; Dey, Dash et al. 2024; Dey et al. 2023; John & Das 2024).

8.3 Conventional and Novel Environmentally Friendly Wastewater Treatment Technologies

8.3.1 Physical Methods

In petroleum wastewater treatment, the following physical treatment process can be used: gravity separation, dissolved air flotation, oil coalescing, air/steam stripping, centrifugation, activated carbon adsorption, evaporation, filtration, and solvent extraction (Hanafy and Nabih 2007). Dissolved air flotation (DAF) and gravity separation (GS) are the two physical oily wastewater treatment technologies that are most frequently used for primary treatment. The theory of GS is based on the fact that water and oil have different densities. Although GS is a relatively simple process, its main drawback is that it is not effective in treating water with emulsified oil (Abuhasel et al. 2021). Dissolved air flotation (DAF) can effectively remove emulsified oil from oily wastewater. In this method, pressurized air is introduced into the vessel, this air forms microscopic bubbles which collide with the oil molecules and bind to them, causing the oil and contaminants to lift to the top and float.

There are significant drawbacks associated with the current DAF method, primarily stemming from its substantial operational costs, which arise due to the continuous generation of bubbles. This is particularly noteworthy because recent research highlights the preference for ultrafine bubbles because of their extensive surface area. Additionally, the method incurs substantial chemical costs and contributes to environmental degradation through the presence of chemical residuals in the wastewater discharged. Furthermore, the DAF process produces sludge that is not biodegradable (Saththasivam et al. 2022).

To address these undesirable consequences, Saththasivam et al. (2022) proposed an environmentally sustainable and economical DAF technique for extracting oil from oil–water emulsions. This innovative approach incorporates an integrated flocculation and coagulation process, the coagulant being ferric chloride and the biodegradable flocculant being chitosan. In comparison to using these substances separately, this hybrid configuration has proven to reduce chemical expenditures by 37%, decrease non-biodegradable ferric chloride usage by 90%, and reduce the impact of flocculation by 50%.

Biodegradable surfactants can be used as an optimisation strategy to increase the effectiveness of the Dissolved Air Flotation (DAF) process. They can enhance the adhesion of oil particles, thus improving the overall process efficiency. Among these surfactants, biosurfactants emerge as a practical, innovative, and environmentally conscious choice, owing to their resilience even under adverse environmental circumstances, minimal surface tension and less toxic nature (Rocha e Silva et al. 2018).

Silva et al. (2019) conducted a study in which they employed a low-cost biosurfactant derived from Pseudomonas aeruginosa to treat oily effluents within a DAF system. Their research showed that using biosurfactants significantly improved the DAF process's ability to separate oil, with improvements ranging from 65 to 95%. Consequently, the incorporation of biosurfactants as a supplementary measure in DAF procedures designed for treating oily effluents stemming from industrial activities presents a promising and sustainable alternative.

8.3.2 Physicochemical Treatment Methods

Among the prominent methods in the realm of physicochemical oily wastewater treatment techniques, flocculation and lately, electrochemical technology have garnered attention. In the flocculation process, the introduction of flocculants into wastewater serves the purpose of neutralizing the negatively charged particles within the oil emulsion, fostering particle bonding and the formation of larger aggregates referred to as flocs, which can be readily separated. Two main groups of flocculants are frequently utilized to treat oily wastewater: inorganic and organic polymeric flocculants. Polymerized ferrous sulphate, poly aluminium chloride and aluminium sulphate are examples of inorganic flocculants that offer simplicity and are cost-effective, however, they exhibit limited efficiency in flocculation. On the other hand, organic polymeric flocculants like polyacrylamide, deliver robust flocculation capabilities at moderate dosages and are applicable across a wide pH range. Nevertheless, it's worth noting that they present concerns for the well-being of humans and the surrounding ecosystems due to their non-biodegradable nature, and they generate a substantial volume of sludge during the treatment process, necessitating additional processing (Abuhasel et al. 2021).

Chitosan, characterized by its non-toxicity, biodegradability, and cationic properties, has become a very desirable option for wastewater treatment involving the use of flocculants. Lü et al. (2019) undertook the synthesis of chitosan-based flocculants, which were employed to facilitate the flocculation of emulsified drops of oil in synthetic oily effluents. Their findings revealed that the flocculation efficiency of chitosan and the widely used cationic polyacrylamide and poly aluminium chloride (PAC) was poor compared to poly dimethyl acryloyloxyethyl benzyl ammonium chloride grafted chitosan (CS-g-PDBC). The CS-g-PDBC flocculant was utilised to treat actual oily wastewater, and it yielded removal efficiencies above 98% for turbidity and oil. This underscores the potential of CS-g-PDBC as a sustainable and eco-friendly flocculant.

Electrochemical technologies encompass various systems, including electroflotation, electrocoagulation, and electrofloculation. These techniques make use of electrochemical cells, where a fixed voltage is applied while electrodes are immersed in oily effluent. Electrocoagulation and electrofloculation integrate oxidation processes and flocculation processes, employing sacrificial electrodes typically composed of materials like aluminium or iron.

Mazumder et al. (2020) proposed an innovative integrated process incorporating membrane separation with electroflotation and electrocoagulation. Their research findings revealed the remarkable efficiency and cost-effectiveness of this technology, boasting an impressive oil rejection rate of 94–96% and a substantial reduction in flux drop ranging from 43 to 72%. Additionally, the operational energy consumption remained minimal, rendering this integrated process a practical and economical approach for treating oily wastewater that also exhibited a reduced propensity for membrane fouling.

In the field of coagulation-flocculation treatment, there has been growing interest in exploring natural coagulants as more environmentally friendly substitutes to commercial coagulants. These natural coagulants have been used in wastewater treatment, either as principal coagulants or as coagulant aids in conjunction with chemical coagulants (Mohd-Salleh et al. 2019). Nonetheless, it's worth noting that the effectiveness of natural coagulants is still in its infant stages of study, and a more in-depth investigation is required to fully comprehend their coagulation performance.

8.3.3 Chemical Treatment Methods

Techniques that produce hydroxyl radicals are known as advanced oxidation processes (AOPs). These processes are employed to generate highly reactive free radicals and employ non-selective agents to eliminate complex organic pollutants present in various media, including but not limited to wastewater, soil, marine environments, and more (Elmobarak et al. 2021).

Among the most widely researched AOP techniques for wastewater treatment, notable ones include heterogeneous photocatalysis (Ani et al. 2018), ozonation processes (Chen et al. 2015), electrochemical oxidation (Jawad and Abbar 2019) and Fenton and photo-Fenton reactions. These methods offer distinct advantages when dealing with wastewater containing a diverse array of contaminants. They exhibit the ability to rapidly degrade persistent pollutants in wastewater, even when present in small quantities. Importantly, these AOP techniques demonstrate exceptional efficacy in the removal of organic compounds from effluents and do so without producing environmentally detrimental secondary waste streams (Ani et al. 2018).

Photocatalysis serves as an attractive choice for water treatment and has garnered substantial attention in the past few years due to its ability to fully oxidize organic compounds even at relatively low energy levels, its swiftness, effectiveness, environmental friendliness, and cost-efficiency (Aljuboury et al. 2018). However, some of the disadvantages of heterogeneous photocatalysts when treating petroleum refinery wastewater are the challenge of recovering the photocatalysts for reuse, the catalyst's incapacity to break down highly concentrated pollutants and its challenges with coping with the complexity of real petroleum wastewater (Ani et al. 2018).

Gawad et al. (2023) proved that photocatalytic oxidation by $UV/TiO_2/H_2O_2$ as a viable method for treating simulated oily wastewater. Through this study, the optimal process parameters were identified as a 1000 mg/L starting oil concentration, TiO_2 concentration of 1500 mg/L, agitation speed at 850 rpm, pH level at 8, and the addition of 3 mL hydrogen peroxide. Under these ideal conditions, after a 40-min reaction period, an impressive 98.43% of the oil was removed successfully from the oily effluent. Tetteh et al. (2020) evaluated the effectiveness of photocatalysis on oil refinery wastewater and found that under ideal conditions of 90 min runtime, 1.225 L/minutes air flow rate of and 8000 mg/L TiO_2, the phenol removal efficiency reached 76% and the soap, oil and grease (SOG) removal efficiency reached 88%.

8.3.4 Biological Treatment Methods

Wastewater from the petroleum sector is increasingly being treated in anaerobic, aerobic and integrated bioreactors. Nevertheless, substantial quantities of petroleum hydrocarbons often necessitate preliminary physicochemical treatments. Bioreactor-based systems offer notable benefits over conventional activated sludge systems when it comes to treating hydrocarbon-rich wastewater from industry. These benefits include enhanced biodegradation parameters control and handling, enhanced transfer of mass within the liquid phase, hydrophobic constituents (such as hydrocarbons), and solid phases as well as heightened resilience to the adverse effects of toxins within immobilized biomass (Kuyukina et al. 2020).

However, their widespread adoption is impeded by the high operational and maintenance costs involved, prompting efforts to explore compact, economically efficient bioreactor models. Stirred tank, fluidized/packed bed, and membrane bioreactors represent sophisticated configurations capable of operating under aerobic, anaerobic, or combined aerobic/anaerobic modes. Research has indicated that the combined action of anaerobic and aerobic microbes in breaking down complex, resistant organic compounds can enhance the operation of dual anaerobic–aerobic processes (Kuyukina et al. 2020).

A membrane (hollow-fibre) bioreactor's capacity to handle actual effluent from the Arak Petroleum Refining Co. was examined by Razavi and Miri (2015). This bench-scale rig was operated for 160 days and comprised a tank of activated sludge (40L) and an immersed bioreactor. The results revealed that the elimination efficiency for TSS, VSS, and turbidity were all greater than 98%. The average chemical oxygen demand (COD) removal was reported to be between 76 and 82%, while the biochemical oxygen demand (BOD) removal effectiveness was determined to be 89% over the course of the study. It is noteworthy that the treated wastewater met the stipulated discharge standards, making it suitable as makeup water for cooling purposes and agricultural irrigation at the plant. These findings hold significant promise for implementation at an industrial-scale refinery wastewater treatment plant (Mishra et al. 2024; Mohanty et al. 2024; Paul et al. 2024; Priyadarsini & Das 2024, 2025).

In the study conducted by Soo et al. (2022), a comprehensive evaluation of anaerobic biofilm reactors for palm oil processing plant effluent treatment was carried out, along with an assessment of the latest advances in this method of treatment. The research findings highlighted that a hybrid anaerobic–aerobic bioreactor can achieve exceptional removal rates, surpassing 99% for total suspended solids (TSS), COD and BOD. Notably, this hybrid system outperformed up-flow anaerobic filter and up-flow anaerobic sludge blanket, which reached 88 to 94%, and 80% respectively, with respect to the COD removal rate. Additionally, the study identified anaerobic ponds as effective in removing COD, achieving a removal efficiency of 97.8%. However, it's important to note that anaerobic ponds necessitate longer retention times and larger land areas. Considering its substantial benefits over traditional biological treatment methods, the anaerobic biofilm reactor emerges as a highly promising technique for palm oil wastewater treatment.

In a similar investigation by Yap et al. (2021), the effectiveness of an integrated anaerobic bioreactor pilot-plant for palm oil wastewater treatment was assessed. The startup process was completed in 24 days, achieving removal efficiencies of at minimum 90% for COD, BOD, and TSS, along with a biogas composition containing 58% to 63% methane. These results underscore the potential and efficiency of integrated anaerobic bioreactors in addressing palm oil wastewater treatment challenges.

The issue of membrane fouling represents a major hurdle in the widespread adoption of membrane-based technology. Fouling has the potential to diminish the functionality of membranes over time, requiring the use of cleaning chemicals for its mitigation. This complicates the operation of the technology and, in turn, leads to increased operational and capital expenses (Banerjee and El-Naas 2022). Furthermore, an additional drawback lies in the extended retention time, coupled with the significant formation of sludge as a byproduct.

8.3.5 Integrated Technologies

Traditional processes for oily effluent treatment, such as DAF, gravity separation, sedimentation, emulsification, flocculation, chemical coagulation, and biological treatment, come with significant drawbacks, including prolonged settling times, extensive land requirements, and challenges in managing generated sludge (Gawad et al. 2023). To address these limitations and develop more environmentally friendly approaches capable of eliminating secondary pollutants, nonbiodegradable chemicals, and sludge formation, it is essential to explore new or integrated technologies.

Golshenas et al. (2020) implemented a photocatalytic membrane reactor system featuring a TiO_2/-Al_2O_3-modified ceramic membrane to treat synthetic oily wastewater. Their efforts yielded highly favourable outcomes, characterized by substantial permeate flow and exceptional removal efficiency of total organic carbon (TOC). Remarkably, TOC removal efficiencies exceeding 90% were achieved with a 1.5 bar transmembrane pressure and a 32.16 ml/minute flow rate during the filtration of a feed containing 500 and 1000 ppm of oil. This technology demonstrates suitability for industrial-scale applications, offering both environmental friendliness and the capability to effectively treat hazardous oily wastewater.

El-Naas et al. (2016) proposed a pilot plant comprising three key components: an electrocoagulation (EC) unit for preliminary treatment, a spouted bed bioreactor (SBBR) housing immobilized Pseudomonas putida within polyvinyl alcohol gel particles, and an adsorption column. Their study explored the optimal configuration of these units and their performance at a pilot scale for treating actual petroleum refinery wastewater. The results revealed the effectiveness of EC as a preliminary treatment step for removing metals, TDS, and TSS. Operating under the most advantageous operating conditions and with the best unit configuration resulted in an overall COD removal of 96% as well as nearly total removal of phenols and cresols. The SSBR stage resulted in the greatest reduction of phenol pollutants, whereas the adsorption process resulted in the greatest decrease in COD.

8.4 Valorisation of Petroleum Wastewater Streams

The wastewater treatment process's cost has been offset by various strategies. The primary goal of Strategy 1 is to treat wastewater to satisfy end-user needs. While strategies 2 and 3 concentrate on energy valorisation, this is a crucial component because fossil fuel supplies are constantly being depleted and energy consumption is increasing exponentially. A schematic design of valorisation has been shown in Fig. 8.1.

- **Strategy I**

When working with oily sludge, techniques like catalytic wet air oxidation (CWAO) or sludge thermal hydrolysis that reduce the particle size cause a large buildup of dissolved organic molecules in the liquid phase. Given the variability and complexity of oily sludge, waste-activated sludge has not typically been subjected to thermal hydrolysis. Therefore, using this strategy is seen to be innovative and attempts to lessen the original residue's considerable solid concentration.

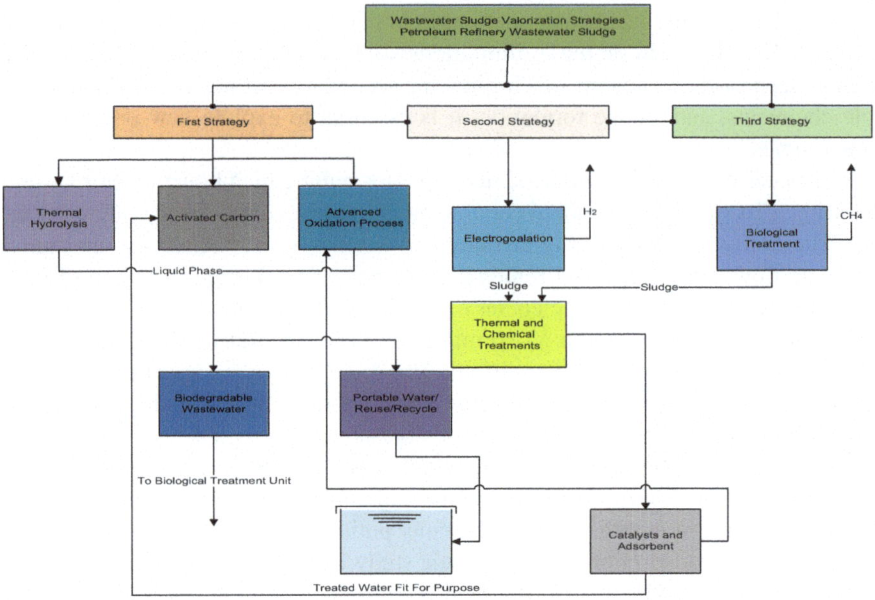

Fig. 8.1 Proposed design strategies for the valorisation of petroleum wastewater

Furthermore, the CWAO method proves valuable in breaking down biologically resistant substances into less complex forms, thereby enhancing the sludge's biodegradability. Consequently, the proposed approach involves combining thermal hydrolysis and CWAO processes to reduce the initial sludge's solid content, enhance its biodegradability and encourage the liquid phase solubilization of organic molecules. Depending on specific circumstances, the utilization of a carbon regeneration furnace, an acid cleaning system, or a filter press system may be necessary. These thermal processes can yield either sludge with the capacity to degrade biologically or an organic liquid effluent with a high concentration which can be processed further through biological means in the refinery as needed.

- **Strategy II**

This technique suggests making use of and valuing the solid and oily components. Following the pre-treatment described in strategy 1, the result would be a potentially compostable liquid effluent. However, the oily phase, which might be recycled in the refinery once it has undergone purification, and the solid fraction, is typically left without additional treatment or utilization. Given the high carbon concentration in this mixture, it may be possible to valorise it by turning it into activated carbon.

Typically, sludge was chemically activated with chemicals like KOH, H_2SO_4, or $ZnCl_2$ and then pyrolyzed to form adsorbent materials. These adsorbent materials are highly porous and have a large surface area, making them useful for a variety of uses. Considering the elevated metal concentration present in the solid phase, a comparable treatment (either thermal or chemical) for this solid material can be proposed. However, the objective of this treatment would differ, aiming to produce catalysts containing active metals such as iron. Hence, the two streams that were not utilized in the strategy I can be used as catalysts or activated carbon in the refining process for a variety of purposes.

- **Strategy III**

The production of various volatile acids, including acetic, formic, and propionic acids, through thermal processes like hydrolysis and WAO (as outlined in Strategy 1), can be effectively reclaimed through anaerobic biodigestion.

Digesting organic materials is a valuable and commonly employed method for deriving value from biosolids, livestock manure, or wet organic substances. Anaerobic processes have been investigated in the petrochemical sector, but only for the purpose of cleaning up contaminated land and water. Methane will be the predominant byproduct of this valorisation, along with a solid residue that might be used to make biochar. This plan would enable meeting a portion of the petroleum refinery's energy needs.

8.5 Recovery of Valuable Materials from Petroleum Wastewater

The solid byproduct from the petroleum industry which is typically comprised of water, crude oil, and solid particulate matter, is called oily sludge. It often contains toxic substances that necessitate treatment for toxicity reduction. It's worth noting that crude oil, a major component of oily sludge, holds significant recycling value, offering substantial economic benefits. Various techniques have been devised for treating oily sludge, typically categorized into two main areas: oil recovery and sludge disposal.

The choice of an oily sludge treatment technology should consider factors such as processing capacity, capital investment, legal requirements for disposal, time constraints and the physical properties of the sludge. Numerous methods, including surfactant-enhanced oil recovery, centrifugation, bio-slurry treatment, froth flotation, and freeze/thaw may be preferable when dealing with oily sludge that has a high-water content. When dealing with oily sludge characterized by high water content, several techniques like centrifugation, surfactant-enhanced oil recovery, freeze/thaw, froth flotation, and bio-slurry treatment may prove more suitable. In contrast, processes like incineration, pyrolysis, and stabilization/solidification are better suited for sludge with lower moisture content, often requiring pre-treatment to reduce its moisture content before processing.

8.5.1 Oil Recovery Methods

8.5.1.1 Solvent Extraction

The process of extracting oil from oily wastes by mixing them with a solvent, ensuring the mixture is fully miscible and then distilling the mixture to separate the oil from the solvent is called solvent extraction. This approach is used to remove organic chemicals from soil and water mixtures.

The extraction of oil from dry and semi-dry petroleum sludge was studied by El Naggar et al. (2010) utilizing petroleum cuts (more precisely, naphtha cut, and kerosene cut). They discovered that the maximum oil recovery within the investigated operating conditions ranged from 80 to 97%, subject to the quantity of oil in the sludge and the efficiency of the solvent used for extraction. These findings are consistent with those of Al-doury (2019), who noted that oil recovery improved from 48 to 95% as the volume of the methyl ethyl ketone solvent was raised from 5 cm^3 to 80 cm^3. According to Al-doury (2019), the most effective solvent for extracting crude oil from oily sludge was methyl ethyl ketone, which had the maximum extraction efficiency for crude oil (95%), solid separation (24%), and water separation (94%) at a temperature of 60 °C.

While solvent extraction is a straightforward and efficient technique that requires a relatively short treatment duration, its application on a large scale has been associated with reduced efficiency and excessive variability (Murungi and Sulaimon 2022). Moreover, solvent recycling necessitates heating, potentially increasing the energy cost of the process. Another significant challenge when using solvent extraction for industrial-scale applications is the substantial volumes of organic solvents required, which can have serious economic and environmental repercussions. Consequently, alternative approaches to enhance solvent extraction performance have been explored.

8.5.1.2 Centrifugation

Centrifugation is a large-scale process which makes use of specialized high-speed rotating equipment for the separation of components (e.g., water, crude oil, and solids) with varying densities. During centrifugal operation, the principle of centrifugal settling is applied, where particles within the oily sludge are driven by the centrifugal force to migrate towards the base of the centrifuge in a circular motion. This technique separates the oily sludge, resulting in the surface-level emulsified oily water being recoverable, while the solid particle pollutants settle at the bottom, requiring further disposal (Hui et al. 2020). To enhance centrifuge operation and lower energy usage, it is essential to reduce the oily sludge viscosity through pre-treatment methods. These pre-treatment approaches may involve the addition of organic solvents, tensioactive chemicals, demulsifying agents, steam injection, or direct heating (Hu et al. 2013).

This pre-treatment step is crucial for enhancing the efficiency of the centrifugation process.

Philemon and Benoît (2013) conducted a study investigating how various centrifugation variables influence the treatment of conditioned oily sludge. According to their research, at 1500 rpm centrifugation speed, 60 °C temperature, 30-min centrifugation period and 25 ppm flocculent concentration, the greatest oil recovery of 27% was reached. However, their analysis indicated that oil was still present in the emulsion, although they reasoned that this issue could be caused by poor centrifugation conditions and heavy hydrocarbons adhering to solid oily sludge particles. In a separate study, Cambiella et al. (2006) employed both a laboratory centrifuge and a bench-scale disc-stack centrifuge to study the effects of a coagulant salt on the breaking down of emulsified oil. Their findings demonstrated that the coagulant salt $CaCl_2$ significantly improved centrifugation and induced emulsion destabilization, resulting in oil removal efficiencies ranging from 92 to 96% in all experiments.

Centrifugation treatment has several drawbacks, it requires high energy consumption, costly equipment investment, limits in small-scale application, frequent part damage, noise, sludge warming for viscosity reduction, and the requirements for surfactants and demulsifying agents (Jahromi and Elektorowicz 2018). These difficulties lead to increased treatment expenses and environmental issues.

8.5.1.3 Surfactant

The chemical cleaning concept, which makes use of the hydrophilic group of the surfactant and the water phase solution of the oil–water emulsion, serves as the foundation for surfactant treatment. Through this procedure, petroleum hydrocarbons become more soluble, lowering the oily sludge's viscosity and surface tension, allowing petroleum hydrocarbons to migrate more easily across the oil and water phases, resulting in demulsified oil and water, the separation of the solid and liquid phases as well as the recovery of oil.

Ramirez et al. (2021) studied the impact of using surfactants on the rate of oil recovery. Their main conclusions showed that, depending on the kind of sludge, the proportion of surfactant volume to oil sludge volume had a considerable impact on oil recovery. In contrast, surfactant concentration showed no substantial impact on the recovery of oil. Notably, oil recovery rates for Triton X-114, Triton X-100, and Rhamnolipid ranged from 40 to 70%. Consequently, it may be easier to selectively remove particular oil hydrocarbon fractions from the oil recovered when a surfactant is used in the washing solution, thus allowing it to be used again as fuel. However, it is worth noting that chemical surfactants that have been used to boost pollutant solubility are hazardous and they are frequently mixed with organic solvents. Organic solvents are difficult to biodegrade and can generate secondary contamination (Hu et al. 2013). To address these concerns, researchers are actively seeking environmentally friendly surfactants characterized by high surface activity, minimal biological toxicity, efficient demulsification, and high selectivity.

Investigations have been undertaken on how to isolate and characterize surfactant-generating bacteria, along with their implementation in environmental rehabilitation (Yan et al. 2012). Biosurfactants, in particular, have the potential for various applications, including cleaning oil storage tanks, enhancing pipeline flow, and improving extraction of oil from reservoirs. However, there hasn't been much research done on using biosurfactants for oil recovery from oily sludge.

The utilization of rhamnolipid biosurfactant as an innovative biodemulsifier for oily sludge destabilization and subsequent oil recovery has been extensively explored both in laboratory and pilot-scale settings. Long et al. (2013) discovered that treatment using rhamnolipids could achieve oil recovery rates exceeding 98% from refractory waste crude oil, accompanied by dewatering efficiencies exceeding 90%. The recovered oil could be used again in the refining process because it had a water content of less than 0.3%, whereas the watery phase could be discarded. These outcomes were in line with those observed by Yan et al. (2012), who discovered that under ideal conditions, which included a 25% proportion of sludge to water, a carbon-to-nitrogen ratio of 10:1 and an inoculum size of 4%, at 35 °C, up to 91.5% of the oil could be recovered. Furthermore, Liu et al. (2018) evaluated several influential factors for recovering oil sludge from tank bottoms utilizing rhamnolipid biosurfactant derived from Pseudomonas aeruginosa. The crude oil recovery rate reached 83.17% under ideal operational conditions, and the recovered oil's moisture content could be decreased to a percentage below 0.42%.

This oil recovery method shows promise as a straightforward, rapid, and efficient method for reclaiming crude oil from oily sludge. It offers a great processing capacity and excellent efficacy without calling for large and complicated apparatus and equipment, in contrast to mechanical centrifugation, solvent extraction, and other procedures.

8.5.1.4 Freeze/Thaw

According to Hui et al. (2020), there are two alternative scenarios that may be used to characterize the freezing/thawing technique for extracting oil from oily sludge. In the first scenario, the oily sludge expands in volume as a result of the water cooling and freezing when its freezing point exceeds that of the oil phase. This will cause the oil phase to begin to amalgamate gradually, which will disrupt the emulsion mixture's equilibrium. Crude oil is separated and recovered as a result of stratification that occurs between the water phase and the oil–water emulsion during the thawing process.

As a result of the oil phase's higher freezing point than the water phase in the second scenario, the oil phase freezes first. The water phase is encased in ice as the oil phase freezes, increasing the amount of the water phase and causing the reservoir to burst. Ice is created as the oil phase freezes. Since the water phase eventually stratifies and separates from the oil phase during the thawing process due to gravity, the oil phase can be recycled (Ghosh and Rousseau 2009).

According to a study by Lin et al. (2008), demulsification was reasonably simple to accomplish in emulsions when the oil phase froze before separating from the water droplet phase during chilling, with a dewatering ratio exceeding 80%. However, in emulsions where the oil phase did not freeze before the water droplet phase, the demulsification process was comparatively more challenging. According to a study by Lin et al. (2008), demulsification was reasonably simple to accomplish in emulsions with a dewatering ratio exceeding 80%, where the freezing of the oil phase happens before that of the water phase during cooling. However, the demulsification procedure was significantly more difficult in emulsions where the water phase froze earlier than the oil phase.

Feng et al. (2017) recovered oil from the waste-cutting fluid using the freeze/thaw technique. The study's major conclusions indicated that − 8 °C and 8 h, respectively, were the ideal freezing temperatures and freezing times without the need for additional salt or pH adjustments. Additionally, 3700 mg/L of oil was recovered from an initial oil concentration of 4500–5000 mg/L, and approximately 80% of the COD was removed from the effluent.

In colder climates where natural freezing is possible, the application of the freeze/thaw process for oil recovery from oily sludge shows considerable promise. This method demands a significant amount of time and energy, making it a more limiting, energy-intensive, and expensive process.

8.5.1.5 Pyrolysis

Pyrolysis is the process of decomposing organic substances at high temperatures in an inert environment, typically between 500 to 1000 °C (Mishra et al. 2021). This method is employed to separate and treat oily sludge by subjecting it to gasification and pyrolysis, where the organic compounds in the sludge undergo thermal degradation under anaerobic conditions with indirect heat transfer. During pyrolysis, the organic compounds are converted into hydrocarbons with smaller molecular weights, resulting in the formation of liquid condensates and non-condensable gases. The solid impurities remaining in the process are carbonized, while the pyrolysis gas is condensed to recover oil. Depending on the process conditions, the final products of pyrolysis consist of char, liquid, and gases (Cheng et al. 2018).

The breakdown of non-volatile minerals and the degradation of volatile chemicals are the two common phases in the pyrolysis of oily sludge. There are three substages to the degradation of volatile chemicals, which largely take place in the temperature range of 100 to 550 °C. Light hydrocarbons vaporize at temperatures between 100 and 200 °C, while heavier hydrocarbons undergo thermal cracking at temperatures between 380 and 550 °C (Li et al. 2021a).

Schmidt and Kaminsky (2001) investigated the applicability of pyrolysis in fluidized bed reactors for oil sludge separation into its fractions and discovered that between 460 and 650 °C, oil separates from oily sludge, and a fluidized bed reactor could separate 70–84% of the oil from sludge. Lin et al. (2018) performed co-pyrolysis experiments using oily sludge and rice husk in a fixed-bed reactor to improve the quality of pyrolysis oil. Research findings indicate that the interaction increased the number of chain hydrocarbons while substantially reducing the number of oxygenated molecules by 46–93%, according to data from gas chromatography/mass spectrometry. Synergy effects on products and sulphur release were investigated. Moreover, because secondary reactions were encouraged, the gas output increased, and more hydrogen gas, carbon monoxide and C_1–C_2 hydrocarbons were produced while H_2S emissions were limited.

To further improve the efficiency of pyrolysis and heat transfer, researchers have explored the addition of biomass and catalysts. Wang et al. (2021) showed that while adding catalysts can speed up reactions, lower reaction temperatures, or restrict the creation and release of hazardous compounds, adding biomass can improve the quantity and quality of light components.

Pyrolysis technologies offer several benefits, including the production of liquid oil that is easy to store and transport. The recovered oil is appropriate for direct use in diesel engines because its quality is comparable to low-grade distillate oil from refineries (Hui et al. 2020). Additionally, pyrolysis generates fewer harmful waste gases compared to incineration. It is a more efficient, faster, and higher-capacity process, making it a popular choice for oily sludge recovery. However, it is associated with high initial, operational, and maintenance costs as well as substantial energy consumption. Pyrolysis is less suitable for handling oily sludge with high water content (Hui et al. 2020).

8.5.1.6 Microwave Irradiation

This method employs electromagnetic waves to heat oily sludge particles and stimulate particle mobility and collision. This method is effective in rapidly increasing the temperature of oil–water mixtures, expediting demulsification processes, and making it easier to separate the molecules of oil from those of water. Heating produces demulsification of the mixture by raising the temperature, decreasing the viscosity, and hastening the settlement of water molecules within the mixtures. Microwave heating is employed at commercial scales and operates within a frequency range spanning from 900 to 2450 MHz.

Demulsification is a crucial stage in treating oily sludge, particularly in the petroleum industry. Lv et al. (2020) investigated the process of demulsifying oily sludge from refineries, using microwave irradiation. The study findings indicate that it is essential to maintain a ≤ 77 °C operational temperature. They also observed that oily sludge demulsification efficiency first climbed and then fell as microwave power and heating temperature increased, peaking at 67.4% efficiency at a 700 W power and 55 °C temperature. Under these microwave irradiation operating parameters, oily sludge demulsification efficiency improved by 25.3% in comparison to conventional heating (55 °C). Additionally, it has been proven that microwave irradiation may reduce the centrifugal speed by 36.7%, the demulsification temperature by greater than 40 °C, and the time by over 95%, indicating that microwave irradiation is a fast, highly efficient, and minimal-energy consumption technology.

Lin et al. (2017) treated oil refinery sludge using microwave pyrolysis. In the experiments, water evaporation and carbonation were coupled by rapid microwave heating, which also sped up the sludge biomass conversion. This process led to several positive outcomes, including increased production of gaseous products, reduced residual solid impurities, and efficient enrichment of trace metals. Notably, when the pyrolysis temperature was maintained at approximately 600 °C, the study achieved the highest oil recovery of 33 and 54% was the maximum energy recovery rate.

Martínez-palou et al. (2013) contrasted traditional oil bath heating and microwave irradiation as demulsification techniques for an oily water emulsion. They found that microwave power improves the separation efficiency in emulsified oil–water mixtures, however, at high power, microwave irradiation produces oil-in-water emulsion overheating, but this results in faster water separation. They also found that using an ionic glycine-based demulsifier had no influence on the process of demulsification when heated in a microwave but significantly boosted it when heated in an oil bath. This showed that irrespective of the addition of a chemical demulsifier, microwave dielectric heating was successful in breaking oil-in-water emulsions.

Microwave irradiation is an energy-efficient technology that can rapidly improve reaction rates. However, its application for adoption to commercial scale treatment is restricted because of the need for specialized equipment, high capital and operating costs, excessive energy consumption, and limited treatment capacity.

8.5.1.7 Froth Flotation

The mining, metallurgical, and mineral industries favour froth flotation due to its high throughput and efficiency. The premise behind this process is to combine oily sludge with a certain quantity of water to create a slurry, and then pressurized air is injected into the vessel, forming microscopic bubbles that collide with and cling to the oil droplets introduced into the mixture, generating tiny bubbles that adhere to and capture the oil droplets. The water–oil mixture's clusters of oil droplets and air bubbles quickly rise to the surface, enabling the gathered oil to be skimmed off and gathered for further processing. The capacity of froth flotation to preferentially recover precious species, including gold, silver, palladium, and more, is another benefit (Ramaswamy et al. 2007).

In order to better understand the flotation mechanism at work in the process and to examine the effects of different flotation parameters on the treatment of oily sludge, extensive research has been done. Li et al. (2021b) found that the optimum reagent for treating oily sludge was Na_2SiO_3 and that improving the pulp shear effect might involve increasing the impeller speed, flotation time, and liquid-to-oil sludge ratio. Furthermore, it was discovered that the impeller speed, which generates shear force and destabilizes the oily sludge in orthogonal experimental data, is the most important factor in the treatment of oily sludge.

Li et al. (2023) analysed the effectiveness of using froth flotation and centrifugation to improve oil recovery from solid materials at room temperature without the usage of additives. According to their results, under specific conditions, including a 25-min flotation period, a 1580 rpm impeller speed, a 1:12 ratio of oily sludge to flotation liquid (tap water), and a temperature of 33 °C, they were able to achieve an oil recovery rate of 93.7% and a residue oil percentage of 3.1%. Importantly, this process generated no wastewater, waste gas, or solid waste, making it an environmentally friendly option for small-scale oily sludge treatment with a modest investment requirement.

8.5.1.8 Ultrasonic Radiation

Ultrasonic radiation alters the oily sludge's structural characteristics and state by using ultrasonic waves. Utilising the cavitation effects and vibrations brought on by auditory radiation, this technique reduces the quantity of crude oil on the particle by dislodging oil droplets that have adhered to solid components. Oil–water emulsions become less viscous after prolonged exposure to ultrasonic waves, and the tiny emulsion droplets' enhanced mobility encourages their collision, and amalgamation, ultimately leading to the oil–water phase separation, thus enabling the recovery of crude oil (Hui et al. 2020).

Sivagami et al. (2019) successfully demonstrated that ultrasound combined with the Fenton process is effective in degrading petroleum oil spill sludge. Their research findings have shown that under specific conditions, including a pH of 3.0, an ultrasonic power of 100W with 40–50% amplitude, and a 10-min treatment period, the highest solubilization was achieved and a removal rate of up to 84.25% for petroleum hydrocarbons. According to the study, the integrated sono-Fenton (SF) approach greatly improved the oil sludge degradation in comparison to using ultrasonic treatment or Fenton oxidation alone.

Almasi et al. (2016) examined how ultrasound affected the oil sludge anaerobic digestion. They found that the combined reactor performed the best, with total COD, TSS, and VSS removal efficiencies of 88.4%, 79%, and 71%, respectively. When ultrasonic treatment (60 min) was used alone, the removal efficiency for total COD, TSS, and VSS was determined to be 82.3%, 61%, and 60%, respectively. This shows that ultrasonic waves can shorten digestion time, reducing the volume of anaerobic digesters and, as a result, expenses.

Ultrasonic treatment is a fast, highly effective, and environmentally friendly method (sometimes known as "green treatment"). However, the implementation of ultrasonic techniques is restricted by their high energy and slow mineralization rate (Sivagami et al. 2019) as well as elements such as treatment time, frequency, the moisture content in the mixture, temperature, the solid component's size, the concentration of the initial petroleum hydrocarbons, the power and intensity of the sonication, salinity, and the presence of surfactant (Murungi and Sulaimon 2022). It is also a very expensive method.

8.5.1.9 Electrokinetic

The electro-kinetic (EK) method's remedial principles can be split into three main classifications: physical properties, electrochemical properties, and electro-kinetic properties. In the context of oil recovery, the EK method uses minimal electrical currents to separate various components from oily sludge. This method is capable of eliminating pollutants from soils, sludges, and sediments that are both organic and inorganic. It uses low-level direct current to transfer, ions, charged particles, organic materials, and water more quickly while also producing less byproducts during post-treatment processes.

In order to destabilise emulsions and recover oil from oily sludge, Gou et al. (2023) tested and proposed an electric field-assisted thermochemical cleaning method. They used an electrolyser to introduce a direct current into a mixture of oil sludge waste from shale gas extraction and a sodium hydroxide solution. Their research revealed that a modest energy requirement of 604.8 J/g may successfully recover more than 60% of the oil. Under optimal conditions, the amount of residual oil in the sludge decreased from 8.6 wt% to 2.8 wt%. The entire treatment process also had low energy usage, was reasonably inexpensive, and was ecologically friendly. This strategy offers a promising alternative for using oily sludge resources.

8.5.2 Oily Sludge Disposal Methods

8.5.2.1 Incineration

Incineration is a method of treating sludge in large refineries that involves the total burning of oily sludge using additional air and fuels. The combustion temperature range in a rotary kiln is between 980 and 1200 °C with a retention period of approximately 30 min and the combustion temperature in a fluidized bed incinerator ranges from 730 to 760 °C with a retention period amounting to days (Islam 2015). Incineration has been shown to be a viable solution for sludge disposal due to its unique ability to reduce volume while recovering energy.

In a fluidized bed boiler, Liu et al. (2009) explored the incineration of oily sludge by co-firing it with a coal-water slurry. They discovered that by changing the rate at which the slurry of coal-water was introduced into the system, they were able to obtain a combustion efficiency of 92.6% and manage the combustion temperature. Importantly, they discovered that during the incineration process, gaseous emissions, and the heavy metals present in the residual ash satisfied environmental standards. Smouldering combustion is a novel method that effectively eliminates organic waste with very modest energy requirements.

For oil sludge treatment with a low addition ratio of filling medium, Gan et al. (2022) examined the viability of smouldering combustion. According to their research, smouldering combustion can effectively eliminate the volatile petroleum hydrocarbons that makeup oil sludge, leaving a non-hazardous carbon-based residue.

This approach successfully degraded extractable oil sludge petroleum hydrocarbons with a removal efficiency of over 99.99% while recovering 51.21% of these hydrocarbons.

Similarly, Zhao et al. (2021) carried out bench-scale studies to investigate the viability of using smouldering combustion for the remediation of refinery oil sludge. According to the study's findings, the solid waste product of the smouldering combustion process was coarse, brick-red in colour, and odourless. A large amount of reclaimed water was found to be appropriate for reuse, and the study also showed that with adequate treatment, the recovered oil amounts to 30.9% of the initial oil sludge sample. The authors came to the conclusion that the ability to reuse both the oil and the water components post-treatment emphasises the resourcefulness of smouldering combustion for oil sludge treatment. The recovered oil had a coking diesel-like composition and had a higher calorific value than kerosene, indicating that it could be reused.

When it comes to treating oily sludge, incineration is a good alternative since it recovers heat energy, effectively removes petroleum hydrocarbons, and leaves barely any oil behind. However, incineration has several limitations, such as poor energy usage, heavy metal leakage, hazardous emissions, high initial equipment and maintenance expenses, and the usage of auxiliary fuels. Fly ash and the gases they produce during combustion may need to be further treated or disposed of. Managing moisture content and successfully dealing with heavy metals in the sludge can also be difficult. Therefore, it is important to consider incineration carefully, taking into account these restrictions as well as any potential financial and environmental consequences.

8.5.2.2 Stabilization/Solidification

Cement-based solidification is a well-known and commonly used technology for immobilizing inorganic hazardous components. This low-cost technique effectively treats a wide range of industrial wastes. However, while attempting to solidify organic wastes, solidification technology has encountered challenges. Hazardous pollutants are either chemically or physically encapsulated in a well-solidified product and the hardened specimen obtained has better structural integrity (Karamalidis and Voudrias 2007).

Boutammine et al. (2020) proposed a promising approach for treating petroleum drill cuttings, a waste product generated in the oil drilling process. They employed a two-step process involving stabilization/solidification using a hydraulic binder such as Portland cement and biological processes to facilitate the petroleum hydrocarbons' degradation in the cementitious matrix.

The study by Li et al. (2015) investigated a novel strategy involving the modification of heavy oily sludge and its potential to serve as a solidification medium in the regulation of heavy metal release. In this investigation, an emulsion that resembled asphalt served as an immobilizing substance, and an ideal proportion of 0.5 (m/m) was found to be the most effective for regulating heavy metal release. It was also discovered that improving the modified oily sludge ratio or using a suitable coating technique might produce leaching tests that performed satisfactorily. According to these findings, heavy metal stabilisation and solidification were improved in the modified oily sludge.

An advantage of the stabilization/solidification method, aside from immobilizing organic pollutants, is its ability to immobilize certain toxic heavy metals found in the oily sludge within the cement matrix. This provides an environmentally beneficial solution for dealing with these contaminants. Karamalidis and Voudrias (2007) investigated the causes of heavy metals leaching from refinery oily sludge that has been solidified and the subsequently generated ash from burning oily sludge using Portland cement. Their results showed that the pH levels were crucial for the heavy metals to become immobilised throughout the stabilization/solidification process. While metals in the solidified oily sludge needed a pH of more than 7 for effective immobilisation, metals in the hardened ash were effectively immobilised at pH levels greater than 6. Notably, only 47% of the heavy metals were successfully immobilised at a pH of 2.5, which is a significant decrease in efficacy. These findings provide important insights for the remediation of oily sludge contaminated with these dangerous contaminants and emphasise the significance of pH control in the stabilization/solidification process when dealing with heavy metals.

There are several benefits of stabilization/solidification which include: rapid and efficient encapsulation of petroleum hydrocarbons in stabilized/solidified products, inexpensive capital investment, and the ability to immobilize heavy metals in oily sludge. However, this method also has a number of disadvantages, including the need to initially lower the oily sludge's water content, the inability to completely eliminate petroleum hydrocarbons, the loss of recyclable energy, and the need for proper management of stabilized/solidified products.

8.5.2.3 Oxidative Treatment

Oxidation treatment, a method employed for breaking down organic contaminants, involves the use of various chemical and oxidation agents to degrade these contaminants. This procedure has made use of a number of oxidation reagents, each with unique benefits and mechanisms for the destruction of organic pollutants.

Ultrasonic irradiation, in particular, has proven to be an effective technique in oxidation treatment. This method harnesses sound waves to incite the movement of oily sludge particles, resulting in the formation of intermediate radicals. These radicals promote the breakdown of complicated hydrocarbons into simple, soluble hydrocarbons. Ultrasound is described as a "green" advanced oxidative process method for removing pollutants without producing secondary pollution (Zhang et al. 2013).

To remediate oily sludge, Zhao et al. (2019) adopted the wet air oxidation approach in two stages. The outcomes of this approach were promising, from the oily sludge, 93.1% of the oil was successfully removed, and the overall quantity of sludge was reduced by an impressive 85.4%. Additionally, this treatment method has the additional benefit of allowing for the rapid separation of water from the oily sludge, facilitating the potential recycling of waste oil. The study also determined the critical elements that affect how effectively oil content is removed from the oily sludge. These variables were the emulsion splitter dosage, duration of the reaction, temperature of the reaction, and the dosage of the oxidant, in that sequence of significance.

Zhang et al. (2013) investigated ultrasonography, the Fenton reaction, and a coupling of the two in order to reduce the total petroleum hydrocarbon content in oily sludge. The results showed that the maximum removal efficiency was 22.6%, 13.8%, and 43.1% respectively. Coupling the process of ultrasonography and Fenton's reaction achieved the highest reductions thus demonstrating that the individual advanced oxidation processes are complementary to one another. This shows that the combined approach may be a more efficient way of treating oily sludge.

Raising water's temperature and pressure over its critical points (375 °C and 22.1 MPa) results in a chemical reaction known as supercritical water oxidation. In this state, most organic molecules containing hydrogen (H), carbon (C), and nitrogen (N) are efficiently converted into water, CO_2, and N_2. Cui et al. (2009) applied supercritical water oxidation to treat oily sludge. Their findings were quite promising, as they observed that 92% of the COD was successfully eliminated after a treatment period of 10 min. This finding emphasises how supercritical water oxidation works to remediate oily sludge efficiently and effectively, demonstrating its potential for rapidly reducing organic contaminants and improving the overall environmental impact of such waste. The major benefits of oxidation treatment are the elimination of petroleum hydrocarbons from oily sludge quickly and completely as well as its relative insensitivity to external disturbances. The high quantities of chemical reagents used, the high capital investment costs, the loss of recyclable energy, and the limited impact on heavy metals are the disadvantages.

8.6 Bioremediation of Petroleum Sludge

Bioremediation is the technique of using microorganisms to remove environmental toxins, and it is frequently used to repair oil-polluted ecosystems by hastening the microbial degradation of petroleum hydrocarbons (Das et al. 2015; Lahiri et al. 2021; Behera et al. 2022a). The most thoroughly explored bioremediation treatments include land treatment, biopile/composting, and bio-slurry treatment.

8.6.1 Land Farming Treatment

Land farming treatment is the process of degrading oily sludge pollutants by combining them with soil. Due to its comparatively low cost, low energy consumption, capacity to accept huge amounts of sludge and straightforward mode of operation, land treatment is chosen over alternative disposal methods (Liu et al. 2010).

In the investigation by Marin et al. (2005), the assessment was made of the potential of landfarming as a means of reducing the overall hydrocarbon content derived from refinery sludge when introduced into the soil. This investigation was carried out within an environment characterized by limited precipitation and elevated ambient temperatures. The study's outcomes revealed that over the course of eleven months, approximately 80% of the hydrocarbons were successfully mitigated. Notably, a substantial portion of this reduction, around 50%, was achieved within the initial three-month period.

Similarly, Hejazi and Husain (2004) explored the application of land treatment to address oily sludge concerns over a 12-month period, particularly in arid conditions. Their research findings highlighted the significance of tilling, which involves the introduction of water and nutrients to the soil. The rate of removal of 76% of petroleum hydrocarbons (PHCs) throughout the process of treating oily sludge land was attributed to tilling, which was identified as the most important factor. These investigations collectively underscore the efficacy of landfarming as an effective approach for diminishing hydrocarbon content in oily sludge, with consideration for specific environmental conditions and the implementation of appropriate soil management practices (Ray et al. 2024; Tripathy et al. 2022).

In low rainfall and hot ambient conditions, Marin et al. (2005) assessed the possibility of land farming in lowering the total hydrocarbon content of soil-absorbed refinery sludge. In eleven months, 80% of the hydrocarbons were eliminated, with the first three months accounting for half of this reduction, according to the study's findings. Additionally, a 12-month study on oily sludge land treatment under arid conditions revealed that tilling (the addition of water and nutrients) was the critical factor that, with a clearance rate of 76%, caused the greatest PHCs reduction.

In Liu et al. (2010), investigation, the scientists successfully bioremediated oily sludge-contaminated soil in situ. They did this by using biostimulation techniques, which involved using manure to increase the activity of local bacteria. Over the course of a year, the treated soil plot's total petroleum hydrocarbons (TPH) decreased by 58.2%. Notably, the biostimulation-treated soil had lower toxicity, higher microbial activity, and a greater variety of soil microorganisms than the untreated control soil. Total petroleum hydrocarbons (TPH)-contaminated soil was the subject of an investigation by Di Marcantonio et al. (2023). Biostimulation, bioaugmentation, and a combination of these methods were employed (Behera et al. 2022b). To simulate the conditions of a landfarming system, the researchers treated actual soil samples in mesocosm reactors the results obtained demonstrated that biostimulation was the most successful method, resulting in a 35% reduction in remediation time although bioaugmentation also had a favourable impact, resulting in a 24% reduction in remediation time. The study emphasises the value of using bioremediation methods, particularly biostimulation, to hasten the remediation of TPH-contaminated soil in agricultural applications, resulting in more environmentally sound and long-lasting soil restoration procedures.

8.6.2 Biopile/Compositing

Petroleum waste biopile/composting has gained popularity as an alternative to landfarming, which requires a vast amount of land. A "biopile" is the process of arranging waste materials into mounds or windrows in the context of managing petroleum waste, typically reaching heights of 2–4 m, with the intention of facilitating their decomposition through the activity of either native or introduced microorganisms. To increase a composting system's effectiveness, composting involves adding organic materials to the system. The biotreatment can be improved using a variety of techniques, such as modifying moisture levels, encouraging aeration, and adding necessary nutrients. This technique frequently incorporates bulking agents including straw, sawdust, bark, wood chips, and other organic materials. These bulking agents increase the soil-sludge mixture's porosity, which improves how evenly air and moisture are distributed throughout the matrix (Prakash et al. 2015). This, in turn, fosters more efficient decomposition and bioremediation of the waste materials.

In their research, Wang et al. (2012) made an intriguing finding, they discovered that adding a bulking agent to the waste layers dramatically increased microbial metabolic activity and diversity in addition to improving uniformity. It also resulted in a rise in the population of two particular bacterial species. Large nutrient additions, however, seemed to have a suppressive impact on these microbes (A. Prasad. Das & Ghosh 2023, A.Prasad Das, Narayan Prasad Das, Ipsita Dipamitra Behera 2024). The pile containing the bulking agent consequently demonstrated an impressive elimination of 49.62% of the total petroleum hydrocarbons from the middle layer over a period of 220 days. The control pile, in contrast, only succeeded in removing 20.44% of the total petroleum hydrocarbons from its inner layer.

According to a study by Atagana (2008), the process of composting sewage sludge and wood chips alongside petroleum hydrocarbons contaminated soil proved effective for bioremediation purposes. According to the study, adding compost that has been inoculated with sewage sludge to soil that has been contaminated with hydrocarbons can hasten the removal of contaminants from the soil matrix under controlled circumstances. Over a 19-month incubation period, the TPH levels were significantly reduced, with a 99% reduction observed in the sewage sludge compost-treated samples compared to a 17% reduction in the control experiments. Additionally, most chosen hydrocarbon component concentrations declined by up to 100% throughout the same time frame.

Numerous benefits come with biopile/composting treatment, including sustainability, a sizable treatment capacity, quicker processing than land treatment, compatibility for cold regions, and suitability for various terrains. In contrast to land treatment, it nevertheless needs a substantial quantity of land, tends to be more expensive, and has a lower treatment capacity.

8.6.3 Bio-slurry Treatment

Contaminants are dissolved into a liquid as part of the bioslurry treatment process to make them solubilized. Typically, 5–50% water by volume of sludge-associated particles are used in this procedure. These pollutants are subsequently transformed by microbial degradation into less harmful intermediates like organic acids and aldehydes which are then converted into the final products of the degradation process, carbon dioxide and water. Slurry-phase biodegradation is often carried out in specially built slurry bioreactors that enhance interaction with microbes, PHCs, nutrients, and oxygen (Webber Jr and Kim 2005). It has been demonstrated that slurry-phase bio-treatment can remove pollutants more quickly than solid-phase treatments like land farming and composting.

Jasmine and Mukherji (2019) used 10% water by volume oily sludge in a 90-day slurry-phase treatment and employed several biodegradation strategies on aged oily sludge. They used a method that involved the intermittent addition of microbes, fertilisers, and surfactants, and the outcome was aged oily sludge that had a maximum oil degradation of 68.4%. This treatment also resulted in the elimination of maltenes (86%) and asphaltenes (54%). This procedure dramatically accelerated the oil decline from ageing sludge, although treating fresh sludge would be quicker and less expensive. These findings are consistent with those of Ward et al. (2003), who used three different microbial consortiums that produce bio-surfactants to investigate the biodegradation of oily sludge in the slurry phase with sludge concentrations ranging from 1.55 to 12.8%. They discovered that in just 10–12 days, the overall amount of petroleum hydrocarbons degraded was between 80 and 99%.

Ayotamuno et al. (2007) bioremediated oily sludge using bioaugmentation. Their study revealed total hydrocarbon reductions ranging from 40.7 to 53.2% within two weeks and 63.7 to 84.5% within six weeks of applying bioaugmentation. In contrast, the control reactor only managed to reduce the overall number of hydrocarbons by 12.8%. This highlights the effectiveness of microbial augmentation in enhancing the breakdown of oily sludge implying that bioaugmentation is an efficient bioremediation approach for dealing with oily sludge.

Although bio-slurry degradation provides a quick and efficient way to get rid of oily sludge, it has significant drawbacks when used on a large scale. In comparison to other disposal options, it is more expensive, has a smaller treatment capacity, involves precise slurry waste management, and requires skilled operation, maintenance, and monitoring. Oily sludge slurry might present operational difficulties due to its non-homogeneous and clay-like composition, necessitating frequent pre-treatment. Additionally, the process could produce compounds that are volatile gases that need to be properly handled.

8.7 Conclusion

In order to effectively handle the substantial issues posed by the wastewater and oily sludge produced by the petroleum industry, the investigation of novel methodologies and environmentally friendly approaches for the remediation of environmental petroleum contaminants constitutes a key step forward. The findings presented in this literature review underscore the urgency of mitigating the adverse environmental and health impacts of untreated petroleum effluents. Convention petroleum wastewater treatment approaches have proven inefficient, expensive, and tend to generate secondary pollutants. Emerging technologies and sustainable practices offer a glimpse into a more environmentally responsible future for the industry, as they offer a more sustainable solution to the adverse environmental and health impacts of untreated petroleum effluents.

This review categorizes innovative solutions into oil recovery, sludge disposal, and remediation methods, highlighting their unique benefits and limitations. It emphasizes continuous improvements to enhance environmental sustainability and the need for integrated approaches that minimize waste, reduce energy consumption, and maximize resource recovery. The review also highlights the potential for valorising waste streams during petroleum pollutant remediation, transforming challenges into opportunities which help offset treatment costs, contribute to the circular economy, and promote a sustainable industry. In the face of increasing environmental concerns and regulatory pressures, the need for the petroleum sector to embrace these novel and eco-friendly techniques cannot be overstated.

References

Abuhasel K, Kchaou M, Alquraish M, Munusamy Y, Jeng YT (2021) Oily wastewater treatment: overview of conventional and modern methods, challenges, and future opportunities. Water 13:980

Al-doury MMI (2019) Treatment of oily sludge using solvent extraction. Pet Sci Technol 37(2):190–196. https://doi.org/10.1080/10916466.2018.1533859

Aljuboury DADA, Palaniandy P, Abdul Aziz HB, Feroz S (2017) Treatment of petroleum wastewater by conventional and new technologies—a review. Glob Nest J 19(3):439–452. https://doi.org/10.30955/gnj.002239

Aljuboury D, Al deen A, Palaniandy P, Feroz S (2018) Advanced oxidation processes (AOPs) to treat the petroleum wastewater. In: Advanced Oxidation Processes (AOPs) in Water and Wastewater Treatment (Issue July). https://doi.org/10.4018/978-1-5225-5766-1.ch005

Almasi A, Mosavi A, Mohammadi M, Azemnia S, Godini K, Zarei A, Mohammadi S, Saleh E (2016) Efficiency of integrated ultrasonic and anaerobic digestion of oil refinery wastewater sludge. Global NEST J 18(4):771–777

Ani IJ, Akpan UG, Olutoye MA, Hameed BH (2018) Photocatalytic degradation of pollutants in petroleum refinery wastewater by TiO_2- and ZnO-based photocatalysts: recent development. J Clean Prod 205:930–954. https://doi.org/10.1016/j.jclepro.2018.08.189

Atagana HI (2008) Compost bioremediation of hydrocarbon-contaminated soil inoculated with organic manure. Afr J Biotech 7(10):1516–1525

Ayotamuno MJ, Okparanma RN, Nweneka EK, Ogaji SOT, Probert SD (2007) Bio-remediation of a sludge containing hydrocarbons. Appl Energy 84:936–943. https://doi.org/10.1016/j.apenergy.2007.02.007

Bala K, Anshul, Verma S, Dogra A, Das AP, Kumar R, Nandini (2024). Sustainable management for urban plastic waste generation. In: Sustainable Management of Urban Plastic Waste Through Circular Economic Approaches, pp 211–236. https://doi.org/10.1201/9781003465447-13

Bala K, Dogra A, Lata P, Das AP (2024) Discussing sources and biological-chemical remediation approaches for microplastic pollution. In: Sustainable Microbial Technology for Synthetic and Cellulosic Microfiber Bioremediation, pp 37–59. https://doi.org/10.1007/978-3-031-62600-5_3

Banerjee A, El-Naas MH (2022) Petroleum industry wastewater: advanced and sustainable treatment methods. Susa Dennis

Behera ID, Basak G, Kumar RR, Sen R, Meikap BC (2020) Treatment of petroleum refinery sludge by petroleum degrading bacterium *Stenotrophomonas pavanii* IRB19 as an efficient novel technology. J Environ Sci Health Part A Toxic/Hazard Subst Environ Eng 56(2):1–13. https://doi.org/10.1080/10934529.2020.1866924

Behera ID, Nayak M, Mishra A, Meikap BC, Sen R (2022) Strategic implementation of integrated bioaugmentation and biostimulation for efficient mitigation of petroleum hydrocarbon pollutants from terrestrial and aquatic environment. Mar Pollut Bull 177:113492. https://doi.org/10.1016/j.marpolbul.2022.113492

Behera ID, Meikap BC, Sen R (2022a) Performance evaluation of bacterial consortium for biodegradation of total petroleum hydrocarbon: a comparative strategic biostimulation study. Research Square.

Boutammine H, Salem Z, Khodja M (2020) Petroleum drill cuttings treatment using stabilization/solidification and biological process combination. Soil Sediment Contam Int J 00(00):1–15. https://doi.org/10.1080/15320383.2020.1722982

Cambiella A, Benito JM, Pazos C, Coca J (2006) Centrifugal separation efficiency in the treatment of waste emulsified oils. Chem Eng Res Des 84(1A):69–76. https://doi.org/10.1205/cherd.05130

Chen C, Yoza BA, Wang Y, Wang P, Li QX, Guo S, Yan G (2015) Catalytic ozonation of petroleum refinery wastewater utilizing Mn–Fe–Cu/Al_2O_3 catalyst. Environ Sci Pollut Res 22(7):5552–5562. https://doi.org/10.1007/s11356-015-4136-0

Cheng S, Chang F, Zhang F, Huang T, Yoshikawa K, Zhang H (2018) Progress in thermal analysis studies on the pyrolysis process of oil sludge. Thermochimica Acta 663(September 2017):125–136. https://doi.org/10.1016/j.tca.2018.01.006

Cui B, Cui F, Jing G, Xu S, Huo W, Liu S (2009) Oxidation of oily sludge in supercritical water. J Hazard Mater 165:511–517. https://doi.org/10.1016/j.jhazmat.2008.10.008

Das A, Dey S, Bala K, Gnanasekaran R, Das AP (2024) Microbial bioremediation technology for sustainable treatment and management of synthetic microfiber waste. In: Sustainable Microbial Technology for Synthetic and Cellulosic Microfiber Bioremediation, pp 221–238. https://doi.org/10.1007/978-3-031-62600-5_12

Das A, Dey S, Das AP (2024) Ecotoxicological impacts of synthetic microfiber pollutants and development of sustainable mitigation strategies. Environ Chem Ecotoxicol. https://doi.org/10.1016/J.ENCECO.2024.12.003

Das AP, Ghosh S, Mohanty S, Sukla LB (2015) Advances in manganese pollution and its bioremediation. In: Sukla LB, Pradhan N, Panda S, Mishra BK (eds) Environmental microbe biotechnology, 1st edn. Springer, Cham, pp 313–328. https://doi.org/10.1007/978-3-319-19018-1_16

Das A. Prasad, Ghosh, Shreya. (2023). Manganese mining microorganisms.

Dey S, Bissoyi A, Koteswara Reddy G, Das AP (2024) Synthetic microfibers: An urban emerging pollutant demanding for sustainable environmental solutions. In: Sustainable Microbial Technology for Synthetic and Cellulosic Microfiber Bioremediation, pp 107–121. https://doi.org/10.1007/978-3-031-62600-5_7

Dey S, Dash N, Ray SK, Priyadarsini S, Chhotaray C, Das AP (2024) Occurrence and quantification of tire wear plastics pollutants from highway city dust of Bhubaneswar City of Odisha. In:

Sustainable Microbial Technology for Synthetic and Cellulosic Microfiber Bioremediation, pp 77–89. https://doi.org/10.1007/978-3-031-62600-5_5

Dey S, Tripathy B, Kumar MS, Das AP (2023) Ecotoxicological consequences of manganese mining pollutants and their biological remediation. Environ Chem Ecotoxicol. https://doi.org/10.1016/J.ENCECO.2023.01.001

Di Marcantonio C, Chiavola A, Noce A, Straccamore E, Giannuzzi A, Jirillo J, Gallo F, Boni MR (2023) A sustainable approach to enhance heavy hydrocarbons removal in landfarming treatment. Biodegradation. https://doi.org/10.1007/s10532-023-10025-6

Diyauddeen BH, Daud WMAW, Aziz ARA (2010) Treatment technologies for petroleum refinery effluents: a review. Process Saf. Environ. Protect. 9(November):95–105. https://doi.org/10.1016/j.psep.2010.11.003

El Gawad HA, Ebrahiem EE, Ghaly MY, Afify AA, Mohamed RM (2023) An application of advanced oxidation process on industrial crude oily wastewater treatment. Sci Rep 0123456789:1–10. https://doi.org/10.1038/s41598-023-29263-y

El Naggar AY, Saad EA, Kandil AT, Elmoher HO (2010) Petroleum cuts as solvent extractor for oil recovery from petroleum sludge. J Petrol Technol Altern Fuels 1(1):10–19

Elmobarak WF, Hameed BH, Almomani F, Abdullah AZ (2021) A review on the treatment of petroleum refinery wastewater using advanced oxidation processes, pp 1–29

El-Naas MH, Surkatti R, Al-Zuhair S (2016) Petroleum refinery wastewater treatment: a pilot scale study. J Water Process Eng 14:71–76. https://doi.org/10.1016/j.jwpe.2016.10.005

Feng W, Yin Y, de Lourdes Mendoza M, Wang L, Chen X, Liu Y, Cai L, Zhang L (2017) Freeze-thaw method for oil recovery from waste cutting fluid without chemical additions. J Clean Prod 148:84–89. https://doi.org/10.1016/j.jclepro.2017.01.156

Gan Z, Zhao C, Li Y, Chen G, Song Z, Zhang Z, Ran W (2022) Experimental investigation on smoldering combustion for oil sludge treatment: influence of key parameters and product analysis. Fuel 316:123354. https://doi.org/10.1016/j.fuel.2022.123354

Ghosh S, Rousseau D (2009) Freeze—thaw stability of water-in-oil emulsions. J Colloid Interf Sci 339(1):91–102. https://doi.org/10.1016/j.jcis.2009.07.047

Golshenas A, Sadeghian Z, Ashrafizadeh SN (2020) Performance evaluation of a ceramic-based photocatalytic membrane reactor for treatment of oily wastewater. J Water Process Eng 36(February):101186. https://doi.org/10.1016/j.jwpe.2020.101186

Gou X, Zheng G, Tang J, Tao C, Liu R, Liu Z (2023) Electric field efficiently enhanced thermochemical cleaning for oil recovery from oily sludge. Chem Eng Process Process Intens 185(December):109314. https://doi.org/10.1016/j.cep.2023.109314

Hanafy M, Nabih HI (2007) Treatment of oily wastewater using dissolved air flotation technique. Energy Sources Part A Recov Util Environ Effects 29(2):143–159. https://doi.org/10.1080/009083190948711

Hejazi RF, Husain T (2004) Landfarm performance under arid conditions. 2. Evaluation of parameters. Environ Sci Technol 38(8):2457–2469. https://doi.org/10.1021/es026045c

Hu G, Li J, Zeng G (2013) Recent development in the treatment of oily sludge from petroleum industry: a review. J Hazard Mater 261:470–490. https://doi.org/10.1016/j.jhazmat.2013.07.069

Hui K, Tang J, Lu H, Xi B, Qu C, Li J (2020) Status and prospect of oil recovery from oily sludge: a review. Arab J Chem 13(8):6523–6543. https://doi.org/10.1016/j.arabjc.2020.06.009

Islam B (2015) Petroleum sludge, its treatment and disposal: a review. Int J Chem Sci 13(4):1584–1602

Jafarinejad S, Jiang SC (2019) Current technologies and future directions for treating petroleum refineries and petrochemical plants (PRPP) wastewaters. J Environ Chem Eng 7(5):103326. https://doi.org/10.1016/j.jece.2019.103326

Jahromi AF, Elektorowicz M (2018) Electrokinetically assisted oil-water phase separation in oily sludge with implementing novel controller system. J Hazard Mater 358(December):434–440. https://doi.org/10.1016/j.jhazmat.2018.07.032

Jasmine J, Mukherji S (2019) Impact of bioremediation strategies on slurry phase treatment of aged oily sludge from a refinery. J Environ Manage 246(May):625–635. https://doi.org/10.1016/j.jenvman.2019.06.029

Jawad SS, Abbar AH (2019) Treatment of petroleum refinery wastewater by electrochemical oxidation using graphite anodes. Al-Qadisiyah J Eng Sci 12:144–150. https://doi.org/10.30772/qjes.v12i3.609

John A, Das AP (2024) Government strategies for achieving circular economy for plastic waste management in Southeast Asia. In: Sustainable Management of Urban Plastic Waste Through Circular Economic Approaches, pp 72–82. https://doi.org/10.1201/9781003465447-6/GOVERNMENT-STRATEGIES-ACHIEVINGCIRCULAR-ECONOMY-PLASTIC-WASTE-MANAGEMENT-SOUTHEAST-ASIA-AKBAR-JOHN-ALOK-PRASAD-DAS

Kanungo J, Sahoo T, Swain LP, Behera ID (2023) Toxicity of persistent hydrocarbon pollutants, sources and sustainable remediation process. In: Behera ID, Das AP (eds) Impact of petroleum waste on environmental pollution and its sustainable management through circular economy. Springer Nature Switzerland, Cham, pp 39–65. https://doi.org/10.1007/978-3-031-48220-5_2

Kanungo J, Sahoo T, Bal M, Behera ID (2024) Performance of bioremediation strategy in waste lubricating oil pollutants: a review. Geomicrobiol J 41(4):360–373. https://doi.org/10.1080/01490451.2023.2245395

Karamalidis AK, Voudrias EA (2007) Cement-based stabilization/solidification of oil refinery sludge: leaching behavior of alkanes and PAHs. J Hazard Mater 148:122–135. https://doi.org/10.1016/j.jhazmat.2007.02.032

Kuokkanen V, Kuokkanen M, Hynynen I, Kuokkanen T (2021) Electrocoagulation treatment of metallurgical industry wastewater—a laboratory scale batch and pilot scale continuous study. Hydrometallurgy 202(September):105596. https://doi.org/10.1016/j.hydromet.2021.105596

Kuyukina MS, Krivoruchko AV, Ivshina IB (2020) Advanced bioreactor treatments of hydrocarbon-containing wastewater. Appl Sci 10(3):831

Lahiri D, Nag M, Dey A, Sarkar T, Joshi S, Pandit S, Das AP, Pati S, Pattanaik S, Tilak VK, Ray RR (2021) Biofilm mediated degradation of petroleum products. Geomicrobiol J. https://doi.org/10.1080/01490451.2021.1968979

Li G, Guo S, Ye H (2015) Thermal treatment of heavy oily sludge: resource recovery and potential utilization of residual asphalt-like emulsion as a stabilization/ solidification material. RSC Adv 5:105299–105306. https://doi.org/10.1039/c5ra14574k

Li J, Lin F, Li K, Zheng F, Yan B, Che L, Tian W, Chen G, Yoshikawa K (2021) A critical review on energy recovery and non-hazardous disposal of oily sludge from petroleum industry by pyrolysis. J Hazard Mater 406(October):124706. https://doi.org/10.1016/j.jhazmat.2020.124706

Li W, Lin H, Yang Y, Shang Z, Li Q, Ma Y, Liu A, Jiang M (2021b) Enhanced separation of oil and solids in oily sludge by froth flotation at normal temperature. Processes 9:2163

Li W, Ma Y, Feng X, Alheshibri M, Bu X, Ma G, Ni M, Li Q, Liu A, Niu X (2023) Optimization of an oil recovery process from oily sludge using a combined technique of froth flotation and centrifugal treatment. J Clean Prod 400(October):136752. https://doi.org/10.1016/j.jclepro.2023.136752

Lin C, He G, Dong C, Liu H, Xiao G, Liu YF (2008) Effect of oil phase transition on freeze/thaw-induced demulsification of water-in-oil emulsions. Langmuir 24(10):5291–5298. https://doi.org/10.1021/la704079s

Lin K, Lai N, Zeng J, Chiang H (2017) Temperature influence on product distribution and characteristics of derived residue and oil in wet sludge pyrolysis using microwave heating. Sci Total Environ 584–585:1248–1255. https://doi.org/10.1016/j.scitotenv.2017.01.195

Lin B, Huang Q, Chi Y (2018) Co-pyrolysis of oily sludge and rice husk for improving pyrolysis oil quality. Fuel Process Technol 177(April):275–282. https://doi.org/10.1016/j.fuproc.2018.05.002

Liu J, Jiang X, Zhou L, Wang H, Han X (2009) Co-firing of oil sludge with coal–water slurry in an industrial internal circulating fluidized bed boiler. J Harzard Mater 167:817–823. https://doi.org/10.1016/j.jhazmat.2009.01.061

Liu W, Luo Y, Teng Y, Li Z, Ma LQ (2010) Bioremediation of oily sludge-contaminated soil by stimulating indigenous microbes. Environ Geochem Health 32(1):23–29. https://doi.org/10.1007/s10653-009-9262-5

Liu C, Zhang Y, Sun S, Huang L, Yu L, Liu X, Lai R, Luo Y, Zhang Z, Zhang Z (2018) Oil recovery from tank bottom sludge using rhamnolipids. J Petrol Sci Eng 170(June):14–20. https://doi.org/10.1016/j.petrol.2018.06.031

Long X, Zhang G, Shen C, Sun G, Wang R, Yin L, Meng Q (2013) Application of rhamnolipid as a novel biodemulsifier for destabilizing waste crude oil. Biores Technol 131:1–5. https://doi.org/10.1016/j.biortech.2012.12.128

Lü T, Luo C, Qi D, Zhang D, Zhao H (2019) Efficient treatment of emulsified oily wastewater by using amphipathic chitosan-based flocculant. React Funct Polym 139(March):133–141. https://doi.org/10.1016/j.reactfunctpolym.2019.03.019

Lv X, Song Z, Yu J, Su Y, Zhao X, Sun J, Mao Y, Wang W (2020) Study on the demulsification of refinery oily sludge enhanced by microwave irradiation. Fuel 279(February):118417. https://doi.org/10.1016/j.fuel.2020.118417

Mallick K, Sahu A, Dubey NK, Das AP (2023) Harvesting marine plastic pollutants-derived renewable energy: a comprehensive review on applied energy and sustainable approach. J Environ Manage 348:119371. https://doi.org/10.1016/j.jenvman.2023.119371

Marin JA, Hernandez T, Garcia C (2005) Bioremediation of oil refinery sludge by landfarming in semiarid conditions: influence on soil microbial activity. Environ Res 98(2):185–195. https://doi.org/10.1016/j.envres.2004.06.005

Martínez-palou R, Cerón-camacho R, Chávez B, Vallejo AA, Villanueva-negrete D, Castellanos J, Karamath J, Reyes J, Aburto J (2013) Demulsification of heavy crude oil-in-water emulsions: a comparative study between microwave and thermal heating. Fuel 113:407–414. https://doi.org/10.1016/j.fuel.2013.05.094

Mazumder A, Chowdhury Z, Sen D, Bhattacharjee C (2020) Electric field assisted membrane separation for oily wastewater with a novel and cost-effective electrocoagulation and electroflotation enhanced membrane module (ECEFMM). Chem Eng Process 151:107918. https://doi.org/10.1016/j.cep.2020.107918

Mishra S, Dash D, Das AP (2024) Microbial degradation of polyester microfibers using indigenously isolated bacterial strain Exiguobacterium Sp. CLEAN – Soil, Air, Water, 52(12):e202300343. https://doi.org/10.1002/CLEN.202300343

Mishra A, Siddiqi H, Kumari U, Behera ID, Mukherjee S, Meikap BC (2021) Pyrolysis of waste lubricating oil/waste motor oil to generate high-grade fuel oil: a comprehensive review. Renew Sustain Energy Rev 150:111446. https://doi.org/10.1016/j.rser.2021.111446

Mohanty M, Tripathy B, Koteswara Reddy G, Das AP (2024) Greener technologies for establishing restraint over microplastic pollution. In: Sustainable Microbial Technology for Synthetic and Cellulosic Microfiber Bioremediation, pp 239–258. https://doi.org/10.1007/978-3-031-62600-5_13

Mohd-Salleh SNA, Mohd-Zin NS, Othman N (2019) A review of wastewater treatment using natural material and its potential as aid and composite coagulant. Sains Malaysiana 48(1):155–164

Murungi PI, Sulaimon AA (2022) Petroleum sludge treatment and disposal techniques: a review. Environ Sci Pollut Res 29(27):40358–40372. https://doi.org/10.1007/s11356-022-19614-z

Paul A, Priyadarsini S, Karakuş S, Behera JK, Gnanasekaran R, Das AP (2024) Fate and accumulation of synthetic microfibers in the environment. In: Sustainable Microbial Technology for Synthetic and Cellulosic Microfiber Bioremediation, pp 61–76. https://doi.org/10.1007/978-3-031-62600-5_4

Philemon ZB, Benoît NM (2013) Treatment of conditioned oily sludge from Cameroon petroleum refinery by centrifugation. Int J Environ Sci. https://doi.org/10.6088/ijes.2013030500007

Priyadarsini S, Das AP (2024) Analytical and structural characterization of waste lithium-ion batteries for their effective recycling strategy. Environ Chem Ecotoxicol.

Priyadarsini S, Das AP (2025) Lithium bioleaching: A review on microbial-assisted sustainable technology for lithium bio-circularity. J Water Process Eng, 69:106744. https://doi.org/10.1016/j.jwpe.2024.106744

Prakash V, Saxena S, Sharma A, Singh S, Singh SK (2015) Treatment of oil sludge contamination by composting. J Bioremed Biodegrad. https://doi.org/10.4172/2155-6199.1000284

Ramaswamy B, Kar DD, De S (2007) A study on recovery of oil from sludge containing oil using froth flotation. J Environ Manage 85:150–154. https://doi.org/10.1016/j.jenvman.2006.08.009

Ramirez D, Shaw LJ, Collins CD (2021) Oil sludge washing with surfactants and co-solvents: oil recovery from different types of oil sludges. Environ Sci Pollut Res 28(5):5867–5879. https://doi.org/10.1007/s11356-020-10591-9

Ray SK, Dey S, Das AP, John A (2024) Microplastic waste management through greener technologies: A review. In: Sustainable Management of Urban Plastic Waste Through Circular Economic Approaches, pp 237–261. https://doi.org/10.1201/9781003465447-14. RENEWABLE ENERGY GENERATION AND VALUE ADDITION FROM ENVIRONMENTAL (2024)

Razavi SMR, Miri T (2015) A real petroleum refinery wastewater treatment using hollow fiber membrane bioreactor (HF-MBR). J Water Process Eng 8:136–141. https://doi.org/10.1016/j.jwpe.2015.09.011

Rocha e Silva FCP, Rocha e Silva NMP, Luna JM, Rufino RD, Santos VA, Sarubbo LA (2018) Dissolved air flotation combined to biosurfactants: a clean and efficient alternative to treat industrial oily water. Rev Environ Sci Biotechnol 17(4): 591–602. https://doi.org/10.1007/s11157-018-9477-y

Saththasivam J, Ogunbiyi O, Lawler J, Al-rewaily R, Liu Z (2022) Evaluating dissolved air flotation for oil/water separation using a hybridized coagulant of ferric chloride and chitosan. J Water Process Eng 47:102836. https://doi.org/10.1016/j.jwpe.2022.102836

Schmidt H, Kaminsky W (2001) Pyrolysis of oil sludge in a fluidised bed reactor. Chemosphere 45:5–10

Silva EJ, Silva IA, Brasileiro PPF, Correa PF, Almeida DG, Rufino RD, Luna JM, Santos VA, Sarubbo LA (2019) Treatment of oily effluent using a low-cost biosurfactant in a flotation system. Biodegradation 30(4):335–350. https://doi.org/10.1007/s10532-019-09881-y

Sivagami K, Anand D, Divyapriya G, Nambi I (2019) Treatment of petroleum oil spill sludge using the combined ultrasound and Fenton oxidation process. Ultrasonics Sonochemistry 51(December):340–349. https://doi.org/10.1016/j.ultsonch.2018.09.007

Soo PL, Bashir MJKB, Wong L (2022) Recent advancements in the treatment of palm oil mill effluent (POME) using anaerobic biofilm reactors: challenges and future perspectives. J Environ Manage 320(June):115750. https://doi.org/10.1016/j.jenvman.2022.115750

Tetteh EK, Rathilal S, Naidoo DB (2020) Photocatalytic degradation of oily waste and phenol from a local South Africa oil refinery wastewater using response methodology. Sci Rep 10(1):1–12. https://doi.org/10.1038/s41598-020-65480-5

Tripathy B, Dash A, Das AP (2022) Detection of environmental microfiber pollutants through vibrational spectroscopic techniques: Recent advances of environmental monitoring and future prospects. Crit Rev Anal Chem, pp 1–11. https://doi.org/10.1080/10408347.2022.2144994

Varjani S, Joshi R, Srivastava VK, Ngo HH, Guo W (2020) Treatment of wastewater from petroleum industry: current practices and perspectives. Environ Sci Pollut Res 27:27172–27180

Wang X, Wang Q, Wang S, Li F, Guo G (2012) Effect of biostimulation on community level physiological profiles of microorganisms in field-scale biopiles composed of aged oil sludge. Biores Technol 111:308–315. https://doi.org/10.1016/j.biortech.2012.01.158

Wang Z, Gong Z, Wang Z, Li X, Chu Z (2021) Application and development of pyrolysis technology in petroleum oily sludge treatment. Environ Eng Res 26(1):1–15. https://doi.org/10.4491/eer.2019.460

Ward O, Singh A, Hamme JV (2003) Accelerated biodegradation of petroleum hydrocarbon waste. Indus Microbiol Biotechnol. https://doi.org/10.1007/s10295-003-0042-4

Webber WJ Jr, Kim HS (2005) Optimizing contaminant desorption and bioavailability in dense slurry systems. 1. Rheology, mechanical mixing, and PAH desorption. Environ Sci Technol 39(7):2267–2273

Yan P, Lu M, Yang Q, Zhang HL, Zhang ZZ, Chen R (2012) Oil recovery from refinery oily sludge using a rhamnolipid biosurfactant-producing *Pseudomonas*. Biores Technol 116:24–28. https://doi.org/10.1016/j.biortech.2012.04.024

Yap CC, Chan YJ, Loh SK, Supramaniam CV, Soh AC, Chong MF, Lim LK (2021) Pilot-scale investigation of the integrated anaerobic–aerobic bioreactor (IAAB) treating palm oil mill effluent (POME): startup and performance evaluation. Ind Eng Chem Res 60(10):3839–3859. https://doi.org/10.1021/acs.iecr.0c05878

Zhang J, Li J, Thring R, Liu L (2013) Application of ultrasound and Fenton's reaction process for the treatment of oily sludge. Proc Environ Sci 18:686–693. https://doi.org/10.1016/j.proenv.2013.04.093

Zhao Y, Yan X, Zhou J, Li R, Yang S, Wang B, Deng R (2019) Treatment of oily sludge by two-stage wet air oxidation. J Energy Inst 92(5):1451–1457. https://doi.org/10.1016/j.joei.2018.08.006

Zhao C, Li Y, Gan Z, Nie M (2021) Method of smoldering combustion for refinery oil sludge treatment. J Hazard Mater 409(November):124995. https://doi.org/10.1016/j.jhazmat.2020.124995

Chapter 9
Fungal Bioremediation: A Sustainable Solution to Petroleum Hydrocarbon Contamination

Lipika Parida

Abstract Petroleum hydrocarbon contamination is a significant environmental issue that requires innovative and sustainable remediation strategies. Fungal bioremediation is a promising approach that utilizes the metabolic capabilities of diverse fungal species to efficiently degrade hydrocarbons. This eco-friendly approach is based on the selection of hydrocarbon-degrading fungi, which are then isolated and screened for their specific hydrocarbon-degrading capabilities. Genetic engineering techniques are used to introduce genes encoding enzymes crucial for breaking down complex hydrocarbon structures, enhancing their remediation potential. Biostimulation and bioaugmentation strategies optimize environmental conditions and nutrients to promote fungal growth and metabolic activity. Bioremediation enhances hydrophobic hydrocarbon bioavailability, while mycorrhizal fungi improve remediation outcomes. Ex situ and in situ applications target contamination at the source, with real-time monitoring techniques guiding adaptive optimization strategies. Hybrid systems, combining fungal bioremediation with phytoremediation, capitalize on the strengths of both approaches. The diverse array of methods presented in this chapter reflects the dynamic and versatile nature of fungal bioremediation, offering a holistic perspective on potential applications and advancements in this field. Further research is required to optimize bioremediation strategies and develop novel solutions for challenging contamination scenarios.

Keywords Petroleum-hydrocarbon · Fungal bioremediation · Sustainable environment

L. Parida (✉)
Department of Chemical Engineering, Veer Surendra Sai University of Technology, Burla, India
e-mail: lparida_chemical@vssut.ac.in

I. D. Behera and A. P. Das (eds.), *Environmental Hydrocarbon Pollution and Zero Waste Approach Towards a Sustainable Waste Management*, Environmental Science and Engineering, https://doi.org/10.1007/978-3-031-81615-4_9

9.1 Introduction

Petroleum hydrocarbons are common contaminants in soil, water, and the air due to a variety of reasons such as industrial operations, transportation-related accidents, and leakage. These contaminants like heavy metals, petroleum, microplastics pose a danger to both the environment and public health, making their remediation crucial (Das et al. 2015, 2023; Paul et al. 2023; Tripathy et al. 2024). Petroleum hydrocarbon pollutants comprise a significant number of organic substances and byproducts. These are considered as priority environmental contaminants because of their persistence and recalcitrance (Bala, Anshul et al. 2024; Bala, Dogra et al. 2024; A. Das, Dey, & Das 2024). They are typically stable and durable; therefore, they last a long time in the environment and do not readily degrade (Ossai et al. 2020; Mallick et al. 2023). However, the remediation of contamination and pollution caused by petroleum hydrocarbons is a highly challenging task. It includes several in situ or ex-situ treatments, including confinement, separation, and degradation, which include chemical, biological, and physicochemical, thermal and heat, electric and electromagnetic, acoustic, and ultrasonic treatments (Biswas et al. 2020; Behera et al. 2022a; Murmu et al. 2023; Pradhan et al. 2023).

Traditional methods for cleaning up petroleum hydrocarbon-contaminated sites often involve physical or chemical techniques to remove or contain contaminants and have been utilized for decades. Excavation and removal involve digging up contaminated soil or sediment and removing it for disposal or treatment, which can disrupt soil structure and ecosystems, generate large volumes of contaminated waste, and be costly and labor-intensive. Soil washing involves washing contaminated soil with water or chemical solutions to remove soluble contaminants, but it produces contaminated wastewater or effluent that requires treatment and proper disposal (Dey et al. 2023; Sahu et al. 2024). It is inefficient for removing hydrophobic or tightly bound contaminants, may cause soil erosion and loss of fertility, and has limited applicability for highly contaminated or clay-rich soils. Pump-and-treat systems involve extracting contaminated groundwater from wells, treating it to remove or neutralize contaminants, and then discharging or reinjecting the treated water. This energy-intensive process requires continuous pumping and treatment, leading to high operational costs and greenhouse gas emissions. It may only address groundwater contamination, leaving soil and sediment contamination untreated. Chemical oxidation involves injecting oxidizing agents into contaminated soil or groundwater to chemically degrade contaminants, but it has high costs, may generate toxic byproducts or secondary contaminants, and can be variable depending on factors like soil composition, contaminant type, and oxidant delivery methods. However, these techniques commonly have drawbacks, such as low efficacy, high costs, and environmental impacts and may not always be feasible, especially in sensitive or remote areas (Lim et al. 2016; Bal et al. 2020). As a result, there is growing interest in developing and implementing innovative and sustainable remediation technologies addressing these constraints and providing more efficient and environmentally friendly solutions for cleaning up petroleum hydrocarbon-contaminated sites (Wei et al. 2023).

Bioremediation is a feasible approach to restoring petroleum-contaminated areas (Lahiri et al. 2021). This strategy uses microorganisms' metabolic ability to break down or change pollutants into less hazardous compounds. Fungi, among the varied array of microorganisms capable of bioremediation, have gained much interest for their capacity to thrive in hydrocarbon-rich environments and efficiently break down a variety of petroleum compounds (Behera et al. 2020; Lahiri et al. 2021). Fungal bioremediation is a process that employs fungus to decompose and remove contaminants from the environment, such as petroleum hydrocarbons, industrial chemicals, pesticides and heavy metals (Timmis 2010; Kumar et al. 2019). Fungi generate enzymes and metabolites that degrade complicated chemicals into simpler, less harmful forms. In the case of petroleum hydrocarbon contamination, fungi use mechanisms like enzymatic degradation, oxidation–reduction reactions, and assimilation into fungal biomass to convert harmful pollutants into harmless byproducts like carbon dioxide, water, and microbial biomass (Wilson and Jones 1993, A. Das, Dey, Bala et al. 2024; Dey, Bissoyi et al. 2024; Dey, Dash et al. 2024; Dey et al. 2023). There are two ways to carry out fungal bioremediation: ex-situ, which involves removing contaminated material and treating it in controlled environments like composting piles or bioreactors, or in situ, which involves introducing fungus directly into polluted habitats. Additionally, fungal bioremediation techniques can be enhanced through the use of microbial consortia, bioaugmentation (adding specific fungi to the contaminated site), biostimulation (providing nutrients or other growth-promoting factors), and the application of surfactants or biosurfactants to enhance hydrocarbon solubility (Hasani Zadeh et al. 2018).

9.2 Fungal Diversity

Exploring fungi diversity helps identify new species and strains with desirable traits for specific applications, contributing to sustainable environmental solutions. This diversity also offers insights into potential bioremediation applications, such as degrading petroleum hydrocarbons.

Fungi are eukaryotic microorganisms with a wide range of sizes, shapes, and complexity. They are further classified into five true phyla, including Ascomycota, Basidiomycota, Zygomycota, Chytridiomycota and Glomeromycota, each containing numerous orders, families, genera, and species. Fungi exhibit remarkable morphological diversity, ranging from microscopic unicellular yeasts to large, multicellular mushrooms. Their structures include hyphae (thread-like filaments), mycelium (network of hyphae), and reproductive structures such as spores and fruiting bodies. Fungi occupy diverse ecological places, thriving in soil, water, air, and various organic substrates (Wu et al. 2019). According to an recent estimatation of fungal diversity, there are between 2.2 and 3.8 million fungal species worldwide (Hawksworth and Lücking 2017). They play crucial roles in nutrient cycling, decomposition, symbiotic relationships (e.g., mycorrhizae), and pathogenesis (e.g., plant pathogens, opportunistic human pathogens) (Alexopoulos et al. 1996). Fungi display extensive metabolic capacities, allowing them to use a variety of carbon sources, which include complex organic molecules such as petroleum hydrocarbons. They generate a broad variety of extracellular enzymes involved in the degradation of lignocellulose, chitin, lipids, and other substrates (Timmis 2010). Fungi have evolved adaptations to thrive in diverse and extreme environments, including high temperatures, acidic or alkaline conditions, high salinity, and polluted sites. Some extremophilic fungi exhibit unique enzymatic activities and metabolic pathways suitable for bioremediation (Sahay 2022). Fungi interact with other organisms in complex ecosystems, forming mutualistic, commensal, or parasitic relationships with plants, animals, bacteria, and other fungi. These interactions influence fungal distribution, diversity, and ecological functions. Certain regions of the world, such as tropical rainforests, temperate forests, and marine environments, have rich fungal biodiversity (Luo et al. 2014). These biodiversity hotspots represent valuable reservoirs of fungal species with potential biotechnological applications, including bioremediation (John & Das 2024; Mishra et al. 2024; Mohanty et al. 2024; Paul et al. 2024; Priyadarsini & Das 2024, 2025).

9.3 Fungi with Potential for Hydrocarbon Degradation

Fungi exhibit diverse metabolic capabilities, making them valuable candidates for hydrocarbon degradation in bioremediation applications. Apart from bacteria, fungi are among the most excellent microorganisms for breaking down oil. Several researches have established that various fungal species, including *Gliocladium*, *Fusarium*, *Rhodotorula*, *Cladosporium*, *Geotrichum*, *Aspergillus*, and *Penicillium*, can use crude oil as their only source of carbon and energy (Al-Hawash et al. 2018a, b). There are several groups of fungi known for their potential in hydrocarbon degradation: White-rot fungi, brown-rot fungi, ascomycete fungi, zygomycete fungi, Endophytic fungi, marine fungi. White-rot fungi, such as species belonging to the genera *Phanerochaete*, *Trametes*, and *Pleurotus*, are renowned for their ability to degrade lignin, a complex polymer similar in structure to many petroleum hydrocarbons. They produce extracellular enzymes that oxidise and degrade crude oil and petroleum molecules, including a variety of aromatic hydrocarbons. Laccases, manganese peroxidases, and lignin peroxidases are a few examples of such enzymes. While brown-rot fungus generally breaks down cellulose and hemicellulose in wood, certain species, like *Gloeophyllum trabeum*, have enzymes that may break down simple hydrocarbons. They can utilize components of crude oil, including aliphatic hydrocarbons, as carbon sources. Ascomycete fungi like Aspergillus, Penicillium, and Fusarium are well-known for their metabolic diversity and ability to synthesise a wide range of extracellular enzymes (Timmis 2010; Kumar et al. 2019). These fungi use enzymes such as cytochrome P450 monooxygenases, esterases, and dioxygenases to degrade both aliphatic and aromatic hydrocarbons. Some zygomycete fungi, including species of *Mucor* and *Rhizopus*, have demonstrated hydrocarbon-degrading

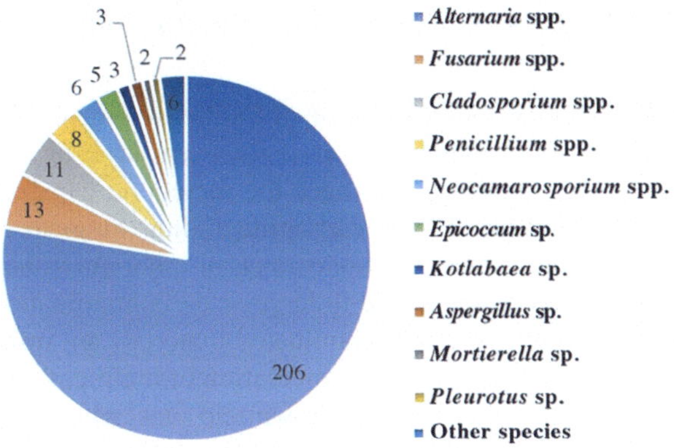

Fig. 9.1 The distribution of fungal species in soil samples from the Yadavaran oilfield in Khuzestan, Iran (Fallahi et al. 2023)

capabilities. They produce enzymes involved in hydrocarbon oxidation and assimilation, contributing to their potential use in bioremediation. Endophytic fungus colonise plant interior tissues without causing noticeable damage. Certain endophytes, such as *Cladosporium* and *Alternaria* species, have been found to possess hydrocarbon-degrading enzymes and may contribute to phytoremediation efforts by enhancing plant tolerance to hydrocarbon contaminants.

Fungi inhabiting marine environments, such as *Trichoderma*, *Aspergillus*, and *Cladosporium* species, are separated from marine sediments contaminated with oil, and they are capable of degrading hydrocarbons in saline environments (Timmis 2010). Their adaptation to marine environments makes them promising candidates for the treatment of oil spills and other marine oil pollution incidents. These fungi, among others, offer potential solutions for bioremediation of hydrocarbon-polluted environments (Kumar et al. 2019). Enzyme activity and metabolic pathways are used to develop techniques for reducing the environmental impact of petroleum spills and other hydrocarbon pollutants. An investigation conducted in Iran identified 265 fungal species from soils subjected to major oil spills to identify native species with significant biodegradation ability for petroleum hydrocarbon contaminants (Fig. 9.1). The most common genera were *Alternaria spp.* (78%), *Fusarium spp.* (5%), and *Cladosporium spp.* (4%). 35% of these isolates showed promising tolerance to crude oil. Two top-performing isolates, 5c-12 and 3b-1, effectively decomposed more than 70% of long-chain hydrocarbons in oil-contaminated soil after 60 days (Fallahi et al. 2023).

9.4 Types of Fungal Bioremediation

There are two distinct approaches used to remediate contaminated environments using fungi: In situ and ex situ fungal bioremediation. Each approach offers unique advantages and is selected based on parameters such as the properties of the pollutants, site characteristics, and remediation objectives (Marykensa 2011).

9.4.1 In Situ Fungal Bioremediation

In-situ fungal remediation involves the treatment of contaminated sites directly at their location without the need for excavation or removal of contaminated materials. This approach leverages the metabolic capabilities of indigenous or exogenously introduced hydrocarbon-degrading fungi to remediate hydrocarbon pollution in soil, sediment, groundwater, or other environmental matrices. In-situ fungal remediation is applicable to large-scale contamination scenarios where excavation and transportation of contaminated materials are impractical or cost-prohibitive. It is commonly used for the remediation of hydrocarbon-polluted sites, such as industrial facilities, landfills, and brownfield sites, where contamination is widespread and dispersed. In-situ fungal remediation involves techniques such as biostimulation and bioaugmentation (Megharaj et al. 2014).

In-situ biostimulation involves the addition of oxygen or nutrients to stimulate the growth and activity of indigenous hydrocarbon-degrading fungi in contaminated environments. This approach enhances microbial metabolism and promotes the natural degradation of hydrocarbon contaminants. In-situ bioaugmentation involves the introduction of exogenous hydrocarbon-degrading fungal strains or consortia into contaminated sites to supplement or enhance the existing microbial community. Selected fungi with specialized metabolic capabilities are inoculated to accelerate the degradation of target hydrocarbon contaminants, improving remediation efficiency.

In-situ remediation offers several advantages over ex-situ methods. The remediation process is cost-effective and ecologically advantageous, with minimal disturbance to sites and ecosystem damage, making it excellent for large-scale pollution issues. It also preserves soil structure and fertility by preserving its physical and chemical properties. However, it faces challenges such as limited control over environmental conditions, longer remediation times for recalcitrant or complex contaminants, and regulatory compliance, as it must adhere to environmental regulations to ensure human health and safety (Ray et al. 2024; Tripathy et al. 2022; A. Prasad. Das & Ghosh 2023).

9.4.2 Ex-Situ Fungal Bioremediation

Ex situ fungal bioremediation is the process of treating contaminated materials outside the contaminated site, typically in controlled environments such as bioreactors or composting facilities. Fungi, known for their versatile metabolic capabilities, are utilized to degrade, transform, or sequester contaminants, including hydrocarbons. Ex situ bioremediation is suitable for contaminated materials that can be excavated, transported, and treated in centralized facilities. It is commonly used for the treatment of contaminated soil, sediment, sludge, wastewater, and other matrices where in-situ treatment is impractical or ineffective (Megharaj et al. 2014).

Bioreactors: Contaminated materials, such as soil, sediment, or wastewater, are placed in bioreactors where fungal biodegradation occurs under controlled conditions. Depending on the metabolic requirements of the fungi and the pollutants being treated, these bioreactors can be either aerobic or anaerobic. Fungi are introduced into the bioreactor, and environmental conditions like temperature, pH, moisture, and nutrient levels are optimized to enhance fungal activity and contaminant degradation.

Fermentation: Fermentation involves the growth of hydrocarbon-degrading fungi on solid substrates, such as agricultural residues, sawdust, or peat, in the absence of freely-flowing water. Fungi colonize on solid substrate and produce extracellular enzymes for hydrocarbon degradation, allowing for the remediation of hydrocarbon-contaminated soil or solid waste materials.

Composting: To aid in the breakdown of organic contaminants, especially hydrocarbons, contaminated organic materials such as yard trash, municipal solid waste, or agricultural residues are composted with fungal inoculants. Fungi accelerate the composting process and contribute to the breakdown of organic contaminants, converting them into stabilized, nutrient-rich compost.

Ex situ fungal bioremediation offers advantages such as controlled conditions, rapid remediation, and versatility in degrading hydrocarbon contaminants. It allows for precise control of environmental parameters, optimizing fungal activity and efficiency. However, it may involve higher capital and operational costs due to infrastructure, equipment, and energy inputs. Site disturbance and logistical challenges can arise from excavation and transportation of contaminated materials. Additionally, post-treatment disposal or further processing may be required to meet regulatory requirements and environmental standards (Pal and Sen 2023).

9.4.3 Fungal Phytoremediation

Fungal phytoremediation is a bioremediation method that uses fungi and plants together to remediate contaminated environments. In this approach, fungi colonize the rhizosphere (the soil surrounding plant roots) and form symbiotic associations with plants, enhancing the degradation, immobilization, or sequestration of contaminants, including hydrocarbons (Kumar et al. 2019). For instance, mycorrhizal fungi live in symbiotic relationships with plants, colonizing their roots to aid in nutrient intake in exchange for the plants' carbon molecules. This enhances the overall phytoremediation efficiency of contaminated sites. Hydrocarbon-degrading fungi, such as oyster mushrooms and white-rot fungi, possess the enzymatic machinery to degrade hydrocarbon contaminants in the soil (Ossai et al. 2020). When associated with plants in the rhizosphere, these fungi can metabolize contaminants, breaking them down into less toxic or non-toxic compounds, enhancing the overall remediation of hydrocarbon-contaminated soil or water.

Fungi also improve the phytoremediation capacity of plants through several methods, such as increased nutrient absorption, synthesis of plant growth-promoting compounds and induced systemic resistance (Timmis 2010; Mohanta et al. 2024). However, fungal phytoremediation may have a slower remediation rate compared to other remediation techniques due to its reliance on natural processes and ecological interactions. Success of fungal phytoremediation depends on site-dependent factors, such as type of soil, climate, contaminant type, and plant-fungal interactions, which may vary from site to site.

9.5 Mechanisms of Fungal Bioremediation

Fungi have unique adaptations that enable them to thrive in hydrocarbon-contaminated environments, enabling them to efficiently degrade hydrocarbons and play a crucial role in fungal bioremediation efforts. These adaptations include physiological, biochemical, and genetic mechanisms that enable fungi to withstand the challenges posed by hydrocarbon exposure. Fungal bioremediation is a crucial process for enhancing treatment efficiency, predicting remediation outcomes, minimizing environmental risks, and ensuring long-term monitoring and management. Figure 9.2 illustrates various strategies adopted by fungi to degrade hydrocarbon-contaminated pollutants.

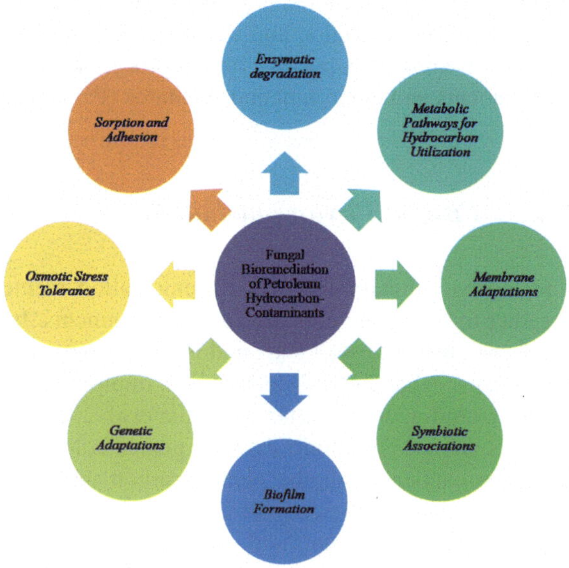

Fig. 9.2 Various strategies adopted by fungi to degrade hydrocarbon-contaminated pollutants

9.5.1 Enzymatic Degradation of Hydrocarbons by Fungi

Fungi produce a variety of extracellular enzymes, including oxidoreductases, hydrolases, and dioxygenases, that are responsible for breaking down hydrocarbon compounds (Yadav et al. 2019). These enzymes are crucial in fungal bioremediation, as they help break down hydrocarbon contaminants into less harmful or non-toxic forms. Key hydrocarbon-degrading enzymes produced by fungi include cytochrome P450 monooxygenases, peroxidases, dioxygenases, laccases, esterases and lipases, alcohol and aldehyde dehydrogenases, and hydroxylases.

Cytochrome P450 monooxygenases are heme-containing enzymes that catalyze the addition of an oxygen atom into hydrocarbon substrates, leading to hydroxylated metabolites. Aromatic hydrocarbons are mostly broken down by peroxidases, including lignin peroxidases, manganese peroxidases, and multifunctional peroxidases. Furthermore, Dioxygenases catalyze the incorporation of molecular oxygen into hydrocarbon molecules, facilitating their degradation (Vipotnik et al. 2021). Laccases are multicopper oxidases that can oxidize a wide range of substrates, including aromatic hydrocarbons and phenolic compounds. Esterases and lipases hydrolyze hydrocarbon esters and triglycerides, releasing hydrocarbon substrates for further degradation. Alcohol and aldehyde dehydrogenases oxidize aliphatic and aromatic alcohols and aldehydes, converting them into carboxylic acids for further degradation. Hydroxylases increase solubility and reactivity by adding hydroxyl groups to hydrocarbon molecules. The synthesis of enzymes by fungi that degrade hydrocarbons is regulated by environmental factors like hydrocarbon substrate availability, nutrient availability, pH, temperature, and oxygen levels (Hyde et al. 2019).

9.5.2 Metabolic Pathways for Hydrocarbon Utilization

Fungi have developed metabolic pathways that allow them to use hydrocarbons as both carbon and energy sources. These pathways involve the activation, oxidation, and degradation of hydrocarbon molecules, leading to their conversion into simpler compounds for fungal growth and metabolism. Fungi utilize diverse metabolic pathways, such as aerobic oxidation, alkene and alkyne degradation, aromatic ring cleavage, and lignin degradation. Alkane degradation involves aerobic oxidation pathways, where alkanes are hydroxylated by alkane monooxygenases, producing primary alcohols. Alkenes and alkynes can be metabolized through similar oxidation pathways, but these compounds may require specific oxygenases or desaturases. Aromatic ring cleavage is facilitated by specialized enzymes known as aromatic ring-cleaving dioxygenase. They catalyze the incorporation of molecular oxygen into aromatic rings, which produces dihydroxylated intermediates. Lignin degradation involves fungi possessing enzymes that degrade complex aromatic polymers in lignocellulosic materials, facilitating the breakdown of lignin into simpler compounds.

Fungi assimilate hydrocarbon-derived carbon and energy into central metabolic pathways, such as glycolysis and the TCA cycle, generating ATP and reducing equivalents (NADH and FADH2) for cellular energy and biosynthesis. Assimilated hydrocarbon carbon can be incorporated into various cellular components to support fungal growth and maintenance. Fungi can co-metabolize hydrocarbons with other sources of carbon, like sugars, organic acids, or amino acids, to meet their metabolic demands (Timmis 2010). This allows fungi to adapt to diverse environmental conditions.

9.5.3 Membrane Adaptations

Fungi have unique membrane adaptations that enable them to survive in hydrocarbon-contaminated environments. These adaptations include adjusting the ergosterol content of their membranes, altering the composition and structure of membrane lipids, regulating the expression of membrane transporters involved in the uptake, efflux, and sequestration of hydrocarbon molecules, forming extracellular vesicles, producing extracellular polymers, activating enzymes associated with the metabolism and detoxification of hydrocarbons and regulating membrane permeability (Suyal and Soni 2021).

Ergosterol is a sterol component of fungal cell membranes that plays a role in membrane fluidity and stability. Fungi can adjust the ergosterol content of their membranes in response to hydrocarbon exposure, modulating membrane properties to adapt to changes in environmental conditions. Membrane lipids can also be altered in response to hydrocarbon exposure, affecting the composition and structure of fatty acid composition, degree of unsaturation, and lipid bilayer organization. Fungi also regulate the expression of transporters in the membrane that participate in the uptake, efflux, and sequestration of hydrocarbon molecules, facilitating the movement of hydrophobic compounds across the cell membrane (Timmis 2010). These transporters regulate intracellular concentrations and protect cells from toxic effects.

9.5.4 Symbiotic Associations

Symbiotic associations involving fungi for hydrocarbon degradation are less common but can play significant roles in environmental remediation efforts. Examples include rhizosphere associations, where fungi provide plants with water and nutrients, and endophytic associations where fungi can degrade hydrocarbons in the soil, enhancing the effectiveness of hydrocarbon degradation and plant growth in polluted environments. Endophytic fungi have hydrocarbon-degrading capabilities, colonizing plant tissues without harm and contributing to hydrocarbon degradation within the host plant. Symbiotic relationships exist between fungus and algae or cyanobacteria, and these relationships are known as lichen associations. Recent studies have identified lichen-associated fungi capable of degrading hydrocarbons, contributing to hydrocarbon degradation within lichen thalli, especially in polluted environments. Biofilm consortia, where fungi form symbiotic associations with bacteria and other microorganisms, can efficiently degrade a wide range of hydrocarbon contaminants by working together (Robertson et al. 2007). These symbiotic associates can benefit from synergistic interactions and metabolic cooperation, enhancing the effectiveness of hydrocarbon degradation and plant growth in polluted environments.

9.5.5 Biofilm Formation

Fungi can form biofilms on surfaces exposed to hydrocarbon contamination, providing a protective microenvironment for microbial growth and metabolic activity. This enhances fungal resistance to adverse environmental conditions, promotes cell-to-cell communication, and facilitates the exchange of genetic material and metabolic products. Biofilm formation in fungi plays a significant role in hydrocarbon remediation by enhancing the degradation efficiency and resilience of fungal communities in contaminated environments (Lahiri et al. 2021).

Fungi in biofilms adhere to surfaces near hydrocarbon contaminants, facilitating the colonization of hydrocarbon-rich environments. Within biofilms, fungal cells aggregate and form structured communities, creating microenvironments with high cell densities and metabolic activity. Synergistic interactions between different fungal species and microbial consortia within biofilms can enhance hydrocarbon degradation through metabolic cooperation, resource sharing, and cross-feeding mechanisms.

Fungi in biofilms produce and secrete the extracellular enzymes that degrade hydrocarbon pollutants, such as cytochrome P450 monooxygenases, peroxidases, dioxygenases, and laccases. The extracellular Polymeric Substances (EPS) matrix surrounding fungal cells in biofilms provides structural support and protection against environmental stresses, including hydrocarbon toxicity, desiccation, and predation.

Fungal biofilms contribute to nutrient cycling and biogeochemical transformations in hydrocarbon-contaminated environments, optimizing the degradation pathways and enhancing the overall efficiency of hydrocarbon remediation processes. They also exhibit increased resilience to environmental stresses compared to planktonic cells, providing mechanical stability and protection against shear forces. Understanding the ecological roles of fungal biofilms and their interactions with hydrocarbon pollutants is essential for developing sustainable bioremediation approaches and restoring ecosystem health.

9.5.6 Genetic Adaptations

Fungi exhibit genetic adaptations to thrive in hydrocarbon-contaminated environments and efficiently degrade hydrocarbon pollutants. These adaptations involve changes at the level of individual genes, gene expression patterns, and genomic structures, allowing fungi to develop specialized metabolic capabilities, stress responses, and tolerance mechanisms. Key genetic adaptations include gene regulation and expression, enzyme evolution and diversification, horizontal gene transfer, metabolic pathway optimization, stress response genes, and evolution of transporters and membrane proteins (Singh 2019).

In response to environmental signals, such as the presence of hydrocarbon pollutants, fungi can modify the expression of genes involved in hydrocarbon metabolism. Regulatory mechanisms, including transcription factors, signal transduction pathways, and epigenetic modifications, control the activation or repression of genes encoding hydrocarbon-degrading enzymes, transporters, and stress response proteins. Fungi undergo genetic adaptations to optimize the composition and activities of enzymes involved in hydrocarbon degradation. Gene duplication events, point mutations, and recombinational events can lead to the evolution and diversification of enzyme families, resulting in variants with enhanced catalytic efficiency, substrate specificity, and stability in hydrocarbon-contaminated environments.

Fungi can acquire genes involved in hydrocarbon metabolism through horizontal or lateral gene transfer from other microorganisms in their environment. The genetic diversity and adaptive capacity of fungal populations are enhanced by lateral gene transfer events that are made possible by mobile genetic elements like viruses, plasmids, and transposons. This enables them to gain unique metabolic capacities for hydrocarbon degradation. Fungi optimize metabolic pathways for hydrocarbon degradation through genetic adaptations that streamline and enhance pathway efficiency. Gene clusters encoding enzymes and regulatory proteins involved in hydrocarbon metabolism may undergo evolutionary changes to improve pathway flux, substrate affinity, and product yield, enabling fungi to metabolize hydrocarbons more effectively as carbon and energy sources.

Fungi possess stress response genes that are activated in response to hydrocarbon-induced stressors, such as oxidative stress, osmotic stress, and DNA damage. Genetic adaptations in stress response pathways, including DNA repair mechanisms, antioxidant systems, and heat shock proteins, enhance fungal tolerance to hydrocarbon toxicity and promote cell survival under adverse conditions. Fungi undergo genetic adaptations to modify the composition and activity of membrane transporters and proteins involved in hydrocarbon uptake, efflux, and sequestration. Changes in transporter specificity, substrate affinity, and membrane permeability contribute to fungal adaptation to hydrocarbon-contaminated environments, facilitating the uptake and excretion of hydrocarbon substrates and metabolites.

9.5.7 *Osmotic Stress Tolerance*

Hydrocarbon-contaminated environments may experience osmotic stress due to the presence of hydrophobic compounds and alterations in soil water content. Fungi have evolved mechanisms to tolerate osmotic stress, such as the synthesis of osmolytes (e.g., trehalose, glycerol) and the regulation of ion transporters, enabling them to maintain cellular homeostasis and metabolic activity under challenging conditions.

9.5.8 *Sorption and Adhesion*

Fungi can adsorb hydrocarbon molecules onto their cell surfaces or mycelial structures through physical interactions, including adsorption, absorption, and adhesion. Fungal cell walls often possess hydrophobic properties, allowing them to adhere to hydrocarbon molecules and facilitate their uptake and utilization by fungal cells. This hydrophobicity enhances the attachment of fungi to hydrophobic surfaces, such as oil droplets or contaminated soil particles, promoting colonization and degradation of hydrocarbon substrates (Xu et al. 2013).

Understanding the process of contaminants degradation by fungi helps in the optimization of bioremediation strategies, such as selecting appropriate fungal strains, designing efficient treatment protocols, and manipulating environmental conditions. Customized approaches are also possible due to the unique characteristics of each contaminated site. Mechanistic insights can be used to develop strategies for enhanced treatment efficiency, such as genetic engineering, co-cultivation with other microorganisms, or the use of additives. Long-term monitoring and management strategies are also informed by understanding factors influencing fungal activity and contaminant degradation. The development of novel fungal strains, optimization of treatment technologies, and integration of bioremediation with other methods further enhance efficacy. Thus, understanding fungal bioremediation mechanisms is vital for developing, implementing, and operating the remediation methods aimed to mitigate environmental contamination and promote ecosystem health. Table 9.1 provides a brief overview of selected fungi known for their potential to degrade specific hydrocarbon compounds with their mechanisms of degradation.

9.6 Factors Influencing Fungal Bioremediation

Numerous factors can impact the efficacy and efficiency of cleanup operations using fungal bioremediation (Wei et al. 2023). These factors include the microbial population present, nutrient availability, soil composition, temperature, pH levels, and moisture content. Additionally, characteristics such as the nature, concentration, and accessibility of pollutants, as well as the genetic competence and metabolic activity

Table 9.1 An overview of selected fungi with potential degrading specific hydrocarbon compounds

Fungi name	Hydrocarbon compound	Mechanism of degradation	References
Trichoderma spp.	Polycyclic aromatic hydrocarbons such as naphthalene, phenanthrene, chrysene, pyrene, and benzo pyrene	Alternate enzyme systems such as multicooper laccases, peroxidases, and ring-cleavage dioxygenases	Zafra and Cortés-Espinosa (2015)
Aspergillus niger	n-alkanes, aromatic nitrogen, sulfur and oxygen containing compounds, and poly-cyclic aromatic hydrocarbons	Intra- and extracellular enzymes degradation	Mittal and Singh (2009) and Al-Hawash et al. (2018a, b)
Candida tropicalis	Phenanthrene, pyrene, diesel oil	Entrapment and physical adsorption, intra- and extracellular enzymes degradation	Kashyap et al. (2020) and Chandran and Das (2011)
Pseudomonas putida	Gasoline, jet fuel, aliphatic, aromatic, poly-aromatic hydrocarbons, naphthalene, phenanthrene	Aerobic metabolism, degradation by mono-oxygenases and dioxygenases	Zylstra et al. (1988) and Palleroni et al. (2010)
Penicillium spp.	Diesel, petroleum diesel, crude oil	Enzymatic degradation, hydrophobicity	Gupta and Rodriguez-Couto (2018)
Pleurotus ostreatus	Polycyclic aromatic hydrocarbons (PAHs)	Enzymatic degradation: production of ligninolytic enzymes	Bezalel et al. (1996)

of the fungi involved, play crucial roles in determining the success of fungal bioremediation efforts as described in Fig. 9.3. These factors interact intricately and must be carefully considered to optimize fungal bioremediation strategies for specific contaminated sites (Kumar et al. 2019).

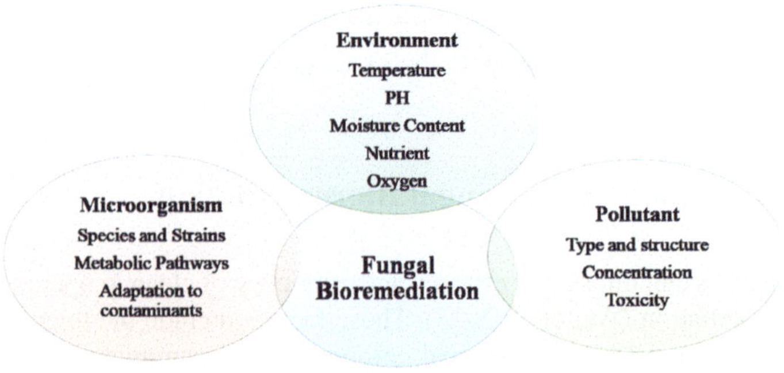

Fig. 9.3 Various factors influencing fungal bioremediation

9.6.1 Temperature

Temperature has a significant impact in fungal bioremediation processes, affecting growth, enzyme activity, and metabolic rates. The optimal temperature range for fungal growth is typically between 25 and 30 °C, with some fungi thriving at lower temperatures. Thermophilic fungi are adapted to high-temperature environments and can efficiently degrade hydrocarbons under elevated conditions, making them suitable for remediation in hot environments like compost piles or industries. Enzymes produced by fungi for hydrocarbon degradation have specific optimal temperatures, and their activity may vary based on temperature. Fungal metabolic rates, including respiration and substrate utilization, are also affected by temperature. Higher temperatures generally increase metabolic rates, leading to faster degradation of hydrocarbon contaminants. However, extreme temperatures can inhibit fungal growth and reduce bioremediation efficiency. Temperature affects other environmental parameters, including as moisture content, pH, and nutrients availability, to influence fungal bioremediation processes. Optimal temperature ranges may vary depending on the specific environmental conditions of the contaminated site. Seasonal variations in temperature can impact fungal bioremediation activities, particularly in outdoor or field-scale remediation projects (Al-Dossary et al. 2020).

9.6.2 Moisture Content

Adequate moisture is crucial for fungal growth and enzymatic activity. Proper aeration and moisture management are essential for maintaining aerobic conditions during bioremediation. Furthermore, moisture content significantly affects substrate availability, and microbial interactions (Prasad 2018). Adequate moisture promotes biosynthesis, hydrolytic enzyme release, substrate solubility, and microbial diversity. Moisture stress, like drought or waterlogging, can inhibit fungal growth, enzyme production, and hydrocarbon degradation efficiency.

9.6.3 pH

Fungal species thrive in neutral to slightly acidic pH ranges, while extremes can hinder growth and efficiency. Optimal pH conditions are necessary for hydrolytic enzymes involved in hydrocarbon degradation, promoting the dissolution and diffusion of hydrophobic contaminants, enhancing microbial diversity, metabolic activity, and synergistic interactions in hydrocarbon-contaminated environments. pH also affects the solubility and bioavailability of metals, affecting their toxicity and enzymatic activity. Some fungal species exhibit pH tolerance, allowing them to grow and degrade hydrocarbons over a wide pH range (Al-Dossary et al. 2020).

9.6.4 Nutrient and Oxygen Availability

Fungi require carbon for energy and growth, and hydrocarbon contaminants can serve as a carbon source for many hydrocarbon-degrading fungi. However, additional nutrients like nitrogen, phosphorus, sulfur, and trace elements are also essential for microbial growth and metabolism (Prasad 2018). Thus can enhance fungal biomass production and enzymatic activity, promoting efficient hydrocarbon degradation. Micronutrient deficiencies can limit enzyme activity and hinder contaminant degradation.

Organic amendments, such as compost or organic fertilizers, can supplement nutrient availability in contaminated soils, stimulating fungal growth and bioremediation activity. Oxygen availability is crucial for aerobic conditions, where fungi utilize oxygen as a terminal electron acceptor in respiratory pathways. Aerobic metabolism is more efficient in generating energy, allowing fungi to sustain higher metabolic rates and enzymatic activities during hydrocarbon degradation. Bioventing techniques can enhance oxygen delivery to contaminated soils, promoting fungal bioremediation by increasing oxygen availability in anaerobic or low-oxygen environments (Al-Dossary et al. 2020).

9.6.5 Fungal-Associated Factors

Fungal bioremediation processes are influenced by factors such as species and strains selection, metabolic pathways and physiological adaptations to pollutants. Species selection is crucial for efficient bioremediation, as different fungal species have varying capabilities for degrading specific contaminants. Enzymes produced by fungi, such as oxidases, peroxidases, and lipases, influence their capacity to degrade pollutants effectively. Fungi utilize diverse metabolic pathways to metabolize contaminants into less toxic or non-toxic byproducts.

Physiological adaptations enable fungi to thrive in diverse environmental conditions, including contaminated sites. Tolerance to environmental stressors enhances their resilience during bioremediation. Growth rates, biomass production, and reproductive strategies vary among fungal species, affecting the rate and extent of contaminant degradation. Competitive interactions with other microorganisms can impact fungal community dynamics and bioremediation consequences. Genetic diversity within fungal populations contributes to their adaptability and resilience in response to changing environmental conditions. High genetic diversity increases the likelihood of encountering strains with optimal traits for contaminant degradation. Regulatory mechanisms coordinate gene expression and enzyme production in response to environmental cues and contaminant availability.

9.6.6 Contaminant-Specific Factors

Contaminant-specific factors include the chemical structure of contaminants, their complexity and molecular weight, solubility, toxicity and concentration. The chemical structure of contaminants determines their susceptibility to fungal degradation, and fungi may exhibit specificity in degrading certain chemical bonds or functional groups within contaminants (Wang and Loh 2001). Fungi may have varying abilities to degrade aliphatic hydrocarbons, aromatic compounds, or complex mixtures based on their enzymatic pathways. Complex contaminants with high molecular weights may require more extensive enzymatic pathways or longer degradation times. High concentrations of contaminants can inhibit fungal growth and metabolic activities. Lowering contaminant concentrations through dilution or using fungal consortia can enhance bioremediation efficiency. Solubility affects bioavailability, with highly soluble contaminants being more accessible to fungi. Toxicity can inhibit fungal growth, enzyme production, or metabolic activity, and high levels of toxicity can reduce bioremediation efficiency. Concentration of pollutants influences fungal bioremediation rates and efficiency.

9.7 Strategies for Enhancing Fungal Bioremediation

The objective of fungal bioremediation is to increase the effectiveness and efficiency of hydrocarbon breakdown. This is achieved by selecting efficient fungal strains, optimizing growth conditions, combining fungal strains with other microorganisms, using bioaugmentation and biostimulation, using surfactants and biosurfactants, genetic engineering, and using natural or engineered sorbents. In situ and ex situ remediation approaches are also used to treat contaminated sites and restore ecosystem health (Ossai et al. 2020).

9.7.1 Selection and Optimization of Fungal Strains

The selection and optimization of fungal strains are crucial for developing effective bioremediation strategies for hydrocarbon-contaminated environments (Barzee et al. 2021). A structured approach involves screening for hydrocarbon-degrading ability, taxonomic identification and characterization, optimization of growth conditions, genetic engineering and strain improvement, microbial consortia and co-cultivation, field testing and validation, and continuous monitoring and optimization (Fennell et al. 2011).

Screening involves assessing fungal isolates for their growth on hydrocarbon substrates, measuring hydrocarbon degradation rates, and identifying strains with high hydrocarbon-degrading potential using biochemical assays. Taxonomic identification and characterization help determine the phylogenetic relationship, genetic diversity, and ecological niche of selected fungal strains. Optimization of growth conditions, including temperature, pH, carbon and nitrogen sources, and oxygen availability, optimizes the growth and metabolic activity of selected fungal strains. Genetic engineering techniques can be employed to enhance hydrocarbon-degrading capabilities, introduce metabolic pathways, or improve stress tolerance and nutrient utilization efficiency.

Microbial consortia and co-cultivation approaches can enhance hydrocarbon degradation efficiency by combining selected fungal strains with other microorganisms. Field testing and validation experiments assess the performance of optimized fungal strains in real-world bioremediation scenarios, monitoring key parameters such as hydrocarbon degradation rates, microbial community dynamics, and ecosystem recovery indicators. Continuous monitoring and optimization strategies are implemented to refine fungal bioremediation processes over time, adjusting operational parameters or introducing amendments as needed to optimize remediation performance and achieve remediation goals.

9.7.2 Application of Microbial Consortia for Synergistic Degradation

Microbial consortia is a promising approach to enhance the efficiency, effectiveness, and sustainability of bioremediation efforts targeting hydrocarbon-contaminated environments. It involves the cooperative interactions between different microorganisms, including fungi, bacteria, and algae, to enhance the degradation of hydrocarbon contaminants. Microbial consortia otherwise called microbial community consist of multiple species with diverse metabolic capabilities, allowing them to collectively target a wide range of hydrocarbon compounds. By combining these species, they can degrade hydrocarbons more efficiently and comprehensively than individual species (Wójcik et al. 2023).

Synergistic interactions occur when different microorganisms cooperate to enhance hydrocarbon degradation processes. For example, fungi may produce extracellular enzymes that break down complex hydrocarbon molecules into smaller intermediates, which can then be further metabolized by bacteria or algae. This results in the accelerated degradation of hydrocarbon contaminants and the conversion of pollutants into less toxic forms. Nutrient cycling and reciprocal beneficial relationships are also possible through microbial consortia. Fungi may release organic compounds as metabolic byproducts, which serve as carbon and energy sources for bacteria or algae within the consortium. In return, bacteria and algae may produce metabolites or excrete enzymes that support fungal growth and enhance hydrocarbon degradation activities.

Microbial consortia exhibit redundancy and resilience, providing backup mechanisms for hydrocarbon degradation even if some members are inhibited or compromised by environmental factors or stressors (Behera et al. 2022b). They can adapt to diverse environmental conditions, maintaining optimal metabolic activities across a range of parameters. In both in situ and ex situ bioremediation approaches, microbial consortia can be used to treat contaminated material of the site.

9.7.3 Bioaugmentation and Biostimulation Approaches

Bioaugmentation and biostimulation are two complementary approaches used in bioremediation to enhance the degradation of hydrocarbon contaminants by microbial communities, including fungi.

9.7.3.1 Bioaugmentation

Bioaugmentation is the process of introducing microbial strains or consortia into contaminated environments to improve the degradation of specific pollutants, such as hydrocarbons. These microorganisms augment the existing microbial community, providing additional metabolic activities and enzymatic pathways for hydrocarbon degradation. This allows for targeted and controlled remediation efforts, especially in sites with specific pollutant profiles or conditions. However, the introduction of these strains may require compatibility with the site-specific environmental conditions, long-term survival, and potential competition with indigenous microbial populations. Bioaugmentation can be used in conjunction with in situ or ex situ bioremediation techniques to accelerate hydrocarbon degradation processes, especially in contaminated sites with low microbial activity or limited natural degradation capacity.

9.7.3.2 Biostimulation

Biostimulation is a technique that involves altering environmental conditions or adding nutrients to stimulate the growth and activity of indigenous microbial populations capable of degrading hydrocarbon contaminants (Verma and Sharma 2024; Behera et al. 2021). It aims to optimize key environmental parameters like temperature, pH, moisture content, and nutrient availability. Common amendments include nitrogen, phosphorus, potassium, micronutrients, and electron acceptors. Biostimulation promotes the growth of indigenous microbial communities, reducing the need for external inoculants and enhancing hydrocarbon degradation. It is cost-effective and environmentally sustainable compared to bioaugmentation. However, it requires careful selection of amendments and monitoring of environmental parameters. It can be applied in various bioremediation scenarios, including in situ and ex situ treatments, and combined with other techniques to achieve remediation goals. By combining bioaugmentation and biostimulation, bioremediation techniques may be specifically designed for contaminated locations, maximizing hydrocarbon degradation and regaining ecosystem health and environmental quality.

9.7.4 Use of Surfactants and Biosurfactants to Enhance Hydrocarbon Solubility

The use of surfactants and biosurfactants is a common strategy employed in bioremediation to enhance the solubility, dispersion, and bioavailability of hydrocarbon contaminants in aqueous environments (Yesankar et al. 2022).

9.7.4.1 Surfactants

Surfactants are amphiphilic molecules with hydrophilic and hydrophobic regions that reduce the surface tension between water and hydrophobic substances, allowing them to solubilize and disperse in aqueous solutions. They adsorb onto hydrophobic hydrocarbon molecules, forming micelles or emulsions that increase their solubility in water. This process enhances hydrocarbon dispersal and facilitates their uptake and utilization by microorganisms, including fungi. Surfactants can be classified into anionic, cationic, nonionic, and zwitterionic types, each with unique characteristics influencing their effectiveness in solubilizing hydrocarbons and promoting microbial degradation. They are commonly used in bioremediation techniques, such as soil washing, surfactant-enhanced bioremediation, and pump-and-treat systems, to mobilize and remediate hydrocarbon-contaminated soil and groundwater.

9.7.4.2 Biosurfactants

Biosurfactants are surface-active compounds produced by microorganisms like fungi, bacteria, and yeasts, which are biodegradable and environmentally friendly alternatives to synthetic surfactants. They function by reducing the interfacial tension between water and hydrophobic substances, enhancing the solubility and emulsification of hydrocarbon contaminants. Biosurfactants can be synthesized from renewable resources and come in various chemical classes, such as glycolipids, lipopeptides, phospholipids, and polysaccharides. Each type has unique surface-active properties and may have specific applications in bioremediation. Biosurfactants are used in bioremediation to improve the bioavailability and degradation of hydrocarbon contaminants in various environmental matrices, such as soil, water, and sediments.

9.8 Conclusions

Fungal bioremediation is a promising method for removing petroleum hydrocarbons from contaminated sites. It utilizes fungi's diverse enzymatic repertoire, metabolic pathways, and physiological adaptations to degrade various contaminants. Their ability to produce enzymes, adapt to environmental conditions, and form symbiotic associations enhances their efficacy. However, the success of fungal bioremediation depends on factors like nutrient and oxygen availability, pH, temperature, moisture content, and contaminant-specific characteristics. Optimizing these factors is crucial for maximizing fungal activity and biodegradation rates. Further research is needed to understand the complex interactions between fungi, contaminants, and environmental conditions, and to develop innovative bioremediation strategies. Integrating fungal bioremediation with other remediation technologies like bioaugmentation, biostimulation, and surfactant-enhanced remediation can help address challenging contamination scenarios and achieve sustainable remediation outcomes.

References

Al-Dossary MA, Abood SA, Al-Saad HT (2020) Factors affecting polycyclic aromatic hydrocarbon biodegradation by Aspergillus flavus. Remediat J 30(4):17–25. https://doi.org/10.1002/rem.21658

Alexopoulos CJ, Mims CW, Blackwell MM (1996) Introductory mycology. Wiley

Al-Hawash AB, Alkooranee JT, Abbood HA, Zhang J, Sun J, Zhang X, Ma F (2018a) Isolation and characterization of two crude oil-degrading fungi strains from Rumaila oil field, Iraq. Biotechnology Reports 17:104–109. https://doi.org/10.1016/j.btre.2017.12.006

Al-Hawash AB, Zhang X, Ma F (2018b) Removal and biodegradation of different petroleum hydrocarbons using the filamentous fungus Aspergillus sp. RFC-1. Microbiol Open 8(1):e00619. https://doi.org/10.1002/mbo3.619

Bal M, Behera ID, Kumari U, Biswas S, Meikap BC (2020) Hydrodynamic study and particulate matter removal in a self priming venturi scrubber. Environ Technol Innov 20:101167. https://doi.org/10.1016/j.eti.2020.101167

Bala K, Anshul, Verma S, Dogra A, Das AP, Kumar R, Nandini (2024) Sustainable management for urban plastic waste generation. In: Sustainable Management of Urban Plastic Waste Through Circular Economic Approaches, pp 211–236. https://doi.org/10.1201/9781003465447-13

Bala K, Dogra A, Lata P, Das AP (2024) Discussing sources and biological-chemical remediation approaches for microplastic pollution. In: Sustainable Microbial Technology for Synthetic and Cellulosic Microfiber Bioremediation, pp 37–59. https://doi.org/10.1007/978-3-031-62600-5_3

Barzee TJ, Cao L, Pan Z, Zhang R (2021) Fungi for future foods. J Futur Foods 1(1):25–37. https://doi.org/10.1016/j.jfutfo.2021.09.002

Behera ID, Basak G, Kumar RR, Sen R, Meikap BC (2020) Treatment of petroleum refinery sludge by petroleum degrading bacterium Stenotrophomonas pavanii IRB19 as an efficient novel technology. J Environ Sci Health Part A Toxic/Hazard Subst Environ Eng 56(2):1–13. https://doi.org/10.1080/10934529.2020.1866924

Behera ID, Nayak M, Biswas S, Meikap BC, Sen R (2021) Enhanced biodegradation of total petroleum hydrocarbons by implementing a novel two-step bioaugmentation strategy using

indigenous bacterial consortium. J Environ Manag 292:112746. https://doi.org/10.1016/j.jenvman.2021.112746
Behera ID, Meikap BC, Sen R (2022a) Performance evaluation of bacterial consortium for biodegradation of total petroleum hydrocarbon: a comparative strategic biostimulation study. *Research Square*
Behera ID, Nayak M, Mishra A, Meikap BC, Sen R (2022b) Strategic implementation of integrated bioaugmentation and biostimulation for efficient mitigation of petroleum hydrocarbon pollutants from terrestrial and aquatic environment. Mar Pollut Bull 177:113492. https://doi.org/10.1016/j.marpolbul.2022.113492
Bezalel L, Hadar Y, Cerniglia CE (1996) Mineralization of polycyclic aromatic hydrocarbons by the White rot fungus pleurotus ostreatus. Appl Environ Microbiol 62(1):292–295. https://doi.org/10.1128/aem.62.1.292-295.1996
Biswas S, Diwakar RK, Behera ID, Meikap BC, Sen TK, Khiadani M (2020) Aqueous phase phenol removal from synthetic and real steel plant effluents through a batch and semifluidized bed column operation: experimental and model analysis. J Environ Chem Eng 8(5):104441
Chandran P, Das N (2011) Degradation of diesel oil by immobilized Candida tropicalis and biofilm formed on gravels. Biodegradation 22(6):1181–1189. https://doi.org/10.1007/s10532-011-9473-1
Das A, Dey S, Bala K, Gnanasekaran R, Das AP (2024) Microbial bioremediation technology for sustainable treatment and management of synthetic microfiber waste. In: Sustainable Microbial Technology for Synthetic and Cellulosic Microfiber Bioremediation, pp 221–238. https://doi.org/10.1007/978-3-031-62600-5_12
Das A, Dey S, Das AP (2024) Ecotoxicological impacts of synthetic microfiber pollutants and development of sustainable mitigation strategies. Environ Chem Ecotoxicol. https://doi.org/10.1016/J.ENCECO.2024.12.003
Das AP, Ghosh S, Mohanty S, Sukla LB (2015) Advances in manganese pollution and its bioremediation. In: Sukla LB, Pradhan N, Panda S, Mishra BK (eds) Environmental microbe biotechnology, 1st edn. Springer, Cham, pp 313–328. https://doi.org/10.1007/978-3-319-19018-1_16
Das AP, Mohanty AK, van Hullebusch ED, Figueiredo G, Das AP, Mohanty AK, van Hullebusch ED, Figueiredo G (2023) Bioremediation as an emerging technology for the removal of synthetic microplastic pollutants from marine ecosystem. MarPB 194:115297. https://doi.org/10.1016/J.MARPOLBUL.2023.115297
Das A. Prasad, Ghosh, Shreya (2023) Manganese mining microorganisms
Dey S, Bissoyi A, Koteswara Reddy G, Das AP (2024) Synthetic microfibers: An urban emerging pollutant demanding for sustainable environmental solutions. In: Sustainable Microbial Technology for Synthetic and Cellulosic Microfiber Bioremediation, pp 107–121. https://doi.org/10.1007/978-3-031-62600-5_7
Dey S, Das A, Mallick K, Sahu A, Das AP (2023) Environmental petroleum waste: pollution, toxicity, sustainable remediation. In: Behera ID, Das AP (eds) Impact of petroleum waste on environmental pollution and its sustainable management through circular economy. Environmental science and engineering. Springer, Cham. https://doi.org/10.1007/978-3-031-48220-5_7
Dey S, Dash N, Ray SK, Priyadarsini S, Chhotaray C, Das AP (2024) Occurrence and quantification of tire wear plastics pollutants from highway city dust of Bhubaneswar City of Odisha. In: Sustainable Microbial Technology for Synthetic and Cellulosic Microfiber Bioremediation, pp 77–89. https://doi.org/10.1007/978-3-031-62600-5_5
Dey S, Tripathy B, Kumar MS, Das AP (2023) Ecotoxicological consequences of manganese mining pollutants and their biological remediation. Environ Chem Ecotoxicol. https://doi.org/10.1016/J.ENCECO.2023.01.001
Fallahi M, Sarempour M, Mirzadi Gohari A (2023) Potential biodegradation of polycyclic aromatic hydrocarbons (PAHs) and petroleum hydrocarbons by indigenous fungi recovered from crude oil-contaminated soil in Iran. Sci Rep 13:22153. https://doi.org/10.1038/s41598-023-49630-z

Fennell DE, Du S, Liu F, Liu H, Häggblom MM (2011) Dehalogenation of polychlorinated dibenzo-p-dioxins and dibenzofurans, polychlorinated biphenyls, and brominated flame retardants, and potential as a bioremediation strategy. In: Environmental biotechnology and safety, vol 6. Elsevier Inc, pp 135–149. https://doi.org/10.1016/B978-0-08-088504-9.00369-X

Gupta VG, Rodriguez-Couto S (2018) New and future developments in microbial biotechnology and bioengineering. Elsevier

Hasani Zadeh P, Moghimi H, Hamedi J (2018) Biosurfactant production by Mucor circinelloides: environmental applications and surface-active properties. Eng Life Sci 18(5):317–325

Hawksworth DL, Lücking R (2017) Fungal diversity revisited: 2.2 to 3.8 million species. Microbiol Spectr 5(4). https://doi.org/10.1128/microbiolspec.funk-0052-2016

https://doi.org/10.1002/elsc.201700149

https://doi.org/10.1080/21501203.2019.1614106

Hyde KD, Xu J, Rapior S, Jeewon R, Lumyong S, Niego AGT, Abeywickrama PD, Aluthmuhandiram JVS, Brahamanage RS, Brooks S, Chaiyasen A, Chethana KWT, Chomnunti P, Chepkirui C, Chuankid B, de Silva NI, Lom M, Faulds C, Gentekaki E et al (2019) The amazing potential of fungi: 50 ways we can exploit fungi industrially. Fungal Divers 97(1):1–136. https://doi.org/10.1007/s13225-019-00430-9

John A, Das AP (2024) Government strategies for achieving circular economy for plastic waste management in Southeast Asia. In: Sustainable Management of Urban Plastic Waste Through Circular Economic Approaches, pp 72–82. https://doi.org/10.1201/9781003465447-6/GOVERNMENT-STRATEGIES-ACHIEVINGCIRCULAR-ECONOMY-PLASTIC-WASTE-MANAGEMENT-SOUTHEAST-ASIA-AKBAR-JOHN-ALOK-PRASAD-DAS

Kashyap N, Roy K, Moholkar VS (2020) Mechanistic investigation in Co-biodegradation of phenanthrene and pyrene by Candida tropicalis MTCC 184. Chem Eng J 399:125659. https://doi.org/10.1016/j.cej.2020.125659

Kumar V, Kumar M, Prasad R (2019) Microbial action on hydrocarbons. Springer. https://doi.org/10.1007/978-981-13-1840-5

Lahiri D, Nag M, Dey A, Sarkar T, Joshi S, Pandit S, Das AP, Pati S, Pattanaik S, Tilak VK, Ray RR (2021) Biofilm mediated degradation of petroleum products. Geomicrobiol J 39(3–5):389–398. https://doi.org/10.1080/01490451.2021.1968979

Lim MW, Lau EV, Poh PE (2016) A comprehensive guide of remediation technologies for oil contaminated soil—present works and future directions. Mar Pollut Bull 109(1):14–45. https://doi.org/10.1016/j.marpolbul.2016.04.023

Luo J, Walsh E, Naik A, Zhuang W, Zhang K, Cai L, Zhang N (2014) Temperate pine barrens and tropical rain forests are both rich in undescribed fungi. PLoS One 9(7):e103753. https://doi.org/10.1371/journal.pone.0103753

Mallick K, Sahu A, Dubey NK, Das AP (2023) Harvesting marine plastic pollutants-derived renewable energy: a comprehensive review on applied energy and sustainable approach. J Environ Manage 348:119371. https://doi.org/10.1016/j.jenvman.2023.119371

Marykensa V (2011) Bioremediation—an overview. J Ind Pollut Control 27:161–168

Megharaj M, Venkateswarlu K, Naidu R (2014) Bioremediation. In: Wexler P (ed) Encyclopedia of toxicology, 3rd edn. Academic Press, Oxford, pp 485–489

Mishra S, Dash D, Das AP (2024) Microbial degradation of polyester microfibers using indigenously isolated bacterial strain Exiguobacterium Sp. CLEAN – Soil, Air, Water, 52(12):e202300343. https://doi.org/10.1002/CLEN.202300343

Mittal A, Singh P (2009) Studies on biodegradation of crude oil by Aspergillus niger. S Pac J Nat Appl Sci 27(1):57. https://doi.org/10.1071/sp09010

Mohanta S, Pradhan B, Behera ID (2024) Impact and remediation of petroleum hydrocarbon pollutants on agricultural land: a review. Geomicrobiol J 41(4):345–359. https://doi.org/10.1080/01490451.2023.2243925

Mohanty M, Tripathy B, Koteswara Reddy G, Das AP (2024) Greener technologies for establishing restraint over microplastic pollution. In: Sustainable Microbial Technology for Synthetic and

Cellulosic Microfiber Bioremediation, pp 239–258. https://doi.org/10.1007/978-3-031-62600-5_13
Murmu AK, Parida L, Prakash V (2023) Petroleum hydrocarbon waste recycling, reusing, repairing, and recovering value added products. In: Behera ID, Das AP (eds) Impact of petroleum waste on environmental pollution and its sustainable management through circular economy. Environmental science and engineering. Springer, Cham. https://doi.org/10.1007/978-3-031-48220-5_9
Ossai IC, Ahmed A, Hassan A, Hamid FS (2020) Remediation of soil and water contaminated with petroleum hydrocarbon: a review. Environ Technol Innov 17:100526. https://doi.org/10.1016/j.eti.2019.100526
Pal D, Sen S (2023) In-depth coverage of petroleum waste sources, characteristics, environmental impact, and sustainable remediation process. In: Behera ID, Das AP (eds) Impact of petroleum waste on environmental pollution and its sustainable management through circular economy. Environmental science and engineering. Springer, Cham. https://doi.org/10.1007/978-3-031-48220-5_1
Palleroni NJ, Pieper DH, Moore ERB (2010) Microbiology of hydrocarbon-degrading pseudomonas. In Timmis KN (eds) Handbook of hydrocarbon and lipid microbiology. Springer, Berlin, Heidelberg. https://doi.org/10.1007/978-3-540-77587-4_129
Paul A, Dey S, Ram DK, Das AP (2023) Hexavalent chromium pollution and its sustainable management through bioremediation. Geomicrobiol J, 1–11. https://doi.org/10.1080/01490451.2023.2218377
Paul A, Priyadarsini S, Karakuş S, Behera JK, Gnanasekaran R, Das AP (2024) Fate and accumulation of synthetic microfibers in the environment. In: Sustainable Microbial Technology for Synthetic and Cellulosic Microfiber Bioremediation, pp 61–76. https://doi.org/10.1007/978-3-031-62600-5_4
Pradhan B, Behera ID, Sahoo K, Mohanta S (2023) Applicability of the teetered bed separator for beneficiating Indian iron ore fines: an experimental study. J Mines Metals Fuels 71(11):1817–1825. https://doi.org/10.18311/jmmf/2023/31100
Prasad R (2018) Mycoremediation and environmental sustainability. Springer
Priyadarsini S, Das AP (2024) Analytical and structural characterization of waste lithium-ion batteries for their effective recycling strategy. Environ Chem Ecotoxicol. https://doi.org/10.1016/j.enceco.2024.12.004
Priyadarsini S, Das AP. (2025). Lithium bioleaching: A review on microbial-assisted sustainable technology for lithium bio-circularity. J Water Process Eng, 69:106744. https://doi.org/10.1016/j.jwpe.2024.106744
Ray SK, Dey S, Das AP, John A (2024) Microplastic Waste management through greener technologies: A review. In: Sustainable Management of Urban Plastic Waste Through Circular Economic Approaches, pp 237–261. https://doi.org/10.1201/9781003465447-14 RENEWABLE ENERGY GENERATION AND VALUE ADDITION FROM ENVIRONMENTAL. (2024)
Robertson SJ, McGill WB, Massicotte HB, Rutherford PM (2007) Petroleum hydrocarbon contamination in boreal forest soils: a mycorrhizal ecosystems perspective. Biol Rev 82(2):213–240. https://doi.org/10.1111/j.1469-185x.2007.00012.x
Sahay S (2022) Extremophilic fungi. Springer Nature. https://doi.org/10.1007/978-981-16-4907-3
Sahu A, Mallick K, Das AP (2024) Environmental impact of single-use synthetic face mask and its recycling: a sustainable approach. In: Das AP, Mishra S (eds) Impact of COVID-19 waste on environmental pollution and its sustainable management. Environmental science and engineering. Springer, Cham. https://doi.org/10.1007/978-3-031-50840-0_11
Singh BP (2019) Advances in endophytic fungal research. Springer. https://doi.org/10.1007/978-3-030-03589-1
Suyal DC, Soni R (2021) Bioremediation of environmental pollutants. Springer Nature. https://doi.org/10.1007/978-3-030-86169-8

Timmis KN (2010) Handbook of hydrocarbon and lipid microbiology. Springer. https://doi.org/10.1007/978-3-540-77587-4

Tripathy B, Dash A, Das AP (2022) Detection of environmental microfiber pollutants through vibrational spectroscopic techniques: Recent advances of environmental monitoring and future prospects. Crit Rev Anal Chem, pp 1–11. https://doi.org/10.1080/10408347.2022.2144994

Tripathy B, Sahoo PP, Sundaray H, Das AP (2024) In-vitro biodegradation of discarded marine microplastics across the eastern coast of the Bay of Bengal, India using Exiguobacterium sp. Environ Chem Ecotoxicol 6:236–247. https://doi.org/10.1016/j.enceco.2024.07.003

Verma N, Sharma A (2024) Biological interventions in bioremediation of cadmium poisoning. In: Jha AK, Kumar N (eds) Cadmium toxicity mitigation. Springer, Cham. https://doi.org/10.1007/978-3-031-47390-6_5

Vipotnik Z, Michelin M, Tavares T (2021) Ligninolytic enzymes production during polycyclic aromatic hydrocarbons degradation: effect of soil pH, soil amendments and fungal co-cultivation. Biodegrad 32(2):193–215. https://doi.org/10.1007/s10532-021-09933-2

Wang SJ, Loh KC (2001) Biotransformation kinetics of **Pseudomonas putida** for cometabolism of phenol and 4-chlorophenol in the presence of sodium glutamate. Biodegrad 12:189–199. https://doi.org/10.1023/A:1013170322269

Wei J, Wang X, Tu C, Long T, Bu Y, Wang H, Jeyakumar P, Jiang J, Deng S (2023) Remediation technologies for neonicotinoids in contaminated environments: current state and future prospects. Environ Int 178:108044. https://doi.org/10.1016/j.envint.2023.108044

Wilson SC, Jones KC (1993) Bioremediation of soil contaminated with polynuclear aromatic hydrocarbons (PAHs): a review. Environ Pollut 81(3):229–249. https://doi.org/10.1016/0269-7491(93)90206-4

Wójcik M, Koper P, Żebracki K, Marczak M, Mazur A (2023) Genomic and metabolic characterization of plant growth-promoting rhizobacteria isolated from nodules of clovers grown in non-farmed soil. Int J Mol Sci 24(23):16679. https://doi.org/10.3390/ijms242316679

Wu B, Hussain M, Zhang W, Stadler M, Liu X, Xiang M (2019) Current insights into fungal species diversity and perspective on naming the environmental DNA sequences of fungi. Mycology 10(3):127–140

Xu N, Bao M, Sun P, Li Y (2013) Study on bioadsorption and biodegradation of petroleum hydrocarbons by a microbial consortium. Biores Technol 149:22–30. https://doi.org/10.1016/j.biortech.2013.09.024

Yadav AN, Singh S, Mishra S, Gupta A (eds) (2019) Recent advancement in White biotechnology through fungi. Fungal Biology. https://doi.org/10.1007/978-3-030-25506-0

Yesankar PJ, Pal M, Patil A, Qureshi A (2022) Microbial exopolymeric substances and biosurfactants as 'bioavailability enhancers' for polycyclic aromatic hydrocarbons biodegradation. Int J Environ Sci Technol 20(5):5823–5844. https://doi.org/10.1007/s13762-022-04068-0

Zafra G, Cortés-Espinosa DV (2015) Biodegradation of polycyclic aromatic hydrocarbons by Trichoderma species: a mini review. Environ Sci Pollut Res 22(24):19426–19433. https://doi.org/10.1007/s11356-015-5602-4

Zylstra GJ, McCombie WR, Gibson DT, Finette BA (1988) Toluene degradation by Pseudomonas putida F1: genetic organization of the tod operon. Appl Environ Microbiol 54(6):1498–1503. https://doi.org/10.1128/aem.54.6.1498-1503.1988

Chapter 10
Sustainable Technologies in the Development of Novel Processes and Products from Environmental Hydrocarbon Pollutants

Deeptimayee Pal and Sujit Sen

Abstract Integrating sustainable technologies has become a crucial aspect in the evolution of industries, specifically in advancing novel processes and products derived from environmental hydrocarbon sources. Novel developments in catalytic conversion, biotechnological applications, and green chemistry approaches characterize the vanguard of this paradigm shift. These technologies aim to tackle the environmental issues of hydrocarbon pollutants by applying precise engineering techniques and sustainable methodologies. Researchers are currently investigating efficient and selective conversion strategies that facilitate the synthesis of valuable chemicals from hydrocarbon waste streams via catalytic processes. Ecologically appropriate options for the degradation and transformation of hydrocarbon pollutants may be found in biotechnological technologies such as microbial remediation and bioconversion. Moreover, using green chemistry principles enables the establishment of procedures that prioritize the efficient use of resources, reduce the production of harmful by-products, and comply with sustainable practices. Combining these sustainable solutions can safeguard the environment and provide the groundwork for developing circular and ecologically friendly industrial processes.

Keywords Hydrocarbon pollutants · Sustainable technologies · Green chemistry · Catalytic conversion · Biotechnological applications

---

D. Pal (✉) · S. Sen
Department of Chemical Engineering, National Institute of Technology Rourkela, Rourkela, Odisha 769008, India
e-mail: deeptimayeenitrkl@outlook.com

I. D. Behera and A. P. Das (eds.), *Environmental Hydrocarbon Pollution and Zero Waste Approach Towards a Sustainable Waste Management*, Environmental Science and Engineering, https://doi.org/10.1007/978-3-031-81615-4_10

10.1 Introduction

Sustainable technologies signify a fundamental change in individual approaches to industrial operations, resource utilization, and environmental stewardship. Amidst rising concerns about climate change, resource scarcity, and environmental deterioration, embracing sustainable technology to attain lasting socio-economic well-being and protect the planet's health is crucial. Sustainable technologies fundamentally include environmental stewardship, equitable society, and economic feasibility (Soini and Birkeland 2014). They include many inventions and practices that strive to reduce the environmental impact, save resources, and encourage sustainable growth in many areas, such as energy, industry, agriculture, transportation, and waste management. The adoption of sustainable technology is motivated by acknowledging the interdependence between human actions and the natural ecosystem. The current industrial methods, which include using too many resources, generating pollution, and accumulating waste, are no longer sustainable in a society dealing with tremendous environmental issues. Sustainable technologies provide a means to achieve more effective, robust, and restorative systems that prioritize the preservation of the environment and the well-being of humans (Dey, Bissoyi et al. 2024; Dey, Dash et al. 2024; Dey et al. 2023).

Sustainable technologies enhance resource efficiency by minimizing waste, maximizing recycling and reuse, and lowering energy usage (Mohanty et al. 2024; Raj and Das 2023). Their primary focus is optimizing raw materials, water, and energy resources to reduce the adverse effects on the environment and improve economic competitiveness. Sustainable technologies strive to reduce or eliminate pollution directly by using cleaner industrial techniques, pollution control measures, and waste reduction initiatives. They actively preserve the environment and protect public health by implementing emission mitigation strategies, reducing harmful discharges, and promoting environmentally friendly behaviours. Sustainable technologies enable the shift towards renewable energy sources, including solar, wind, biomass, and hydroelectric power, to decrease dependence on fossil fuels and alleviate the impact of greenhouse gas emissions (Bala, Anshul et al. 2024; Bala, Dogra et al. 2024; A. Das, Dey, & Das 2024; A. Das, Dey, Bala et al. 2024). They promote the development of decentralized, robust energy systems that use renewable and plentiful energy sources. Sustainable technologies embody the circular economy principles, prioritizing resource recovery, reuse, and regeneration (Velenturf and Purnell 2021; Mallick et al. 2023). They advocate for closed-loop solutions that facilitate the continuous circulation of resources throughout the economy, reducing waste production and maximizing value creation. By considering their needs and interests, sustainable technologies emphasize social fairness and the involvement of all individuals, particularly those from marginalized groups, workers, and future generations. They aim to promote fair and equal access to resources, opportunities, and benefits while addressing socioeconomic inequities and inequalities.

10.1.1 Sustainable Development Principles

The concepts of sustainable development impart a framework for attaining a balance between economic advancement, social fairness, and environmental preservation (Roseland 2000). They guide making decisions and taking action to build a more resilient and equitable society while protecting the integrity of natural systems for current and future generations. Some significant principles of sustainable development are as follows:

- Sustainable development acknowledges social, economic, and environmental interdependence. It recognizes that activities carried out in one area may have a widespread impact on other regions, emphasizing the need to use comprehensive methods to solve problems and make decisions.
- The precautionary principle promotes using precautionary measures where there is uncertainty or possible damage to human health or the environment. It highlights the significance of foreseeing and averting negative consequences, especially in the lack of definitive scientific data.
- Sustainable development seeks fair and impartial access to resources, opportunities, and advantages for everyone, irrespective of race, gender, socioeconomic level, or geographic location. It aims to tackle inequities and inequalities by implementing inclusive policies and practices.
- Intergenerational equality acknowledges future generations' entitlement to inherit a healthy and sustainable world. It requires appropriate management of natural resources and ecosystems to guarantee their ongoing availability and integrity for future use.
- Sustainable development aims to keep ecosystems, biodiversity, and ecological processes intact (Blicharska et al. 2019). Ecosystem resilience in the face of environmental difficulties is one of its primary goals, along with the conservation and restoration of natural ecosystems and the prevention of biodiversity loss.
- Resource efficiency refers to optimizing the use of natural resources to extract the most value from them while reducing waste, pollution, and adverse environmental effects. It promotes the use of renewable resources, the practice of recycling, and the adoption of sustainable production and consumption methods to maximize resource utilization and reduce ecological impact.
- Sustainable development fosters the capacity to withstand and adjust to environmental, social, and economic challenges. It promotes the formation of resilient and versatile systems capable of enduring shocks, foreseeing change, and swiftly recovering from disturbances.
- To foster cooperation, sustainable development prioritizes inclusive decision-making, including many stakeholders, such as governments, corporations, civil society, and local communities. The statement acknowledges the significance of using inclusive and transparent governance procedures to attain sustainable results.
- Sustainable development recognizes the interdependence of global concerns and all countries' joint obligations to confront them. International collaboration, unity,

and reciprocal assistance are necessary to tackle global challenges such as climate change, poverty, and inequality (Brodhag and Taliere 2006).

10.1.2 Sustainable Technologies' Significance for Environmental Concerns

Sustainable technologies are crucial in tackling the many environmental difficulties our world is experiencing. As the world's population increases and industrialization speeds up, there is a greater need for resources and energy, resulting in more strain on ecosystems, biodiversity, and the climate. Sustainable technologies provide effective solutions for mitigating the release of greenhouse gas emissions, which are the main catalysts of climate change. Renewable energy technology, energy-efficient practices, and carbon capture and storage (CCS) technologies aid in the shift away from fossil fuels, effectively reducing carbon emissions and lessening the effects of climate change. Sustainable technologies enhance resource efficiency and conservation by optimizing resource utilization, minimizing waste generation, and advocating for circular economy principles (Aithal and Aithal 2023; Behera and Nayak et al. 2022). These technologies, ranging from sustainable agricultural and forestry practices to eco-friendly industrial processes and waste reduction programs, contribute to preserving natural resources and ecosystems for future generations (John & Das 2024; Mishra et al. 2024; Mohanty et al. 2024; Paul et al. 2024). Biodiversity and ecosystem health are greatly endangered by habitat loss, pollution, and climate change. Sustainable technologies provide novel solutions for restoring habitats, conserving animals, and preserving ecosystems. These technologies protect Earth's diverse range of life by advocating for sustainable land use practices, habitat restoration programs, and biodiversity conservation activities. Air and water pollution harm human health, ecosystems, and biodiversity. Sustainable technologies, such as state-of-the-art water treatment systems, pollution control devices, and clean energy technologies, are crucial in mitigating air and water pollution. These technologies preserve clean air and water, guaranteeing their availability for communities worldwide, as listed in Table 10.1.

Table 10.1 Sustainable technologies' significance in addressing environmental concerns

Sustainable technology	Environmental significance
Circular economy practices	Reduction of resource depletion and generation of waste
Green chemistry	Minimization of hazardous materials during manufacturing processes
Renewable energy	Lowering greenhouse gas emissions from energy generation
Carbon capture and storage	Reduction of carbon dioxide (CO_2) emissions from industrial activities
Bioremediation	Restoration of ecosystems in polluted water and soil

Environmental deterioration often has a more significant impact on marginalized populations, making socioeconomic disparities and injustices worse. Sustainable technologies prioritize environmental justice and fairness by focusing on the needs and concerns of disadvantaged people, encouraging community involvement and empowerment, and pushing for inclusive and participatory decision-making. Sustainable technologies enhance the capacity to withstand and adjust to environmental uncertainties and obstacles, promoting resilience and adaptability (Nelson et al. 2007). By promoting sustainable agriculture, implementing disaster preparation and response techniques, and developing climate-resilient infrastructure, these technologies assist communities and ecosystems in enduring and recovering from environmental shocks and disturbances. Using sustainable technology presents prospects for economic expansion, employment generation, and inventive advancements. By implementing renewable energy projects, constructing green infrastructure, and supporting sustainable agriculture and eco-tourism efforts, these technologies stimulate economic growth while advancing environmental sustainability. Environmental concerns are beyond the limits of individual countries and need to coordinate worldwide efforts and collaboration. Sustainable technologies promote global cooperation, information exchange, and technology transfer, allowing nations to cooperatively tackle shared environmental issues and construct a more sustainable and resilient future for humanity (Pandey et al. 2022).

10.2 Environmental Hydrocarbon Pollutants

Environmental hydrocarbon pollutants are chemical substances that originate from petroleum and other sources of fossil fuels. These pollutants harm the environment and human health (Pal and Sen 2023a; Behera et al. 2021). Hydrocarbon pollutants refer to a wide variety of substances that originate from petroleum and other sources of fossil fuels. The following are prevalent categories of hydrocarbon pollutants:

(a) **Polyaromatic Hydrocarbons (PAHs)**: Organic molecules of multiple fused aromatic rings are known as PAHs. They originate from incomplete combustion processes involving organic materials, including wood, tobacco, and fossil fuels. Some examples are Benzo[a]pyrene, phenanthrene, anthracene, and naphthalene. Long-term environmental persistence and carcinogenic and mutagenic qualities are well-known characteristics of PAHs.

(b) **BTEX**: Benzene, toluene, ethylbenzene, and xylene are referred to as BTEX chemicals. These are the volatile organic compounds (VOCs) that are often present in products made from petroleum. BTEX chemicals are emitted during the manufacturing, transportation, and use of petrol, diesel, and other hydrocarbons. They can potentially impart health hazards to people, such as neurological and respiratory diseases.

(c) **Aliphatic Hydrocarbons**: Alkyl and hydrogen atoms combine to form linear or branched chains known as aliphatic hydrocarbons. They may be found in natural gas, petroleum products, and crude oil. Chemicals known as alpha-halocarbons may pollute land and water when they spill, leak, or runoff from industrial areas. Alkanes, alkenes, and alkynes are a few examples.
(d) **Chlorinated Hydrocarbons**: Organic molecules with chlorine atoms bound to carbon atoms are chlorinated hydrocarbons. They are used as industrial compounds, insecticides, and solvents. Chlorinated hydrocarbons can persist in the environment and accumulate inside living things. Polychlorinated biphenyls (PCBs), dichloromethane, and chloroform are a few examples (Erickson and Kaley 2011).
(e) **Aromatic Hydrocarbons:** The cyclic molecules known as aromatic hydrocarbons have one or more benzene rings. They are often found in petroleum-based goods, including lubricants, fuel, and petrol. In addition to being hazardous, aromatic hydrocarbons may harm the environment and human health. Benzene, toluene, ethylbenzene, and xylenes are a few examples.

10.2.1 Sources of Hydrocarbon Pollutants

Hydrocarbon contamination arises from several causes, such as industrial operations, transportation, natural processes, and urbanization, as shown in Fig. 10.1. Below are some of the prevalent causes of hydrocarbon pollution:

(a) **Industrial Activities**: Emissions, leaks, and effluents release hydrocarbons during refining, which turns crude oil into different petroleum products. Numerous organic chemicals, including hydrocarbons, which may pollute land, water, and air, are produced at chemical production plants. Hydrocarbons are

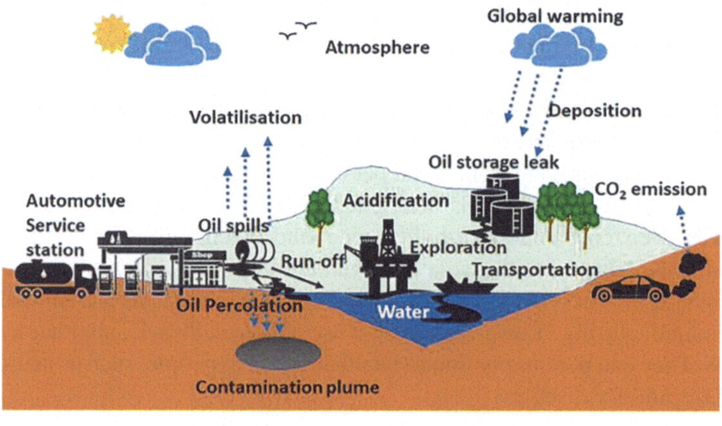

Fig. 10.1 Sources of hydrocarbon pollutants, as adapted from (Chukwunonso Ossai et al. 2022)

released as emissions or byproducts of industrial operations such as metal fabrication, making plastics, and manufacturing textiles. Hydrocarbon pollutants are released into the atmosphere during the combustion of fossil fuels, particularly by power production facilities that use coal, oil, and natural gas.

(b) **Transportation**: During gasoline burning, vehicles such as cars, trucks, buses, ships, airplanes, and other means of transportation release hydrocarbons as exhaust emissions. Unintentional leaks that occur when transporting, storing, or refueling petrol, diesel, and other hydrocarbon fuels have the potential to pollute land and water. Petroleum products stored in underground and aboveground storage tanks have the potential to leak gradually, resulting in the release of hydrocarbons into the environment (Pal and Sen 2023b).

(c) **Oil and Gas Operations**: Oil and gas exploration and drilling operations can potentially cause spills, leaks, and the release of hydrocarbons into the environment, both on land and in the ocean. Due to leakage, venting, and flaring, hydrocarbons may be released from oil and gas production infrastructure, such as wellheads, pumps, separators, and pipelines. Refineries and petrochemical facilities convert crude oil and natural gas into fuels, chemicals, and polymers, releasing hydrocarbons as pollutants and effluents (Behera et al. 2020; Lahiri et al. 2021; Mohanta et al. 2024).

(d) **Natural Seepage**: Hydrocarbons are released into the atmosphere, soil, and water via geological structures, including fissures, fractures, and oil and gas reserves. Hydrocarbons are released into marine ecosystems via submarine seeps, contributing to the hydrocarbon pollution of deep-sea habitats and coastal regions at baseline.

(e) **Urbanization and Development**: Hydrocarbons from parking lots, industrial sites, residential areas, and roadways may enter rivers, lakes, and coastal waterways via stormwater runoff of metropolitan regions. Excavation, building, and clearing land may disrupt the soil and release hydrocarbons that are buried pollutants or trapped in silt. Hydrocarbon pollution of soil and groundwater may result from improper disposal of waste substances, such as industrial waste, home chemicals, and automobile fluids.

(f) **Natural Processes**: Plants emit hydrocarbons, mainly VOCs, due to their inherent metabolic processes, influencing the amounts of hydrocarbons in the atmosphere. Using their metabolic activities, microorganisms in soil and water can break down organic substances, such as hydrocarbons. This contributes to the natural reduction of these substances, known as attenuation processes.

10.2.2 Environmental Impact of Hydrocarbon Pollutants

Hydrocarbon pollutants have a complex and wide-ranging environmental effect, which may significantly affect ecosystems, biodiversity, human health, and the general functioning of natural systems (Behera and Das 2023). The following are essential factors to consider about the environmental consequences of hydrocarbon pollutants:

(a) **Air pollution**: Hydrocarbon pollutants exacerbate air pollution by releasing VOCs, nitrogen oxides (NOx), particulate matter, and other contaminants. These pollutants can undergo chemical reactions with sunlight and other atmospheric substances, forming smog and ground-level ozone. These pollutants may cause irritation to the respiratory system, worsen asthma and other respiratory illnesses, and contribute to the development of cardiovascular diseases. Prolonged exposure to air pollution caused by hydrocarbon emissions may result in chronic respiratory conditions, lung cancer, and cardiovascular disorders in humans.
(b) **Water pollution**: Hydrocarbon contaminants have the potential to pollute both surface water and groundwater as a result of spills, leaks, runoff, and industrial discharges. Oil spills can cover water surfaces and asphyxiate aquatic animals, disturb marine ecosystems, and cause damage to birds, mammals, and fish. Hydrocarbon pollution can deteriorate water quality, hinder the development of aquatic habitats, diminish biodiversity, and disturb the functioning of ecosystems. Benzene and other aromatic hydrocarbons can seep into groundwater, polluting drinking water sources and presenting potential hazards to human health (Pal and Sen 2023b).
(c) **Soil contamination**: Leaks and spills of hydrocarbons can pollute soil, lowering its fertility, preventing plant development, and upsetting soil ecosystems. Long-term soil persistence of petroleum hydrocarbons may be hazardous to plants, soil organisms, and groundwater quality. Remediation techniques such as soil excavation, bioremediation, or soil vapor extraction may be necessary to restore the health and productivity of contaminated soil (Ossai et al. 2020).
(d) **Ecological Disruption**: Hydrocarbon pollution can disturb land-based and water-based ecosystems, causing changes in the physical makeup of habitats, the interconnectedness of food chains, and the dynamics of species relationships. Oil spills can cover shorelines, mangroves, and coral reefs, suffocating plants and animals and causing disturbances in the equilibrium of marine ecosystems (John & Das 2024; Mishra et al. 2024; Mohanty et al. 2024; Paul et al. 2024). Prolonged exposure to hydrocarbon pollutants may result in a decrease in population, extinction of species, and deterioration of ecosystems, ultimately diminishing the overall resilience and stability of the ecosystem.
(e) **Bioaccumulation and Biomagnification**: Hydrocarbon contaminants can accumulate in organisms, resulting in elevated levels in the tissues of apex predators in the food chain. Biomagnification is when hydrocarbon pollutants get more concentrated as they go up the food chain, resulting in increased dangers for apex predators, including humans. Persistent organic pollutants, like PAHs and chlorinated hydrocarbons, can increase in concentration as they go up the food chain in aquatic ecosystems. This may raise health concerns for creatures at higher levels of the food chain.

(f) **Human Health Risks**: Exposure to hydrocarbon pollutants may provide substantial hazards to human health, such as respiratory diseases, neurological disorders, reproductive issues, and cancer. Benzene, an aromatic hydrocarbon with carcinogenic properties, is present in gasoline and crude oil and has been associated with the development of leukemia and other types of cancer. Specific groups at higher risk, including children, the elderly, and persons with pre-existing health disorders, are more prone to experiencing the health consequences of hydrocarbon pollution (Pal and Sen 2023a).

10.3 Challenges in Hydrocarbon Pollution Management

Dealing with hydrocarbon pollution is challenging because of the intricate characteristics of the pollutants, their extensive origins, and the many environmental and socioeconomic consequences. Hydrocarbon pollutants arise from many sources, such as industrial activity, transportation, natural seepage, and urbanization, which complicates the task of monitoring and controlling emissions. Hydrocarbon compounds have unique characteristics and behaviours, resulting in diverse environmental destinies and travel routes, complicating pollution control efforts. Hydrocarbon pollutants have the potential to negatively affect the quality of air, water bodies, soil health, and ecosystems. This may result in several environmental outcomes, including air pollution, water contamination, soil degradation, and disturbance of ecological balance. The interconnection between different environmental compartments adds complexity to evaluating and controlling the effects of hydrocarbon contamination on ecosystems and human health. Several hydrocarbon pollutants, including PAHs and chlorinated hydrocarbons, have a lasting presence in the environment and may accumulate in organisms, posing long-term threats to ecosystems and human health (Alegbeleye et al. 2017). The inherent durability of hydrocarbons in soil, water, and living organisms poses challenges to the cleanup process and requires extended monitoring and management approaches. Hence, various policies and regulations have to be maintained; an overview has been illustrated in Fig. 10.2.

Physical limits do not limit hydrocarbon contamination and may have cross-border effects, impacting ecosystems, air quality, and human health in other countries. Oil spills can go great distances via ocean currents, impacting coastal towns, marine environments, and animals in faraway areas. Identifying and monitoring hydrocarbon pollutants in the environment presents technological difficulties owing to the diverse compounds, stringent detection thresholds, and the need for advanced analytical methods. Enhancing real-time monitoring and early identification of hydrocarbon contamination incidents requires remote sensing technology, sensor networks, and improved analytical methodologies. The remediation of areas polluted by hydrocarbons, such as oil spills and industrial zones, is complex because of the pollutants' intricate characteristics, the site's unique circumstances, and the extent of the contamination. To obtain good results, rigorous planning, execution, and monitoring are necessary for effective remediation approaches, such as bioremediation, chemical

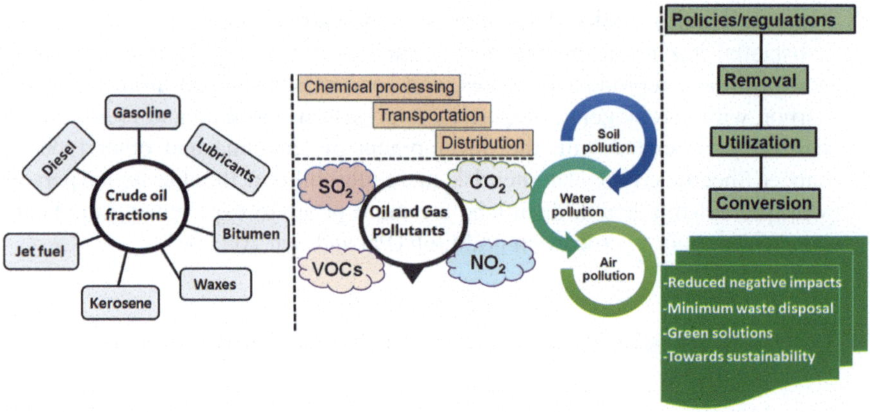

Fig. 10.2 Schematic representation of hydrocarbon pollution generation and management, adapted from (Haruna et al. 2023)

treatment, and soil vapor extraction (Labianca et al. 2020). Hydrocarbon pollution has a more significant impact on disadvantaged communities, indigenous people, and marginalized groups, resulting in environmental injustices and unequal levels of exposure to health hazards associated with pollution. To prevent hydrocarbon pollution effectively, it is necessary to consider socioeconomic variables, connect with the community, and include equity concerns in pollution management plans to achieve fair and equitable results for all parties involved. Effective regulation and policy development to manage hydrocarbon contamination requires cooperation between governments, industry, NGOs, and communities. Establishing consistent standards, ensuring compliance, and filling gaps in monitoring and enforcement systems are crucial to enhancing pollution management and achieving environmental sustainability objectives.

10.3.1 Compliance Concerns and Regulatory Framework

Regulatory frameworks are essential for effectively controlling hydrocarbon pollution. They set standards, guidelines, and enforcement mechanisms to limit emissions, prevent pollution, and safeguard human health and the environment. Nevertheless, implementing and enforcing these standards might provide many problems and compliance concerns. The regulatory frameworks for managing hydrocarbon pollution may be complex and fragmented, including several government bodies, jurisdictions, and overlapping missions.

Hydrocarbon operations, including exploration, production, transportation, refining, and waste disposal, are subject to different rules, which may result in regulatory overlap and inconsistency (Ite et al. 2013). Ineffective implementation and enforcement of hydrocarbon pollution legislation may be hindered by a lack of cooperation among regulatory bodies and stakeholders.

Industries and stakeholders may need clarification and compliance issues due to inconsistent enforcement tactics, contradictory rules, and jurisdictional conflicts. The limited resources, competence, and capability of regulatory authorities might hinder their ability to efficiently monitor, enforce, and guarantee compliance with legislation regarding hydrocarbon pollution. Lack of adequate funds, workforce, and technological skills may impede regulatory bodies' capacity to conduct inspections, impose fines, and promptly react to pollution occurrences.

Industries often must work on time to comply with monitoring and reporting requirements for hydrocarbon pollutant emissions (Ahmad et al. 2008). This involves constant monitoring, data collecting, and reporting emissions data to regulatory authorities. Different sectors may have varying levels of adherence to monitoring and reporting obligations, resulting in discrepancies in the quality, accuracy, and openness of data. Rapid technological progress in extracting, producing, and controlling hydrocarbon emissions poses problems for regulatory systems in keeping up with new technologies and industry standards. Regulatory frameworks may encounter difficulties adjusting to emerging technology, inventive methodologies, and altering industrial norms, resulting in deficiencies in regulation and adherence. Lack of sufficient involvement from the public and stakeholders in creating and enforcing rules on hydrocarbon pollution may weaken the credibility and efficiency of such regulations. The absence of transparency, public availability of information, and chances for substantial involvement may undermine confidence in regulatory procedures and decision-making. Non-uniform implementation methods, moderate consequences, and insufficient deterrent measures might be ineffective in preventing non-compliance and illicit actions within petroleum operations.

Robust enforcement measures, such as imposing penalties, fines, and punishments, are crucial to guarantee adherence to rules and discourage pollution occurrences.

10.3.2 Technological Constraints in the Treatment and Remediation of Pollution

Technological constraints in pollution treatment and cleanup might provide considerable difficulties in efficiently tackling hydrocarbon contamination (Kuppusamy et al. 2017). Although there have been improvements in remediation technology, several limitations still affect pollution treatment initiatives' efficiency, efficacy, and viability. Hydrocarbon contaminants possess a wide range of chemical compositions, physical characteristics, and environmental behaviors, which complicates and presents challenges in their rehabilitation. Hydrocarbons of various kinds, including aliphatic, aromatic, chlorinated, and volatile chemicals, need customized remediation methods designed according to their distinct properties and how they behave in the environment. Several hydrocarbon pollutants exhibit persistence in the environment since they defy degradation and stay in soil, water, and sediments for extended durations (Varjani et al. 2017). The persistence of hydrocarbons, exceptionally resistant chemicals such as hydrocarbons, also known as PAHs, and chlorinated solvents might impede their elimination and remediation from polluted areas. The available choices for addressing hydrocarbon pollution are restricted due to considerations such as the site's characteristics, the levels of contaminants present, and the regulations that must be followed. Remediation approaches, such as physical, chemical, and biological procedures, may need to be improved in their efficacy, application, and cost-effectiveness based on the unique features of the site and attributes of the contaminants.

Challenges arise when attempting to achieve equal dispersion of treatment agents and sufficient interaction with pollutants in subterranean settings during in situ remediation procedures, such as soil vapor extraction, bioremediation, and chemical oxidation. Soil permeability, heterogeneity, and matrix composition might affect the efficiency of in situ treatment methods and may need customized adjustments and optimization for each site. Ex situ remediation procedures, such as excavation, soil washing, and thermal desorption, include extracting and treating polluted soil or water at facilities away from the original site. Logistical constraints, transportation expenses, and disposal concerns might restrict the use of ex-situ treatment procedures, especially for pollution sites of significant size or in distant areas. Implementing remediation technology for hydrocarbon contamination may incur substantial costs, necessitating a significant financial commitment, allocation of resources, and specialized skills (Adesipo et al. 2020). The substantial expenses related to cleanup might present difficulties for parties with limited resources, such as government agencies, companies, and communities, restricting their capacity to tackle pollution effectively. To ensure the long-term efficacy of remediation activities, it is necessary to continuously monitor, maintain, and evaluate the performance of remediated sites. Ensuring a project's long-term monitoring and maintenance may need help securing continuous financing, enough workforce, and institutional backing. These obstacles could undermine the effectiveness and durability of the remediation efforts.

10.3.3 Social and Financial Obstacles to Initiatives to Minimize Hydrocarbon Pollution

Challenges in the form of economic and societal factors may greatly hinder attempts to reduce pollution, particularly in tackling hydrocarbon contamination. These obstacles often interconnect and engage, resulting in intricate difficulties for governments, companies, communities, and other individuals with an ownership stake. Implementing hydrocarbon pollution control measures and adopting cleaner technologies generally require companies, corporations, and governments to invest a substantial amount of money initially. Adhering to environmental regulations, such as hydrocarbon emission standards and pollution control measures, may necessitate investments in equipment upgrades, process alterations, and hydrocarbon pollution reduction technologies, which can present financial difficulties, especially for small and medium-sized enterprises (SMEs). Industries may be motivated to engage in hydrocarbon pollution reduction measures via economic incentives, such as subsidies, tax breaks, and grants for pollution control and clean technology. Governments, particularly in developing nations, may have challenges in efficiently implementing and enforcing hydrocarbon pollution control measures due to constraints such as limited financial resources, technical experience, and infrastructural capability. Limited resources may also impede investment in research and development of cutting-edge hydrocarbon pollution reduction technology and sustainable practices, thus slowing down progress in mitigating pollution. Nevertheless, the lack of compelling economic motivations or counterproductive incentives, such as government support for fossil fuel development or lenient enforcement of environmental legislation, may discourage attempts to reduce hydrocarbon pollution and sustain ecologically detrimental activities.

Hydrocarbon pollution mitigation strategies and environmental legislation may disparately affect various socioeconomic groups, sectors, and geographies (Shonkoff et al. 2011). Vulnerable populations, such as low-income and marginalized communities, may endure a disproportionate share of hydrocarbon pollution-related health hazards and environmental injustices, compounding socioeconomic inequities and disparities in pollution exposure. Industries heavily involved in hydrocarbon pollution, such as fossil fuel extraction, refining, and manufacturing, often provide substantial job prospects and significant economic contributions to local economies. Policymakers attempting to balance environmental and economic goals may face obstacles when they shift from polluting sectors to cleaner, sustainable alternatives. These effects might include job losses, economic disruption, and opposition from impacted communities. Public support for hydrocarbon pollution reduction programs and community involvement in environmental stewardship may be impeded by resistance to change, skepticism towards environmental rules, and little knowledge or comprehension of pollution concerns. Adopting and enforcing hydrocarbon pollution control measures may be undermined by political interests, lobbying activities, and regulatory capture by entrenched interests, particularly those in polluting sectors. Political resistance, the reversal of legislation, and the industry's impact on governmental choices may undermine environmental regulations and impede attempts to reduce hydrocarbon pollution.

10.4 Novel Methods for Mitigating Hydrocarbon Pollution

Novel methods for mitigating hydrocarbon pollution use state-of-the-art technology, multidisciplinary partnerships, and innovative tactics to tackle the intricate issues linked to hydrocarbon contamination. These techniques improve pollution prevention, remediation, and sustainable management practices while promoting environmental protection and socio-economic growth.

(a) **Advanced Remediation Techniques:** Cleanup techniques for hydrocarbon-contaminated soil and water are made possible by innovative remediation technologies, including electrokinetic remediation, phytoremediation, and nanoremediation, which are effective and affordable (Kuppusamy et al. 2020). For instance, modified nanoparticles are used in remediation approaches based on nanoparticles to improve the breakdown and removal of contaminants, hence increasing remediation efficiency and lowering the time and cost of cleaning.
(b) **Bioremediation and Microbial Remediation:** Bioremediation utilizes the inherent metabolic capacity of microorganisms to break down hydrocarbon contaminants in soil and water. Biotechnological advancements, such as the use of genetically modified microorganisms and groups of bacteria, allow for the focused and improved breakdown of specific hydrocarbon molecules, hence speeding up the cleanup process (Kanungo et al. 2024; Das et al. 2024).
(c) **Green infrastructure and environmentally friendly solutions:** Natural alternatives to control stormwater runoff and purify water tainted with hydrocarbons include built wetlands, biofilters, and green roofs. Green infrastructure initiatives use naturally occurring ecosystems and constructed systems to improve water quality, lower pollution, and lessen the effects of urbanization on hydrocarbon runoff and contamination.
(d) **Technologies for Remote Sensing and Monitoring:** Remote sensing technologies, such as satellite imaging, airborne drones, and sensor networks, enable the immediate monitoring and surveillance of hydrocarbon pollution sources, spills, and emissions. Predictive modeling and early identification of pollution incidents are made possible by advanced data analytics and machine learning algorithms. This allows for quick reaction and mitigation measures.
(e) **Resource Recovery and the Circular Economy:** The concepts of a circular economy advocate for the efficient use of resources, the minimization of waste, and the implementation of sustainable strategies for managing resources in hydrocarbon industries. Novel methodologies, such as waste-to-energy technologies, bioconversion processes, and chemical recycling, facilitate the retrieval and utilization of hydrocarbon-based waste streams, therefore diminishing environmental consequences and advocating for the preservation of resources.
(f) **Initiatives from the Community and Citizen Science:** Community involvement and citizen science activities enable local communities to monitor, report, and mitigate hydrocarbon contamination in their neighborhoods. Collaborative alliances among communities, academics, NGOs, and government agencies enable the exchange of information, skills development, and implementation of local initiatives to tackle pollution issues successfully.
(g) **Regulatory Reform and Policy Innovation:** Novel policy tools, such as pollution levies, cap-and-trade systems, and green finance methods, provide incentives for reducing pollution, stimulating investments in clean technology,

and fostering sustainable habits. Adopting adaptive management techniques, engaging stakeholders, and implementing co-regulatory frameworks promote regulatory innovation and enhance the ability to respond to changing pollution hazards and difficulties.

10.4.1 Bioremediation for Mitigating Hydrocarbon Pollution

Bioremediation approaches use indigenous microbes, plants, or their enzymes to break down or render harmless contaminants in the environment. Bioremediation is a successful and eco-friendly method for cleaning up soil and water in areas polluted by hydrocarbons (Azubuike et al. 2020). Biostimulation refers to improving the development and activity of naturally occurring microorganisms that can break down hydrocarbons. This is achieved by supplying them with nutrition, oxygen, and other substances that promote their growth. Typical biostimulation methods include introducing nitrogen, phosphorous, and potassium fertilizers, as well as oxygenation via aeration or bioventing, to enhance microbial metabolism and speed up the breakdown of hydrocarbons. Bioaugmentation introduces certain groups of genetically modified microorganisms into polluted settings to improve their ability to break down and remove contaminants. Microbial strains chosen for their unique metabolic pathways in hydrocarbon degradation, such as bacteria and fungi that can break down hydrocarbons, are introduced into the polluted site to enhance microbial populations and rapidly remove pollutants. Phytoremediation uses plants to absorb, process, or isolate hydrocarbon contaminants from soil, water, or air. Poplar trees, willows, and certain grass species, which have a high capacity for producing a large amount of plant material and have deep root systems, are cultivated in polluted areas. These plants take hydrocarbons from the soil via their roots and then transport them to their above-ground parts to break them down or store them (Dietz and Schnoor 2001).

Rhizoremediation is a technique that combines the concepts of phytoremediation with the activity of microorganisms that are connected with plant roots. Its purpose is to improve the decomposition of hydrocarbons in the rhizosphere, which is the soil area impacted by plant roots. The interactions between plants and microorganisms in the rhizosphere enhance the development and activity of microbes, provide favorable circumstances for the breakdown of hydrocarbons, and stimulate processes that aid in the cleanup of pollutants with the help of plants (Dietz and Schnoor 2001). Mycoremediation uses fungi, including white-rot fungus and mycorrhizal fungi, to break down hydrocarbon pollutants via enzymatic degradation and metabolic processes. Fungi secrete extracellular enzymes, such as ligninases and peroxidases, which can decompose complex hydrocarbon molecules into simpler compounds that may be used as carbon and energy sources, as shown in Fig. 10.3. Biochar, a carbonaceous substance derived from the thermal decomposition of biomass, may be used as a soil amendment to augment microbial activity and facilitate the breakdown of hydrocarbons in polluted soils. Biochar enhances soil composition, moisture retention, and nutrient accessibility, fostering optimal microbial proliferation and hydrocarbon

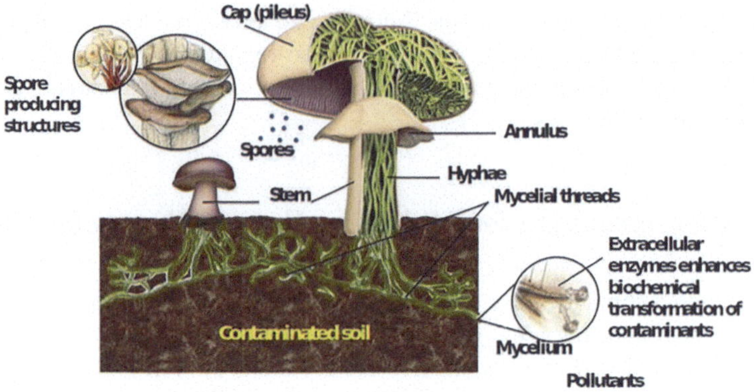

Fig. 10.3 Mycoremediation of petroleum-contaminated hydrocarbon soil, as adapted from (Chukwunonso Ossai et al. 2022)

decomposition. Microbial electroremediation employs electrochemical methods to boost microbial activity and improve the breakdown of hydrocarbons in polluted settings. Electrodes are placed into the polluted soil or water, producing an electric field that enhances electron transfer processes, microbial metabolism, and the breakdown of hydrocarbons (Muttaleb and Ali 2022).

10.4.2 Advanced Water Treatment Technologies for Mitigating Hydrocarbon Pollution

Advanced water treatment technologies are essential for removing hydrocarbons from polluted water sources, safeguarding the environment, and providing a safe drinking water supply. These technologies use advanced methods and systems to efficiently isolate, break down, or absorb hydrocarbon contaminants from water. Activated carbon adsorption is a commonly used technique for eliminating hydrocarbons, such as volatile organic compounds (VOCs) and non-polar pollutants, from water. Activated carbon has a large surface area and porosity, which enables it to adsorb hydrocarbons onto its surface via physical adsorption and chemical interactions. Advanced oxidation processes (AOPs), such as ozone oxidation, ultraviolet (UV) photolysis, and advanced oxidation using hydrogen peroxide (H_2O_2), produce very reactive hydroxyl radicals (•OH) that break down hydrocarbon pollutants into smaller, less harmful substances. Advanced oxidation processes (AOPs) efficiently treat stubborn hydrocarbons and intricate combinations, providing a fast and thorough breakdown of pollutants in water. Membrane filtration methods such as reverse osmosis (RO), nanofiltration (NF), and ultrafiltration (UF) may effectively eliminate hydrocarbons and other impurities from water by using size exclusion, electrostatic interactions, and molecular sieving. Membrane systems have excellent removal efficiency and provide purified water suited for various uses, such as drinking water treatment and industrial process water purification (Lee et al. 2016).

Biological treatment methods, such as aerobic and anaerobic bioreactors, use microbial activity to metabolize and break down hydrocarbon contaminants found in water.

Aerobic bioreactors provide an environment with abundant oxygen to support the breakdown of hydrocarbons by aerobic microbes into CO_2 and water. On the other hand, anaerobic bioreactors promote the microbial degradation of hydrocarbons in an environment with low oxygen levels. Coagulation and flocculation consist of introducing coagulants, such as aluminum sulfate (alum) or ferric chloride, and then combining flocculants to disrupt and gather hydrocarbon particles, resulting in the formation of bigger flocs that may be efficiently eliminated using sedimentation or filtering. Chemical coagulation and flocculation processes successfully eliminate suspended hydrocarbons, emulsified oils, and other particle matter from water, enhancing water purity and quality. Electrocoagulation (EC) is an electrochemical method that disrupts and combines hydrocarbon pollutants in water by producing coagulant ions and gas bubbles at the electrodes (Lu et al. 2021). Electrocoagulation (EC) systems have many benefits, including fast treatment, decreased chemical use, and less sludge generation. As a result, they are well-suited for removing hydrocarbons from generated water, industrial waste, and polluted groundwater. Graphene-based materials, carbon nanotubes, and metal–organic frameworks (MOFs) are advanced sorption materials that can effectively remove hydrocarbons from water due to their high adsorption capabilities and selectivity. These materials have favorable prospects for developing next-generation adsorbents with improved performance, stability, and recyclability for water treatment applications.

10.4.3 Nanotechnology for Mitigating Hydrocarbon Pollution

Nanotechnology applications are emerging as viable solutions for hydrocarbon pollution management by using the distinctive features of nanoparticles to tackle environmental concerns successfully (Patil et al. 2016). Nanotechnology employs methods that allow for accurate control and modification of materials on a size of 1 to 100 nm. This results in improved performance, efficiency, and adaptability in reducing pollution. Nanostructured adsorbents, including carbon nanotubes, graphene, and mesoporous silica nanoparticles, possess large surface areas and adjustable surface characteristics, effectively allowing them to adsorb contaminants from air, water, and soil. These materials demonstrate an improved ability to attract and hold substances, quick rates of reaction, and the ability to selectively eliminate impurities, such as heavy metals, organic pollutants, and hydrocarbons, enhancing the overall effectiveness of pollutant removal. Nanocatalysts, such as metal nanoparticles (e.g., titanium dioxide, iron oxide) and metal–organic frameworks (MOFs), improve the effectiveness of advanced oxidation processes (AOPs) in breaking down persistent pollutants. When exposed to light or electron beam irradiation, the nanocatalysts produce reactive oxygen species, such as hydroxyl radicals. This allows for the quick and thorough breakdown of organic pollutants, medicines, and new contaminants during water and wastewater treatment. Incorporating nanomaterials, such as carbon nanotubes and graphene oxide, into membrane matrices has significantly transformed membrane filtration technologies, particularly in nanofiltration (NF) and reverse osmosis (RO). Nano-enhanced membranes provide better separation efficiency, resistance to fouling, and permeability, effectively eliminating contaminants, pathogens, and microorganisms from water sources, such as saltwater desalination and wastewater reuse (Giwa et al. 2016).

Nanobiosensors identify and track environmental hydrocarbon pollutants, toxins, and diseases, exhibiting exceptional sensitivity, specificity, and selectivity. These small sensing devices use nanomaterials like quantum dots, carbon nanotubes, and metallic nanoparticles as sensors to identify and measure small amounts of contaminants in air, water, and soil samples. This helps to detect and respond to pollution incidents quickly. Nanoremediation methods use nanoparticles, surfactant-stabilized emulsions, or nanostructured materials to improve the cleaning of polluted soil and water. Nanoparticles, such as zero-valent iron (nZVI) and nanoscale zero-valent iron (NZVI), enhance the degradation and immobilization of hydrocarbon pollutants, chlorinated solvents, heavy metals, and organic pollutants by facilitating reductive processes. This leads to improved efficiency in remediation and reduced treatment duration. Nano-enhanced fertilizers, nanopesticides, and nanoscale soil additives provide precise transportation of nutrients, insecticides, and agrochemicals to crops, decreasing agricultural runoff and soil erosion. These formulations, based on nanomaterials, promote the efficiency of nutrient absorption, the effectiveness of pest management, and soil fertility. As a result, they limit the adverse effects on the environment and increase crop yield. Additionally, they minimize the dangers of hydrocarbon pollution linked with traditional agricultural methods. Nanotechnology-driven air purification methods, such as photocatalytic air filters, nanostructured coatings, and nanofiber membranes, efficiently eliminate airborne contaminants, including VOCs, particulate matter (PM), and NOx. Nanostructured materials, such as titanium dioxide nanoparticles, facilitate the breakdown of hydrocarbon contaminants when exposed to light, allowing for the effective and environmentally friendly cleansing of indoor and outdoor air environments (Kumar et al. 2021).

10.5 Circular Economy

The circular economy theory represents a fundamental change in our production, consumption, and resource management approach (Desing et al. 2020). Its goal is to separate economic development from the depletion of resources and the deterioration of the environment. A circular economy aims to maximize resource value, minimize waste, and promote sustainable resource usage throughout the product lifespan, in contrast to the linear economy's "take-make-dispose" production and consumption paradigm. The circular economy is based on three fundamental principles:

- The circular economy prioritizes the development of goods, processes, and systems that prevent the creation of waste and pollution from the beginning. Design strategies include several aspects, such as ensuring the longevity of goods, their potential for reuse, their capacity to be recycled, and the use of renewable materials. These strategies enable things to be repurposed, repaired, or recycled after they end their useful life.
- The circular economy seeks to extend the lifespan of items and resources by promoting their reuse, refurbishment, remanufacturing, and sharing rather than simply disposing of them after use. By prolonging product lifespan and optimizing resource use, we can minimize the need to extract new resources and decrease the environmental impact linked to manufacturing and consumption.
- The circular economy aims to rejuvenate natural systems and enhance ecosystem health by promoting sustainable resource management, conservation, and restoration initiatives. The circular economy promotes long-term environmental sustainability and resilience by completing nutrient and material loops, minimizing resource depletion, and restoring ecosystems' capability to offer ecosystem services.

The essential factors that drive the circular economy are innovation, cooperation, digitization, and regulatory assistance. Innovation propels the development of novel technologies, business strategies, and materials that facilitate circularity across many industries. Collaboration among stakeholders, such as enterprises, governments, academia, and civil society, promotes the exchange of information, efficient use of resources, and joint efforts towards achieving circularity. Digitalization simplifies monitoring resources, enhancing visibility in supply chains, optimizing resource utilization, and promoting more effective resource management and circular business strategies. Policy support, in the form of legislation, incentives, and standards, establishes a favorable environment for the shift toward a circular economy. It offers explicit indications and motivations for enterprises to embrace circular practices (Suárez-Eiroa et al. 2019).

10.5.1 Circular Economy Strategies for Hydrocarbon Industries

The hydrocarbon sector, which includes oil and gas extraction, refining, and the manufacture of petrochemicals, substantially impacts worldwide energy output and economic growth. Nevertheless, the industry is increasingly under pressure to decrease its environmental impact, minimize the effects of climate change, and shift towards sustainable ways of managing resources. Implementing circular economy techniques provides a means for the petroleum sector to overhaul its operations, reduce waste production, and optimize resource utilization, all while remaining competitive and profitable and implementing resource-efficient methodologies and technologies to minimize the consumption of raw materials, utilization of energy, and creation of waste throughout the whole hydrocarbon value chain and implementing cutting-edge process technologies, such as integrated refining and petrochemical complexes, to maximize the efficient use of resources and minimize the environmental impact of hydrocarbon production and processing processes and enacting waste reduction strategies and waste-to-value projects to transform wastes of the petroleum sector, including refinery leftovers, flaring gases, and waste plastic, into valuable commodities and materials. Allocating resources towards waste valorization technologies, such as pyrolysis, gasification, and chemical recycling, to convert waste streams from hydrocarbons into raw materials for producing new goods, including fuels, chemicals, and materials, and establishing interconnected systems and supply chains that facilitate the reuse, recycling, and repurposing of materials and products obtained from hydrocarbons. Implementing reverse logistics networks and product take-back programs to retrieve and incorporate end-of-life items and materials into the production cycle decreases the need for new resources and minimizes waste disposal.

Increasing the utilization of bio-based feedstocks, renewable energy sources, and sustainable alternatives to fossil fuels in hydrocarbon production and processing activities.

Allocating resources towards bio-refining technologies, such as biomass conversion and biofuel generation, to broaden the range of materials used, decrease carbon emissions, and facilitate the shift towards a low-carbon circular economy. Increasing the longevity of hydrocarbon-based goods and equipment by refurbishing, remanufacturing, and maintenance services. Adopting servitization models, such as leasing, renting, and performance-based contracts, encourages the extended lifespan of products, promotes the sharing of resources, and makes it easier to manage products at the end of their life. Utilizing digital technology, data analytics, and predictive maintenance solutions to enhance process efficiency, allocate resources effectively, and control waste in hydrocarbon operations. Digital twins, IoT sensors, and real-time monitoring systems can be used to improve the visibility of the supply chain, monitor the movement of resources, and identify potential solutions for a circular economy (Lieder and Rashid 2016). Engaging in partnerships with industry partners, governments, research institutions, and NGOs to promote innovation, exchange best

Table 10.2 Circular economy practices for hydrocarbon pollution mitigation

Circular economy practice	Description	Advantages	Examples
Recycling	Reusing materials based on hydrocarbons	Minimizes the need for virgin resources	Recycling waste plastic for manufacturing novel products
Reuse	Utilizing hydrocarbon products again	Increases the lifecycle of the product	Restoring utilized oil barrels for use as storage
Remanufacturing	Restoring products based on hydrocarbons	Reduction in energy consumption and waste generation	Automobile components refurbishing for future selling
Closed loop systems	Incorporating recyclable materials	Minimizes waste and the exploitation of resources	Plastic packaging systems with closed loops

practices, and tackle shared difficulties in implementing circular economy principles in the hydrocarbon sector. Interacting with local communities, indigenous groups, and other stakeholders to guarantee inclusive decision-making, permission to operate, and conformity with sustainable development objectives. Various circular economy practices for hydrocarbon pollution mitigation are listed in Table 10.2.

10.6 Green Chemistry

Green chemistry concepts provide novel and sustainable approaches to address hydrocarbon contamination, reduce environmental consequences, and encourage resource preservation throughout the whole hydrocarbon lifetime. Green chemistry techniques focus on the development, creation, and use of chemicals and materials that are environmentally friendly (Mulvihill et al. 2011). The goal is to minimize pollution, improve energy efficiency, and minimize the dangers associated with hydrocarbon production, processing, and usage. The objective is to create and produce hydrocarbon-based goods, fuels, and chemicals using chemical processes that are safer and less dangerous, hence minimizing the production of toxic byproducts and emissions. Alternative feedstocks, such as bio-based and renewable resources, should be utilized instead of fossil fuels to reduce the environmental effects and carbon footprint of chemical manufacturing. Catalytic methods, such as heterogeneous and biocatalysis, can facilitate specific and efficient changes in hydrocarbon raw materials while minimizing waste production. Researching innovative catalytic materials and optimizing reaction conditions to boost process efficiency, reduce energy use, and maximize product yields in hydrocarbon refining and petrochemical synthesis. In extracting, purifying, and synthesizing hydrocarbons, replacing dangerous solvents with more environmentally friendly options, such as water, supercritical fluids, and ionic liquids. Adopting solvent-free or low-solvent reaction protocols helps reduce the solvent used, decrease solvent waste, and optimize chemical processes without compromising product quality and output (Anastas and Eghbali 2010).

Researching the development of environmentally friendly substances that can dissolve and disperse hydrocarbon contaminants in water. Developing environmentally sustainable formulations for oil spill dispersants, soil washing agents, and industrial cleaning solutions to aid in the cleanup and restoration of hydrocarbon-contaminated locations without causing further environmental damage. Promoting the development and use of bio-based polymers, bioplastics, and renewable materials obtained from biomass or agricultural leftovers as substitutes for traditional plastics and packaging materials made from hydrocarbons.

The objective is to develop bio-based additives and composites with enhanced characteristics such as better qualities, biodegradability, and recyclability (Kanungo et al. 2023; Behera et al. 2023; Mishra et al. 2021). These materials are intended for various sectors, including packaging, construction, and automotive. The goal is to create cutting-edge recycling technologies and closed-loop systems that can effectively recover and reuse hydrocarbon-based materials and products as they end their useful life. Utilising chemical depolymerization, pyrolysis, and solvent extraction techniques to transform post-consumer plastics, rubber, and waste oils into valuable raw materials for producing new goods and uses. Performing life cycle assessments (LCAs) and analyzing green metrics to analyze the environmental effects and sustainability of hydrocarbon-based goods, processes, and technology. Integrating green chemistry indicators, such as atom economy, E-factor, and environmental footprint, into decision-making processes facilitates the choice of more environmentally friendly options and optimization tactics (Chen et al. 2020) 0.6.1 Principles of green chemistry and its applications.

The concepts of green chemistry provide guidelines for designing, synthesizing, and utilizing chemicals and materials that reduce environmental harm and encourage sustainability at every stage of their existence. The twelve principles of green chemistry (Abdussalam-mohammed et al. 2020) are:

a. **Prevention:** Preventing waste is more advantageous than treating or remedying trash after its formation.
b. **Atom economy:** The design of synthetic techniques should aim to optimize the integration of all elements used in the process into the result.
c. **Less hazardous chemical syntheses:** Synthetic processes should be developed to produce materials without environmental and human health toxicity.
d. **Designing safer chemicals:** Toxicological risks to humans and the environment should be reduced in the design of chemical goods without sacrificing their effectiveness.
e. **Safer solvents or auxiliaries:** Auxiliary compounds (e.g., solvents, separation agents, etc.) should be avoided if feasible and dealt with safely.
f. **Design for energy efficiency:** Chemical processes should use as little energy as possible to minimize consumption and environmental effects.

g. **Use of renewable feedstocks:** Raw materials and feedstocks must be derived from sustainable, biobased, or renewable sources wherever feasible.
h. **Reduce derivatives:** Minimise or prevent unnecessary derivatization, such as blocking groups, protection/deprotection, or temporary alterations, if feasible.
i. **Catalysis:** Catalytic reagents exhibit a high degree of selectivity and are more advantageous than stoichiometric reagents.
j. **Design for degradation:** To reduce their environmental impact, chemical items should break down into harmless byproducts after usage.
k. **Real-time analysis for pollution prevention:** Developing analytical techniques is necessary to identify and stop the emergence of dangerous compounds during chemical reactions.
l. **Inherently safer chemistry for accident prevention:** Preventing chemical mishaps, such as releases, explosions, and fires, requires carefully selecting substances and their forms for use in chemical processes.

10.6.1 Sustainable Synthesis Routes for Hydrocarbon-Based Products

Sustainable synthesis routes for hydrocarbon-based products are designed to reduce environmental impact, resource consumption, and carbon emissions throughout manufacturing. Sustainable synthesis routes allow the generation of hydrocarbon-based products with a lower environmental impact and greater sustainability by using green chemistry principles and utilizing renewable feedstocks, energy-efficient methods, and novel catalytic technologies.

Renewable feedstocks sourced from biomass, such as plant oils, lignocellulosic biomass, and algae, are substitutes for fossil-based hydrocarbons. Bio-based feedstocks have the benefit of being renewable, carbon–neutral, and possibly biodegradable. This reduces dependence on limited fossil resources and helps to alleviate greenhouse gas emissions.

- Utilizing ecologically friendly solvents, such as water, supercritical fluids, and ionic liquids, in hydrocarbon synthesis operations to reduce environmental harm and enhance process safety. Improving the reaction parameters, such as temperature, pressure, and catalyst amount, decreases energy use, increases the reaction's selectivity, and minimizes waste production (Sheldon 2005).
- Implementing catalytic conversion techniques, including heterogeneous, homogeneous, and enzymatic catalysis, to facilitate specific and compelling changes in hydrocarbon raw materials. Researching and developing innovative catalysts and catalytic systems with exceptional activity, selectivity, and stability. These catalysts will efficiently manufacture hydrocarbon-based products while minimizing the formation of unwanted byproducts and waste.
- Biocatalysts, including enzymes and microbial cells, should be considered to produce hydrocarbon-based compounds and intermediates. Biocatalytic techniques have benefits like mild reaction conditions, precise substrate specificity,

and compatibility with living organisms, allowing for synthesizing intricate hydrocarbons with minimized environmental harm.

- Minimizing carbon emissions and energy use, using green hydrogen sources, such as renewable power or hydrogen from biomass, to carry out hydrogenation and oxidation processes. Researching and developing durable catalytic systems for hydrogenation and oxidation processes, including transition metal catalysts, MOFs, and supported nanoparticles.
- Utilizing energy-efficient methods such as membrane filtration, distillation, and adsorption to separate and purify hydrocarbon-based products. Developing cohesive process flows and separation systems to minimize energy consumption, solvent use, and environmental effects linked to product purification.
- Performing LCAs and green metrics analysis to examine environmental and environmental effects and sustainability of hydrocarbon synthesis methods. Integrating green chemistry indicators, such as atom economy, E-factor, and carbon footprint, into decision-making processes to inform the choice of sustainable synthesis pathways and optimization tactics (Kralisch et al. 2015).
- Utilizing process intensification methods, such as microreactor systems, continuous flow reactors, and intensified reaction conditions, increases efficiency, decreases space requirements, and minimizes waste production. Developing modular and scalable methods for synthesis that allow for flexible production, quick expansion, and effective resource utilization, hence promoting the shift towards sustainable manufacturing practices.

10.6.2 Designing Eco-Friendly Methods for Employing Hydrocarbons

Designing environmentally friendly techniques for using hydrocarbons entails creating sustainable approaches that minimize the adverse effects on the environment, reduce the use of resources, and decrease carbon emissions at every stage of the lifetime of hydrocarbon-based goods. Environmentally benign processes strive to mitigate environmental hazards associated with hydrocarbon utilization and promote sustainability and resource conservation by combining green chemistry concepts, renewable feedstocks, energy-efficient technology, and waste minimization methods. Renewable feedstocks, such as biomass, bio-based oils, and hydrocarbons from waste, can be used as substitutes for fossil fuels in hydrocarbon conversion processes. Renewable feedstocks are sustainable, carbon–neutral, and biodegradable (Srivastava et al. 2021). This reduces our dependence on limited fossil resources and helps to decrease greenhouse gas emissions. Incorporating eco-friendly solvents, such as water, supercritical fluids, and ionic liquids, in hydrocarbon utilization procedures to reduce environmental harm and enhance process safety. Enhancing reaction conditions, such as adjusting temperature, pressure, and catalyst choice, decreases energy use, increases reaction specificity, and minimizes waste production. Employing catalytic processes heterogeneous, homogeneous, and enzymatic catalysis, among others, to effectively and selectively transform hydrocarbon feedstocks into value-added products and developing innovative catalysts and catalytic systems with high selectivity, activity, and stability to produce fuels, chemicals, and materials generated from hydrocarbons with the least waste and byproducts possible.

Implementing biological systems, such as enzymes, biocatalysts, and microorganisms, to transform hydrocarbon feedstock into the necessary products. Optimizing microbial pathways, increasing product yields, and boosting process efficiency in hydrocarbon utilization via fermentation, synthetic biology, and metabolic engineering and capturing and converting CO_2 emissions into valuable goods and commodities by integrating carbon capture and utilization technology into hydrocarbon utilization processes (Mikulčić et al. 2019) and investigating CO_2 utilization strategies to convert CO_2 into building materials, chemicals, and fuels to promote carbon neutrality and lower greenhouse gas emissions. These pathways include carbonation reactions, electrochemical reduction, and biological fixation. Establishing procedures that reduce waste production and increase the value of coproducts and byproducts produced using hydrocarbons. Establishing waste-to-value techniques into practice, such as upcycling, recycling, and using byproducts, to transform waste streams into useful resources while lessening their adverse effects on the environment and improving resource efficiency, and performing LCAs and sustainability evaluations to measure hydrocarbon utilization processes' environmental impact and sustainability performance and integrating green chemistry indicators, such as atom economy, process efficiency, and environmental footprint, into the design and optimization of processes to inform decision-making and encourage sustainable behaviors. The objective is to integrate process units and optimize energy flows in hydrocarbon utilization processes to achieve maximum energy efficiency, minimum energy consumption, and reduced greenhouse gas emissions. Utilizing heat integration, cogeneration, and renewable energy sources to decrease dependence on fossil fuels, improve the sustainability of processes, and lessen the effects of climate change (Bagherian and Mehranzamir 2020).

10.7 Hydrocarbon Waste Stream Conversion into Value-Added Products

The conversion of hydrocarbon waste streams into valuable products provides a sustainable approach to waste management since it offers economic possibilities and reduces environmental effects. Diverse methods and procedures may convert hydrocarbon waste streams, including industrial wastes, refinery leftovers, and post-consumer plastics, into useful goods, fuels, chemicals, and materials. Pyrolysis is a thermally induced breakdown process that disintegrates organic substances into pyrolysis oil, syngas, and char, including hydrocarbon waste streams like plastics, rubber, and biomass. Pyrolysis oil may undergo further processing to produce biofuels, specialized chemicals, or lubricants (Jahirul et al. 2012). Syngas, conversely, can be used as a fuel for generating electricity or as a raw material for chemical synthesis. Char generated by pyrolysis may serve as a soil supplement, activated carbon, or carbon black for diverse uses. Catalytic upgrading employs catalysts to transform hydrocarbon waste streams into more valuable products via hydrotreating, hydrocracking, and catalytic reforming. Catalysts expedite the decomposition of intricate hydrocarbons into less complicated molecules, enhancing the end product's quality, quantity, and worth. Catalytic upgrading methods may convert hydrocarbon waste streams, such as heavy oils, tars, and refinery wastes, into transportation fuels, petrochemical feedstocks, or specialty chemicals. Chemical recycling, called feedstock recycling or depolymerization, transforms plastic waste into monomers or intermediate compounds using chemical methods, including pyrolysis, hydrolysis, or glycolysis. Plastics, such as polyethylene (PE), polypropylene (PP), and polystyrene (PS), may undergo depolymerization to break down into their monomers. These monomers can then create fresh plastics or other chemical products. Chemical recycling presents a potential alternative for addressing the issue of plastic waste and completing the cycle of plastic production, allowing for the retrieval of valuable materials from plastic trash that is mixed or contaminated. Some value-added products and their applications obtained from hydrocarbon waste are listed in Table 10.3.

Biological conversion methods, such as anaerobic digestion and fermentation, use microorganisms to decompose organic hydrocarbon waste streams into biogas, bioethanol, or biochemicals. Anaerobic digestion is a process that transforms organic waste into biogas, consisting of methane and CO_2, as well as digestate, which may be used as a source of renewable energy or organic fertilizer (Manyi-Loh et al. 2019). Fermentation techniques generate bioethanol, biodiesel, and other biochemicals from organic waste streams, including agricultural leftovers, food waste, and lignocellulosic biomass. Waste-to-energy methods, such as incineration, gasification, and pyrolysis, transform hydrocarbon waste into heat, electricity, or fuels. Incineration procedures include combusting waste materials at high temperatures and producing heat and steam that may be harnessed to produce electricity or provide district heating. Gasification and pyrolysis technologies transform hydrocarbon waste into syngas, bio-oil, or char, which may be used as fuels or chemical raw materials. Carbon capture and utilization methods extract CO_2 emissions from hydrocarbon waste streams and

Table 10.3 Value-added products from hydrocarbon waste

Waste stream	Value-added product	Applications
Used cooking oil	Biodiesel	Transportation fuel, renewable energy
Plastic waste	Recycled plastic	Packaging, construction materials, consumer goods
Oil refinery byproducts	Asphalt binder	Road construction, roofing materials
Organic waste	Biogas	Electricity generation, heating, cooking

transform them into valuable goods such as fuels, chemicals, or construction materials. CO_2 may be extracted from industrial flue gases or straight from the environment and transformed into methanol, formic acid, or carbonate minerals using chemical or biological methods.

CCU technologies can provide advantages such as reducing CO_2 emissions, recovering valuable resources, and integrating with the circular economy. Hydrothermal processing methods, such as hydrothermal liquefaction (HTL) and hydrothermal carbonization (HTC), transform moist hydrocarbon waste streams, such as sewage sludge, food waste, and algae, into biofuels, charcoal, or biochemical (Marzbali et al. 2021). HTL utilizes elevated temperatures and pressures to transform moist biomass into bio-crude oil, while HTC generates hydrochar, a solid material rich in carbon, from organic waste substances.

10.7.1 Renewable Energy Sources' Incorporation into Hydrocarbon Operations

Integrating renewable energy technology with hydrocarbon sectors offers a significant chance to decrease carbon emissions, improve energy efficiency, and encourage sustainable behaviors. Hydrocarbon businesses may effectively reduce their carbon footprint and address greenhouse gas emissions by substituting fossil fuel-based energy sources with renewable options like solar, wind, or biomass. This action supports international efforts to address climate change and attain carbon neutrality. Renewable energy sources often provide reduced operating expenses compared to conventional fossil fuels, significantly as the cost of renewable technology decreases (Ellabban et al. 2014). Incorporating renewable energy technologies like solar photovoltaics (PV), wind turbines, or biomass boilers may assist hydrocarbon industries in decreasing their energy costs and enhance their overall cost competitiveness. Integrating renewable energy sources helps diversify the energy supply, improving energy security by reducing reliance on limited fossil fuel supplies and unstable global energy markets. Renewable energy systems are also more resistant to supply outages and geopolitical threats, guaranteeing a dependable and continuous supply for hydrocarbon activities. Renewable energy integration optimizes resource utilization by exploiting plentiful natural energy sources like sunshine, wind, or biomass wastes. This enhances the efficient use of resources and reduces the adverse effects on the environment, which aligns with sustainable development and responsible management of resources.

Using renewable energy technology in hydrocarbon sectors promotes technical advancement and encourages collaborative research (Sen and Ganguly 2017). This encompasses advancing hybrid energy systems, energy storage solutions, and smart grid technologies designed to meet hydrocarbon facilities' distinct needs and operating demands. As regulatory pressure to decrease emissions and shift towards renewable energy sources grows, incorporating renewables into hydrocarbon sectors allows corporations to comply with regulations and showcase their commitment to corporate responsibility. Implementing renewable energy solutions improves the standing of a company, reinforces connections with stakeholders, and reduces the potential harm to its reputation due to environmental effects. Integrating renewable energy may improve community involvement and relationships with stakeholders by showing a dedication to sustainability, environmental responsibility, and local economic growth. Projects integrating renewable energy technology often receive backing from local communities, governments, and environmental organizations, cultivating favorable connections and societal approval to function. Incorporating renewable energy into hydrocarbon sectors helps organizations achieve long-term sustainability and adaptability in a quickly changing energy environment. Hydrocarbon industries may maintain their competitiveness and resilience in the global marketplace by diversifying their energy portfolios and shifting towards cleaner and more sustainable energy sources. This allows them to react to changing market dynamics, regulatory frameworks, and social expectations (Cherepovitsyn et al. 2023).

Integrating renewable energy sources into hydrocarbon operations offers a strategic opportunity to decrease greenhouse gas emissions, improve energy efficiency, and encourage sustainable practices across the sector. Hydrocarbon facilities can include renewable energy-generating technologies on their premises, such as PV panels, wind turbines, or biomass boilers, to fulfill a part of their energy requirements. Renewable energy systems may be incorporated into existing infrastructure, such as roofs, open land, or offshore platforms, to reduce land use and optimize energy generation. Hybrid energy systems integrate various renewable energy sources with conventional energy sources or energy storage technologies to optimize the supply and demand of energy. By using supplementary renewable energy technologies, such as solar and wind, hydrocarbon operations may improve energy dependability, decrease irregularity, and attain greater levels of energy self-sufficiency. Energy storage systems, such as batteries, pumped hydro storage, or compressed air energy storage, may enhance renewable energy generation by storing surplus energy for future use or supplying backup power during low renewable energy production. Energy storage technologies bolster grid stability, promote energy dependability, and facilitate the seamless integration of renewable energy into hydrocarbon operations. Hydrocarbon firms can engage in power purchase agreements with renewable energy producers to get renewable electricity from off-site renewable energy projects, such as large-scale solar or wind farms. Power Purchase Agreements (PPAs) enable hydrocarbon businesses to use affordable renewable energy without requiring investment in on-site infrastructure. This helps to decrease carbon emissions and decrease power expenses. Before incorporating renewable energy sources, hydrocarbon operations may use energy efficiency methods to optimize energy usage and minimize total energy demand (Igogo et al. 2021).

Energy efficiency activities, such as upgrading equipment, optimizing processes, and implementing energy management systems, promote operational efficiency, reduce energy expenses, and improve the efficacy of integrating renewable energy. Microgrid technologies allow hydrocarbon operations to function autonomously or in tandem with the main power grid by using renewable energy production, energy storage, and energy management systems to optimize the balance between energy supply and demand. Microgrids provide robustness, dependability, and adaptability in energy provision, especially in distant or isolated areas where the availability of conventional grid infrastructure may be restricted (Khare and Chaturvedi 2023). By incorporating renewable energy sources into hydrocarbon operations, companies may ensure compliance with environmental regulations and reduce greenhouse gas emissions. This approach also helps minimize the environmental effects of their activities and enables them to achieve their sustainability goals. Integrating renewable energy supports worldwide climate objectives, regulatory mandates, and industry standards while improving business standing and reducing regulatory vulnerabilities. Integrating renewable energy showcases a company's commitment to social responsibility and promotes active involvement from stakeholders by solving environmental issues, assisting local communities, and contributing to sustainable development. Interacting with stakeholders, such as workers, investors, communities, and governments, guarantees that the organization meets the expectations of these parties and encourages their support for renewable energy projects.

10.7.2 Transitioning the Hydrocarbon Industry Toward a Sustainable Energy System

The process of shifting towards sustainable energy systems in the hydrocarbon industry includes adopting renewable energy sources, improving energy efficiency, and implementing carbon mitigation methods to decrease environmental harm and foster long-term sustainability. The hydrocarbon industry can integrate renewable energy technologies, such as solar, wind, and biomass, into its activities to decrease dependence on fossil fuels and diminish carbon emissions (Abolhosseini et al. 2014). One way to do this is by investing in producing renewable energy on-site, forming PPAs with renewable energy companies, or engaging in renewable energy initiatives within the community. Enhancing energy efficiency is vital for decreasing energy use and optimizing the utilization of resources in hydrocarbon activities. This may be accomplished by implementing energy management systems, using energy-efficient technology, and adopting operational best practices focusing on reducing waste and maximizing energy production. CCS systems are designed to absorb CO_2 emissions from industrial operations, such as power generation and petrochemical manufacture. These emissions are then stored underground to prevent their release into the atmosphere. By incorporating CCS into hydrocarbon operations, it is possible to decrease greenhouse gas emissions and address climate change significantly, all while still being able to use fossil fuels. Hydrocarbon firms can shift their focus towards using low-carbon fuels, such as biofuels, hydrogen, and synthetic fuels derived from renewable sources. These alternative fuels provide cleaner and more sustainable options than traditional fossil fuels, aiding in reducing carbon emissions in transportation, heating, and industrial activities.

Enhancing resilience and flexibility in hydrocarbon operations may be achieved by diversifying energy portfolios via investments in renewable energy sources, energy storage technology, and demand-side management systems. This allows organizations to adjust to changing market situations, regulatory mandates, and technical progress in the energy industry. Investment in research and innovation is crucial for developing and implementing sustainable energy solutions in the hydrocarbon industry (Chu and Majumdar 2012). This includes providing financial resources for research initiatives, partnerships with academic institutions and research organizations, and assistance to technological incubators and businesses that prioritize the development of sustainable energy solutions. Effective cooperation among industry players, government agencies, research institutions, and non-governmental organizations is crucial for facilitating the shift towards sustainable energy systems in the hydrocarbon sector. Collaborations may expedite the dissemination of information, transfer of technology, and formulation of policies to hasten the implementation of clean energy technologies and practices. Regulatory assistance and incentives, such as carbon pricing schemes, tax credits, and subsidies for renewable energy research and deployment, might motivate hydrocarbon businesses to use sustainable energy solutions. Well-defined and uniform regulatory frameworks provide assurance and steadiness for enduring investments in sustainable energy.

The emerging developments in sustainable technologies for hydrocarbon pollution control include novel methods to minimize environmental damage, improve remediation procedures, and advance circular economy concepts. Bioremediation methods are advancing, allowing for the more effective and precise breakdown of hydrocarbon contaminants in soil and water. This encompasses the advancement of genetically modified microbes, techniques for enhancing biodegradation and pollution removal via bioaugmentation, and the utilization of bioelectrochemical systems. Nanotechnology is used to clean areas polluted by hydrocarbons (Mehndiratta et al. 2013). Nanomaterials, including nano zero-valent iron (nZVI), carbon nanotubes, and nanoparticles, possess improved sorption, catalytic, and reactive characteristics that effectively treat soil and groundwater polluted with hydrocarbons. Electrokinetic remediation approaches use electric currents to mobilize and eliminate hydrocarbon pollutants from soil and groundwater. Novel methods, such as electrokinetic-enhanced bioremediation and electrochemical oxidation, provide effective and economical alternatives for addressing difficult hydrocarbon contamination sites. Situ Chemical Oxidation (ISCO) methods use chemical oxidants, such as hydrogen peroxide, ozone, and persulfate, to break down hydrocarbon pollutants in their original location. The progress in ISCO approaches, including the integration of chemical and biological treatments, the use of reactive barriers, and the development of controlled release formulations, is enhancing the effectiveness of treatment and shortening the time required for remediation. Phytoremediation strategies use plants to absorb, process, or isolate hydrocarbon contaminants from soil and water. Phytoremediation is a field that is seeing the emergence of many new trends. These include the careful selection of plants that can accumulate high levels of pollutants, the application of genetic engineering techniques to boost the ability of plants to absorb impurities, and the use of microbial-assisted methods for phytoremediation (Cristaldi et al. 2017).

Hydrocarbon pollution control solutions increasingly use circular economy concepts to encourage the recovery of resources, the valorization of waste, and the sustainable use of materials. This includes developing closed-loop systems to recycle polluted soils, extracting valuable hydrocarbons from waste streams, and using byproducts for energy production or industrial purposes. Remote sensing and monitoring technologies, such as satellite imagery, unmanned aerial vehicles (UAVs), and sensor networks, can identify and track hydrocarbon pollution sources and their environmental consequences in real time. Fusing remote sensing data with geographic information systems (GIS) facilitates enhanced decision-making and administration of polluted locations. The concepts of green chemistry are directing the creation of sustainable methods for producing hydrocarbon-based goods, aiming to minimize waste production and decrease the adverse effects on the environment. This encompasses utilizing sustainable raw materials, reactions that do not need solvents and processes that employ catalysts to produce chemicals, fuels, and materials that have a lower impact on carbon emissions. Efforts are being made to investigate using blockchain technology and traceability solutions to improve transparency and accountability in managing hydrocarbon contamination. Blockchain technology allows for safe and decentralized data administration, which may be used to monitor the origins of pollutants, ensure compliance with regulations, and verify the effectiveness of remediation efforts. Public–private partnerships and collaborative activities promote information sharing, developing skills, and transferring technology to control sustainable hydrocarbon pollution (Ning et al. 2023). Collaborations involving several stakeholders, consortia formed by industries, and networks focused on research are crucial in promoting innovation and expediting the implementation of sustainable solutions to prevent and remedy pollution.

The future of sustainable hydrocarbon development encounters several obstacles and elements to consider, which demonstrate the intricate relationship between energy demand, environmental issues, and socioeconomic aspects. Tackling climate change is a crucial and pressing obstacle for the hydrocarbon business, considering the substantial release of greenhouse gases linked to fossil fuel extraction, production, and use. To meet climate change mitigation objectives, a quick transition to low-carbon energy sources, the use of CCS technology, and the integration of renewable energy. The primary focus of sustainable hydrocarbon development should be minimizing the adverse environmental effects such as land degradation, water pollution, and habitat loss linked to extraction and production operations (Brittingham et al. 2014). Enforcing strict environmental legislation, implementing effective pollution prevention and repair methods, and investing in cutting-edge technology are crucial for minimizing environmental damage. Ensuring long-term sustainability requires carefully balancing developing hydrocarbon resources and their management and protection. To achieve sustainable hydrocarbon development, it is essential to responsibly manage natural resources such as oil, gas, and minerals. This involves preventing their depletion, maximizing their use, and minimizing any negative environmental impact. Ensuring widespread availability of inexpensive, dependable, and environmentally friendly energy is crucial in promoting sustainable hydrocarbon development. Encouraging clean cooking solutions, investing in renewable energy infrastructure in underserved regions, and increasing the availability of modern energy services are all critical steps towards addressing energy poverty and advancing energy justice.

The future of sustainable hydrocarbon production relies on adopting technical innovation and shifting towards cleaner and more efficient methods (Dovì et al. 2009). Allocating resources to research and development (R&D) projects, promoting the development of clean energy technologies, and facilitating cooperation among business, academia, and government is crucial for promoting technical advancement and facilitating the shift towards sustainable energy. Establishing trust, encouraging communication, and encouraging community involvement are essential for obtaining social approval to operate and establishing sustainable hydrocarbon production. Engaging in partnerships with local communities, upholding the rights of indigenous people, and tackling social issues such as health effects and livelihoods are crucial for establishing meaningful connections and attaining mutually advantageous results. Geopolitical considerations, such as conflicts between nations, the desire to control resources, and worries about the security of energy sources, provide obstacles to the sustainable development of hydrocarbon resources. Diversifying energy sources, strengthening energy resilience, and fostering international collaboration to reduce geopolitical risks and guarantee a stable and secure energy supply are crucial. Developing strong legislative and regulatory frameworks is essential for directing the sustainable extraction of hydrocarbons and ensuring responsible management of natural resources. Establishing a favorable environment for sustainable energy investments is crucial to implementing efficient regulatory monitoring, enforcing environmental standards, and ensuring that policies align with climate goals.

Achieving a low-carbon economy requires significant changes in energy infrastructure, industrial processes, and consumer habits. Coordinating with wider decarbonization initiatives is crucial to ensure hydrocarbon production's long-term viability. This includes gradually eliminating government support for fossil fuels, implementing a system to charge for carbon emissions, and actively encouraging the use of energy-efficient practices and the adoption of renewable energy sources. Developing the capacity to withstand and recover from external disruptions, such as political instability, unpredictable market changes, and natural calamities, is crucial for guaranteeing the long-term viability of hydrocarbon production. It is essential to improve adaptive ability, diversify income sources, and implement risk management methods to navigate uncertainties and mitigate the implications of external disruptions effectively.

The hydrocarbon sector may benefit significantly from innovation and cooperation since they provide substantial prospects for promoting sustainable practices and tackling various difficulties. Collaborative research and development activities have the potential to stimulate innovation in sustainable energy technologies, CCS, and environmental remediation approaches. Through the collaboration and use of collective resources and specialized knowledge, stakeholders can expedite the development and implementation of innovative technologies that may effectively decrease the release of greenhouse gases, improve energy efficiency, and minimize environmental harm. Using big data analytics, artificial intelligence (AI), and digital technology can enhance the efficiency of hydrocarbon exploration, production, and distribution systems (Waqar et al. 2023a). Cooperative endeavors in exchanging data, using predictive modeling, and implementing decision support systems may enhance operational effectiveness, save expenses, and mitigate environmental impact across the hydrocarbon value chain. Collaboration between industry players, governments, and research institutes may facilitate the implementation of circular economy ideas in the hydrocarbon sector. Advancements in converting trash into energy, recovering materials and byproducts, and implementing closed-loop systems can reduce waste production, optimize the recovery of resources, and generate value from hydrocarbon waste streams. The collective efforts to shift towards low-carbon energy sources, such as renewable energy and hydrogen, provide prospects for innovation and investment. Collaborations among hydrocarbon corporations, renewable energy innovators, and financial establishments may expedite the implementation of eco-friendly energy infrastructure, bolster initiatives for energy diversification, and assist in reducing carbon emissions in the energy sector.

Cooperation in carbon management and efforts to reduce emissions may effectively tackle climate change and advance sustainable development. Carbon capture, utilization, and storage (CCUS) initiatives, carbon offset programs, and emissions trading systems provide prospects for inventive solutions and cooperation in reducing greenhouse gas emissions from hydrocarbon activities. By actively involving local communities, indigenous groups, and other stakeholders, collaborative initiatives may cultivate societal acceptance of operations and facilitate inclusive development. Collaborations in community development projects, efforts to boost skills and abilities, and programs to promote long-term economic well-being may effectively tackle social issues, strengthen stakeholder trust, and provide mutual benefits in locations where hydrocarbon production occurs. Collaboration among governments, industry groups, and civil society organizations may facilitate sustainable hydrocarbon development, which can help establish effective policy and regulatory frameworks (Das, A. Prasad., & Ghosh, Shreya. 2023). Dialogue platforms, working groups, and public–private partnerships may help foster agreement, promote information sharing, and ensure that rules align with environmental and social goals.

Cooperative education, training, and skills development efforts can generate a workforce with the necessary knowledge and experience to shift toward sustainable hydrocarbon operations. Collaborations among academic institutions, industry, and vocational training institutions may facilitate skills growth, encourage diversity and inclusivity, and stimulate innovation in the hydrocarbon industry. Cooperation within innovation ecosystems, such as technological clusters, incubators, and accelerators, may accelerate entrepreneurship and innovation in the hydrocarbon business. Collaborations among startups, scale-ups, investors, and existing enterprises may expedite the acceptance of technology, entrance into the market, and commercialization of novel solutions for sustainable hydrocarbon production. Cooperation on global platforms, initiatives, and collaborations may facilitate joint efforts in attaining worldwide sustainability objectives, such as the United Nations Sustainable Development

Goals (SDGs). Collaboration across different sectors, information exchange, and technology transfer may accelerate the advancement toward a more sustainable and resilient future for the hydrocarbon industry and society (Waqar et al. 2023b).

10.8 Conclusion and Future Perspectives

The future of sustainable hydrocarbon development relies on a comprehensive strategy that effectively manages environmental conservation, social accountability, and economic feasibility. Despite the several issues confronting the hydrocarbon business, such as climate change, environmental degradation, and the need to shift to other energy sources, there are substantial prospects for fostering innovation and cooperation to bring about beneficial transformations. The hydrocarbon industry may adopt sustainable practices, use technology and data-driven solutions, and collaborate with many sectors and stakeholders to achieve a more sustainable and resilient future. There are several strategies to drive innovation and achieve common sustainability objectives, such as pushing renewable energy technology, implementing circular economy concepts, fostering community participation, and creating social impact. Furthermore, the need for the sustainable development of hydrocarbons goes beyond the limits of the sector, necessitating collaborative efforts from governments, enterprises, civil society, and academia. By collaborating to tackle environmental issues, foster inclusive growth, and shift towards low-carbon energy systems, stakeholders can lay the foundation for a fair, prosperous, and sustainable future.

Adopting sustainability as a fundamental concept and prioritizing investing in research, development, and implementation of cutting-edge technology to minimize environmental harm, improve energy efficiency, and support circular economy principles. Establishing and enforcing legislative and regulatory frameworks that incentivize sustainable behaviors, promote technological innovation, and foster cooperation among industry stakeholders. Engaging in innovative research, fostering partnerships with industry, and educating future professionals in sustainable technology, environmental management, and methods for transitioning to clean energy. Promoting transparency, accountability, and responsible practices in the hydrocarbon business. Engaging with stakeholders to increase knowledge of the environmental and social consequences. Foster communication and cooperation to achieve common sustainability objectives. Participating in decision-making processes, expressing concerns and objectives regarding hydrocarbon development, and participating in productive discussions with industry stakeholders to ensure that development projects are socially and ecologically accountable. Integrating environmental, social, and governance (ESG) issues into investment choices promotes sustainable projects and initiatives and encourages enterprises to adopt optimal practices and achieve quantifiable sustainability results.

Developing and promoting cutting-edge approaches to the hydrocarbon sector's pollution, carbon, and integration of renewable energy needs; working with industry partners to investigate and expand new technologies; and showing the practicality and efficacy of sustainable substitutes, and promoting the sharing of information, development of skills, and cooperation across nations and regions to tackle global issues associated with hydrocarbon development, encourage technology transfer, and assist in achieving sustainable development objectives. Advocating for openness, accountability, and sustainability from hydrocarbon firms, endorse programs that advance cleaner energy sources and responsible resource management, and make well-informed decisions that foster a more sustainable energy future. Promoting sustainable technologies in the hydrocarbon industry and engaging in collective action and collaboration among different sectors and stakeholders is crucial. This involves committing to common sustainability objectives and working together to establish a more resilient, fair, and sustainable energy system for future generations.

References

Abdussalam-mohammed W, Qasem A, Errayes AO (2020) Green chemistry: principles, applications, and disadvantages. Chem Methodol 4(4):408–423. https://doi.org/10.33945/sami/chemm.2020.4.4

Abolhosseini S, Heshmati A, Altmann J (2014) A review of renewable energy supply and energy efficiency technologies. SSRN Electron J 8145.https://doi.org/10.2139/ssrn.2432429

Adesipo AA, Freese D, Nwadinigwe AO (2020) Prospects of in-situ remediation of crude oil contaminated lands in Nigeria. Scient African 8:e00403. https://doi.org/10.1016/j.sciaf.2020.e00403

Ahmad M, Bajahlan AS, Hammad WS (2008) Industrial effluent quality, pollution monitoring and environmental management. Environ Monit Assess 147(1–3):297–306. https://doi.org/10.1007/s10661-007-0121-5

Aithal S, Aithal PS (2023) Importance of circular economy for resource optimization in various industry sectors—a review-based opportunity analysis. Int J Appl Eng Managem Lett 1:191–215. https://doi.org/10.47992/IJAEML.2581.7000.0182

Alegbeleye OO, Opeolu BO, Jackson VA (2017) Polycyclic aromatic hydrocarbons: a critical review of environmental occurrence and bioremediation. Environ Manage 60(4):758–783. https://doi.org/10.1007/s00267-017-0896-2

Anastas P, Eghbali N (2010) Green chemistry: principles and practice. Chem Soc Rev 39(1):301–312. https://doi.org/10.1039/B918763B

Azubuike CC, Chikere CB, Okpokwasili GC (2020) Bioremediation: an eco-friendly sustainable technology for environmental management. In: Saxena G, Bharagava RN (eds) Bioremediation of industrial waste for environmental safety. Springer, Singapore. https://doi.org/10.1007/978-981-13-1891-7

Bagherian MA, Mehranzamir K (2020) A comprehensive review on renewable energy integration for combined heat and power production. Energy Conver Managem 224:113454. https://doi.org/10.1016/j.enconman.2020.113454

Bala K, Anshul, Verma S, Dogra A, Das AP, Kumar R, Nandini (2024) Sustainable management for urban plastic waste generation. In: Sustainable Management of Urban Plastic Waste Through Circular Economic Approaches, pp 211–236. https://doi.org/10.1201/9781003465447-13

Bala K, Dogra A, Lata P, Das AP (2024) Discussing sources and biological-chemical remediation approaches for microplastic pollution. In: Sustainable Microbial Technology for Synthetic and Cellulosic Microfiber Bioremediation, pp 37–59. https://doi.org/10.1007/978-3-031-62600-5_3

Behera ID, Das AP (2023) Impact of petroleum waste on environmental pollution and its sustainable management through circular economy. Springer

Behera ID, Basak G, Kumar RR, Sen R, Meikap BC (2020) Treatment of petroleum refinery sludge by petroleum degrading bacterium Stenotrophomonas pavanii IRB19 as an efficient novel technology. J Environ Sci Health—Part A Toxic/Hazardous Substances and Environ Eng 56(2). https://doi.org/10.1080/10934529.2020.1866924

Behera ID, Nayak M, Biswas S, Meikap BC, Sen R (2021) Enhanced biodegradation of total petroleum hydrocarbons by implementing a novel two-step bioaugmentation strategy using indigenous bacterial consortium. J Environ Managem 292:112746. https://doi.org/10.1016/j.jenvman.2021.112746

Behera ID, Nayak M, Mishra A, Meikap BC, Sen R (2022) Strategic implementation of integrated bioaugmentation and biostimulation for efficient mitigation of petroleum hydrocarbon pollutants from terrestrial and aquatic environment. Marine Pollut Bulletin 177:113492. https://doi.org/10.1016/j.marpolbul.2022.113492

Blicharska M, Smithers RJ, Mikusiński G, Rönnbäck P, Harrison PA, Nilsson M, Sutherland WJ (2019) Biodiversity's contributions to sustainable development. Nature Sustain 2(12):1083–1093. https://doi.org/10.1038/s41893-019-0417-9

Brittingham MC, Maloney KO, Farag AM, Harper DD, Bowen ZH (2014) Ecological risks of shale oil and gas development to wildlife, aquatic resources and their habitats. Environ Sci Technol 48(19):11034–11047. https://doi.org/10.1021/es5020482

Brodhag C, Taliere S (2006) Sustainable development strategies: tools for policy coherence. Nat Res Forum 30(2):136–145. https://doi.org/10.1111/j.1477-8947.2006.00166.x

Chen T-L, Kim H, Pan S-Y, Tseng P-C, Lin Y-P, Chiang P-C (2020) Implementation of green chemistry principles in circular economy system towards sustainable development goals: Challenges and perspectives. Sci Total Environ 716(1):136998. https://doi.org/10.1016/j.scitotenv.2020.136998

Cherepovitsyn A, Kazanin A, Rutenko E (2023) Strategic priorities for green diversification of oil and gas companies. Energies 16(13):1–17. https://doi.org/10.3390/en16134985

Chu S, Majumdar A (2012) Opportunities and challenges for a sustainable energy future. Nature 488(7411):294–303. https://doi.org/10.1038/nature11475

Chukwunonso Ossai I, Shahul Hamid F, Hassan A (2022) Biological treatments for petroleum hydrocarbon pollutions: the eco-friendly technologies. In: Hazardous waste management. IntechOpen. https://doi.org/10.5772/intechopen.102053

Cristaldi A, Conti GO, Jho EH, Zuccarello P, Grasso A, Copat C, Ferrante M (2017) Phytoremediation of contaminated soils by heavy metals and PAHs. a brief review. Environ Technol Innov 8:309–326. https://doi.org/10.1016/j.eti.2017.08.002

Das A, Dey S, Bala K, Gnanasekaran R, Das AP (2024) Microbial bioremediation technology for sustainable treatment and management of synthetic microfiber waste. In: Sustainable Microbial Technology for Synthetic and Cellulosic Microfiber Bioremediation, pp 221–238. https://doi.org/10.1007/978-3-031-62600-5_12

Das A, Dey S, Das AP (2024) Ecotoxicological impacts of synthetic microfiber pollutants and development of sustainable mitigation strategies. Environ Chem Ecotoxicol. https://doi.org/10.1016/J.ENCECO.2024.12.003

Das A Prasad, Ghosh, Shreya (2023) Manganese mining microorganisms

Das AP, Nayak B, Behera ID (2024) Applications of microorganisms in advanced geomicrobiological and biotechnological exploration. Geomicrobiol J 41(4):309–311. https://doi.org/10.1080/01490451.2024.2327059

Desing H, Brunner D, Takacs F, Nahrath S, Frankenberger K, Hischier R (2020) A circular economy within the planetary boundaries: towards a resource-based, systemic approach. Resour Conserv Recycling 155:104673. https://doi.org/10.1016/j.resconrec.2019.104673

Dey S, Bissoyi A, Koteswara Reddy G, Das AP (2024) Synthetic microfibers: An urban emerging pollutant demanding for sustainable environmental solutions. In: Sustainable Microbial Technology for Synthetic and Cellulosic Microfiber Bioremediation, pp 107–121. https://doi.org/10.1007/978-3-031-62600-5_7

Dey S, Dash N, Ray SK, Priyadarsini S, Chhotaray C, Das AP (2024) Occurrence and quantification of tire wear plastics pollutants from highway city dust of Bhubaneswar City of Odisha. In: Sustainable Microbial Technology for Synthetic and Cellulosic Microfiber Bioremediation, pp 77–89. https://doi.org/10.1007/978-3-031-62600-5_5

Dey S, Tripathy B, Kumar MS, Das AP (2023) Ecotoxicological consequences of manganese mining pollutants and their biological remediation. Environ Chem Ecotoxicol. https://doi.org/10.1016/J.ENCECO.2023.01.001

Dietz AC, Schnoor JL (2001) Advances in phytoremediation. Environ Health Perspect 109(SUPPL1):163–168. https://doi.org/10.1289/ehp.01109s1163

Dovì VG, Friedler F, Huisingh D, Klemeš JJ (2009) Cleaner energy for sustainable future. J Clean Prod 17(10):889–895. https://doi.org/10.1016/j.jclepro.2009.02.001

Ellabban O, Abu-Rub H, Blaabjerg F (2014) Renewable energy resources: current status, future prospects and their enabling technology. Renew Sustain Energy Rev 39:748–764. https://doi.org/10.1016/j.rser.2014.07.113

Erickson MD, Kaley RG (2011) Applications of polychlorinated biphenyls. Environ Sci Pollut Res 18(2):135–151. https://doi.org/10.1007/s11356-010-0392-1

Giwa A, Akther N, Dufour V, Hasan SW (2016) A critical review on recent polymeric and nano-enhanced membranes for reverse osmosis. RSC Adv 6(10):8134–8163. https://doi.org/10.1039/C5RA17221G

Haruna A, Tanimu G, Ibrahim I, Garba ZN, Yahaya SM, Musa SG, Merican ZMA (2023) Mitigating oil and gas pollutants for a sustainable environment—critical review and prospects. J Cleaner Prod 416:137863. https://doi.org/10.1016/j.jclepro.2023.137863

Igogo T, Awuah-Offei K, Newman A, Lowder T, Engel-Cox J (2021) Integrating renewable energy into mining operations: opportunities, challenges, and enabling approaches. Appl Energy 300:117375. https://doi.org/10.1016/j.apenergy.2021.117375

Ite AE, Ibok UJ, Ite M, Petters SW (2013) Petroleum exploration and production: past and present environmental issues in the Nigeria's Niger Delta. Amer J Environ Protect 1(4):78–90. https://doi.org/10.12691/env-1-4-2

John A, Das AP (2024) Government strategies for achieving circular economy for plastic waste management in Southeast Asia. In: Sustainable Management of Urban Plastic Waste Through Circular Economic Approaches, pp 72–82. https://doi.org/10.1201/9781003465447-6/GOVERNMENT-STRATEGIES-ACHIEVING-CIRCULAR-ECONOMY-PLASTIC-WASTE-MANAGEMENT-SOUTHEAST-ASIA-AKBAR-JOHN-ALOK-PRASAD-DAS

Jahirul M, Rasul M, Chowdhury A, Ashwath N (2012) Biofuels production through biomass pyrolysis—a technological review. Energies 5(12):4952–5001. https://doi.org/10.3390/en5124952

Kanungo J, Sahoo T, Bal M, Behera ID (2024) Performance of bioremediation strategy in waste lubricating oil pollutants: a review. Geomicrobiol J 41(4):360–373. https://doi.org/10.1080/01490451.2023.2245395

Kanungo J, Sahoo T, Swain LP, Behera ID (2023) Toxicity of persistent hydrocarbon pollutants, sources and sustainable remediation process. In: Behera ID, Das AP (eds) Impact of petroleum waste on environmental pollution and its sustainable management through circular economy. Springer Nature Switzerland, pp 39–65. https://doi.org/10.1007/978-3-031-48220-5_2

Khare V, Chaturvedi P (2023) Design, control, reliability, economic and energy management of microgrid: a review. E-Prime—Adv Electri Eng Electron Energy 5:100239. https://doi.org/10.1016/j.prime.2023.100239

Kralisch D, Ott D, Gericke D (2015) Rules and benefits of life cycle assessment in green chemical process and synthesis design: a tutorial review. Green Chem 17(1):123–145. https://doi.org/10.1039/C4GC01153H

Kumar L, Ragunathan V, Chugh M, Bharadvaja N (2021) Nanomaterials for remediation of contaminants: a review. Environ Chem Lett 19(4):3139–3163. https://doi.org/10.1007/s10311-021-01212-z

Kuppusamy S, Thavamani P, Venkateswarlu K, Lee YB, Naidu R, Megharaj M (2017) Remediation approaches for polycyclic aromatic hydrocarbons (PAHs) contaminated soils: technological constraints, emerging trends and future directions. Chemosphere 168:944–968. https://doi.org/10.1016/j.chemosphere.2016.10.115

Kuppusamy S, Maddela NR, Megharaj M, Venkateswarlu K (2020) Approaches for remediation of sites contaminated with total petroleum hydrocarbons abstract. In: Total petroleum hydrocarbons: environmental fate, toxicity, and remediation. Springer International Publishing. https://doi.org/10.1007/978-3-030-24035-6

Labianca C, De Gisi S, Picardi F, Todaro F, Notarnicola M (2020) Remediation of a petroleum hydrocarbon-contaminated site by soil vapor extraction: a full-scale case study. Appl Sci 10(12):4261. https://doi.org/10.3390/app10124261

Lahiri D, Nag M, Dey A, Sarkar T, Joshi S, Pandit S, Das AP, Pati S, Pattanaik S, Tilak VK, Ray RR (2021) Biofilm mediated degradation of petroleum products. Geomicrobiol J 1–10. https://doi.org/10.1080/01490451.2021.1968979

Lee A, Elam JW, Darling SB (2016) Membrane materials for water purification: design, development, and application. Environ Sci: Water Res Technol 2(1):17–42. https://doi.org/10.1039/C5EW00159E

Lieder M, Rashid A (2016) Towards circular economy implementation: a comprehensive review in context of manufacturing industry. J Clean Prod 115:36–51. https://doi.org/10.1016/j.jclepro.2015.12.042

Lu J, Zhang P, Li J (2021) Electrocoagulation technology for water purification: an update review on reactor design and some newly concerned pollutants removal. J Environ Managem 296:113259. https://doi.org/10.1016/j.jenvman.2021.113259

Mallick K, Sahu A, Dubey NK, Das AP (2023) Harvesting marine plastic pollutants-derived renewable energy: a comprehensive review on applied energy and sustainable approach. J Environ Managem 348:119371. https://doi.org/10.1016/J.JENVMAN.2023.119371

Manyi-Loh CE, Mamphweli SN, Meyer EL, Okoh AI (2019) Microbial anaerobic digestion: process dynamics and implications from the renewable energy, environmental and agronomy perspectives. Int J Environ Sci Technol 16(7):3913–3934. https://doi.org/10.1007/s13762-019-02380-w

Marzbali MH, Kundu S, Halder P, Patel S, Hakeem IG, Paz-Ferreiro J, Madapusi S, Surapaneni A, Shah K (2021) Wet organic waste treatment via hydrothermal processing: a critical review. Chemosphere 279:130557. https://doi.org/10.1016/j.chemosphere.2021.130557

Mehndiratta P, Jain A, Srivastava S, Gupta N (2013) Environmental pollution and nanotechnology. Environ Pollut 2(2):49–58. https://doi.org/10.5539/ep.v2n2p49

Mikulčić H, Ridjan Skov I, Dominković DF, Wan Alwi SR, Manan ZA, Tan R, Duić N, Hidayah Mohamad SN, Wang X (2019) Flexible carbon capture and utilization technologies in future energy systems and the utilization pathways of captured CO_2. Renew Sustain Energy Rev 114:109338. https://doi.org/10.1016/j.rser.2019.109338

Mishra S, Dash D, Das AP (2024) Microbial degradation of polyester microfibers using indigenously isolated bacterial strain Exiguobacterium Sp. CLEAN – Soil, Air, Water, 52(12):e202300343. https://doi.org/10.1002/CLEN.202300343

Mishra A, Siddiqi H, Kumari U, Behera ID, Mukherjee S, Meikap BC (2021) Pyrolysis of waste lubricating oil/waste motor oil to generate high-grade fuel oil: a comprehensive review. Renew Sustain Energy Rev 150:111446. https://doi.org/10.1016/j.rser.2021.111446

Mohanta S, Pradhan B, Behera ID (2024) Impact and remediation of petroleum hydrocarbon pollutants on agricultural land: a review. Geomicrobiol J 41(4):345–359. https://doi.org/10.1080/01490451.2023.2243925

Mohanty M, Mohanty J, Dey S, Dutta K, Shah MP, Das AP (2024) The face mask: a tale from protection to pollution and demanding sustainable solution. Emerg Contaminants 10(2):100298. https://doi.org/10.1016/J.EMCON.2023.100298

Mohanty M, Tripathy B, Koteswara Reddy G, Das AP (2024) Greener technologies for establishing restraint over microplastic pollution. In: Sustainable Microbial Technology for Synthetic and Cellulosic Microfiber Bioremediation, pp 239–258. https://doi.org/10.1007/978-3-031-62600-5_13

Mulvihill MJ, Beach ES, Zimmerman JB, Anastas PT (2011) Green chemistry and green engineering: a framework for sustainable technology development. Annu Rev Environ Resour 36(1):271–293. https://doi.org/10.1146/annurev-environ-032009-095500

Muttaleb WH, Ali ZH (2022) Bioremediation an eco-friendly method for administration of environmental contaminants. MINAR Int J Appl Sci Technol 04(02). https://doi.org/10.47832/2717-8234.11.3

Nelson DR, Adger WN, Brown K (2007) Adaptation to environmental change: contributions of a resilience framework. Annu Rev Environ Resour 32(1):395–419. https://doi.org/10.1146/annurev.energy.32.051807.090348

Ning L, Abbasi KR, Hussain K, Alvarado R, Ramzan M (2023) Analyzing the role of green innovation and public-private partnerships in achieving sustainable development goals: a novel policy framework. Environ Sci Pollut Res.https://doi.org/10.1007/s11356-023-26414-6

Ossai IC, Ahmed A, Hassan A, Hamid FS (2020) Remediation of soil and water contaminated with petroleum hydrocarbon: a review. Environ Technol Innov 17:100526. https://doi.org/10.1016/j.eti.2019.100526

Pal D, Sen S (2023a) emerging petroleum pollutants and their adverse effects on the environment. In: Impact of petroleum waste on environmental pollution and its sustainable management through circular economy. pp 103–137. https://doi.org/10.1007/978-3-031-48220-5_5

Pal D, Sen S (2023b) In-depth coverage of petroleum waste sources, characteristics, environmental impact, and sustainable remediation process. In: Impact of petroleum waste on environmental pollution and its sustainable management through circular economy. pp 1–38. https://doi.org/10.1007/978-3-031-48220-5_1

Pandey N, de Coninck H, Sagar AD (2022) Beyond technology transfer: innovation cooperation to advance sustainable development in developing countries. Wires Energy Environ 11(2):1–25. https://doi.org/10.1002/wene.422

Patil SS, Shedbalkar UU, Truskewycz A, Chopade BA, Ball AS (2016) Nanoparticles for environmental clean-up: a review of potential risks and emerging solutions. Environ Technol Innov 5:10–21. https://doi.org/10.1016/j.eti.2015.11.001

Paul A, Priyadarsini S, Karakuş S, Behera JK, Gnanasekaran R, Das AP (2024) Fate and accumulation of synthetic microfibers in the environment. In: Sustainable Microbial Technology for Synthetic and Cellulosic Microfiber Bioremediation, pp 61–76. https://doi.org/10.1007/978-3-031-62600-5_4

Priyadarsini S, Das AP (2024) Analytical and structural characterization of waste lithium-ion batteries for their effective recycling strategy. Environ Chem Ecotoxicol. https://doi.org/10.1016/j.enceco.2024.12.004

Priyadarsini S, Das AP (2025) Lithium bioleaching: A review on microbial-assisted sustainable technology for lithium bio-circularity. J Water Process Eng, 69, 106744. https://doi.org/10.1016/j.jwpe.2024.106744

Raj K, Das AP (2023) Lead pollution: Impact on environment and human health and approach for a sustainable solution. Environ Chem Ecotoxicol 5:79–85. https://doi.org/10.1016/J.ENCECO.2023.02.001

Ray SK, Dey S, Das AP, John A (2024) Microplastic waste management through greener technologies: A review. In: Sustainable Management of Urban Plastic Waste Through Circular Economic Approaches, pp 237–261. https://doi.org/10.1201/9781003465447-14

Roseland M (2000) Sustainable community development: integrating environmental, economic, and social objectives. Prog Plan 54(2):73–132. https://doi.org/10.1016/S0305-9006(00)00003-9

Sen S, Ganguly S (2017) Opportunities, barriers and issues with renewable energy development—A discussion. Renew Sustain Energy Rev 69:1170–1181. https://doi.org/10.1016/j.rser.2016.09.137

Sheldon RA (2005) Green solvents for sustainable organic synthesis: state of the art. Green Chem 7(5):267. https://doi.org/10.1039/b418069k

Shonkoff SB, Morello-Frosch R, Pastor M, Sadd J (2011) The climate gap: environmental health and equity implications of climate change and mitigation policies in California-a review of the literature. Clim Change 109(SUPPL1):485–503. https://doi.org/10.1007/s10584-011-0310-7

Soini K, Birkeland I (2014) Exploring the scientific discourse on cultural sustainability. Geoforum 51:213–223. https://doi.org/10.1016/j.geoforum.2013.12.001

Srivastava RK, Shetti NP, Reddy KR, Kwon EE, Nadagouda MN, Aminabhavi TM (2021) Biomass utilization and production of biofuels from carbon neutral materials. Environ Pollution 276:116731. https://doi.org/10.1016/j.envpol.2021.116731

Suárez-Eiroa B, Fernández E, Méndez-Martínez G, Soto-Oñate D (2019) Operational principles of circular economy for sustainable development: Linking theory and practice. J Clean Prod 214:952–961. https://doi.org/10.1016/j.jclepro.2018.12.271

Tripathy B, Dash A, Das AP (2022) Detection of environmental microfiber pollutants through vibrational spectroscopic techniques: Recent advances of environmental monitoring and future prospects. Crit Rev Anal Chem, pp 1–11. https://doi.org/10.1080/10408347.2022.2144994

Varjani SJ, Gnansounou E, Pandey A (2017) Comprehensive review on toxicity of persistent organic pollutants from petroleum refinery waste and their degradation by microorganisms. Chemosphere 188:280–291. https://doi.org/10.1016/j.chemosphere.2017.09.005

Velenturf APM, Purnell P (2021) Principles for a sustainable circular economy. Sustain Prod Consum 27:1437–1457. https://doi.org/10.1016/j.spc.2021.02.018

Waqar A, Othman I, Shafiq N, Mansoor MS (2023) Applications of AI in oil and gas projects towards sustainable development: a systematic literature review. Artif Intell Rev 56(11):12771–12798. https://doi.org/10.1007/s10462-023-10467-7

Chapter 11
Sustainable Utilization of Waste Motor Oil: A Hazardous Hydrocarbon Pollutant

Asmita Mishra and Nirmalendu Sekhar Mishra

Abstract The rapid population, industrialization, and urbanization increase have led to a multi-fold increase in automobiles, thereby increasing the demand for lubricating oil. Among the available lubricants, automotive lubricating oil (motor oil) accounts for 57% of global lubricant demand, which is used to lubricate vehicular engine's metal parts, reducing wear and increasing engine efficiency. However, with usage, it degrades due to (i) oxidation, (ii) thermal breakdown, (iii) micro-dieseling, (iv) additive depletion, and demands alteration with fresh motor oil. This chapter aims to discuss the characteristics of fresh and waste motor oil. Besides, it extensively explains the impact of waste motor oil (WMO) on water bodies, marine life, soil, and human health due to illegal disposal to overcome high waste management costs. Furthermore, it also focuses on the WMO hierarchy and suggests different treatment methods based on the hierarchy. However, among the available treatment methods, pyrolysis has drawn notice due to its adaptability and product diversification.

Keywords Lubricant · Waste motor oil · Waste management · Waste-to-energy · Pyrolysis

11.1 Introduction

Transport vehicles have increased drastically as a result of quick urbanization and population growth. The vehicle increase has led to an upsurge in demand for lubricants. The size of the lubricant market worldwide is estimated to be 157.6 billion USD in 2025 and is expected to grow to 182.6 billion USD. Out of all the lubricants

A. Mishra (✉)
Department of Chemical Engineering, Parala Maharaja Engineering College, Berhampur, Odisha 761003, India
e-mail: asmita.asmitamishra@gmail.com

N. S. Mishra
Department of Environmental Science and Engineering, SRM University, Amaravati, Andhra Pradesh 522240, India

I. D. Behera and A. P. Das (eds.), *Environmental Hydrocarbon Pollution and Zero Waste Approach Towards a Sustainable Waste Management*, Environmental Science and Engineering, https://doi.org/10.1007/978-3-031-81615-4_11

currently on the market, 57% of the demand is for motor oil for automobiles (A. Das, Dey, & Das 2024; A. Das, Dey, Bala et al. 2024; Dey, Bissoyi et al. 2024). This motor oil lubricates the vehicular engine's metal parts, reduces wear, and increases engine efficiency. However, with use, the motor oil degrades. Thus, new motor oil replaces this degraded oil to improve its performance. Hence, around 50% of this lubricating oil is a waste (Mishra et al. 2020). Due to high management costs, only 25–45% of these waste oils are collected globally; the remaining 55–75% are either disposed of or used improperly in the environment (Mishra et al. 2021). The US Environmental Protection Agency (EPA) defines waste motor oil (WMO) as "any petroleum-based or synthetic oil that has been used for vehicle lubrication, and as a result of routine use, motor oil becomes contaminated with various impurities, such as dirt, water, chemicals, or metals from the vehicle engine." If this WMO is incorrectly disposed of into the environment, it may contaminate streams through runoff and soil. Additionally, these waste hydrocarbons may be used to produce energy. As a result, a financially reasonable, socially acceptable, and environmentally sustainable solution is required.

11.2 Lubricant Market

After China and the United States, India is the third-largest lubricant market in the world, according to Castrol India Ltd (Sushil Finance). It is also among the lubricant markets with the fastest growth rates. There are three segments in the lubricants market: process, industrial, and automotive oils. According to industry sources, the Indian lubricants industry is a 3-billion-litre business, with the automotive market accounting for 42%, the industrial segment accounting for 23%, transformer and white oils accounting for 23%, process oils accounting for 8%, and greases accounting for 4% as shown in Fig. 11.1a Of the more than 20 companies, the three public sector organizations, BPCL, HPCL, IOCL, and the largest private sector company, Castrol, control about 55% of the market (Fig. 11.1b). Industry reports predict that over the next several years, the Indian lubricants market will expand at a rate of 4–5% annually.

The Indian Oil/Lubricant Market, which is classified into goods such as engine oil, greases, general industrial oils, hydraulic fluid, gear oils, and process oils, is expected to grow at a Compound-Annual Growth Rate (CAGR) of 4.64% between 2019 and 2024. The expanding infrastructure and building activity are vital contributors to the growth.

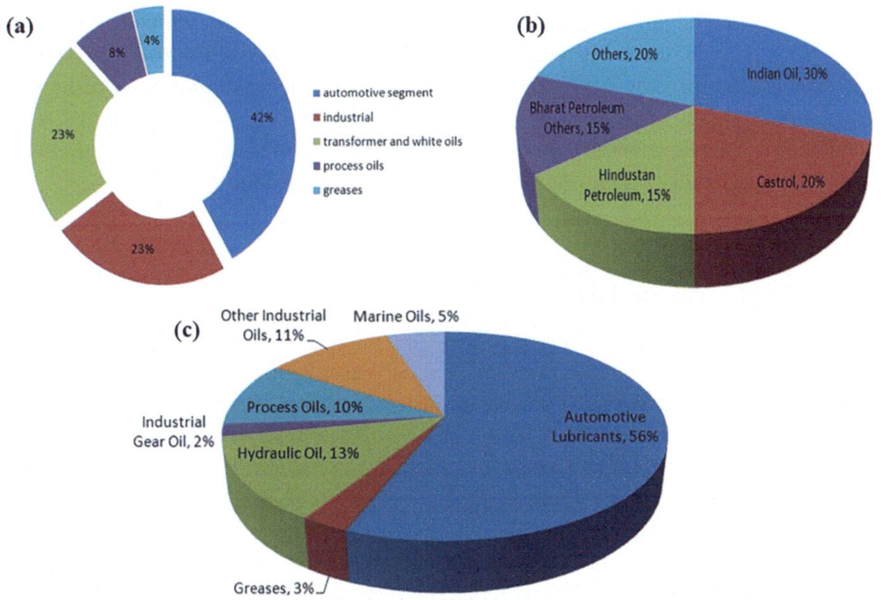

Fig. 11.1 **a** Market share of the Indian lubricants sector, **b** Market share analysis, and **c** Percentage contribution of waste oils

11.3 Motor Oil

Motor oil, commonly known as engine oil or engine lubricant, was discovered on 6 September 1866 by an American named John Ellis as a breakthrough because it worked effectively at high temperatures, posing lower gum content. These motor oils are an amalgamation of base oil and esters. The base oil comprises petroleum-based hydrocarbons and polyalphaolefins, whereas the esters are the additives for enhancing the lubricating properties. Additives are incorporated into the base oil to (i) minimise wear, (ii) control viscosity, (iii) prevent oxidation, and (iv) minimize the impact of water on machinery.

The motor oil is a film between the vehicles moving parts, thereby reducing friction and engine wear, improving engine efficiency, and retarding degradation. Besides engine protection, it also cools the engine by transferring heat from moving parts, neutralizes acids originating from fuel and lubricant oxidation, cleans the engine from sludge, and improves the sealing of piston rings. However, with usage, these motor oils degrade due to (i) oxidation, (ii) thermal breakdown, (iii) micro-dieseling, and (iv) additive depletion.

Lubricant oxidation increases viscosity, thereby generating varnish, sludge, and sediment. This leads to an increase in the acid number promoting corrosion. Thermal breakdown occurs due to heat dissipation, deteriorating oil's viscometric properties. Micro-dieseling is the pressure-induced thermal degradation where an air bubble transits from low/negative pressure to a high-pressure zone. The temperature at the bubble interface is sufficient to carbonize oil, forming carbon byproducts (sludge and varnish). Additive depletion reduces the oil's life. Oil contamination occurs due to dirt, water, air, etc. The fine metallic particles and dirt act as a catalyst that sparks and speeds up the degradation process (Bala, Anshul et al. 2024; Bala, Dogra et al. 2024). Air and water contamination leads to lubricant oxidation.

11.4 Waste Motor Oil

According to the Schedule IV of the Hazardous Waste Rules, 2016, "*Waste oil refers to crude oil spills, emulsions, tank bottom sludge, and slop oil produced by petroleum refineries, installations, or ships that can be utilized as a fuel in energy recovery furnaces if it fits the standards set out in Part-B of Schedule V, either as is or after reprocessing.*" Reprocessing entails recovering wasted oil and converting it into high-quality oil that may be utilized in manufacturing operations. The process can result in a significant decline in the amount of spent oil that is regarded as trash. In terms of business, acquiring new oils results in substantial savings. The oil's life cycle is prolonged by reintroducing reprocessed oil into the same manufacturing cycle.

11.4.1 Sources of Waste Motor Oil

The sources of used oil include the oil generated from crude oil or synthetic oil mixes (used engine oil, wasted oil, hydraulic oil, turbine oil, gear oil, compressor oil, industrial gear oil, transformer oil, heat transfer oil, and tank bottom sludges), according to the Hazardous Waste (Management, Handling, and Transboundary Movement) Rules, 2016. Additionally, the sources of waste oil include crude oil spills, tank bottom sludge, emulsions, and slop oil produced by petroleum refineries, installations, or ships (Behera et al. 2020, 2021; Mohanta et al. 2024). Figure 11.1c depicts the percentage contribution of waste oils. Some facts about used oil:

One million gallons of fresh water can be contaminated by used oil from a single oil change.

- One cup of old motor oil is required to create an oil sheen on a one-acre pond.
- Recycled used motor oil that is converted into fresh oil, processed into fuel oils, and used as raw materials in the petroleum sector.
- One gallon of WMO has the same amount of lubricant as 42 gallons of crude oil.

11.4.2 Characterization of Used/Waste Oil

The additives present in motor oil breaks down with usage. The subsequent interaction of these simpler compounds formed during the process of chemical decomposition of additives results in the formation of undesirable chemicals and corrosive acids. These unwanted chemicals and the corrosive acids that form are the primary contaminants, demanding the replacement of used oil with fresh motor oil. Used oil is classified as a hazardous waste under the Basel Convention on Control of Transboundary Movements of Hazardous Wastes and Their Disposal. Heavy metals, lubricant additives, and polyaromatic hydrocarbons (PAHs) are the three main toxins in used oil (Kanungo et al. 2023). Lead is generally present in high percentages among the metals in engines that use leaded gasoline. The usage of leaded gasoline and the breakdown of lead scavengers leads to the inclusion of bromine and chlorine, which leads to substantial amounts of chlorinated solvents. Other lubrication additives, such as molybdenum disulfide, zinc dialkyl dithiophosphates, and other organo-metallic compounds, are harmful to the environment and human health (Dey, Dash et al. 2024; Dey et al. 2023; John & Das 2024; Mishra et al. 2024; Mohanty et al. 2024).

PAHs, in WMO, are produced when fuel is burned during the operation of engines or other machinery. With operational time, the PAH concentration in crankcase oil keeps enhancing. In addition, polyaromatic hydrocarbons (PAHs) are a significant concern for their proven carcinogenicity. The wetlands, rivers, aquatic life, air pollution, and human health can all be significantly impacted by its poor management. However, the quantity of pollutants in WMO is dependent on numerous parameters, such as (i) the category of detergents and dilutants introduced to the virgin oil, (ii) the site of WMO storage, and (iii) the management techniques involved. For instance, suppose leaded gasoline is progressively phased out, then the lead concentrations in waste oil will drastically decrease. Besides, it will also reduce the quantity of bromine and chlorine additions required, and, as a result, halogenated hydrocarbons in waste oil will be reduced even more.

11.4.3 Impact of WMO on Ecology

According to the EPA web archive, nearly two hundred million gallons of WMO are improperly disposed of yearly. When the waste motor oil is spilled or disposed of in the water bodies, it floats on water due to the lighter oil density. This oil layer hinders the available amount of sunlight, thus inhibiting photosynthesis and disrupting the oxygen cycle, consequently promoting the growth and reproduction of the oil-feeding microbes. This uncontrolled growth of oil-feeding microorganisms in the water bodies results in decreased dissolved oxygen required for aquatic organisms, leading to eutrophication (Das et al. 2023a, b; Das et al. 2024; Tripathy et al. 2024). In addition, the larval stages of marine species are sensitive to the harmful chemicals found in these waste oils. At the bottom of the food chain, the chances of toxin build-up in plankton and other microscopic animals could eventually reach humans. Ingesting waste-oil-contaminated water poses various health risks, ranging from minor signs of toxic chemical accumulation in the liver to severe physiological impairment and, subsequently, death (El-Fadel and Khoury 2001; Mishra and Meikap 2024; Moodley et al. 2022). Thus, illegal disposal in water bodies jeopardizes marine life by inhibiting the oxygen supply. Further, one million litres of water taste are hampered with just one litre of used oil. Around 3784 m^2 of soil can be polluted by one litre of WMO, making it non-functional for farming or plant development for hundred years. Illegal disposal of WMO into the lithosphere will deteriorate soil fertility by causing a reduction in pH, immobilization of nutrients, and poor aeration. These changes may cause large-scale animal extinctions in the upper layers of soil, which would upset the ecological balance of the soil. Long-term exposure may increase the probability of developing cancer, renal or liver illness, damage to the bone marrow, and other conditions due to polyaromatic hydrocarbons (PAHs) and heavy metals (Mishra et al. 2021; Mishra and Meikap 2023). Thus, these WMO are one of the mainstream groups of environmental pollutants with the highest risk to human health and the environment, demanding sustainable management (Mishra and Meikap 2024).

11.4.4 Biodegradability of Oils/Lubricants

As the industry grows more ecologically conscious, tremendous gains have been achieved in lubricant management and disposal. Lubricants are being developed to comply with even the most rigorous environmental requirements. Lubricant biodegradability is frequently evaluated and used as a reference when choosing a lubricant for a particular environment (Paul et al. 2024; Priyadarsini & Das 2024, 2025; Ray et al. 2024; Tripathy et al. 2022). A lubricant's biodegradability relates to how quickly it may be transformed to carbon dioxide and water by naturally occurring microbes. The primary biodegraded quantity range for vegetable oils, polyols and diesters, white oils, minerals, PAG, PAO, and Polyether are 70–100, 55–100, 25–45, 15–35, 10–20, 5–30, and 0–25% respectively. The expanding market for biodegradable lubricants is expected to provide a better opportunity shortly.

11.5 Hazardous Waste Management in India

According to the Hazardous Wastes (Management, Handling and Transboundary Movement) Rules, 2008 "*Recycling' is defined as reclamation or reprocessing of hazardous waste in an environmentally sound manner for the original purpose or for other purposes. 'reuse' means the use of a hazardous waste for a purpose*

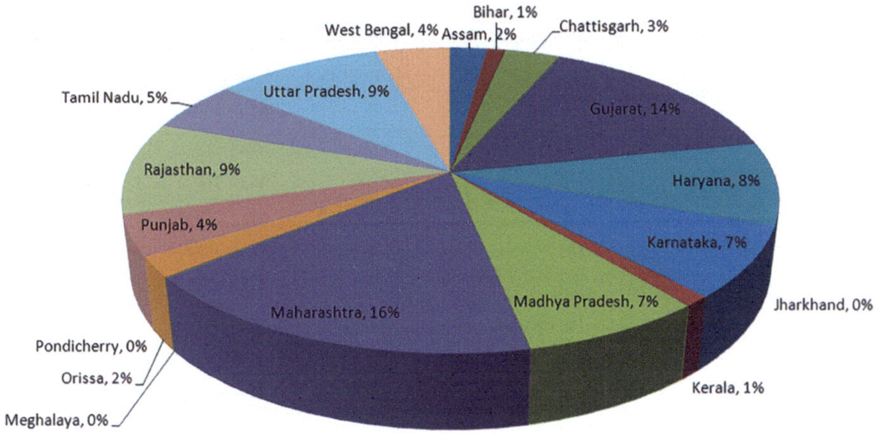

Fig. 11.2 State-by-state percentage share of total recyclers

of its original use or other use." The Rules outline the obligation of any 'occupier' of hazardous waste for the safe and ecologically sound treatment of environmental waste. Hazardous waste stored by a 'occupier' must be transported or sold to a recycler or reprocessor authorized to dispose of it properly. Furthermore, the holder responsible for hazardous waste must take all necessary precautions to control contaminating chemicals and prevent adverse effects on human health and the environment. The state-by-state percentage share of total recyclers is illustrated in Fig. 11.2, with the highest recyclers percentage in Maharashtra, followed by Gujarat. According to HWM Rules, 2016, for the management of hazardous and other wastes, an occupier shall follow the steps in hierarchy (i) prevention, (ii) minimization, (iii) reuse, (iv) recycling, (v) recovery, utilization including co-processing, and then (vi) safe disposal.

11.6 Environmentally Sound Management of Used/Waste Oil

In 2008, approximately 5.6 million metric tonnes of waste lubricating oils (WOs) were produced in the European Union (EU). According to the European Waste Catalogue (code 13), these WOs are hazardous and may cause ecotoxic, damaging, and carcinogenic consequences. These WOs have been categorized as hazardous waste, requiring correct and safe treatment. The rule on waste oil control was issued on 21 January 2004 in the Official Gazette under the number 25353, and it follows European Directive 75/439/EEC on waste oil disposal. The rule demands appropriate waste oil management throughout the recycling process and forbids unauthorized disposal. The categories are structured in a hierarchical structure and are dependent

Table 11.1 Waste oil categories (MoEF 2008)

Pollutants (ppm)	Limit values		
	Category I	Category II	Category III
As	< 5	Max 5	> 5
Cd	< 2	Max 2	> 2
Cr	< 10	Max 10	> 10
Cl^-	Max 200	Max 2000	> 2000
Pb	< 100	Max 100	> 100
TX	Max 200	Max 2000	> 2000
PCBs	Max 10	Max 50	> 50
Flash point	Min 38 °C	Min 38 °C	–
Application	**Recycling**	**Recovery**	**Disposal**

on whether or not the oil is polluted with chemical contaminants. The pollutants and category limit values are shown in Table 11.1.

11.7 Importance of Used/Waste Oil Recycling

A wide variety of waste oils can be recycled and recovered in several methods, either immediately or after separation and refining. The first alternative, according to the waste management hierarchy (Fig. 11.3), is to preserve the original properties of the oil, allowing for direct reuse. Other possibilities include recovering its heating value and reusing it in lower-level applications. Certain waste oils, particularly lubricants, can be recycled and reused directly (Diphare et al. 2013). The disposal and recycling of waste lubricating oils has grown significantly due to the growing need for environmental protection and increasingly strict environmental laws. Recycling waste lubricating oils may be achieved in three primary ways: reprocessing, re-refining, and incineration (Mishra et al. 2022a, 2021), as shown in Table 11.2. Other common/generic treatment procedures for old lubricating oil include acid-clay, vacuum distillation (VD), and solvent extraction (SE).

11.8 Industrial Recycling Processes

A wide range of physical and chemical techniques for regenerating, re-refining, and processing industrial and automotive lubricants have been developed. Some of these treatments aim to improve the oil to be used in the same equipment again. Many industrial enterprises that use a large amount of lubricating oil have their own purification and refurbishment systems. The advantages and disadvantages of some common recycling methods are tabulated in Table 11.3.

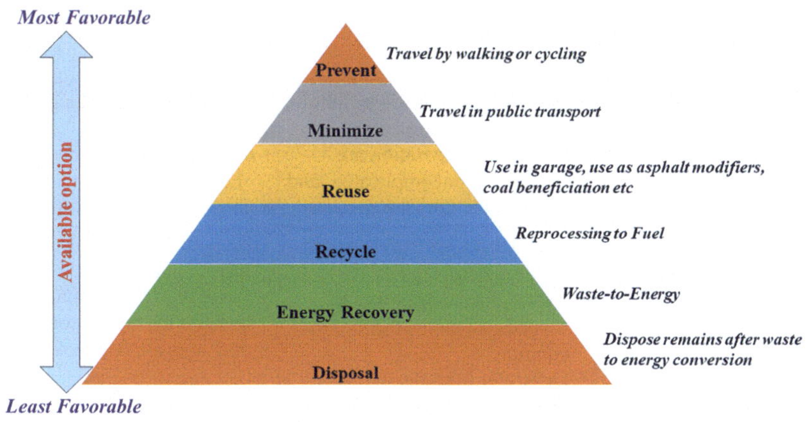

Fig. 11.3 WMO hierarchy

Table 11.2 Advantages and disadvantages of waste oil management techniques (Diphare et al. 2013)

Technique	Advantage	Disadvantage
Re-refining	• Long-term environmentally sound solution • Reduces the quantity of lubricating oil imported	• A well-developed collection mechanism is required • A well-developed market is required for re-refined lubricant oil • Extensive capital investment is required • The re-refining method necessitates using a professional recycling firm to assure the product's marketability • Proper disposal of end-of-life waste leftovers is expensive
Reprocessing	• An excellent alternative for low-grade fuels • Reduces the harmful impacts of the practise of uncontrolled waste burning • The purchaser monitors the quality control of the reprocessed fuel oil	• A well-developed collection mechanism is required • Extensive capital investment is required • Proper disposal of end-of-life waste leftovers is expensive
Incineration	• Financially viable at lower processing volumes • Cement plants are eager to acquire waste oil • Requires less capital than the preceding choices • Concentrates waste oil disposal to fewer, more readily monitored and supervised places	• Air emissions, although minor, must nonetheless be handled • Opposition from regulatory and government agencies

Table 11.3 Advantages and Disadvantages of some recycling methods

Recycling methods	Advantages	Disadvantages
Acid-clay (Udonne and Bakare 2013)	• It is a widely utilised technique • Lower production costs • Process and operation are simple and uncomplicated, and trained operators are necessary	• It creates waste, which causes pollution • Corrosive to equipment and shortens its life • Insufficient oil recovery
Vacuum distillation (Hamawand et al. 2013)	• Appropriate for high-capacity plants • It does not cause pollution • It needs sophisticated equipment • Generates high-quality base oils	• Requires a significant capital investment • Because of the advanced technology, highly competent operators are required
Solvent extraction (Oladimeji et al. 2018)	• Solvent is a recyclable material • It does not cause pollution • High recovery of base oil	• Only economical for high-capacity plants • It necessitates the use of highly qualified operators

Acid Clay Process: The acid clay technique has dominated the re-refining business. The majority of the time, this procedure is utilized to recover car oils. According to an assessment of the various recycling procedures, the acid-clay recycling approach poses the most significant environmental danger while being the least expensive. The method involves treating used/waste oil with acid and clay. The clay absorbs the odour and dark colour by acting as an adsorbent. The following features distinguish acid-clay from other methods: simple technique, minimal capital investment, low operational cost, and no need for trained operators (Udonne and Bakare 2013).

Vacuum Distillation: The essential procedures in Vacuum Distillation include pretreatment of oil to eliminate contaminants that might cause fouling and corrosion of the equipment, distillation to separate water and light hydrocarbons, and vacuum distillation utilizing a thin film or a typical vacuum column. Following vacuum distillation, the distillate is hydrotreated at high pressure and temperature in the presence of a catalyst to remove chlorine, sulphur, nitrogen, and organic components. Under high vacuum, hydrotreated oil is further fractionated into industrial, hydraulic, and motor oil components. The residue from the vacuum distillation procedure can be utilized to make road and roof bitumen (Diphare et al. 2013).

Solvent Extraction: Selective extraction can be used to separate particular chemicals. However, the characteristics of the produced oil are frequently changed. It is goal is to save natural resources while also recovering lubricating oil, making it a more appealing cleaner technology. There have been a variety of solvents studied for the regeneration of Waste Oils using liquid extraction for the removal of contaminants and parameters proposed for the choice of solvent or solvents mixture (Udonne and Bakare 2013). Still, no proposals that provide information on determining the optimal mass ratio of solvents for the maximum removal of pollutants have been made.

11.9 Modern Technology for the Treatment of WMO

Membrane technology employs different polymer hollow fibre membranes to treat waste lubricating oils through filtering in order to remove metal particles and carbon soot contained in the waste oil and restore some of the oil's lubricating characteristics (Das, A. Prasad., & Ghosh, Shreya. (2023)). Polyethersulfone (PES), polyvinylidene fluoride (PVDF), and polyacrylonitrile (PAN) are examples of polymers used in membrane filters (Lam et al. 2016; Oladimeji et al. 2020).

Another modern option is the pyrolysis process, which has recently been applied as an alternate method of effectively converting old lubricants to refined ones (Mishra et al. 2022b, 2020). Pyrolysis is a thermal degradation process that heats and decomposes substances at high temperatures (300–1000 °C) in an inert, oxygen-free atmosphere. The pyrolysis process is not yet widely used, but it is gaining popularity because of its ability to manufacture energy-dense goods from waste materials.

References

Bala K, Anshul, Verma S, Dogra A, Das AP, Kumar R, Nandini (2024) Sustainable management for urban plastic waste generation. In: Sustainable Management of Urban Plastic Waste Through Circular Economic Approaches, pp 211–236. https://doi.org/10.1201/9781003465447-13

Bala K, Dogra A, Lata P, Das AP (2024) Discussing sources and biological-chemical remediation approaches for microplastic pollution. In: Sustainable Microbial Technology for Synthetic and Cellulosic Microfiber Bioremediation, pp 37–59. https://doi.org/10.1007/978-3-031-62600-5_3

Behera ID, Basak G, Kumar RR, Sen R, Meikap BC (2020) Treatment of petroleum refinery sludge by petroleum degrading bacterium *Stenotrophomonas pavanii* IRB19 as an efficient novel technology. J Environ Sci Health Part A 56(2):226–239. https://doi.org/10.1080/10934529.2020.1866924

Behera ID, Nayak M, Biswas S, Meikap BC, Sen R (2021) Enhanced biodegradation of total petroleum hydrocarbons by implementing a novel two-step bioaugmentation strategy using indigenous bacterial consortium. J Environ Manage 292:112746. https://doi.org/10.1016/j.jenvman.2021.112746

Das A, Dey S, Bala K, Gnanasekaran R, Das AP (2024) Microbial bioremediation technology for sustainable treatment and management of synthetic microfiber waste. In: Sustainable Microbial Technology for Synthetic and Cellulosic Microfiber Bioremediation, pp 221–238. https://doi.org/10.1007/978-3-031-62600-5_12

Das A, Dey S, Das AP (2024) Ecotoxicological impacts of synthetic microfiber pollutants and development of sustainable mitigation strategies. Environ Chem Ecotoxicol. https://doi.org/10.1016/J.ENCECO.2024.12.003

Das AP, Dutta K, Khatun R, Behera ID, Singh S, Mishra S (2023) Microfiber pollution and its microbial mitigation: a review on current trends and future prospects. J Taiwan Inst Chem Eng 166:105104. https://doi.org/10.1016/j.jtice.2023.105104

Das AP, Mohanty AK, van Hullebusch ED, Figueiredo G, Das AP, Mohanty AK, van Hullebusch ED, Figueiredo G (2023b) Bioremediation as an emerging technology for the removal of synthetic microplastic pollutants from marine ecosystem. MarPB 194:115297. https://doi.org/10.1016/J.MARPOLBUL.2023.115297

Das AP, Nayak B, Behera ID (2024) Applications of microorganisms in advanced geomicrobiological and biotechnological exploration. Geomicrobiol J 41(4):309–311. https://doi.org/10.1080/01490451.2024.2327059

Das A. Prasad, Ghosh, Shreya (2023) Manganese mining microorganisms

Dey S, Bissoyi A, Koteswara Reddy G, Das AP (2024) Synthetic microfibers: An urban emerging pollutant demanding for sustainable environmental solutions. In: Sustainable Microbial Technology for Synthetic and Cellulosic Microfiber Bioremediation, pp 107–121. https://doi.org/10.1007/978-3-031-62600-5_7

Dey S, Dash N, Ray SK, Priyadarsini S, Chhotaray C, Das AP (2024) Occurrence and quantification of tire wear plastics pollutants from highway city dust of Bhubaneswar City of Odisha. In: Sustainable Microbial Technology for Synthetic and Cellulosic Microfiber Bioremediation, pp 77–89. https://doi.org/10.1007/978-3-031-62600-5_5

Dey S, Tripathy B, Kumar MS, Das AP (2023) Ecotoxicological consequences of manganese mining pollutants and their biological remediation. Environ Chem Ecotoxicol. https://doi.org/10.1016/J.ENCECO.2023.01.001

Diphare MJ, Pilusa J, Muzenda E, Mollagee M (2013) A review of waste lubricating grease management. In: 2nd International Conference on Environment, Agriculture and Food Sciences, pp 131–134

El-Fadel M, Khoury R (2001) Strategies for vehicle waste-oil management: a case study. Resour Conserv Recycl 33:75–91. https://doi.org/10.1016/S0921-3449(01)00058-1

Hamawand I, Yusaf T, Rafat S (2013) Recycling of waste engine oils using a new washing agent. Energies 6:1023–1049. https://doi.org/10.3390/en6021023

John A, Das AP (2024) Government strategies for achieving circular economy for plastic waste management in Southeast Asia. In: Sustainable Management of Urban Plastic Waste Through Circular Economic Approaches, pp 72–82. https://doi.org/10.1201/9781003465447-6/GOVERNMENT-STRATEGIES-ACHIEVING-CIRCULAR-ECONOMY-PLASTIC-WASTE-MANAGEMENT-SOUTHEAST-ASIA-AKBAR-JOHN-ALOKPRASAD-DAS

Kanungo J, Sahoo T, Swain LP, Behera ID (2023) Toxicity of persistent hydrocarbon pollutants, sources and sustainable remediation process. In: Behera ID, Das AP (eds) Impact of petroleum waste on environmental pollution and its sustainable management through circular economy. Springer Nature Switzerland, Cham, pp 39–65

Lam SS, Liew RK, Jusoh A, Chong CT, Ani FN, Chase HA (2016) Progress in waste oil to sustainable energy, with emphasis on pyrolysis techniques. Renew Sustain Energy Rev 53:741–753. https://doi.org/10.1016/j.rser.2015.09.005

Mishra S, Dash D, Das AP (2024) Microbial degradation of polyester microfibers using indigenously isolated bacterial strain Exiguobacterium Sp. CLEAN – Soil, Air, Water, 52(12):e202300343. https://doi.org/10.1002/CLEN.202300343

Mishra A, Kumari U, Turlapati VY, Siddiqi H, Meikap BC (2020) Extensive thermogravimetric and thermo-kinetic study of waste motor oil based on iso-conversional methods. Energy Convers Manag 221:113–194. https://doi.org/10.1016/j.enconman.2020.113194

Mishra A, Meikap BC (2024) Optimization and screening of process parameters for the robust co-pyrolytic study of waste motor oil and rice stubble toward sustainable waste-to-fuel generation. Chemosphere. https://doi.org/10.1016/j.chemosphere.2024.141450

Mishra A, Meikap BC (2023) Optimization of process parameters for waste motor oil pyrolysis towards sustainable waste-to-energy utilizing a combinatorial approach of response surface methodology and desirability criteria. Fuel 353:129226

Mishra A, Siddiqi H, Kumari U, Behera ID, Mukherjee S, Meikap BC (2021) Pyrolysis of waste lubricating oil/waste motor oil to generate high-grade fuel oil: a comprehensive review. Renew Sustain Energy Rev 150:111446. https://doi.org/10.1016/j.rser.2021.111446

Mishra A, Siddiqi H, Meikap B (2022a) Elucidating sustainable waste management approaches along with waste-to-energy pathways: a critical review. In: Energy from Waste: Production and Storage. CRC Press, Boca Raton, p 83. https://doi.org/10.1201/9781003178354-8

Mishra A, Sonowal M, Turlapati VY, Maiti P, Meikap BC (2022b) A comprehensive thermo-kinetics devolatilization analysis of waste motor oil: thermal degradation kinetics, kinetic model, thermodynamic analysis, and ANN. Int J Green Energy 00:1–13. https://doi.org/10.1080/15435075.2022.2155825

Mohanta S, Pradhan B, Behera ID (2024) Impact and remediation of petroleum hydrocarbon pollutants on agricultural land: a review. Geomicrobiol J 41(4):345–359. https://doi.org/10.1080/01490451.2023.2243925

Mohanty M, Tripathy B, Koteswara Reddy G, Das AP (2024) Greener technologies for establishing restraint over microplastic pollution. In: Sustainable Microbial Technology for Synthetic and Cellulosic Microfiber Bioremediation, pp 239–258. https://doi.org/10.1007/978-3-031-62600-5_13

Moodley P, Lokhat D, Ahmad A, Mishra A, Meikap BC (2022) Kinetic studies on the potential use of citrus-based green and low-cost demulsifying agents for the oil-in-water emulsions ‘ treatment. J Environ Chem Eng 10:107127. https://doi.org/10.1016/j.jece.2021.107127

Oladimeji TE, Oguntuashe KM, Emetere ME, Efeovbokhan VE, Odunlami OA, Obanla OR (2020) Industrial- and automotive-used lubricating oils recycling cum acidic sludge treatment. Int J Adv Manuf Technol 106:4157–4167. https://doi.org/10.1007/s00170-019-04751-6

Paul A, Priyadarsini S, Karakuş S, Behera JK, Gnanasekaran R, Das AP (2024) Fate and accumulation of synthetic microfibers in the environment. In: Sustainable Microbial Technology for Synthetic and Cellulosic Microfiber Bioremediation, pp 61–76. https://doi.org/10.1007/978-3-031-62600-5_4

Priyadarsini S, Das AP (2024) Analytical and structural characterization of waste lithium-ion batteries for their effective recycling strategy. Environ Chem Ecotoxicol. https://doi.org/10.1016/j.enceco.2024.12.004

Priyadarsini S, Das AP (2025) Lithium bioleaching: A review on microbial-assisted sustainable technology for lithium bio-circularity. J Water Process Eng, 69:106744. https://doi.org/10.1016/j.jwpe.2024.106744

Ray SK, Dey S, Das AP, John A (2024) Microplastic waste management through greener technologies: A review. In: Sustainable Management of Urban Plastic Waste Through Circular Economic Approaches, pp 237–261. https://doi.org/10.1201/9781003465447-14

Oladimeji TE, Sonibare JA, Omoleye JA, Emetere ME, Elehinafe FB (2018) A review on treatment methods of used lubricating oil. Int J Civ Eng Technol 9:506–514

Tripathy B, Dash A, Das AP (2022) Detection of environmental microfiber pollutants through vibrational spectroscopic techniques: Recent advances of environmental monitoring and future prospects. Crit Rev Anal Chem, pp 1–11. https://doi.org/10.1080/10408347.2022.2144994

Tripathy B, Sahoo PP, Sundaray H, Das AP (2024) In-vitro biodegradation of discarded marine microplastics across the eastern coast of the Bay of Bengal, India using *Exiguobacterium* sp. Environ Chem Ecotoxicol 6:236–247. https://doi.org/10.1016/j.enceco.2024.07.003

Udonne JD, Bakare OA (2013) Recycling of used lubricating oil using three samples of acids and clay as a method of treatment. Int Arch Appl Sci Technol 4:8–14

Chapter 12
Bioremediation Strategies Against Hydrocarbon Wastes and Synthesis of Novel Recycled Products

Snigdha Ghosh, Snehanjana Patra, and Priyanka Jha

Abstract Petrochemical industries, extraction of crude oil, automobiles, mining, oil spills, drill cuttings and other metallurgy based industries release enormous hazardous solid as well as liquid hydrocarbon containing wastes in our environment. Such pollutants (flue gas, oil sludge, refinery sludge, effluents) are detrimental to our natural resources including vast flora and fauna, microbial community and human beings. Different physical and chemical methods are conducted for the remediation and restoration of hydrocarbon-contaminated regions, yet, they aren't eco-friendlier and cost-effective. Bioremediation technique utilizes (microbial species as bacteria, fungi, algae and also plants) in order to remove total petroleum waste hydrocarbons from water resources, soil and even air. Bio-removal of hydrocarbon waste residues is quite promising towards its sustainability, eco-friendliness, and it also stands ecologically as well as economically efficient. This paper highlights the productiveness of bioremediation and its corresponding advantages, group of diverse microflora and plants that are beneficial for absorbing hydrocarbon pollutants, rehabilitation of petrochemical polluted sites and recycled products that can be reused further.

Keywords Sustainability · Fossil fuels · Bioremediation · Petroleum oil sludge · Hydrocarbons · Environment

---

S. Ghosh · S. Patra
Amity Institute of Biotechnology, Amity University, Kolkata Plot No: 36, 37 and 38, Major Arterial Road, Action Area II, Kadampukur Village, Rajarhat, Newtown, Kolkata, West Bengal 700135, India

P. Jha (✉)
Department of Biotechnology, School of Bioengineering and Biosciences, Lovely Professional University, Phagwara, Punjab 144411, India
e-mail: priyanka.bt.jha@gmail.com

I. D. Behera and A. P. Das (eds.), *Environmental Hydrocarbon Pollution and Zero Waste Approach Towards a Sustainable Waste Management*, Environmental Science and Engineering, https://doi.org/10.1007/978-3-031-81615-4_12

12.1 Introduction

Nowadays, 'pollution' and its wide range of origination followed by devastating consequences which is witnessed by our planet is neither unknown nor neglected (Mishra et al. 2019, 2021a, b). This alarming situation has inflated the rage of developing emergency precautions and a sense of high-time importance. In order to shield our green planet as well as our future generations from the lap of destruction and concurrent disease progression, researchers are critically in search of novel products and techniques to outcast environmental pollutants like hydrocarbon waste, microfiber (Das et al. 2023; Mishra and Das 2021), heavy metals (Das et al. 2011; Raj and Das 2023), etc. Among these, hydrocarbon wastes are the most obnoxious substances present in our surroundings, components of our ecosystem (air, water, and soil), between food chains, and last but not least, hampering human health and other living species (Behera et al. 2020; Lahiri et al. 2021).

Organic compounds like hydrocarbons are exclusively built with hydrogen and carbon molecules that are generally found in fossil fuels (coal, petroleum, natural gas) and their corresponding derivatives. Whether it be aliphatic or aromatic hydrocarbon compounds, burning of such non-renewable energy sources emits PAHs, PCBs, and greenhouse gases too. Most prominently, PAHs (Polycyclic aromatic hydrocarbons), PCBs (Polychlorinated biphenyls), and PCDD/FS (Polychlorinated dibenzo-p-dioxins and dibenzofurans) are released by various thermal power stations, oil refineries plants, petrochemical-based industries, pesticide and fertilizer manufacturing factories, automobiles, etc. Besides this, oil spill accidents or chronic leakages of land and marine pipelines, fuel tanks, domestic and industrial waste effluents, partial combustion of fuels, and wood burning are also some of the major reasons for adding hydrocarbon contents in the atmosphere, marine systems and soil sediments (Bala, Anshul et al. 2024; Bala, Dogra et al. 2024; A. Das, Dey, & Das 2024).

Crude oil, along with other petroleum-oriented products like—diesel (n-paraffins, isoparaffins, naphthenes, and aromatics), kerosene, gasoline, lubricants, and grease, falls under liquid hydrocarbons. Sludges, incinerator ash, bottom ash, boiler slag, fly ash, flue gases comes under hydrocarbon wastes, that are equally relevant to hydrocarbon pollution (Haider et al. 2021).

Post-leakage of oil tankers and pipelines while transportation and oil spillage over soil and oceanic regions extensively reduces the survival rate of marine species. In case of oil contaminated soil or hydrocarbon efflunts released from petrochemical industries, seeps into the lower layers of soil, resulting to low seed germination and nutrients translocation, induced oxidative stress, disturbed plant metabolic activity (A. Das, Dey, Bala et al. 2024; Dey, Bissoyi et al. 2024; Dey, Dash et al. 2024; Dey et al. 2023; John & Das 2024). Ultimately leading to dimished plant yield and survival (Haider et al. 2021). However, the ejection of wastewater from chemical manufacturing units directly to nearby rivers or lakes also leads to the same repercussions. Coming to human lives, they deal with multiple types of carcinomas (liver, skin, kidney etc.), lung and respiratory sickness along with cardiovascular, kidney as well as liver disorders, when such hydrocarbon toxicants are inhaled, consumed or comes in dermal contact (Mishra et al. 2024; Mohanty et al. 2024; Paul et al. 2024; Priyadarsini & Das 2024, 2025).

In reponse to such detrimental outcomes, different physical (excavation of contaminated soil, flotation, microwave-assisted heating), chemical (photocatalytic oxidation, soil flushing, usage of synthetic surfactants or detergents), and bioremediation techniques are already into action for complete or maximum eradication of hydrocarbon pollutants. On that note, bioremediative strategies by the incorporation of selective microbial strains and plant species are quiet sustainable, eco-friendly, efficient and cost-effective in their mode of quenching those toxic particles. Moreover, recycling of waste derived from coal gasification and petroleum oil sludges, proves to be a beneficial alternative to fossil resources combined with lesser hydrocarbon footprint.

This chapter highlights fundamental concepts of hydrocarbon pollution, affecting ecological balance and living entities. It also sheds light upon mechanism related to various hydrocarbon reduction strategies and their advantages, especially bioremoval procedures by uplifting its sustainability standard. Additionally, it also showcases novel recycled products and their efficacy as a substitute to fossil-based energy sources.

12.2 Sources of Hydrocarbon Waste Substances

12.2.1 Natural

Increased anthropogenic activities, rapid industrialisation and urbanization are the prime reasons for global warming as well as frequent natural disasters to take place. Forest fires significantly start due to high atmospheric temperature and dryness, and as per estimation, between $(530 - 555 \times 106)$ hectare of land, which is involved with fires annually on a global scale. It also depends on temperature, humidity, nature and quantity of dead fuel, wind speed, and topography of the area (Faboya et al. 2020). Carbon dioxide emissions along with PAHs, polychlorinated naphthalenes (PCNs), and polychlorinated dibenzo-p-dioxins/dibenzofurans (PCDD/Fs) are produced from biomass dissociation during fire combustion. Likewise, traces of persistent organic pollutants (POPs) and polycyclic aromatic hydrocarbons (PAHs), volatile organic compounds (VOCs), NOx (nitrogen oxides), CO (carbon monoxide), SO_2 (sulphur dioxide), are also emitted or re-evaporated during forest fires (Gong and Wang 2021).

On the other hand, mud volcanoes are superficial geological phenomenon in which a complex mixture of gas, liquid and solid material is continuously or explosively evolved due to pressurized fluids at depth. They are variable in shape and size located both in sub-aerial and sub-marine environments. It liberates methane, carbon-dioxide, PAHs and other higher hydrocarbons into the atmosphere hence, it alters the soil, marine ecosystem and air quality. Such volcanic events are controlled by tectonic activity (compression or seismicity), sedimentary or tectonic loading (rapid sedimentation or overthrusting), the presence of substantial thicknesses of fine-grained plastic sediments, and continuous hydrocarbon generation.

12.2.2 Anthropogenic

12.2.2.1 Petroleum Oil Spill

Total Petroleum Hydrocarbon (TPH) or the crude oil accounts for 50–98% of carbon, hydrogen, oxygen, nitrogen, sulphur, heavy metals (mercury, arsenic, cadmium, chromium etc.) blend consisting diverse alkanes, cycloalkanes, aromatics, polyaromatic hydrocarbons (PAHs) and other compounds. Their toxicity towards environment and carcinogenic properties makes them the most outragious source of hydrocarbon pollution around the globe.

Loading and discharging of crude oil, routine operations, transportation, oil processing and refining events deliberatly releases aromatic organic compounds and phenolic substances that are barely degradable by nature. Some refined petroleum products (petrol, gasoline, diesel, jet fuel, asphalt, wax, lubricating oil, tar, kerosene, and liquefied petroleum gas) and petrochemical intermediates (solvents, dye stuffs, pharmaceuticals, polymers) are also quiet common in discharging hydrocarbon material.

Moreover, oil spill accidents due to malfunctioning of ship operations, poor maintainance, transportation and shipment of oil, offshore oil exploration, corossion, ruptures of tanks, leakages of underwater pipes, fire and expolsion are well noticed. North-eastern USA, the Gulf of Mexico, and the Mediterranean Sea are regarded as the hotspots for such accident to occur. Aftereffects of oil spillage gets more worse when spilled oil reaches coastal lines and seashore regions, damaging soil productivity, food security, marine life and species (Asif et al. 2022).

12.2.2.2 Combustion of Fossil Fuels

For extensive power generation and balance the supply–demand chain, coal is mostly preferred than any other substance. Fossils (coal, natural gas, petroleum and its products like gasoline as well as diesel fuel) are currently used for electricity prodution, heating purposes, transportation etc. Combustion of coal in thermal power plants, releases carbon dioxide (CO_2) and carbon monoxide (CO), sulfur dioxide (SO_2) and sulfur trioxide (SO_3), and nitrogen dioxide (NO_2) and nitric oxide (NO), respectively. Similarly, coal-fired plants generates flue gas containing (carbon dioxide, sulfur dioxide, nitrogen dioxide, fly ash (PM), and heavy metals), while oil refineries produces NOx, CO, CO_2, SOx, and total hydrocarbon in the environment.

Black carbon, polycyclic aromatic hydrocarbons (PAH), volatile organic chemicals that form ground-level ozone (O_3), phenols, and aliphatic compounds are also emitted to the atmosphere as a result of incomplete combustion of coal or coal-biomass fuel mixtures that are usually added to run coal stoves and coal-fired power plants (Czaplicka et al. 2021). In an experiment conducted by Marković et al. (2023) showcased particulate matter, trace elements (except Cd and Pb), benzo[a]pyrene, and benzo[b]fluoranthene levels to exceed the upper limit standards. Along with a comparison of PAHs produced through fly ashes, when combustion of lignite and coal waste is conducted in thermal power plants and semi-industrial fluidized bed boilers.

12.2.2.3 Chemical Industries

Pesticides as well as fertilizers enhance agricultural production, growth of cultivated crops by maximizing their yield and defensing from pests and microorganisms for further diseases. Consequently, excessive usage of such chemical compounds causes serious environmental impact and health concerns. Pesticide manufacturing factories produces variety of solvents comprising VOCs, that are profusely ejected through wastewater or in gaseous form. On top of that, organic and inorganic raw materials, organic solvents, and catalysts are used to synthesize pesticides. Aromatic hydrocarbons, esters, ethers, ketones, and aldehydes, are some of them. As a result, PAHs are also liberated into the atmosphere as shown by Wang et al. (2023) in their experimental findings. Pesticide industries ejecting 52 such $PM_{2.5}$-bound PAHs during the daytime and nighttime, especially high molecuar weight PAHs, was observed. These toxic substances can be found in several food products (pineapple, cucumber, catfish, honey, herbs, potato, tomato etc.) as analysed by GC–MS/MS and other techniques to enumerate various classes of PAHs, PCBs pollutants (Borowiec et al. 2022).

12.2.2.4 Automobile Exhaust

Transportation and travelling is vital for industrial supply chain and collection of raw materials, logistics and exchange of goods and services, home-based deliveries related to medicine, food, clothes etc., and last but not the least for personal use and visiting tourist spots. Diesel or petrol-running vehicles's exhausts emit black smoke and other air-polluting substances like Carbon Monoxide (CO) and/or Particulate Matter (PM), Hydrocarbons (HC), and Carbon Dioxide (CO_2), respectively. Incomplete combustion of fuels results to elevated fuel consumption and degraded engine performance. Therefore, releasing soot, NOx, green house gases and other ultrafine particles directly to air. Lung diseases and their corresponding respiratory tract infections, cancers, asthma etc., are common these days.

12.2.2.5 Incineration of Wastes

Municipal solid wastes, plastics and polythene debris, and crop residue, are a few examples that are burned inside furnaces and in the open air. Despite the fact that, crop waste residues can be used as feedstock yet, those are incinerated to get rid off. Lack of education, awareness and technological advancements leads to such practices.

While agicultural waste residues combustion process, it eliminates volatile organic compounds (VOCs), carbon monoxide, hydrocarbons (PAHs). Aside from greenhouse gasses (CH_4, NO_2, CO_2), air contaminants (non-methane hydrocarbons, smoke, SOx and NOx are also given out. Cheng et al. (2022) observed polycyclic aromatic hydrocarbons (PAHs) exposed to air, and PM2.5 samples were collected from the open burning of different types of municipal solid waste (MSW), especially wood and bamboo wastes, to generate the highest toxicants.

Industrial waste incinerator plants also releases Non-ortho PCBs and mono-ortho PCBs. However, drinking water or wastewater chlorination, chlorine bleaching of cellulose pulp, pigment or dye production, are also some sources (Gabryszewska and Gworek 2020). In addition, PCDD/Fs (polychlorinated dibenzo-p-dioxins and furans) dl-PCBs (dioxin-like polychlorinated biphenyls), chlorophenols (CPs), and chlorobenzenes (CBz) in the municipal waste incinerator emmisions are also visible (Hsu et al. 2021) (Fig. 12.1).

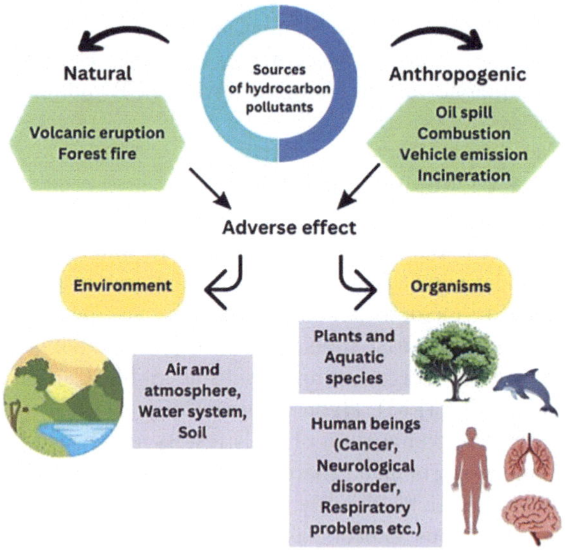

Fig. 12.1 Contributors to hydrocarbon waste and their impact on environment and different species

12.3 Range of Hydrocarbon Pollutants

12.3.1 PAHs

Polycyclic aromatic hydrocarbons (PAHs) are basically colourless or pale yellow, posessing high melting and boiling points (therefore they are solid), low vapor pressure, and very low aqueous solubility. These results to increased lipophilic organic compounds with two or more single/fused aromatic rings. Having upto six aromatic rings can be called as ‘small PAHs’ whereas, more than six as ‘large PAHs’ respectively. Besides that, light sensitivity, heat resistance, conductivity, corrosion resistance, and UV absorbance spectrum are also showcased as general characteristics of PAHs. PAHs can also be distinguished as low molecular weight (LMW) if they have two or three fused rings, and as high molecular weight (HMW) with four or more fused rings. Generally, low molecular PAHs are more susceptible to degradation and volatilisation. Furthermore, hydrophobicity/lipophilicity is directly proportional to molecular weight, on the other hand, its inversely proportional to water solubility and vapour pressure.

Tranformation of PAHs particles in environement produces containing oxygen (O-PAHs), nitrogen (N-PAHs and aazarenes AZA) or sulfur (PAHs) in the aromatic ring. Combination of oxygen, nitrogen and sulphur in aromatic rings elevates the toxicity of this particular hydrocarbon pollutant (Krzyszczak and Czech 2021). Such derivatives generate in two ways—either by direct combustion of fossil-based substances (primary emission) or by undergoing photochemical reactions (secondary emission) between PAHs and other entities (nitrogen oxides, sulfur oxides) or ozone in the atmosphere.

PAHs and its derivatives are persistant in environment causing adverse health conditions as they are mutagenic, teratogenic and genotoxic to humans. Infact, their deposition and presence in air, oceanic regions and sediments of soil builds a path for acute toxicity and interference between food chains. Later, it affects other aqautic as well as terrestrial species which is discussed below. Few common examples are (Naphthalene, Acenaphthylene, Acenaphthene, Anthracene, Phenanthrene, Fluorene, Fluoranthene, Pyrene, Benzo(a)-anthracene, Benzo(a)-pyrene and more) (Kim et al. 2013).

According to the list of 16 most dreadful PAHs compounds mentioned by The United States Environmental Protection Agency (USEPA), benzo [a] anthracene (BaA), benzo [a] pyrene (BaP), benzo [b] fluoranthene (BbF), benzo [k] fluoranthene (BkF), chrysene (Chr), dibenzo [a,h] anthracene (DBahA), and indeno [1,2,3-c,d] pyrene (IcP) are denoted as B2 class of carcinogens (Sankar et al. 2023). Moreover, the four major PAHs found in foods: benz[a]anthracene (BaA), chrysene (Chr), benzo[b]fluoranthene (BbF), and benzo[a]pyrene (BaP). Hence, they are denoted as mutagenic, cancerous and genotoxic by IARC (The International Agency for Research on Cancer), the Joint Food and Agriculture Organization/World Health Organization (FAO/WHO), Scientific Committee on Food (SCF), and European Food Safety Authority (EFSA (Sampaio et al. 2021).

12.3.2 PCBs

Polychlorinated Biphenyls (PCBs) can be classified as POPs (persistent organic pollutants) and around 1929, their commercial production started. With reference to Stockholm Convention, established by the United Nations Environment Programme in 2001, PCB-containing products and their further synthesis was aimed to completely shut down by 2028. They are another sort of aromatic compounds having biphenyl structure bonded to hydrogen and chlorine atoms according to the chemical formula $C_{12}H_xCl_y$, where x and y range from 1 to 10, and $x + y = 10$. IUPAC states, 209 different PCB congeners based on hydrogen substitution with chlorine atoms and their positioning on the biphenyl rings (Othman et al. 2022). They are light to deep yellow, oily liquid with insulating ability, thermal stability, resistance to acids, oxidation, hydrolysis, and flame resistance properties. Therefore, PCBs are suitable for acting as plasticizers in rubber and resins, carbonless copy paper, in adhesives, wax extenders, in dedusting agents, paints/inks, and also for capacitors and in transformers.

PCBs are lipophilic, posessing high boiling point, tolerant to alkali and acids, super thermal conductivity, yet poor electrical transmission is seen. Its solubility standards decreases with escalation of chlorine content. Apart from that, as the degree of chlorination increases, the viscosity of PCBs also increases (i.e., light yellow oily liquids and then waxy solids). Even at low temperatures, PCBs do not crystallize but turn into solid resins (Ododo and Wabalo 2019).

Moreover, oxidation of PCBs also generates Hydroxylated polychlorinated biphenyls (OH-PCBs), which causes inhibition of mitochondrial respiration, generation of reactive oxygen species, oxidative damage to DNA, and endocrine disrupting effect in humans as well as more damages to our ecosystem. Along with that, skin diseases, immunotoxic effects, hormonal imbalance, genotoxicity, reproductive disorders and more occurs to human when they are inhaled or ingested.

12.3.3 PCDD/Fs

(Polychlorinated-dibenzo-p-dioxins, PCDDs) and (polychlorinated-dibenzofurans, PCDFs), are together called as PCDD/Fs, which is another type of hydrocarbon pollutant arising from several natural as well as industrial processes. They are semi-volatile, aromatic hydrocarbons formed by a triple-ring structure of two benzene rings interconnected by a third oxygenated ring. As per their structural configuration and attachment of chlorine in positions (especially 2, 3, 7, 8), leads to the formation of different congeners (Dopico and Gómez 2015). A family of 210 congeners, are mainly released during combustion fumes and thermal processes involving carbon and chlorine. POP (persistant oganic pollutants) like PCDD/Fs also generates from older-generation municipal solid waste (MSW) incinerators, household or hospital incinerators, paper-pulp industries, chlorine bleaching processes, pesticide production, metal smelting, cement plants and disel-running vehicles (Vernez et al. 2023; Gonzále and Domingo 2021).

Settlement of these hydrocarbons over topsoil and further absorption by roots, leads to damaged crop yield, development and reproduction. Furthermore, sea foods (fishes, crabs, shrips), meat (beef, pork, chicken), eggs, cereals (rice), vegetables, fruits, dairy products containing PCDD/Fs levels disrupts the human dietary schedule, is also mentioned by Gonzále and Domingo (2021). It causes severe illness, including (growth retardation in fetuses and infants, neurodevelopmental dysfunction, thyroid deficiency, impaired immune system, reproductive disorders, and cancer. World Health Organisation (WHO) indicated PCDD/Fs has the potential to elevate the risk of breast cancer in women. Likewise, they are also related to liver, kidney, and pancreatic carcinomas in humans, because of their presence in such organ, as observed by scientific research techniques (Themba et al. 2023).

Besides PCDD/Fs, Polybrominated dibenzo-p-dioxins and dibenzofurans (PBDD/Fs) are quiet similar to PCDDs as per their physical structure. Talking about its chemical properties, PBDD/Fs are having higher molecular weights and melting points, lower water solubility, and lower vapor pressures. PBDDs are widely produced through industrial thermal processes, brominated flame retardant (BFR) products, e-waste dismantling, metal smelting processes, and waste incineration. Infact, PBDD/Fs proves to bioaccumulate more than PCDD/Fs (Vernez et al. 2023).

2,3,7,8-Tetrachlorodibenzo-p-dioxin (TCDD) is one of the congeners of PCDD/Fs which is extremely toxic and generates as the by-product while herbicide manufacturing takes place, such as 2,4-dichloroacetophenol and 2,4,5-trichloroacetophenol (Takiguchi et al. 2022). Back in 2012, The International Agency for Research on Cancer (IARC) already stated 2,3,7,8-TCDD and 2,3,4,7,8-PCDF as 'group 1 carcinogens', followed by respiratory tract cancers, prostate cancer, and certain types of lymphomas and sarcomas in humans (Vernez et al. 2023).

12.4 Impact of Hydrocarbons

12.4.1 Repercussions on the Environment

12.4.1.1 Soil

Contamination with petroleum hydrocarbons (PHs) degrades the soil quality which in turn proves to be inhabitable for micro-organisms in the soil ultimately affecting its biological variability. A case study on the Loess Plateau in China revealed that this plateau was abundant in oil and gas resources that were overburdened, which primarily resulted in environmental pollution. Upon analysis of the properties of the soil, it displayed increased levels of pollution according to European soil quality standards and lastly, an evaluation depicted the potential risks to cancer in some age groups, mostly women (Wang et al. 2020).

A recent research performed by Jia et al. (2023) confirms the adverse effects of petroleum hydrocarbon contamination. The study illustrates that increased concentrations of PHs in the soil were directly proportional to a decreased microbial community which in turn affected the ecological function. Several occurrences in the world are associated with the dumping of petroleum wastes in a free environment because of which they are a prevalent environmental contagion. PHs contamination in agricultural lands has shown a drastic change in the land's pH and nutritional value affecting the agricultural yield (Mohanta et al. 2023). Several studies have also reported PHs contamination via leakage where the agricultural lands were turning infertile due to leakage from garbage heaps.

Therefore, it is of paramount importance to control the further loss of land, especially agricultural lands. So, for the renewal of such lands, one should take on remediation strategies to control the widespread contamination of petroleum hydrocarbons.

12.4.1.2 Air

Reports since the 2000s have stated the extensive atmospheric hydrocarbon pollution that has led to climate change. Investigations have also reported that indoor air has a higher concentration of harmful volatile organic compounds.

Hydrocarbons are mainly responsible for a plethora of effects, including a reduction in air quality caused by the emission of volatile organic compounds (VOCs) into the atmosphere. These VOCs in turn react with other contaminants present in the air and can cause several adverse effects on human health (Ray et al. 2024; Tripathy et al. 2022; Das, A. Prasad., & Ghosh, Shreya 2023). A cohort study that took place in Taiwan aimed at understanding the association between hydrocarbons in the air and ischemic stroke (IS) and the findings of this study indicated that long-term exposure to hydrocarbons in the air like total hydrocarbons (THC) and nonmethane hydrocarbons (NHMC) posed a threat for ischemic stroke (Zhang et al. 2019).

Wang et al. (2011), in their study demonstrated that PAHs in the air accounted for total genotoxicity. When hydrocarbons are released into the air through processes like vehicle emissions they can contribute to air pollution with the percentages of PAHs derivatives being 8.23 and 27.68% (Krzyszczak and Czech 2021). Hydrocarbons, when released into the atmosphere, can also contribute towards global warming, and these hydrocarbons react with nitrogen oxides in the presence of sunlight to form ground-level ozone, which also has adverse effects on human health, especially children and the elderly.

12.4.1.3 Aquatic System

Worldwide, water resources have been polluted with petroleum hydrocarbons posing a threat to the marine ecosystem. Several sources like agrochemicals, petroleum and different toxic substances released into water streams contribute to increased hydrocarbon pollution. These PHs are classified as fractious pollutants in nature and are mostly released in water during the refining process and are considered to be neurologically harmful to the flora and fauna present underwater (Sun et al. 2020). Oil spill is the major cause of hydrocarbon contamination in water followed by the release of urban storm water. Car parks and roads are contaminated daily by fuel from vehicles and during heavy rains, these contaminants reach the water source polluting them. A study was conducted by Jabbar (2018) on the effect of oil contamination on the aquatic environment in four different locations in the Baher Al-Najaf Depression for a period of one year and the results demonstrated that the plants in the contaminated sites were highly affected and had an accumulation of heavy metals and it could be concluded that contamination by oil posed a threat to the biodiversity underwater.

Several studies (Mfarrej et al. 2022) have reported that the release of PHs contaminants has adversely affected groundwater and the ocean. The major reason contributing to permanent damage is bio-amplification due to the steady release of PHs. Many studies have also focused on the detrimental effects of petroleum hydrocarbons on marine biota. A study was conducted to demonstrate the severe impacts of petroleum hydrocarbons on blood clam, *Tegillarca granosa*. The results depicted that petroleum hydrocarbon contamination exerted notable immunotoxic impacts (Sun et al. 2020). A 2006 study also reported on the toxic effects where the contaminants drastically affected eggs in fish and their growth. Therefore, it is imperative to deploy waste removal techniques to control water pollution by petroleum hydrocarbons.

12.4.2 Repercussions on Living Organisms

12.4.2.1 Human Health

Hydrocarbons are the general category of environmental pollutants and during exploration, production, refining, transport and storage of petroleum and petroleum-based products, there are possibilities for leakage and these hydrocarbons are potentially hazardous to marine ecosystems and also share an impact on human health. These can be potentially toxic if they enter the food chain and can come up with toxic, mutagenic and cancerogenic effects (Abdullah et al. 2020). Many reports have shown the negative effects of hydrocarbons including polynuclear aromatic hydrocarbons (PAHs).

A study on various routes of exposure suggests that people are mainly exposed to PAHs in a few ways: by breathing in the air around them, which includes both indoors and outdoors, eating a portion of food that has PAHs in it, smoking tobacco or breathing in smoke from fires. Tobacco smoke has several PAHs that are considered hazardous to health. Some crops like wheat, rye or lentils can also contain PAH acquired through air, water, or soil. The studies were conclusive that some routes of exposure may entail more than one route simultaneously making the PAH exposure occur regularly (Abdel-Shafy and Mansour 2016).

The observation of PAHs is of particular concern because of their ubiquitous distribution and toxicity and certain PAHs carcinogenicity has been classified by the International Agency for Research on Cancer. Some examples of these include naphthalene, chrysene and benzo[b]fluoranthene (Group 2B). The primary health risk associated with inhalation exposure to these PAHs is an elevated likelihood of developing lung cancer (Kim et al. 2013).

A review on the impact of total petroleum hydrocarbon (TPHs) demonstrated the strong adverse effect these TPHs had on mental health which further induced physiological effects and were considered hazardous to the immune and endocrine systems and further research is required to understand the long-term effects of TPHs in the post-exposure period (Kuppusamy et al. 2019).

Short-term exposure to an increased level of hydrocarbons is also considered to be hazardous and several studies have presented the damaging effects of PAHs on the nervous system which mainly focuses on neurodegeneration and neurodevelopmental toxicity (Olasehinde and Olaniran 2022). Additionally, some adverse symptoms like reproductive toxicity and excess indications of various types of cancer ranging from skin to colon have been identified and have come to the centre stage in recent years (Table 12.1).

12.4.2.2 Plants

The extraction of petroleum hydrocarbons to ensure the increasing world population's fossil fuel demand has wreaked havoc on humans and the ecosystem. Contamination of the environment by petroleum hydrocarbons has been the prime issue. The development of civilization has put a lot of pressure on energy sources and with the increasing use of petroleum hydrocarbons (PHs), these can mix with organic substances, hampering the growth of plants. The use of refinery products has been linked to further deterioration of the physiochemical properties of the soil, which, therefore, cause a damaging effect on the plant growth and yield. Generally, the lipo-protein membrane of the plants accumulates PHs (Mukome et al. 2020) and increased concentrations of these PHs in the soil have been seen to lessen the rate of germination of different plant species and also cause severe deformities in the leaves that ultimately led to tissue and cell necrosis.

More studies suggest that PHs have the ability to tightly attach to proteins, DNA and RNA within the cells interfering with growth, and enzyme functions that lead to a rise in reactivity and an affinity for reacting with other molecules (Bao et al. 2020). The presence of petroleum hydrocarbons has contributed to a significant phytotoxicity and some studies have shown that when *Arabidopsis thaliana* L. Plants were exposed to PHs, there was an initiation of oxidative stress. Some recent research

Table 12.1 Diseases caused by toxic hydrocarbons affecting human health

Pollutants	Source	Targeted organ system	Effects on health	References
Benzene	Petroleum derivatives	• Respiratory • Blood	• Chronic swelling • Potentially carcinogenic • Genetic damage • Epigenetic changes	Cordiano et al. (2022)
Toluene	Partial combustion of liquid fuels	• Central nervous system • Respiratory	• Cognitive decline • Dilation of the ventricles	Cruz et al. (2014)
Polycyclic aromatic hydrocarbons (PAH)	Burning of coal, oil, wood, and garbage	• Immune system • Respiratory • Circulatory	• Lung cancer • Cardiovascular disease (CVD) • Embryo toxicity	Kim et al. (2013)
Polychlorinated biphenyls (PCBs)	Thermal treatments, worn out equipment that used PCBs	• Nervous system • Immune system • Circulatory system • Reproductive system	• Cognitive impairment • Decreased fertility • Cardiovascular disease	Montano et al. (2022)
Total petroleum hydrocarbons (TPHs)	Chemical industry, gasoline pumps	• Endocrine system • Nervous system	• Skin problems • Neurological disorders • Metabolic disorders	Kuppusamy et al. (2019)

has also highlighted *Oryza sativa* L. resistance to stress which has been induced by PHs (Odukoya et al. 2019).

Seed germination is the primary activity in the vegetation cycle, and according to research studies, the soil polluted with certain volatile PHs caused a reduction in the germination rate of several grasses like *Lolium multiform* L. by approximately 50% (Afegbua and Batty 2018). So, a higher concentration of these PHs is directly proportional to a higher toxicity recorded. Similar findings were observed by Eze et al. (2020), who reported that upon adulteration by diesel oil, the alfalfa seeds showed a significant reduction rate of around 90%.

Different studies have also found that PHs contamination is directly linked to a decreased microbial community in the soil as the toxic PHs because of their limited density and higher viscosity could easily pass through the soil disturbing the soil microbes which in turn affected the nutrient uptake (Wang et al. 2020).

Available studies further confirm the devastating effect of PHs on photosynthesis activity.

Studying how certain plants change their chemical processes when exposed to PHs and understanding how plant nutrition affects this at a molecular level is another important area for future research (Haider et al. 2021).

12.4.2.3 Animals

Petroleum hydrocarbons and their derivates have received much attention owing to their severe range of toxicity that increases with the increase in mass. The marine environment is also severely affected because of the oil spills that lead to polycyclic aromatic hydrocarbon (PAHs) contamination, which in turn affects the marine animals on varying levels.

The horizon oil spill accident that affected a wide area of the Gulf of Mexico had adverse effects on marine life and caused acute PAH toxicity ranging from immune system harmfulness to developmental anomalies in several marine vertebrates (Romero et al. 2018). The rise in hydrocarbon contamination has seen a widespread tumour outbreak called epizootic neoplasia, and certain reports have also mentioned the increased carcinogenic effects in fish, with the cancer-causing properties studied in feral fish like English sole (*Parophrys vetulus*) and flounder (*Platichthys stellatus*). Investigations have also reported several toxicities like bone destabilization in Pacific herring, pink salmon, and medaka upon exposure to PAH contaminants (Honda and Suzuki 2020). Through further research, it has been concluded that PAH exposure is not only toxic to the liver but also causes reproductive issues disrupting the process of hatching.

A research work to evaluate the effects of exposure to various types of polycyclic aromatic hydrocarbons was conducted on planarian *Girardia tigrina* and the results depicted an increase in toxicity in a dose-dependent manner (Simão et al. 2020). The invertebrates face a reduction in number followed by severe metabolic alterations on exposure to PAH pollutants. Further experiments have revealed an increase in the accumulation of PAHS in fishes like red snapper, and golden tilefish that further impaired growth, heightened vulnerability to illness. Birds too are affected negatively by oil hydrocarbon contamination and the effects range from feather contamination to metal toxicity (King et al. 2021). Few studies on terrestrial mammals reveal the potential harmfulness of PAH like a reduction in food intake and immune system suppression. There is little information about how amphibians are affected by PAHs exposure, however, they might be especially in danger of the synergistic interactions between PAHs harmfulness (Table 12.2).

12.5 Conventional Waste Removal Techniques

From the above discussion, we can see the impact of petroleum hydrocarbons on the environment and human health. The existence of toxic components in soils contaminated by petroleum waste makes it harder to entirely remove the PHs following the

Table 12.2 Advantages and disadvantages of each technique used for the removal of hydrocarbon pollutants

Sl No.	Remediation technique	Advantages	Disadvantages	References
Physical				
1.	Flotation	• High removal efficacy • Easy to conduct	• Huge constructional cost	Lim et al. (2018)
2.	Combustion	• Quick process • Successful destruction of pollutants	• Causes air pollution • Generation of CO, Nox	Ali et al. (2020)
3.	Thermal Stripping	• Simple • Effective method	• Not so cost efficient	Zhao et al. (2019)
4.	Sonication	• Fast process • No need of chemical reagents	• Equipments are costly • Not for largescale application	Wulandari and Effendi (2018)
5.	Microwave irradication	• High recovery rate • Easy to operate	• Needs special designing of equipment • Largescale application is low	Sivagami et al. (2019)
Chemical				
6.	Oxidative treatment	• Rapid selectivity of PHCs • Can be used in combination with other remediation techniques	• Requires enormous amount of chemicals • Persistent contamination	Besha et al. (2018)
7.	Soil flushing	• Easy detachment of hydrocarbon compounds; easy solubility • Non-intrusive method	• Harmful chemical containing sludges • Disposal issues	Effendi et al. (2022)
8.	Synthetic detergents	• Efficient dispersion • Rapid removal of hydrocarbon polluted sites	• Formation of toxic substances	Rocha e Silva et al. (2018)
9.	Photodecomposition	• Ensures complete decay of hydrocarbons • Does not involve any toxic catalysts	• Generation of strong oxidants and OH radicles • Only applicable for small-scale remediation	Samuel et al. (2023)
Biological				
10.	Microbial	• Environmental friendly • High performance • Economical approach	• Time consuming • Maintanance of optimal condition	Sharma and Prasad (2019)
11.	Plants	• Low cost • Convinient and eco-friendly	• Not so specific • Time constraint	Abdullah et al. (2020)

natural methods of degradation. Therefore, several methods have been established for the removal of petroleum hydrocarbon pollutants, some of which are discussed below.

12.5.1 Physical Methods

Globally, there have been reports of many physical treatment methods to degenerate and remove PHs from the environment, including several methods incorporating separation and recycling. various ex-situ methods like cleansing, and degradation technology were used for the removal of PHs found in soil and groundwater.

12.5.1.1 Flotation

This is a physical method that is commonly utilized in wastewater treatment and oil spill decontamination. This method uses the attachment of gas bubbles to hydrocarbon molecules, allowing them to float to the top where they can be taken off.

Group of (Lim et al. 2018) studied the flotation method that used a gas–liquid–solid system for retrieving oil from polluted sandy soil in laboratory conditions and the effectiveness of the oil recuperation rate was found to be around 70%. A study in 2001 stated the importance of dissolved air flotation (DAF) in the decontamination of hydrocarbons and the results from this study showed a removal efficacy of around 90% (Galil and Wolf 2001).

12.5.1.2 Combustion

The process of combustion is widely applicable in the context of hazardous waste management as it involves the restrained burning of organic materials, including hydrocarbons, at very increased temperatures in specially designed facilities known as incinerators. According to a study by Hao et al. (2020), it was found that there was a 95% reduction in petroleum contaminants upon burning those contaminants at increased temperatures of 1000 °C. However, this process of incineration required a thermal oxidizer which increased the cost of remediation (Ali et al. 2020).

12.5.1.3 Thermal Stripping

When considering ex-situ methods, heat treatments are a must as they remediate the toxic petroleum hydrocarbons. This method involves the use of high temperatures that further increase the vapour tension of the organic pollutants leading to the removal of the pollutant. This technique has been in use for inflammable pollutants like polycyclic aromatic hydrocarbons (PAHs) (Liu et al. 2018). However, the use of high energy for the increased temperature poses a major drawback by increasing the cost.

12.5.1.4 Sonication

This is a method used for the removal of hydrocarbons from various materials which includes water, soil, and industrial equipment. Sonication involves the use of high-frequency sound waves that are generally above 20 kHz to the polluted medium which causes cavitation that further leads to intense agitation and mixing thereby leading to effective dispersion of hydrocarbon pollutants.

A research study on the treatment of wastewater utilized two ultrasound systems which were of different ranges of frequency for the removal of PAH and the results reported that using a higher frequency at a lower power augmented the PAH removal rate efficiently (Ghasemi et al. 2019). In another study by Wulandari and Effendi (2018), they used a stainless-steel reactor for the sanitation process using frequencies of 28 and 48 kHz. The results demonstrated an effective removal of total petroleum hydrocarbon (TPH) at an ideal frequency of 48 kHz.

12.5.1.5 Physical Barriers and Excavation

A provisional solution to decontaminate petroleum hydrocarbon (PHs) from the soil surface involves the method of excavation which incorporates unearthing followed by removal of the unearthed contaminants to remediation sites.

Surface capping is another method that is administered to minimize surface soil and water pollution. The process involved isolating the PHs-contaminated soil from affected entities (Li et al. 2020). Wagner and Yarmak (2017) found that obstructions like vertical barriers, and trenches with a down-gradient installed helped minimise the flow of petroleum hydrocarbons in the soil. A study was performed on the effectiveness of sequenced permeable reactive barriers (PRBs) that facilitated the formation of biofilm and enhanced the remediation rate of petroleum hydrocarbons in Antarctic conditions.

The results showed the limited formation of biofilms, however, when the adsorptive material reached depletion, there was a desorption of hydrocarbons that led to biofilm formation and enhanced degradation (Freidman et al. 2017).

12.5.1.6 Microwave Irradiation

A cost-effective physical method in use has been the removal of petroleum hydrocarbons (PHs) by microwave heating, where microwave energy is converted into heat energy for the removal of contaminants via volatilization.

Sivagami et al. (2019) in their investigation of using waste as a resource found that microwave irradiation using spent graphite for treating petroleum hydrocarbon-polluted soil at a set frequency of 2.45 GHz eliminated around 90% TPH from the soil. However, this method is contained in laboratory settings for small-scale remediation and does not find its applicability for large-scale operations.

12.5.2 Chemical Methods

Several reports talk about the chemical methods of PHs removal which includes a variety of processes described below.

12.5.2.1 Oxidative Treatment

The chemical process includes breaking down of molecules into small parts that help with the hydrolysis. The technique of oxidation uses various oxidisers like hydrogen peroxide (H_2O_2), persulfate and ozone for the treatment of contaminated soil. The transformed organic compounds create stable, less mobile, and inert compounds, which may undergo additional treatment processes (Besha et al. 2018).

Using this method of oxidative treatment requires careful consideration of the site and characterization to ascertain the properties of the impacted site ranging from the mass of the pollutant to the pH of the soil.

Chen et al. (2015) conducted an in-situ chemical oxidation (ISCO) decontamination on soil polluted by diesel. The results from the study revealed various removal efficacies ranging around 48–90% by permanganate, and hydrogen peroxide under varying concentrations within a processing time of 120 days.

12.5.2.2 Soil Scrubbing

Soil scrubbing or flushing has long been in use for the removal of volatile and semi-volatile substances. This process involves the following steps: pre-treatment, separation, coarse-grained treatment, fine-grained treatment, process water treatment and residuals management. It has been seen that a soil-washing system was implemented to remediate the polluted soil. The results indicated a 97% removal of total petroleum hydrocarbon (TPH) and in a review by Tran et al. (2022), the method of soil flushing was mentioned where the petroleum hydrocarbons were divided from the soil particles using cleaning agents.

12.5.2.3 Synthetic Detergents

The use of synthetic detergents is also preferred because these surfactants are particles with both hydrophilic heads and hydrophobic tails. They are utilized to eliminate hydrocarbons from the polluted soil. When added to water, these detergents lower the surface tension of the hydrocarbon-polluted mixture, leading to the formation of colloidal aggregates and subsequent separation (Rocha e Silva et al. 2018). The industry-based surfactants used for decontamination are alkyl benzene sultanate, sodium dodecyl sulphate and Tween 20. However, the drawback that weighs them down is that these surfactants in use have a further adverse effect on the environment.

12.5.2.4 Photodecomposition

A review by Koe et al. (2019) also mentions the method of photodecomposition in which ultraviolet radiation (UV), ozone (O_3) and other metal oxides like titanium oxide act as a catalyst to break down petroleum hydrocarbon pollutants. Increased intensity of UV radiation treatment is in use by employing hydrogen peroxide to oxidize contaminants into water, carbon dioxide and salts. A study by Moradi et al. (2021) mentioned the usage of photochemical $FeTiO_3$/graphene oxide nano-structured material which showed an increased phenol decay under visible light.

The aforementioned techniques in use are efficient to an extent for the removal of petroleum hydrocarbons, however, these processes are not economical and are toxic to the ecosystems. Therefore, researchers are on the lookout for techniques of removal that are eco-friendly.

12.6 Bioremediation

The drawbacks of the physical and chemical methods have encouraged the researchers to develop alternative modes of removal which are low-energy consumption, reduces the operative costs and is also environment friendly. A technique that meets all of the requirements and has come a long way in the process of bioremediation.

Bioremediation is a technique that includes the use of microorganisms to alleviate, break down and reduce the harmfulness of the contaminants without adversely affecting the environment (Behera et al. 2022b).

Organic breakdown, primarily facilitated by petroleum-metabolizing microorganisms, is the prime mechanism that is in use to break down hydrocarbon pollutants further utilizing them as an organic origin of energy and carbon.

Indigenous microorganisms, too, play a pivotal role in the decontamination of hydrocarbons owing to their catalytic activities (Varjani and Upasani 2017; Tripathi and Gaur 2021).

However, according to some studies, the biodegradation kinetics of different compounds vary. According to a study by Sridhar et al. (2020), it was suggested that microorganisms could play a significant role in the soil–plant–atmosphere system.

Various studies have been reported on the degradation of petroleum hydrocarbon contaminants biologically and it was noted that this natural process of degradation of petroleum waste products took a lot of time (Sengupta and Pal 2021). Bioremediation like any other technique has its fair share of limitations, like the inability to completely degrade the petroleum hydrocarbon (PHs) contaminants. Despite this, the method is widely applied in different ways and is categorized into in situ and ex situ techniques which are discussed below.

12.6.1 In Situ Bioremediation

These methods necessitate decontamination of the polluted substances right where they were created. This method is considered to be cost-effective in comparison to ex situ techniques. Bioaugmentation, biosparging, bioventing, and bioslurping are examples of in situ bioremediation techniques.

In situ biorepair practice can be classified into two different types: natural in situ remediation and technical in situ bioremediation.

Natural remediation is the process which makes use of contaminated sites without the intervention of humans and the breakdown process is based on aerobic and anaerobic processes in microorganisms which simultaneously reduces the cost of use.

Natural remediation finds its way as a method of decontamination for the biological treatment of obstructed groundwater and the concentration of the contaminants rapidly reduced from 40 to 5 ppm within a treatment range of 3 years (Yadav et al. 2018).

For technical or engineered in situ bioremediation, there is an introduction of a specific microorganism to the contaminated site which then helps in the process of decontamination. The examples mentioned above have been discussed below.

12.6.1.1 Bioaugmentation

This method involves the addition of certain types of potential bacteria or groups of bacteria that have been collected from other polluted areas or bacteria that have undergone genetic modification to help clean up the contamination (Behera et al. 2022a). Furthermore, the microorganisms have a specific degradation process, which has been proven to enhance the rate of degeneration.

This method is generally used to treat waste generated by municipalities. Microbial cell bioaugmentation is a technique that has been commonly used for the removal of pollutants. Reports have also stated the development of other augmentation methods to help the decontamination processes (Behera et al. 2020, 2022a).

12.6.1.2 Bioventing

A promising new method that makes use of regulated airflow to expand the activity of local bacteria for the process of decontamination by providing oxygen to the under-saturated region. The process is assisted by the addition of nutrient and water content during the bioventing process which helps the bacteria break down the pollutants into harmless substances (Da Silva et al. 2020). Engineered bioventing systems may be installed horizontally or vertically, depending on the geology of the place.

This process of bioventing boosts the local microorganisms in the soil by adding lots of air which helps the microbes degenerate the pollutants better and clean up heavy metals by causing them to settle down (Anekwe and Isa 2021).

12.6.1.3 Biosparging

It is a method that is deployed to clean up soil and groundwater contaminated with hydrocarbon wastes. This involves administering air into the contaminated site below the ground which stimulates the growth of native microorganisms that degenerate the hydrocarbons into non-toxic byproducts.

The effectiveness of biosparging is influenced by soil permeability and pollutant decomposability. Biosparging is an effective way to remediate sites contaminated by oil spills and is also used for the removal of diesel and kerosene from water resources. To make the process of disparaging more effective, oxygen is supplied to the microbes during the increased decontamination process. Moreover, bioventing and biosparging have been merged with several treatment plans like barrier treatment methods and vapour recovery from soil.

12.6.1.4 Bioslurping

Bioslurping is an innovative method that combines elements of vacuum-assisted extraction and bioremediation techniques. In bioslurping, a vacuum is applied to the polluted area to remove the hydrocarbon-laden groundwater and vapours. Bioslurping can be used to fix soils polluted with flammable organic substances (Anekwe and Isa 2021). Given that bioslurping, uses less groundwater, it accelerates the removal of PHs contaminated sites efficiently making it a valuable tool in environmental remediation efforts.

12.6.2 Ex Situ Bioremediation

Ex situ bioremediation method entails the treatment of polluted soil, water, or sediment outside the original location. This approach allows for better control and monitoring of the remediation process and the downside is the increased cost due to excavation and transportation (Abdullah et al. 2020). Examples of ex-situ bioremediation techniques include land farming, composting, biostacking and bio-reactor processes which are discussed below.

12.6.2.1 Land Farming

Land farming is a process that spreads polluted soil in thin layers over a treatment area and promotes microbe-assisted decay through aeration and nutrient addition (Guarino et al. 2017). Land farming can effectively degrade soil contaminated by PWHCs. This method is given preference over landfilling and burning as it requires less energy and also has a reduction in groundwater pollution due to the stationarity of the hydrocarbons in the soil.

12.6.2.2 Biostacking

Biostacks are constructed stacks of polluted soil that is damp and has added nutrients to enhance microbial activity. Biostacking is gaining importance because of its cost efficiency and features like pH and nutrient control.

The flexibility of biostacking allows a reduction in decontamination time by enhancing the microbial function, which further improves the rate of degradation.

There is a mention of bio-available organic carbon (BOC) in a review that plays a significant role in the remediation process through the method of bio stacking. Soil polluted by petroleum hydrocarbons has been decontaminated using moderate temperature conditions using alpha, beta and gamma proteobacteria. Biostacking has also found its way to the treatment of diesel-contaminated soil in the sub-Antarctic area.

Reports have revealed a removal rate of 93% of total petroleum hydrocarbon (TPH) in about a year using the innovative method of biostacking (Jaain and Patel 2019).

12.6.2.3 Biological Reactors

Biological reactors alter raw materials into specific products by providing a favorable growth environment. Upon using the bioreactors for the bioremediation process, it has been seen that there was an accelerated rate of bioremediation as conditions like pH, agitation, temperature, aeration and inoculum concentration were precisely regulated.

Reports have stated the use of bioslurry reactors for an increased process yield using the batch mode (Avona et al. 2022). Some recent studies have shed light on the bioremediation of PHs polluted soil using the bioslurry system and various removal efficacy percentages have been stated with a 74% removal rate of total petroleum hydrocarbons (TPH) in about 210 days (Ramadass et al. 2018) (Fig. 12.2).

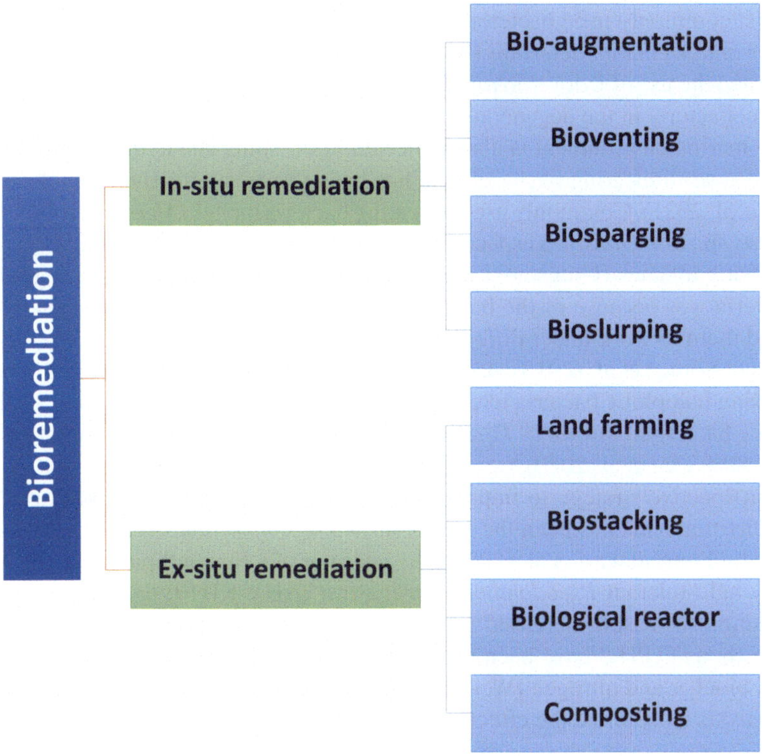

Fig. 12.2 Flow chart on ecologically varying remediation strategies

12.7 Other Bioremediative Technologies

12.7.1 Microorganisms Assisted Bioremediation

As discussed above, bioremediation uses bacteria, fungi, algae and even yeast to eliminate polluted materials from the environment and there is a need to understand the dynamics of the microbial community to enhance these bioremediation techniques further. Microbial resilience and its physiological processes make the microbes an ideal choice for decontamination.

12.7.1.1 Bacteria

Various reports in the past decade have stated the input of microbes in the degeneration of petroleum hydrocarbons (PHs). Hydrocarbon-consuming microbes play a significant role in the cleanup of polluted water and other polluted media and the most commonly used bacterial strains include *Clostridium, Zymomonas, Enterobacter and Methanobacterium* (Nwidee et al. 2016). Dubinsky et al. (2013) have mentioned the use of *Colwellia and Pseudoalteromonas* which belong to the phylum of proteobacteria in the degeneration of aromatic hydrocarbons.

The marine environment is also subjected to oil spills and hydrocarbon contamination so administering bacterial taxa has been the most effective way of removal (Anno et al. 2021). Commonly used oil-eating bacteria include *Alcanovirax* sp., *Pseudomonas* sp., *and Oleispira* especially *Oleispira* antarctica RB-8 for the degradation of aliphatic alkanes (Gregson et al. 2020). The increased yield in the degradation of bacteria was because of the bacteria's adaptation to high-salt environments that allowed them to work under different levels of osmotic stress.

Rezaei Somee et al. (2018) also reported a degradation rate of diesel oil by about 40% using halophilic bacteria like *Halomonas, Dietzia and Arthrobacter*. Similarly, a marine bacterium known as *Paracoccus sp.* has shown a removal rate for diesel of around 80% (Xu et al. 2020).

A prospective strategy to improve the efficiency of bioremediation of contaminated environments has been the use of surface agents synthesized by microbes. In a study by Pourfadakari et al. (2021), the addition of a rhamnolipid analogue generated by halo-tolerant *Pseudomonas aeruginosa* (AHV-KH10) caused around 70% degradation of diesel. Increased removal rates of around 80% for total petroleum hydrocarbons (TPHs) have been reported on the combined usage of surface-active agents, biochar and nitrogen (Wei et al. 2020). Researchers also suggest an accurate risk analysis to examine the effects of biological treatments.

12.7.1.2 Fungi

Reportedly, fungi are the most multifaceted group of single-celled eukaryotes with more than five million species. Although research has primarily focused on bacteria, fungi too have gained increased attention in bioremediation due to their versatile metabolic adaptability, which helps in the absorption of hydrocarbons using low concentrations and acidic environment.

Various genera of fungi including *Cladosporium, Aspergillus, Cunninghamella, Penicillium, Fusarium* and *Mucor* have been involved in the degradation of aliphatic and aromatic hydrocarbons. Recent study reported that fungi like *Phanerochaete chrysosporium, Agaricus bisporus* and *Traemetes versicolor* were successful in remediating contaminated sites (Xu et al. 2020).

Several studies have also provided evidence of the promising use of marine fungi for the degradation of hydrocarbons (Barone et al. 2019).

Barnes et al. (2017) worked on ten fungal cultures belonging to *Aspergillus, Fusarium, Penicillium*, and *Acremonium* that could break down crude oil. The results obtained demonstrated that out of these isolates, *Penicillium citrinum* caused a 70% degeneration in crude oil and about 90% for n-alkane.

Several studies have also focused on lignin-degrading fungi like *Phanerochaete chrysosporium* for the breakdown of PAHs (Li et al. 2020). Fungi found from the northern coast of Brazil like *Aspergillus sclerotiorum* CBMAI 849 and *Mucor racemosus* CBMAI 847 have shown a 90% efficacy in degrading pyrene (Passarini et al. 2011).

However, mycoremediation is still in its earlier stages and entails promising biotechnological applications.

12.7.1.3 Microalgae

Microalgae are microscopic photosynthetic organisms that represent a fundamental part of the treatment of contaminated water. Several studies have examined the biofilm-forming ability of microalgae for the removal of pollutants like total petroleum hydrocarbon (TPHS) and PAHs (Ugya et al. 2021).

Chlorella sp. has been under investigation owing to their unique ability to persist in polluted media and studies by Kuttiyathil et al., investigated the removal of crude oil by using *Chlorella* spp. and the results indicated around 80% crude oil removal rate within a span of 5 days by using this microalga.

Further studies have also mentioned the use of green microalgae *Chlorophyceae* in the removal of crude oil contaminants *and Skeletonema costatum and Nitschia* sp. for the efficient degradation of polycyclic aromatic hydrocarbons (PAHs) (Hong et al. 2008).

A study by Ghasemi et al (2011) reported the use of oceanic blue-green algae like *A. quadruplicate, Microcoleus Chthonoplastes* for the removal of PAHs. Some haptophytes like *Isochrysis galbana* have been administered for the removal of hydrocarbons in polluted oceans and the removal rate stands around 80% (Ammar et al. 2018). Microalgae-assisted bioremediation shows promise in this field of bioremediation and further research is required to optimize cultivation techniques and scale-up processes for further applications.

12.7.2 Plant Assisted Bioremediation

The utilization of plants is a promising alternative approach to bioremediation. This process is a relatively cost-effective mode of treatment that includes the use of plants and microbes in the root environment to degrade petroleum contaminants. Fan et al. (2018) in their study demonstrated that in the root environment, bacteria and fungi break down harmful compounds by utilizing them as a source of nutrients and energy. Several studies have emphasized plant characteristics as key players in plant-assisted bioremediation (Jeevanantham et al. 2019).

Several plants like Fescue (*Lolium arundinaceum* Schreb), Ryegrass (*Lolium multiflorum* L.), Black rush (*Juncus roemeriamus*) have served as phytoremediators in wastewater treatment (Jeevanantham et al. 2019).

According to the reports, around 62% of plants are land plants and 33% are aquatic plants. Saeed and Sun (2012) in their study reported a 100% removal efficacy of benzene from contaminated water using a polishing filter in combination with plants; willows in general. Plant-assisted remediation includes diverse approaches that are discussed below.

12.7.2.1 Plant Assisted Degradation

Plant-assisted degradation is a process that utilizes vegetal enzymes to break down pollutants. Organic materials can be subjected to plant-assisted degradation within the plant itself or in the root environment commonly known as the rhizosphere. This method effectively causes a reduction in pollutants like groundwater, oil and aromatic compounds in the soil.

Researchers have found that some plants use enzymes such as dehalogenase, peroxidase, and nitrilase to destroy the contaminants (Kumar Yadav et al. 2018).

Additionally, investigations by Alanbary et al. (2019) have shown that the long carbon chains in the total petroleum hydrocarbon have been broken down into lower carbon chains after exposure. García-Sánchez et al. (2018) have highlighted that using microorganisms for phytodegradation is an intricate process that turns toxic compounds into less poisonous ones.

12.7.2.2 Rhizoremediation

This is an innovative environmental cleanup technique that utilizes the natural capabilities of plant roots and their associated microorganisms to decontaminate polluted soil or groundwater. Plants are grown in the contaminated area and the microorganisms, in turn, degrade the contaminants present thereby effectively reducing the toxic concentrations. According to Alotaibi et al. (2021), rhizodegradation is a successful way to degrade petroleum hydrocarbons and heavy metals.

Hoang et al. (2021) examined several plants like *Acacia pyrifolia, Banksia seminuda, Hakea prostrata* and *Chloris truncata* for their rhizodegradability and some plants like *Hakea prostrata* showed efficacy in reducing TPHs found in soil polluted by diesel.

12.7.2.3 Plant Mediated Extraction

As the name suggests, plant-mediated extraction involves growing specific plants known as hyperaccumulators which can reduce the toxic concentrations of hydrocarbons without their tissues being harmed. As these plants grow, they take up the pollutants from the site and move them to leaves and stems effectively reducing the concentration of contaminants in the soil. An investigative study by Hidayati et al. (2018) reported that total petroleum hydrocarbons were transported and taken up into the stomata of the swamp forest.

While plant-mediated extraction shows a promising approach for contaminated sites, research is being conducted to enhance its applicability and efficacy.

12.7.2.4 Plant-Mediated Volatilization

This is a process where plants take up pollutants from the soil or water and release them into the atmosphere in a vapour form. This method has a specific use for the removal of volatile organic compounds (VOCs) from polluted sites. This process also depends on the plant size meaning that the larger the plant, the more effective way of transpiration.

The utilization of plants for remediation purposes is an environment-friendly and relatively simple method to reduce further levels of hydrocarbons. However, a few steps should be taken to optimize the process and research should be conducted to understand the aftereffects of the extracted hydrocarbons by the plants (Table 12.3).

12.8 Biologically Engineered Organisms for Enhanced Bioremediation

The use of genetically enhanced organisms for the degradation of contaminants has been on the rise and is considered a cost-effective alternative to physical and chemical methods.

In this process of decontamination, several genetically engineered microorganisms are used to degrade polyaromatic hydrocarbons using their metabolic pathways (Kumar et al. 2018). Various techniques like metabolic engineering, recombinant DNA technology and omics approach have been applied in tandem to enhance the process of. According to recent investigations, researchers have been utilizing the CRISPR-Cas systems in experimental organisms like *Pseudomonas* or E. coli and some organisms like *Achromobacter* sp. HZ01 has also been under consideration because of the ability to express function-oriented genes relevant to bioremediation (Okoli et al. 2021).

Research by Sazanova et al. (2022) has mentioned the effectiveness of genetically modified organisms (GMOs) like *P. putida* KT2440 in the process of degradation of organophosphates and organochlorines the main constituents of insecticides. Kharitonov et al. (2022) mentioned *Rhodococcus opacus* R7's pobA's efficacy in degrading naphthenic acid. In a study by Hara et al. (2021), a PCB-breaking experiment was performed by utilizing a genetically enhanced PCB-degrading bacterium in combination with a wild-type PCB-degrading bacteria and the results demonstrated a reduction in the dioxin-like PCB congeners which further led the researchers to believe that amplification of the bacterial strain can be of help concerning PCB contaminated waters.

Table 12.3 Represents distinct species that are efficient reducers of hydrocarbon pollutants

Sl No	Organisms involved	Species	Type of hydrocarbon pollutant/ substrate	Removal rate (%)	References
1	Bacteria	*Pseudomonas* sp.	Petrol, diesel and waste engine oil	Petrol—76%, diesel—83%, and waste engine oil—69%	Ambaye et al. (2022)
		Ochrobactrum sp.	Crude oil	83%	
		Bacillus soreness D11, *Pseudomonas stutzeri* D13, and *Bacillus cereus* D12	Petroleum–diesel	80%	Anno et al. (2021)
		Alcaligenes aquatilis BU33N	Crude oil and phenanthrene	70%	
		Cycloclasticus sp.strain BG-2	Phenanthrene	–	
		Cycloclasticus sp. 78-ME	Polycyclic aromatic hydrocarbons (PAH)	–	
		Cycloclasticus sp. strain P1	Naphthalene, phenanthrene, pyrene	–	
		Pseudomonas aeruginosa N6P6	Phenanthrene and pyrene	Phenanthrene (80–85%), pyrene (40–45%)	
		Rhodococcus sp.P14	Phenanthrene, pyrene, benzo(a)pyrene	–	Alaidaroos (2023)
		Enterobacter cloacae	Crude oil	–	
		B. stereothermophilus, *Pseudomonas* sp., *Corynebacterium* sp. *Vibrio* sp. *Bacillus* sp. *Achromobacter* sp.	Monocyclic aromatic hydrocarbons	–	
		Corynebacterium sp. *Flavobacterium* sp. *Acinetobacter* sp. *Moraxella* sp. *Flavobacterium* sp.	Xylene, benzene, hexane, crude oil, kerosene, gasoline, diesel and olive oil	–	
		Bacillus sorensis	PAH and Diesel	32% and 39%	Mekonnen et al. (2024)
		Flavobacterium johnsoniae and *Shewanella baltica*	PAH	93%	

(continued)

Table 12.3 (continued)

Sl No	Organisms involved	Species	Type of hydrocarbon pollutant/ substrate	Removal rate (%)	References
		Bacillus subtilis (M16K, and M19F)	Crude oil and Phenanthrene	94% and 85%	
		Gordonia strain (Q8)	Naphthalene, Phenanthrene, Anthracene, Pyrene	100%, 95%, 73% 53%	
		Photobacterium phosphoreum	TPH	85%	
		Geobacillus thermodenitrifican Gordonia sihwensis Desulfosarcina sp. and *Desulfococcus* sp.	Aliphatic hydrocarbons	–	Ahmed et al. (2023)
		Stenotrophomonas sp. *Bacillus licheniformis* and *Bacillus mojavensis*	Aromatic hydrocarbons	–	
		Acinetobacter radioresistens KA2 and KA5	Oil sludge	90%	Radhakrishnan et al. (2023)
		Staphylococcus equorum KA4, *Enterobacter hormaechei* KA3	Heavy oil sludge	89%	
		Enterobacter hormaechei KA6	Oily sludge	81%	
		Acinetobacter baumannii	Crude oil	43%	Mahjoubi et al. (2018)
		Alcanivorax and Thalassolituus		70%	
2	Fungi	*Aspergillus ficuum* *A. fumigatus* *A. flavus* *Cladosporium* sp.	Aromatic hydrocarbons	54–60%	Alaidaroos (2023)
		Coriolopis byrsina *Marasmiellus* sp. *Merulius tremellosus*	PAH	96% 98% 83%	
		Peniophora incarnata *Trichoderma* sp. *Pleurotus pulmonarius*	PAH	82% 78% 99%	
		P. ostreatus *Paecilomyces formosus*	TPH	87% 50%	Mekonnen et al. (2024)
		Aspergillus sp., *A. niger*	n-alkanes, Kerosene	50% and 78%	
		Fusarium solani	Diesel	70%	
		Trichoderma harzianum *Polyporus* sp. S133	Crude oil	77% 93%	

(continued)

Table 12.3 (continued)

Sl No	Organisms involved	Species	Type of hydrocarbon pollutant/ substrate	Removal rate (%)	References
		Aspergillus niger (KY473958), *Talaromyces purpurogenus* (KY488463), *Trichoderma harzianum* (KY488466) and *Aspergillus flavus* (KY488467)	PAH	95%	
		Candida tropicalis	Diesel	83%	Mahjoubi et al. (2018)
		Absidia cylindrospora	Fluorene	90%	
		Aspergillus niger *Candida krusei* *C. glabrata* *Saccharomyces cerevisiae*	Crude oil	–	Bala et al. (2022)
		Aspergillus spp. *Penicillium* sp. RMA1 and RMA2	Crude oil	87% 57%—RMA1, 55% RMA2	Ambaye et al. (2022)
		Aspergillus ochraceus CBMAI 849	Benzo [a] pyrene	99%	
		Aspergillus sydowii NIOSN-SK56C42 *Acremonium sclerotigenum* NIOSN-M109 *Penicillium polonicum* *Penicillium cyclopium* *Penicillium mononematosum*	Crude oil, alkanes	–	Anno et al. (2021) and Barnes et al. (2017)
		Tinctoporellus sp. CBMAI1061	PAH, pyrene	100%—pyrene	
		Aspergillus sclerotiorum CBMAI 849 *Mucor racemosus* CBMAI 847	pyrene and benzo[a]pyrene	99%—pyrene, 76% benzo[a]pyrene	
		Cladosporium sphaerospermum	Saturated hydrocarbons	–	Hamad et al. (2021) and Li et al. (2020)
		Alternaria alternata *Penicillium chrysogenum*	PAH	72% 59%	
		A. ramosus *A. flavus* *A. terreus* *P. canescens* *P. janczewskii* *P. montanense* *P. simplicissimum* *P. restrictum* *P. harzianum* *P. terrestre*	Crude oil, PAH, Pyrene	–	Ghosh et al. (2023)

(continued)

Table 12.3 (continued)

Sl No	Organisms involved	Species	Type of hydrocarbon pollutant/ substrate	Removal rate (%)	References
3	Algae	*Selenastrum capricornutum*	Naphthalene, Phenanthrene, Pyrene	–	Chekroun et al. (2014)
		Scenedesmus obliquus GH2	Crude oil	–	
		Chlorella sorokiniana *Pseudomonas migulae*	Phenanthrene	–	
		Sargassum latifolium *Corallina ofcinalis* *Ulva intestinalis*	n-alkanes and PAH	62–78%	Akl et al. (2023)
		Chlorella vulgaris BS1	PAH	98%	Das and Deka (2019)
		Ankistrodesmus capricornutum	Benzo[a]pyrene	–	Michel et al. (2024)
		Dunaliella salina *Cylindrotheca* sp. *Synedra* sp. *Amphora* sp.	Naphthalene	–	
		Nitzschia sp. *Chlamydomonas reinhardii*	Phenanthrene	–	
		Prototheca zopfii	Saturated aliphatic hydrocarbons	49%	Touliabah et al. (2022)
		Prototheca zopfii	Aromatic compounds	26.50%	
		Oscillatoria sp.	Pyrene	95%	
		Chlorophyceae *Chlorella* spp. *Haematococcus pluvialis*	Crude oil and PAH		Radice et al. (2023)
4	Plants	*Phragmites australis and Juncus maritimus*	Crude petroleum oil	–	
		Triticum aestivum *Zea mays* *Trifolium repens* *Medicago sativa* *Lolium multiflorum*	Petroleum hydrocarbons	35–60%	Ambaye et al. (2022)
		Medicago sativa L			
		Imperata cylindrica Mucuna bracteata *Pteris vittata*	TPH	91%	
		Epipremnum aureum	Crude oil	30–50%	

(continued)

Table 12.3 (continued)

Sl No	Organisms involved	Species	Type of hydrocarbon pollutant/ substrate	Removal rate (%)	References
		Chromolaena odorata	Phenanthrene, Fuoranthene and Benzo(a)pyrene	60%, 38%, 36%	Futughe et al. (2023)
		Alhagi camelorum	TPH	53%	Nemati et al. (2024)
		Mirabilis jalapa *Impatiens balsamina*	TPH	–	Liu et al. (2018)
		Echinacea purpurea *Festuca arundinacea*		46–49%	
		Tagetes patula *I. balsamina*	PCB	79–92%	
		Medicago sativa L *Lolium perenne L*	Oil hydrocarbons	72–90%	Panchenko et al. (2023)
		Eucalyptus camaldulensis	Oil hydrocarbons	82%	
		Juncus roemerianus	n-alkanes and PAH	–	
		Sorghum sudanense *Melilotus offcinalis* *Pinus sylvestris* *Festuca arundinacea* *Lepironia articulate*	PAH	80–82%	Abdullah et al. (2020)
		Phragmites australis *Panicum hemitomon*	Petroleum crude oil	–	
		Festuca rubra *Poa pratensis* *Lolium perenne* *Trifolium repens* *Pisum sativum*	Diesel	–	
		Zea mays *Phaseolus nipponesis OWH1*	TPH	86%	

A recent review in 2023 has focused on degrading polyaromatic hydrocarbons (PAHs) using the mixed microbial system (MMS). This artificial MMS construction has been simplified by organizing the microbes more simply, making sure that each microbe has a specific job to do, thereby making the process more efficient (Cui et al. 2023). Methods like synthetic biology are being used increasingly to deal with contaminated sites thus promoting the remediation strategies.

Following the use of green technology, several plants have also been modified to help with the process of plant-assisted bioremediation. Alfalfa is one such plant that shares a history of adaptability and genetically enhanced alfalfa plants have been utilized for the phytoremediation of persistent organic pollutants (Tussipkan and Manabayeva 2022). In another study by Anees et al. (2020), the increased efficacy of sodium plant-assisted bioremediation by *Pseudomonas, Sanguibacter, Bacillus* and *Bacillus* species was reported.

1,2,3-trichloropropane (TCP) is a well-known groundwater pollutant and genetically enhanced TCP-degrading *Pseudomonas* strains have been increasingly used to combat the growing pollution (Janssen and Stucki 2020). Duque et al. (2022) in their study mentioned the wide usage of gene cloning and the possibility of using model bacterial cells for the treatment of pollutants in soil and water. That being said, the widespread usage of GMOs for bioremediation requires high-throughput technology study to identify any detrimental effect that might destroy the delicate balance of the biological ecosystem.

12.9 Novel Products from Hydrocarbon Waste Residues

Post-combustion residues from coal fired power plants (slag, bottom ash), liquid oily sludge and chemical containing water generated from petroleum refineries as well as fertilizer manufacturing industries, boiler slag, soot, fly ash, flue gas (CO_2 and SO_2) etc. are exceeding and their detrimental impact on our Earthis uncontrollable. Different activities such as drilling, burning of waste dumps, industrial chimneys, transportation, refining and gas production operations in the oil and gas industry produces hydrocarbon entities. Those hydrocarbon waste residues can be modified and treated to manufacture cement, concrete, tiles, composites, adsorbents, fillers, pipes, and so on (Asim et al. 2021; Mishra et al. 2021a, b).

As per the current growth of recycled products and adaptation of sustainable development by utilizing waste substances, several novel techniques are implemented thesedays to reuse hydrocarbon wastes too. Therefore, recovering and manufacturing novel value-added products. The members of World Wide Coal Combustion Products Network, emphasizes the utilization and reuse of coal combustion by-products as raw materials for other sectors. Japan is the leading country having 99% of new procedures to recycle hydrocarbon waste entities followed by Africa and Middle East to be the least. In 2016, Poland took initiative were 20 million tonnes of CCPs were generated, of which approximately 65% was recycled and reused in that year and 35% was landfilled (Bielecka and Kulczycka 2020).

Recovery technologies of oil from oil sludge includes solvent extraction, ultrasonification, surfactants, pyrolysis, gasification to generate syngas, char and pyrolysed oil. Moreover, flue gases can also be treated for the removal of acid gases (H_2S and CO_2) by applying various desulfurization, adsoprtion, and membrane seperation techniques. Recently, it is observed that the potential of petroleum sludge for boosting constructional sector for making bricks, cement, asphalt-concrete mixtures, ceramics and home insulation materials. On top of that, adsorbents made out of petroleum sludges are also evidently used for wastewater treatment (Jagaba et al. 2022).

Global production of CFA (Coal Fly ash) is 750 million tonnes which is expected to evolve in coming years. Several elements like aluminium, titatinum, iron, silica, germanium, gallium, chromium, cadmium and many more can be efficiently recovered from coal fly ash using chemical reagents (Sahoo et al. 2016). Application of fly ash in manufacturing concrete, bricks, road and embankment works, paints, tiles as well as agricultural sector is quiet prominently described by Gorai (2017). Back in 2017, it has been displayed X-Ray diffraction and differential thermal analysis results to prove fly ash from thermal power plants as a promising candidate for filler in cement binding.

Besides this, soot is also known as one of the deadly component for global warming after carbon dioxide. Each year 8 milliom metric tonnes of soot gets released in the atmosphere. Hence, Uttaravalli et al. (2022) showed how soot can be utilised in making car tires, batteries, plastics, coatings, pipes, adsorbents etc. as uplifting the sustainable approach towards wastes produced.

12.9.1 Constructional Sector

12.9.1.1 Bricks

The brick manufacturing sector is one of the largest emitters of greenhouse gases which nearly stands 66–84 million tonnes of CO_2 and over 100,000 tonnes of black carbon every year (Rastogi and Kumar Paul 2020). Effective utilization of fly ash can (i) diminish the occurance of environmental pollution and (ii) preserves the top soil in a cultivated land. As huge amount of clay is needed for making burnt clay brick. Fly ash bricks are durable, light-weight, and prepared in room temperature. Most probably sintering method and autoclaving method are used to manufacture fly ash bricks. Substances such as calcium silicate hydrate generated by fly ash hydration reaction can be used to bond aggregates to improve their mechanical strength. In present scenario, more than 10 types of fly ash bricks are available, such as hollow bricks, wall panels and sound-absorbing bricks (Wang et al. 2023).

Innoative approach comprising CFA (Coal Fly Ash) from coal fired plants is an excellent method to flourish sustainability, as it promotes cost-effectiveness, eco-friendliness and stability in comparison to clay bricks. Central Fuel Research Institute, Dhanbad, India adopts this technique to produce fly ash bricks for buildings (Kelechi et al. 2022).

Furthermore, oil sludges are also relevant in this field of sustainable brick manufactruing process. Scientific studies on oil sludge from Yarakta oil and gas condensate field located in Irkutsk Region of the Russian Federation, mentioned oil sludge as a potential additive for producing bricks. Hence, it eliminates additional economic costs and implements sustainability by reusing waste oil sludge (Gorbaev et al. 2020). Additionally, impurities and organic substances present in oil sludge can decrease the bulk density by up to 25% and, hence, offer thinning and combustible properties, which lead to the production of lightweight bricks (Pauzi et al. 2022).

12.9.1.2 Cement

On the basis of The World of Coal Ash (WOCA) estimation 780 million metric tonnes of CBA (Coal Bottom Ash) gets generated from thermal power stations of which 66% is by Asian countries, followed by Europe and the United States. Along with that, China produces the highest amount of coal ash (395 million metric tons) (Mt), then comes US (118 Mt), India (105 Mt), Europe (52.6 Mt), and Africa (31.1 Mt) respectively. Due to their hazardeous nature, disposal issues are quiet challenging. General properties of bottom ash is that, its not so light like fly ash and gets collected in a hopper after falling through the bottom furnace. It is porous in texture and particular size ranges like sand and gravel particles (Mohammed et al. 2021).

Both bottom ash and fly ash together makes a great combination when used to make a supplementary cementious material of concrete and constituents of Portland cement. This new application of CBA displays an interesting alternative for its disposal. Thus, leading to more sustainable cement production by lowering energy consumption and toxic raw material incorporation (Argiz et al. 2017).

Aim of Xiao et al. (2019) and his group was to utilise oily sludge as a roadbed material with ordinary Portland cement (OPC), fly ash (FA), and silica fume (SF) as binders and phosphogypsum (PG) as a stabilizer, and conduct solidification/stabilization (S/S) technique. Final results were outstanding as it highlighted sufficient strength and good water stability performance, freeze–thaw resistance, enhanced environmental protection and economically profitable.

Apart from CBA, fly ashes are prominently used thesedays for cement production. Some active substances with pozzolanic effect (such as CaO, SiO_2 and Al_2O_3, etc.), are mixed into a raw meal for the production of cement or as a concrete admixture instead of cement. When fly ash is mixed into cement, it establishes a homogenization effect and lubrication effect, thereby increasing the strength and improved impermeability (Wang et al. 2023; Amran et al. 2021).

12.9.1.3 Concrete

Research findings suggests fly ash as an adaptable, affordable, essential cum sustainable component for the production of building materials. Fly ash is used as a cement substitute in blended cement or concrete and is valued for its potential advantages like—greater structural strength, cost efficiency, and environmental sustainability, particle size etc. For manufacturing dry and hard concrete masonry units, fly ash is becoming a popular constituent as a cementing and filler substance. In addtion, manufacturers add fly ash in concrete mixtures to gain required strength, finished molds, sharpened angles, plasticity, and durability of the end product (Shukla et al. 2023).

Recently, Hameed et al. (2022) has observed that fly ash can be used as an substitute to a VMA (viscosity-modifying admixture) to produce cost-effective, self-compacting concrete. Usage of fly ash in concrete is capable of lowering the pollution induced by the cement factories and by decreasing the CO_2 emissions. Moreover, fly ash can act as a significant alternative to the volume of OPC (up to 50%) without influencing concrete durability but 50% FA (fly ash) as a replacement may negatively affect the concrete strength (Amran et al. 2021).

Similarly, Blast Furnace Slag (BFS) and fly ash can also be modulated and mixed further to produce high-performance concrete and composites for buildings. Even boiler slag becomes an excellent coarse aggregate to make concrete. Cenosphere developed from coal fly ash is another material for adding into the mixture of concrete. Especially, lightweight composite material are easily manufactured by the help of cenospheres, resulting to a high-strength and durable concrete.

Furthermore, filler like CBA (coal bottom ash) are known as agglomerated ash particles evolved from coal furnaces which are applied for making concrete and utilised as an substitute to fast depleting natural sand. Likewise, findings of Ganesan et al. (2023) delivers high-strength concrete which can be achieved by using 40% coal bottom ash, as a partial replacement of fine aggregates in combination with 15% Alccofine inclusion as a partial replacement of cement. About 58% improvement in compressive strength was recorded.

12.9.1.4 Ceramics and Tiles

Innumerable studies have conveyed that petroleum oily sludge can be applied for making ceramic items, bricks, aggregates and concrete structures. Additional expansion agents like (carbonates such as $CaCO_3$ and $NaHCO_3$) in combination to oily sludge yield ceramic products. Likewise, by replacing natural kaolin with 5% solid petroleum waste in a porcelain stoneware tile body/composition and then burned at 1240 °C revealed that it reduces flexural strength, linear shrinkage, density. While increasing water absorption in the tile products manufactured and therefore it proves that, oily wastes can be also reused to manufacture porcelain stoneware tiles (Jagaba et al. 2022).

Besides this, less than 50% of fly ash quantity, can be used to build floor as well as wall tiles as initiated by Central Power Research Institute, Bangalore, India (Gorai 2017).

12.9.1.5 Insulation Materials

Advanced research and technologies have revealed that apart from constructional and building material items, coal fly ash can be extensively used for manufacturing lightweight aerospace and automotive composites. This paper showcased a eco-friendly cum green technology for converting fly ash into an advanced and novel aerogel, leading to fabricating thermal and acoustic insulative materials (Duong et al. 2021). Along with that, coal fly ashes are also beneficial in developing recycled polyvinyl chloride (r-PVC) to produce door or window sub-frame (DWSF) composite materials. Hence, estabilising CFA from trash to treasure. Value-added application of CFA to CFA-based DWSFs is simple, robust, and easy to industrialize, which allows large-scale production standards, and from the perspective of "carbon neutrality" as well as production costs (Li et al. 2022).

12.9.2 Agricultural Sector

Talking of the influence of coal combustion residues in cultivation, they are renowned agents for increasing soil quality, growth and development of cultivars, nutrient availability etc. Fly ash when applied on farmlands, promotes macro and micro nutrients (calcium, magnesium, aluminium, sodium, copper, chromium, boron etc.), improves the physical health of soil, pH balance, fertility goes high and water rentention capacity rises. Consequently, 10–40% rise in crop yield is observed and it can also change the soil texture from sandy or clayey to loamy. As per 2018–2019 CEA (Central Electricity Authority) reports, agriculture use of fly ash is 1.38 billion tonnes that sums up to 0.63% of total fly ash generation (Yousuf et al. 2020).

Species of crops like alfalfa, barley, bermuda grass, white clover, field corn, millet, carrots, onion, beans, cabbage, potatoes, tomatoes, brinjal, cabbage, sunflower, groundnuts spectacularly develops in fly ash containing soil and improved yield condition is observed. Moreover, involvement of fly ash along with chemical fertilizer and organic materials can save chemical fertilizer application and enhance efficiency (Gorai 2017).

Flue gas desulfurization (FGD) technique is a set of technology that removes sulphur dioxide (SO_2) present in the flue gas, which is exhausted from the thermal power plant during electricity generation or other fossil based industries. Resulting to calcium sulphite and FGD gypsum also termed as FGD products, which are further utilised in agricultural land. They are a good source of soil nutrients needed for the plant's growth and also used for amelioration of the soil acidity, salinity and alkalinity (Sreshma et al. 2024). Excessive direct application of Flue Gas Desulfurization Gypsum (FGDG), a by-product of scrubbing sulfur from combustion gases can also be harmful for the crop and soil quality. Such toxic trace elements can deaccelerate soil health, crop growth and quality, and impede heavy metal uptake by the plants. Presence of arsenic, mercury, cadmium, lead, and selenium etc. in FGDG has reduced its beneficial value and mass usage in agricultural sector.

Reports of Cross-State Air Pollution Rule (CSAPR), finalized by the U.S. EPA in 2011, that SO_2 emissions from coal combustion plants needs to be reduced and treated prior to being released into the atmosphere Hence, this resulted in U.S. coal-fired power plants for adopting flue gas desulfurization (FGD) technologies. FGDG wide application in agriculture also deals with remediation of soil eutrophication and reduction of soil erosion (Fig. 12.3).

Fig. 12.3 Novel products derived from hydrocarbon waste residues

12.10 Challenges

Environmental pollution to date is considered a major problem, and traditional methods of remediation face significant challenges. Hydrocarbon contaminants are diverse in composition and can persist in the environment for an extended period due to their stability and durability against degradation. Physical methods like flotation need chemicals and there is an issue of sludge handling when coagulant is used. Furthermore, techniques like incineration and sonication require careful monitoring and maintenance which further elevate the operational expenses making these methods less preferred (Kundu and Mishra 2018). Chemical methods like contaminated soil-washing, using synthetic detergents, and oxidative treatments might not be able to completely remove the hydrocarbon contaminants which might lead to persistent contamination and also pose a risk to the soil and water quality. These methods of physical and chemical removal also generate unwanted by-products that pose additional environmental risks followed by their limited applicability in certain environments.

Bioremediation which is considered a sustainable technique has its fair share of challenges. A study on bioremediation of oil spills mentioned hindered degradation of PAHs, poor biodegradability of asphaltenes, and inefficient decay in oxygen-deprived environments as the challenges faced in their study (Behera et al. 2022a, b). For plant-assisted remediation techniques, environmental factors like temperature, weather conditions, and availability of nutrients play a significant role in the remediation efficacy, therefore, implementing plant-assisted remediation techniques can pose a big challenge for any industry or stakeholder.

Given that bioremediation still faces significant challenges, there is ongoing encouragement for innovative approaches to enhance the efficiency and applicability of the remediation strategies for hydrocarbon-contaminated environments.

12.11 Future Perspectives

Genetic engineering has gained prominence in the field of microbial remediation as more and more techniques are being used to develop microbial strains with improved hydrocarbon-decaying properties. Genetically modified plants too have the potential to decontaminate specific pollutants by leveraging discovered enzymes, and genes according to Sarma and Prasad (2019). Several research studies by Kumar et al. (2020) have stated the efficacy of GMOs in cleaning up a wide range of contaminated lands, and when used in combination with assisted bioremediation approaches, can be seen as a promising approach towards reducing pollution from the environment. Leveraging omics technologies can help researchers make informed decisions regarding the design of tailored bioremediation strategies and refining processes like bioaugmentation and biostimulation by using novel strategies will help overcome the environmental constraints.

However, recovery and extraction of oil and other substances (acid gases) from hydrocarbon waste residues (oil sludge, flue gases, boiler slag, botton ash, fly ash) is another prominent strategy to combat demand and supply, detrimental consequences faced by our Earth, cost factor thus, promoting re-utilization. In addition, upscaling the application of Coal Fly Ash (CFA) and Coal Bottom Ash (CBA), boiler slag etc. to manufacture diverse sort of constructional materials as well as in agricultural sector can not only reduce ecological concerns, rather elevate sustainability, affordability and eco-friendliness.

Continued collaboration between policymakers, researchers and practitioners is crucial to expedite the integration of bioremediation, production of recycled products from waste substance and reuseing them with advanced technology and scientifc studies. This will promote restoration of ecosystem and complete erradication of contamination related issues worldwide.

12.12 Conclusion

The generation of petroleum waste has long been an environmental issue. These wastes have a detrimental impact on human lives as well as the environment. Such toxic compounds can enter the human body through various routes of exposure and to combat these several techniques of removal have been deployed out of which bioremediation has gained massive importance. Bioremediation is a cheaper, eco-friendly alternative for the restoration of hydrocarbon polluted sites. This review provides a better understanding on various hydrocarbons and their consecutive after-effects on survivng species including humans on this planet. Along with the environmental hazards and ecological depressions. Parallelly, this review delves into the synthesis of novel recycled products, which needs further investigation and acceptance globally. The knowledge gained so far represents the enforcement of novel research and technologies, the development of fundamental cum sustainable bioremediation techniques, recycling protocols and standardised laws for proper eradication of hydrocarbon pollutants. Thus, preserving and conserving our green planet for future generations.

Declarations

Ethical Approval and Consent to Participate Not applicable.

Competing Interests The authors declare no competing interests.

References

Abdel-Shafy HI, Mansour MSM (2016) A review on polycyclic aromatic hydrocarbons: Source, environmental impact, effect on human health and remediation. Egypt J Pet 25(1):107–123. https://doi.org/10.1016/j.ejpe.2015.03.011

Abdullah SR, Al-Baldawi IA, Almansoory AF, Purwanti IF, Al-Sbani NH, Sharuddin SS (2020) Plant-assisted remediation of hydrocarbons in water and soil: application, mechanisms, challenges and opportunities. Chemosphere 247:125932. https://doi.org/10.1016/j.chemosphere.2020.125932

Afegbua SL, Batty LC (2018) Effect of single and mixed polycyclic aromatic hydrocarbon contamination on plant biomass yield and pah dissipation during phytoremediation. Environ Sci Pollut Res 25(19):18596–18603. https://doi.org/10.1007/s11356-018-1987-1

Ahmed S, Kumari K, Singh D (2023) Different strategies and bio-removal mechanisms of petroleum hydrocarbons from contaminated sites. Arab Gulf J Sci Res. https://doi.org/10.1108/agjsr-09-2022-0172

Akl FM, Ahmed SI, El-Sheekh MM, Makhlof ME (2023) Bioremediation of N-alkanes, polycyclic aromatic hydrocarbons, and heavy metals from wastewater using seaweeds. Environ Sci Pollut Res 30(47):104814–104832. https://doi.org/10.1007/s11356-023-29549-8

Alaidaroos BA (2023) Advancing eco-sustainable bioremediation for hydrocarbon contaminants: challenges and solutions. Processes 11(10):3036. https://doi.org/10.3390/pr11103036

Alanbary SR, Abdullah SR, Al-Baldawi IA, Abu Hasan H, Anuar N, Othman AR, Suja' F (2019) Phytotoxicity of contaminated sand containing crude oil sludge on *Ludwigia octovalvis*. J Ecol Eng 20(11):246–255. https://doi.org/10.12911/22998993/113189

Ali N, Dashti N, Khanafer M, Al-Awadhi H, Radwan S (2020) Bioremediation of soils saturated with spilled crude oil. Sci Rep 10(1). https://doi.org/10.1038/s41598-019-57224-x

Alotaibi F, Hijri M, St-Arnaud M (2021) Overview of approaches to improve rhizoremediation of petroleum hydrocarbon-contaminated soils. Appl Microbiol 1(2):329–351. https://doi.org/10.3390/applmicrobiol1020023

Ambaye TG, Chebbi A, Formicola F, Prasad S, Gomez FH, Franzetti A, Vaccari M (2022) Remediation of soil polluted with petroleum hydrocarbons and its reuse for agriculture: recent progress, challenges, and perspectives. Chemosphere 293:133572. https://doi.org/10.1016/j.chemosphere.2022.133572

Ammar SH, Khadim HJ, Mohamed AI (2018) Cultivation of *Nannochloropsis oculata* and *Isochrysis galbana* microalgae in produced water for bioremediation and biomass production. Environ Technol Innov 10:132–142. https://doi.org/10.1016/j.eti.2018.02.002

Amran M, Fediuk R, Murali G, Avudaiappan S, Ozbakkaloglu T, Vatin N, Karelina M, Klyuev S, Gholampour A (2021) Fly ash-based eco-efficient concretes: a comprehensive review of the short-term properties. Materials 14(15):4264. https://doi.org/10.3390/ma14154264

Anees M, Qayyum A, Jamil M, Rehman F, Abid M, Malik MS, Yunas M, Ullah K (2020) Role of halotolerant and chitinolytic bacteria in phytoremediation of saline soil using spinach plant. Int J Phytoremed 22(6):653–661. https://doi.org/10.1080/15226514.2019.1707160

Anekwe IM, Isa YM (2021) Comparative evaluation of wastewater and bioventing system for the treatment of acid mine drainage contaminated soils. Water-Energy Nexus 4:134–140. https://doi.org/10.1016/j.wen.2021.08.001

Argiz C, Sanjuán MÁ, Menéndez E (2017) Coal bottom ash for Portland cement production. Adv Mater Sci Eng 2017:1–7. https://doi.org/10.1155/2017/6068286

Asif Z, Chen Z, An C, Dong J (2022) Environmental impacts and challenges associated with oil spills on shorelines. J Mar Sci Eng, 10(6):762. https://doi.org/10.3390/jmse10060762

Asim N, Badiei M, Torkashvand M, Mohammad M, Alghoul MA, Gasaymeh SS, Sopian K (2021) Wastes from the petroleum Industries as sustainable resource materials in construction sectors: opportunities, limitations, and directions. J Clean Prod 284:125459. https://doi.org/10.1016/j.jclepro.2020.125459

Avona A, Capodici M, Di Trapani D, Giustra MG, Greco Lucchina P, Lumia L, Di Bella G, Viviani G (2022) Preliminary insights about the treatment of contaminated marine sediments by means of bioslurry reactor: process evaluation and microbiological characterization. Sci Total Environ 806:150708. https://doi.org/10.1016/j.scitotenv.2021.150708

Bala K, Anshul, Verma S, Dogra A, Das AP, Kumar R, Nandini (2024) Sustainable management for urban plastic waste generation. In: Sustainable Management of Urban Plastic Waste Through Circular Economic Approaches, pp 211–236. https://doi.org/10.1201/9781003465447-13

Bala K, Dogra A, Lata P, Das AP (2024) Discussing sources and biological-chemical remediation approaches for microplastic pollution. In: Sustainable Microbial Technology for Synthetic and Cellulosic Microfiber Bioremediation, pp 37–59. https://doi.org/10.1007/978-3-031-62600-5_3

Bala S, Garg D, Thirumalesh BV, Sharma M, Sridhar K, Inbaraj BS, Tripathi M (2022) Recent strategies for bioremediation of emerging pollutants: a review for a green and sustainable environment. Toxics 10(8):484. https://doi.org/10.3390/toxics10080484

Bao H, Wang J, Zhang H, Li J, Li H, Wu F (2020) Effects of biochar and organic substrates on biodegradation of polycyclic aromatic hydrocarbons and microbial community structure in pahs-contaminated soils. J Hazard Mater 385:121595. https://doi.org/10.1016/j.jhazmat.2019.121595

Barnes NM, Khodse VB, Lotlikar NP, Meena RM, Damare SR (2017) Bioremediation potential of hydrocarbon-utilizing fungi from select marine niches of India. 3 Biotech 8(1). https://doi.org/10.1007/s13205-017-1043-8

Barone V, Tangherlini R, Dell'Anno D, Corinaldesi (2019) Marine fungi: biotechnological perspectives from deep-hypersaline anoxic basins. Diversity 11(7):113. https://doi.org/10.3390/d11070113

Behera ID, Basak G, Kumar RR, Sen R, Meikap BC (2020) Treatment of petroleum refinery sludge by petroleum degrading bacterium *Stenotrophomonas pavanii* IRB19 as an efficient novel technology. J Environ Sci Health Part A Toxic Hazard Subst Environ Eng 56(2). https://doi.org/10.1080/10934529.2020.1866924

Behera ID, Meikap BC, Sen R (2022a) Performance evaluation of bacterial consortium for biodegradation of total petroleum hydrocarbon: a comparative strategic biostimulation study. Res Square

Behera ID, Nayak M, Mishra A, Meikap BC, Sen R (2022b) Strategic implementation of integrated bioaugmentation and biostimulation for efficient mitigation of petroleum hydrocarbon pollutants from terrestrial and aquatic environment. Mar Pollut Bull 177:113492. https://doi.org/10.1016/j.marpolbul.2022.113492

Besha AT, Bekele DN, Naidu R, Chadalavada S (2018) Recent advances in surfactant-enhanced in-situ chemical oxidation for the remediation of non-aqueous phase liquid contaminated soils and aquifers. Environ Technol Innov 9:303–322. https://doi.org/10.1016/j.eti.2017.08.004

Bielecka A, Kulczycka J (2020) Coal combustion products management toward a circular economy—a case study of the coal power plant sector in Poland. Energies 13(14):3603. https://doi.org/10.3390/en13143603

Borowiec SM, Szpyrka E, Książek-Trela P, Podbielska M (2022) Simultaneous determination of multi-class pesticide residues and pahs in plant material and soil samples using the optimized QUECHERS method and tandem mass spectrometry analysis. Molecules, 27(7):2140. https://doi.org/10.3390/molecules27072140

Chekroun KB, Sánchez E, Baghour M (2014) The role of algae in bioremediation of organic pollutants. Int Res J Public Environ Health 1(2):19–32

Chen K, Chang Y, Chiou W (2015) Remediation of diesel-contaminated soil using *in situ* chemical oxidation (isco) and the effects of common oxidants on the indigenous microbial community: a comparison study. J Chem Technol Biotechnol 91(6):1877–1888. https://doi.org/10.1002/jctb.4781

Cheng K, Li J-Y, Wang Y, Ji W-W, Cao Y (2022) Characterization and risk assessment of airborne polycyclic aromatic hydrocarbons from open burning of municipal solid waste. Front Environ Sci 10. https://doi.org/10.3389/fenvs.2022.861770

Cordiano R, Papa V, Cicero N, Spatari G, Allegra A, Gangemi S (2022) Effects of benzene: hematological and hypersensitivity manifestations in resident living in oil refinery areas. Toxics 10(11):678. https://doi.org/10.3390/toxics10110678

Cruz SL, Rivera-García MT, Woodward JJ (2014) Review of toluene actions: clinical evidence, animal studies, and molecular targets. J Drug Alcohol Res 3:1–8. https://doi.org/10.4303/jdar/235840

Cui J-Q, He Z-Q, Ntakirutimana S, Liu Z-H, Li B-Z, Yuan Y-J (2023) Artificial mixed microbial system for polycyclic aromatic hydrocarbons degradation. Front Microbiol 14. https://doi.org/10.3389/fmicb.2023.1207196

Czaplicka M, Klyta J, Komosiński B, Konieczny T, Janoszka K (2021) Comparison of carbonaceous compounds emission from the co-combustion of coal and waste in boilers used in residential heating in Poland, Central Europe. Energies 14(17):5326. https://doi.org/10.3390/en14175326

Das A, Dey S, Bala K, Gnanasekaran R, Das AP (2024) Microbial bioremediation technology for sustainable treatment and management of synthetic microfiber waste. In: Sustainable Microbial Technology for Synthetic and Cellulosic Microfiber Bioremediation, pp 221–238. https://doi.org/10.1007/978-3-031-62600-5_12

Das A, Dey S, Das AP (2024) Ecotoxicological impacts of synthetic microfiber pollutants and development of sustainable mitigation strategies. Environ Chem Ecotoxicol. https://doi.org/10.1016/J.ENCECO.2024.12.003

Das A. Prasad, Ghosh, Shreya (2023) Manganese mining microorganisms

Das AP, Sukla LB, Pradhan N, Nayak S (2011) Manganese biomining: a review. Biores Technol 102(16):7381–7387. https://doi.org/10.1016/J.BIORTECH.2011.05.018

Das B, Deka S (2019) A cost-effective and environmentally sustainable process for phycoremediation of oil field formation water for its safe disposal and reuse. Sci Rep 9(1). https://doi.org/10.1038/s41598-019-51806-5

Das AP, Dutta K, Khatun R, Behera ID, Singh S, Mishra S (2023) Microfiber pollution and its microbial mitigation: a review on current trends and future prospects. J Taiwan Inst Chem Eng 105104. https://doi.org/10.1016/J.JTICE.2023.105104

Dell' Anno F, Rastelli E, Sansone C, Brunet C, Ianora A, Dell' Anno A (2021) Bacteria, fungi and microalgae for the bioremediation of marine sediments contaminated by petroleum hydrocarbons in the omics era. Microorganisms 9(8):1695. https://doi.org/10.3390/microorganisms9081695

Dey S, Bissoyi A, Koteswara Reddy G, Das AP (2024) Synthetic microfibers: An urban emerging pollutant demanding for sustainable environmental solutions. In: Sustainable Microbial Technology for Synthetic and Cellulosic Microfiber Bioremediation, pp 107–121. https://doi.org/10.1007/978-3-031-62600-5_7

Dey S, Dash N, Ray SK, Priyadarsini S, Chhotaray C, Das AP (2024) Occurrence and quantification of tire wear plastics pollutants from highway city dust of Bhubaneswar City of Odisha. In: Sustainable Microbial Technology for Synthetic and Cellulosic Microfiber Bioremediation, pp 77–89. https://doi.org/10.1007/978-3-031-62600-5_5

Dey S, Tripathy B, Kumar MS, Das AP (2023) Ecotoxicological consequences of manganese mining pollutants and their biological remediation. Environ Chem Ecotoxicol. https://doi.org/10.1016/J.ENCECO.2023.01.001

Dopico M, Gómez A (2015) Review of the current state and main sources of dioxins around the world. J Air Waste Manag Assoc 65(9):1033–1049. https://doi.org/10.1080/10962247.2015.1058869

Dubinsky EA, Conrad ME, Chakraborty R, Bill M, Borglin SE, Hollibaugh JT, Mason OU, Piceno YM, Reid FC, Stringfellow WT, Tom LM, Hazen TC, Andersen GL (2013) Succession of hydrocarbon-degrading bacteria in the aftermath of the *deepwater horizon* oil spill in the Gulf of Mexico. Environ Sci Technol 47(19):10860–10867. https://doi.org/10.1021/es401676y

Duong HM, Ling NRB, Thai QB, Le DK, Nguyen PTT, Goh XY, Phan-Thien N (2021) A novel aerogel from thermal power plant waste for thermal and acoustic insulation applications. Waste Manage 124:1–7. https://doi.org/10.1016/j.wasman.2021.01.031

Duque E, Udaondo Z, Molina L, de la Torre J, Godoy P, Ramos JL (2022) Providing octane degradation capability to *pseudomonas putida*kt2440 through the horizontal acquisition of *oct* genes located on an integrative and conjugative element. Environ Microbiol Rep 14(6):934–946. https://doi.org/10.1111/1758-2229.13097

Effendi AJ, Ramadan BS, Helmy Q (2022) Enhanced remediation of hydrocarbons contaminated soil using electrokinetic soil flushing—landfarming processes. Bioresour Technol Rep 17:100959. https://doi.org/10.1016/j.biteb.2022.100959

Eze MO, Hose GC, George SC (2020) Assessing the effect of diesel fuel on the seed viability and germination of *Medicago sativa* using the event-time model. Plants 9(9):1062. https://doi.org/10.3390/plants9091062

Faboya OL, Sojinu SO, Oguntuase BJ, Sonibare OO (2020) Impact of forest fires on polycyclic aromatic hydrocarbon concentrations and stable carbon isotope compositions in burnt soils from Tropical Forest, Nigeria. Sci Afr 8. https://doi.org/10.1016/j.sciaf.2020.e00331

Fan M, Xiao X, Guo Y, Zhang J, Wang E, Chen W, Lin Y, Wei G (2018) Enhanced phytoremdiation of *Robinia pseudoacacia* in heavy metal-contaminated soils with rhizobia and the associated bacterial community structure and function. Chemosphere 197:729–740. https://doi.org/10.1016/j.chemosphere.2018.01.102

Freidman BL, Gras SL, Snape I, Stevens GW, Mumford KA (2017) A bio-reactive barrier sequence for petroleum hydrocarbon capture and degradation in low nutrient environments. Int Biodeterior Biodegradation 116:26–37. https://doi.org/10.1016/j.ibiod.2016.09.025

Futughe AE, Jones H, Purchase D (2023) A novel technology of solarization and phytoremediation enhanced with biosurfactant for the sustainable treatment of pah-contaminated soil. Environ Geochem Health 45(6):3847–3863. https://doi.org/10.1007/s10653-022-01460-0

Gabryszewska M, Gworek B (2020) Impact of municipal and industrial waste incinerators on PCBS content in the environment. PLOS ONE 15(11). https://doi.org/10.1371/journal.pone.0242698

Galil NI, Wolf D (2001) Removal of hydrocarbons from petrochemical wastewater by dissolved air flotation. Water Sci Technol 43(8):107–113. https://doi.org/10.2166/wst.2001.0476

Ganesan H, Sachdeva A, Petrounias P, Lampropoulou P, Sharma PK, Kumar A (2023) Impact of fine slag aggregates on the final durability of coal bottom ash to produce sustainable concrete. Sustainability 15(7):6076. https://doi.org/10.3390/su15076076

García-Sánchez M, Košnář Z, Mercl F, Aranda E, Tlustoš P (2018) A comparative study to evaluate natural attenuation, mycoaugmentation, phytoremediation, and microbial-assisted phytoremediation strategies for the bioremediation of an aged pah-polluted soil. Ecotoxicol Environ Saf 147:165–174. https://doi.org/10.1016/j.ecoenv.2017.08.012

Ghasemi Y, Rasoul-Amini S, Fotooh-Abadi E (2011) The Biotransformation, biodegradation, and bioremediation of organic compounds by microalgae1. J Phycol 47(5):969–980. https://doi.org/10.1111/j.1529-8817.2011.01051.x

Ghasemi N, Gbeddy G, Egodawatta P, Zare F, Goonetilleke A (2019) Removal of polycyclic aromatic hydrocarbons from wastewater using dual-mode ultrasound system. Water Environ J 34(S1):425–434. https://doi.org/10.1111/wej.12540

Ghosh S, Rusyn I, Dmytruk OV, Dmytruk KV, Onyeaka H, Gryzenhout M, Gafforov Y (2023) Filamentous fungi for sustainable remediation of pharmaceutical compounds, heavy metal and oil hydrocarbons. Front Bioeng Biotechnol 11. https://doi.org/10.3389/fbioe.2023.1106973

Gong P, Wang X (2021) Forest fires enhance the emission and transport of persistent organic pollutants and polycyclic aromatic hydrocarbons from the central Himalaya to the Tibetan Plateau. Environ Sci Technol Lett 8(7):498–503. https://doi.org/10.1021/acs.estlett.1c00221

González N, Domingo JL (2021) Polychlorinated dibenzo-p-dioxins and dibenzofurans (PCDD/FS) in food and human dietary intake: an update of the scientific literature. Food Chem Toxicol 157:112585. https://doi.org/10.1016/j.fct.2021.112585

Gorai S (2017) Utilization of fly ash for sustainable environment management. J Mater Environ Sci 2018 9(2):385–393. https://doi.org/10.26872/jmes.2018.9.2.42

Gorbaev AV, Gorlenko NV, Timofeeva SS (2020) Use of oil sludge as a combustible additive for the production of bricks. IOP Conf Ser Mater Sci Eng 962(4):042031. https://doi.org/10.1088/1757-899x/962/4/042031

Gregson BH, Metodieva G, Metodiev MV, Golyshin PN, McKew BA (2020) Protein expression in the obligate hydrocarbon-degrading psychrophile *oleispira antarctica* RB-8 during alkane degradation and cold tolerance. Environ Microbiol 22(5):1870–1883. https://doi.org/10.1111/1462-2920.14956

Guarino C, Spada V, Sciarrillo R (2017) Assessment of three approaches of bioremediation (natural attenuation, landfarming and bioagumentation—assistited landfarming) for a petroleum hydrocarbons contaminated soil. Chemosphere 170:10–16. https://doi.org/10.1016/j.chemosphere.2016.11.165

Haider FU, Ejaz M, Cheema SA, Khan MI, Zhao B, Liqun C, Salim MA, Naveed M, Khan N, Núñez-Delgado A, Mustafa A (2021) Phytotoxicity of petroleum hydrocarbons: sources, impacts and remediation strategies. Environ Res 197:111031. https://doi.org/10.1016/j.envres.2021.111031

Hamad AA, Moubasher HA, Moustafa YM, Mohamed NH, Abd-el rhim EH (2021) Petroleum hydrocarbon bioremediation using native fungal isolates and consortia. Sci World J 2021:1–13. https://doi.org/10.1155/2021/6641533

Hameed A, Rasool AM, Ibrahim YE, Afzal MF, Qazi AU, Hameed I (2022) Utilization of fly ash as a viscosity-modifying agent to produce cost-effective, self-compacting concrete: a sustainable solution. Sustainability 14(18):11559. https://doi.org/10.3390/su141811559

Hao X, Chen Q, van Loosdrecht MCM, Li J, Jiang H (2020) Sustainable disposal of excess sludge: incineration without anaerobic digestion. Water Res 170:115298. https://doi.org/10.1016/j.watres.2019.115298

Hara T, Takatsuka Y, Nakata E, Morii T (2021) Augmentation of an engineered bacterial strain potentially improves the cleanup of PCB water pollution. Microbiol Spect 9(3). https://doi.org/10.1128/spectrum.01926-21

Hidayati N, Hamim H, Sulistyaningsih YC (2018) Phytoremediation of petroleum hydrocarbon using three mangrove species applied through tidal bioreactor. Biodiversitas J Biol Divers 19(3):786–792. https://doi.org/10.13057/biodiv/d190305

Hoang SA, Lamb D, Seshadri B, Sarkar B, Choppala G, Kirkham MB, Bolan NS (2021) Rhizoremediation as a green technology for the remediation of petroleum hydrocarbon-contaminated soils. J Hazard Mater 401:123282. https://doi.org/10.1016/j.jhazmat.2020.123282

Honda M, Suzuki N (2020) Toxicities of polycyclic aromatic hydrocarbons for aquatic animals. Int J Environ Res Public Health 17(4):1363. https://doi.org/10.3390/ijerph17041363

Hong Y-W, Yuan D-X, Lin Q-M, Yang T-L (2008) Accumulation and biodegradation of phenanthrene and fluoranthene by the algae enriched from a mangrove aquatic ecosystem. Mar Pollut Bull 56(8):1400–1405. https://doi.org/10.1016/j.marpolbul.2008.05.003

Hsu Y-C, Chang S-H, Chang MB (2021) Emissions of pahs, PCDD/FS, DL-PCBS, chlorophenols and chlorobenzenes from municipal waste incinerator cofiring industrial waste. Chemosphere 280:130645. https://doi.org/10.1016/j.chemosphere.2021.130645

Jaain R, Patel A (2019) Bioremediation of gurugram–faridabad dumpsite at Bandhwari. Waste Valoris Recycl 433–440. https://doi.org/10.1007/978-981-13-2784-1_40

Jabbar SAM (2018) Effect of oil pollution on growing and diversity of aquatic plants

Jagaba AH, Kutty SR, Lawal IM, Birniwa AH, Affam AC, Yaro NS, Usman AK, Umaru I, Abubakar S, Noor A, Soja UB, Yakubu AS (2022) Circular economy potential and contributions of petroleum industry sludge utilization to environmental sustainability through engineered processes—a review. Clean Circul Bioecon 3:100029. https://doi.org/10.1016/j.clcb.2022.100029

Janssen DB, Stucki G (2020) Perspectives of genetically engineered microbes for groundwater bioremediation. Environ Sci Process Impacts 22(3):487–499. https://doi.org/10.1039/c9em00601j

Jeevanantham S, Saravanan A, Hemavathy RV, Kumar PS, Yaashikaa PR, Yuvaraj D (2019) Removal of toxic pollutants from water environment by phytoremediation: a survey on application and future prospects. Environ Technol Innov 13:264–276. https://doi.org/10.1016/j.eti.2018.12.007

Jia W, Cheng L, Tan Q, Liu Y, Dou J, Yang K, Yang Q, Wang S, Li J, Niu G, Zheng L, Ding A (2023) Response of the soil microbial community to petroleum hydrocarbon stress shows a threshold effect: research on aged realistic contaminated fields. Front Microbiol 14. https://doi.org/10.3389/fmicb.2023.1188229

John A, Das AP (2024) Government strategies for achieving circular economy for plastic waste management in Southeast Asia. In: Sustainable Management of Urban Plastic Waste Through Circular Economic Approaches, pp 72–82. https://doi.org/10.1201/9781003465447-6/GOVERNMENT-STRATEGIES-ACHIEVING-CIRCULAR-ECONOMY-PLASTIC-WASTE-MANAGEMENT-SOUTHEAST-ASIA-AKBAR-JOHN-ALOKPRASAD-DAS

Kelechi SE, Adamu M, Uche OA, Okokpujie IP, Ibrahim YE, Obianyo II (2022) A comprehensive review on coal fly ash and its application in the construction industry. Cogent Eng 9(1). https://doi.org/10.1080/23311916.2022.2114201

Kharitonov S, Shchegolkova N, Alexandrova A, Saynchuk A, Michel P, Maciejewski K, Gautier M, Gourdon R, Semenov M, Krasnov G (2022) Taxonomic diversity of fungi and bacteria in Azoé-NP® vertical flow constructed wetlands. Water 14(5):698. https://doi.org/10.3390/w14050698

Kim K-H, Jahan SA, Kabir E, Brown RJC (2013) A review of airborne polycyclic aromatic hydrocarbons (pahs) and their human health effects. Environ Int 60:71–80. https://doi.org/10.1016/j.envint.2013.07.019

King MD, Elliott JE, Williams TD (2021) Effects of petroleum exposure on birds: a review. Sci Total Environ 755:142834. https://doi.org/10.1016/j.scitotenv.2020.142834

Koe WS, Lee JW, Chong WC, Pang YL, Sim LC (2019) An overview of photocatalytic degradation: photocatalysts, mechanisms, and development of photocatalytic membrane. Environ Sci Pollut Res 27(3):2522–2565. https://doi.org/10.1007/s11356-019-07193-5

Krzyszczak A, Czech B (2021) Occurrence and toxicity of polycyclic aromatic hydrocarbons derivatives in environmental matrices. Sci Total Environ 788:147738. https://doi.org/10.1016/j.scitotenv.2021.147738

Kumar RS, Singh D, Bose SK, Trivedi PK (2020) Biodegradation of environmental pollutant through pathways engineering and genetically modified organisms approaches. Microorg Sustain Environ Health 137–165. https://doi.org/10.1016/b978-0-12-819001-2.00007-3

Kumar Yadav K, Gupta N, Kumar A, Reece LM, Singh N, Rezania S, Ahmad Khan S (2018) Mechanistic understanding and holistic approach of phytoremediation: a review on application and future prospects. Ecol Eng 120:274–298. https://doi.org/10.1016/j.ecoleng.2018.05.039

Kundu P, Mishra IM (2018) Treatment and reclamation of hydrocarbon-bearing oily wastewater as a hazardous pollutant by different processes and technologies: a state-of-the-art review. Rev Chem Eng 35(1):73–108. https://doi.org/10.1515/revce-2017-0025

Kuppusamy S, Maddela NR, Megharaj M, Venkateswarlu K (2019) Impact of total petroleum hydrocarbons on human health. Total Petrol Hydrocarb 139–165. https://doi.org/10.1007/978-3-030-24035-6_6

Lahiri D, Nag M, Dey A, Sarkar T, Joshi S, Pandit S, Das AP, Pati S, Pattanaik S, Tilak VK, Ray RR (2021) Biofilm mediated degradation of petroleum products. Geomicrobiol J 1–10. https://doi.org/10.1080/01490451.2021.1968979

Li Q, Liu J, Gadd GM (2020) Fungal bioremediation of soil co-contaminated with petroleum hydrocarbons and toxic metals. Appl Microbiol Biotechnol 104(21):8999–9008. https://doi.org/10.1007/s00253-020-10854-y

Li Z, Wang G, Yan J, Qian Y, Guo S, Liu Y, Li D (2022) Value-added utilization of coal fly ash and recycled polyvinyl chloride in door or window sub-frame composites. Green Process Synth, 12(1). https://doi.org/10.1515/gps-2023-0002

Lim MW, Lau EV, Poh PE (2018) Micro-macrobubbles interactions and its application in flotation technology for the recovery of high density oil from contaminated sands. J Petrol Sci Eng 161:29–37. https://doi.org/10.1016/j.petrol.2017.11.064

Liu J, Xin X, Zhou Q (2018) Phytoremediation of contaminated soils using ornamental plants. Environ Rev 26(1):43–54. https://doi.org/10.1139/er-2017-0022

Mahjoubi M, Cappello S, Souissi Y, Jaouani A, Cherif A (2018) Microbial bioremediation of petroleum hydrocarbon—contaminated marine environments. Recent Insights Petrol Sci Eng. https://doi.org/10.5772/intechopen.72207

Marković JZ, Marinković AD, Savić JZ, Mladenović MR, Erić MD, Marković ZJ, Ristić M (2023) Risk evaluation of pollutants emission from coal and coal waste combustion plants and environmental impact of Fly Ash landfilling. Toxics, 11(4):396. https://doi.org/10.3390/toxics11040396

Mekonnen BA, Aragaw TA, Genet MB (2024) Bioremediation of petroleum hydrocarbon contaminated soil: a review on principles, degradation mechanisms, and advancements. Front Environ Sci 12. https://doi.org/10.3389/fenvs.2024.1354422

Mfarrej MF, Wang X, Fahid M, Saleem MH, Alatawi A, Ali S, Shabir G, Zafar R, Afzal M, Fahad S (2022) Floating treatment wetlands (FTWs) is an innovative approach for the remediation of petroleum hydrocarbons-contaminated water. J Plant Growth Regul 42(3):1402–1420. https://doi.org/10.1007/s00344-022-10674-6

Michel PJ, Banat IM (2024) Unravelling diatoms' potential for the bioremediation of oil hydrocarbons in Marine Environments. Clean Technol, 6(1):93–115. https://doi.org/10.3390/cleantechnol6010007

Mishra S, Das AP (2021) Current treatment technologies for removal of microplastic and microfiber pollutants from wastewater. Wastew Treat Cutting Edge Mol Tools Tech Appl Aspects 237–251. https://doi.org/10.1016/B978-0-12-821881-5.00011-8

Mishra S, Dash D, Das AP (2024) Microbial degradation of polyester microfibers using indigenously isolated bacterial strain Exiguobacterium Sp. CLEAN – Soil, Air, Water, 52(12):e202300343. https://doi.org/10.1002/CLEN.202300343

Mishra S, Rath C, Das AP (2019) Marine microfiber pollution: a review on present status and future challenges. Mar Pollut Bull 140:188–197. https://doi.org/10.1016/J.MARPOLBUL.2019.01.039

Mishra A, Siddiqi H, Kumari U, Behera ID, Mukherjee S, Meikap BC (2021a) Pyrolysis of waste lubricating oil/waste motor oil to generate high-grade fuel oil: a comprehensive review. Renew Sustain Energy Rev 150:111446. https://doi.org/10.1016/j.rser.2021.111446

Mishra S, Rout PK, Das AP (2021b) Emerging microfiber pollution and its remediation. In: Prasad R (ed) Environmental and microbial biotechnology. Springer, Singapore, pp 247–266. https://doi.org/10.1007/978-981-15-5499-5_9

Mohammed SA, Koting S, Katman HY, Babalghaith AM, Abdul Patah MF, Ibrahim MR, Karim MR (2021) A review of the utilization of coal bottom ash (CBA) in the construction industry. Sustainability 13(14):8031. https://doi.org/10.3390/su13148031

Mohanta S, Pradhan B, Behera ID (2023) Impact and remediation of petroleum hydrocarbon pollutants on agricultural land: a review. Geomicrobiol J 1–15. https://doi.org/10.1080/01490451.2023.2243925

Mohanty M, Tripathy B, Koteswara Reddy G, Das AP (2024) Greener technologies for establishing restraint over microplastic pollution. In: Sustainable Microbial Technology for Synthetic and Cellulosic Microfiber Bioremediation, pp 239–258. https://doi.org/10.1007/978-3-031-62600-5_13

Mohd Pauzi NN, Ahmad Shah MS (2022) A review on the treatment of petroleum sludge and use as construction materials. Civil Sustain Urban Eng 2(1):12–20. https://doi.org/10.53623/csue.v2i1.74

Montano L, Pironti C, Pinto G, Ricciardi M, Buono A, Brogna C, Venier M, Piscopo M, Amoresano A, Motta O (2022) Polychlorinated biphenyls (PCBS) in the environment: occupational and exposure events, effects on human health and fertility. Toxics 10(7):365. https://doi.org/10.3390/toxics10070365

Moradi M, Vasseghian Y, Khataee A, Harati M, Arfaeinia H (2021) Ultrasound-assisted synthesis of FETIO3/GO nanocomposite for photocatalytic degradation of phenol under visible light irradiation. Sep Purif Technol 261:118274. https://doi.org/10.1016/j.seppur.2020.118274

Mukome FND, Buelow MC, Shang J, Peng J, Rodriguez M, Mackay DM, Pignatello JJ, Sihota N, Hoelen TP, Parikh SJ (2020) Biochar amendment as a remediation strategy for surface soils impacted by crude oil. Environ Pollut 265:115006. https://doi.org/10.1016/j.envpol.2020.115006

Nemati B, Baneshi MM, Akbari H, Dehghani R, Mostafaii G (2024) Phytoremediation of pollutants in oil-contaminated soils by *Alhagi camelorum*: evaluation and modeling. Sci Rep 14(1). https://doi.org/10.1038/s41598-024-56214-y

Nwidee LN, Theophilus S, Barifcani A, Sarmadivaleh M, Iglauer S (2016) Eor processes, opportunities and technological advancements. Chem Enhanc Oil Recov Pract Overv. https://doi.org/10.5772/64828

Ododo MM, Wabalo BK (2019) Polychlorinated biphenyls (PCBs) and their impacts on human health: a review. J Environ Pollut Human Health 7(2):73–77. https://doi.org/10.12691/jephh-7-2-3

Odukoya J, Lambert R, Sakrabani R (2019) Understanding the impacts of crude oil and its induced abiotic stresses on agrifood production: a review. Horticulturae 5(2):47. https://doi.org/10.3390/horticulturae5020047

Okoli AS, Blix T, Myhr AI, Xu W, Xu X (2021) Sustainable use of CRISPR/Cas in fish aquaculture: the biosafety perspective. Transgen Res 31(1):1–21. https://doi.org/10.1007/s11248-021-00274-7

Olasehinde TA, Olaniran AO (2022) Neurotoxicity of polycyclic aromatic hydrocarbons: a systematic mapping and review of neuropathological mechanisms. Toxics 10(8):417. https://doi.org/10.3390/toxics10080417

Othman N, Ismail Z, Selamat MI, Sheikh Abdul Kadir SH, Shibraumalisi NA (2022) A review of polychlorinated biphenyls (PCBS) pollution in the air: where and how much are we exposed to? Int J Environ Res Public Health 19(21):13923. https://doi.org/10.3390/ijerph192113923

Panchenko L, Muratova A, Dubrovskaya E, Golubev S, Turkovskaya O (2023) Natural and technical phytoremediation of oil-contaminated soil. Life 13(1):177. https://doi.org/10.3390/life13010177

Passarini MRZ, Rodrigues MVN, da Silva M, Sette LD (2011) Marine-derived filamentous fungi and their potential application for polycyclic aromatic hydrocarbon bioremediation. Mar Pollut Bull 62(2):364–370. https://doi.org/10.1016/j.marpolbul.2010.10.003

Paul A, Priyadarsini S, Karakuş S, Behera JK, Gnanasekaran R, Das AP (2024) Fate and accumulation of synthetic microfibers in the environment. In: Sustainable Microbial Technology for Synthetic and Cellulosic Microfiber Bioremediation, pp 61–76. https://doi.org/10.1007/978-3-031-62600-5_4

Pourfadakari S, Ghafari S, Takdastan A, Jorfi S (2021) A salt resistant biosurfactant produced by moderately halotolerant pseudomonas aeruginosa (AHV-KH10) and its application for bioremediation of diesel-contaminated sediment in saline environment. Biodegradation 32(3):327–341. https://doi.org/10.1007/s10532-021-09941-2

Priyadarsini S, Das AP (2024) Analytical and structural characterization of waste lithium-ion batteries for their effective recycling strategy. Environ Chem Ecotoxicol. https://doi.org/10.1016/j.enceco.2024.12.004

Priyadarsini S, Das AP (2025) Lithium bioleaching: A review on microbial-assisted sustainable technology for lithium bio-circularity. J Water Process Eng, 69:106744. https://doi.org/10.1016/j.jwpe.2024.106744

Radhakrishnan A, Balaganesh P, Vasudevan M, Natarajan N, Chauhan A, Arora J, Ranjan A, Rajput VD, Sushkova S, Minkina T, Basniwal RK, Kapardar R, Srivastav R (2023) Bioremediation of hydrocarbon pollutants: recent promising sustainable approaches, scope, and challenges. Sustainability 15(7):5847. https://doi.org/10.3390/su15075847

Radice RP, De Fabrizio V, Donadoni A, Scopa A, Martelli G (2023) Crude oil bioremediation: from bacteria to microalgae. Processes 11(2):442. https://doi.org/10.3390/pr11020442

Raj K, Das AP (2023) Lead pollution: impact on environment and human health and approach for a sustainable solution. Environ Chem Ecotoxicol 5:79–85. https://doi.org/10.1016/J.ENCECO.2023.02.001

Ramadass K, Megharaj M, Venkateswarlu K, Naidu R (2018) Bioavailability of weathered hydrocarbons in engine oil-contaminated soil: Impact of bioaugmentation mediated by pseudomonas spp. on bioremediation. Sci Total Environ 636:968–974. https://doi.org/10.1016/j.scitotenv.2018.04.379

Rastogi A, Kumar Paul V (2020) A critical review of the potential for fly ash utilisation in construction-specific applications in India. Environ Res Eng Manag 76(2):65–75. https://doi.org/10.5755/j01.erem.76.2.25166

Ray SK, Dey S, Das AP, John A (2024) Microplastic waste management through greener technologies: A review. In: Sustainable Management of Urban Plastic Waste Through Circular Economic Approaches, pp 237–261. https://doi.org/10.1201/9781003465447-14 RENEWABLE ENERGY GENERATION AND VALUE ADDITION FROM ENVIRONMENTAL. (2024)

Rezaei Somee M, Shavandi M, Dastgheib SM, Amoozegar MA (2018) Bioremediation of oil-based drill cuttings by a halophilic consortium isolated from oil-contaminated saline soil 3 Biotech 8(5). https://doi.org/10.1007/s13205-018-1261-8

Rocha e Silva NM, Meira HM, Almeida FC, de Soares da Silva R, Almeida DG, Luna JM, Rufino RD, Santos VA, Sarubbo LA (2018) Natural surfactants and their applications for heavy oil removal in industry. Sep Purif Rev 48(4):267–281. https://doi.org/10.1080/15422119.2018.1474477

Romero IC, Sutton T, Carr B, Quintana-Rizzo E, Ross SW, Hollander DJ, Torres JJ (2018) Decadal assessment of polycyclic aromatic hydrocarbons in mesopelagic fishes from the Gulf of Mexico reveals exposure to oil-derived sources. Environ Sci Technol 52(19):10985–10996. https://doi.org/10.1021/acs.est.8b02243

Saeed T, Sun G (2012) A review on nitrogen and organics removal mechanisms in subsurface flow constructed wetlands: dependency on environmental parameters, operating conditions and supporting media. J Environ Manage 112:429–448. https://doi.org/10.1016/j.jenvman.2012.08.011

Sahoo PK, Kim K, Powell MA, Equeenuddin SM (2016) Recovery of metals and other beneficial products from coal fly ash: a sustainable approach for fly ash management. Int J Coal Sci Technol 3(3):267–283. https://doi.org/10.1007/s40789-016-0141-2

Sales da Silva IG, Gomes de Almeida FC, Padilha da Rocha e Silva NM, Casazza AA, Converti A, Asfora Sarubbo L (2020) Soil bioremediation: overview of technologies and trends. Energies 13(18):4664. https://doi.org/10.3390/en13184664

Sampaio GR, Guizellini GM, da Silva SA, de Almeida AP, Pinaffi-Langley AC, Rogero MM, de Camargo AC, Torres EA (2021) Polycyclic aromatic hydrocarbons in foods: biological effects, legislation, occurrence, analytical methods, and strategies to reduce their formation. Int J Mol Sci 22(11):6010. https://doi.org/10.3390/ijms22116010

Samuel O, Othman MH, Kamaludin R, Dzinun H, Imtiaz A, Li T, El-badawy T, Khan AU, Puteh MH, Yuliwati E, Kurniawan TA (2023) Photocatalytic degradation of recalcitrant aromatic hydrocarbon compounds in oilfield-produced water: a critical review. J Clean Prod 415:137567. https://doi.org/10.1016/j.jclepro.2023.137567

Sankar TK, Kumar A, Mahto DK, Das KC, Narayan P, Fukate M, Awachat P, Padghan D, Mohammad F, Al-Lohedan HA, Soleiman AA, Ambade B (2023) The health risk and source assessment of polycyclic aromatic hydrocarbons (pahs) in the soil of industrial cities in India. Toxics 11(6):515. https://doi.org/10.3390/toxics11060515

Sarma H, Prasad MN (2019) Metabolic engineering of Rhizobacteria associated with plants for remediation of toxic metals and metalloids. Transg Plant Technol Remed Toxic Metals Metalloids 299–318. https://doi.org/10.1016/b978-0-12-814389-6.00014-6

Sazanova KV, Zelenskaya MS, Vlasov AD, Bobir SY, Yakkonen KL, Vlasov DY (2022) Microorganisms in superficial deposits on the stone monuments in Saint Petersburg. Microorganisms 10(2):316. https://doi.org/10.3390/microorganisms10020316

Sengupta K, Pal S (2021) A review on microbial diversity and genetic markers involved in methanogenic degradation of hydrocarbons: futuristic prospects of biofuel recovery from contaminated regions. Environ Sci Pollut Res 28(30):40288–40307. https://doi.org/10.1007/s11356-021-13666-3

Shukla BK, Gupta A, Gowda S, Srivastav Y (2023) Constructing a greener future: a comprehensive review on the sustainable use of fly ash in the construction industry and beyond. Mater Today Proc 93:257–264. https://doi.org/10.1016/j.matpr.2023.07.179

Simão FCP, Gravato C, Machado AL, Soares AMVM, Pestana JLT (2020) Toxicity of different polycyclic aromatic hydrocarbons (pahs) to the freshwater planarian *Girardia tigrina*. Environ Pollut 266:115185. https://doi.org/10.1016/j.envpol.2020.115185

Sivagami K, Padmanabhan K, Joy AC, Nambi IM (2019) Microwave (MW) remediation of hydrocarbon contaminated soil using spent graphite—an approach for waste as a resource. J Environ Manage 230:151–158. https://doi.org/10.1016/j.jenvman.2018.08.071

Sreshma CK, Vimalashree H, Lamani HD (2024) Flue gas desulfurization products use on agricultural land use systems. Int J Plant Soil Sci 36(3):199–206. https://doi.org/10.9734/ijpss/2024/v36i34415

Sridhar A, Ponnuchamy M, Kumar PS, Kapoor A (2020) Food preservation techniques and nanotechnology for increased shelf life of fruits, vegetables, beverages and spices: a review. Environ Chem Lett 19(2):1715–1735. https://doi.org/10.1007/s10311-020-01126-2

Sun S, Shi W, Tang Y, Han Y, Du X, Zhou W, Hu Y, Zhou C, Liu G (2020) Immunotoxicity of petroleum hydrocarbons and microplastics alone or in combination to a bivalve species: synergic impacts and potential toxication mechanisms. Sci Total Environ 728:138852. https://doi.org/10.1016/j.scitotenv.2020.138852

Takiguchi T, Vu HT, Nishino Y (2022) Effects of polychlorinated dibenzo-p-dioxins, polychlorinated dibenzofurans, and dioxin-like PCBS on teeth and bones in animals and humans. Toxics 11(1):7. https://doi.org/10.3390/toxics11010007

Themba N, Sibali LL, Chokwe TB (2023) A review on the formation and remediations of polychlorinated dibenzo p-dioxins and dibenzo-furans (PCDD/FS) during thermal processes with a focus on MSW process. Air Qual Atmos Health 16(10):2115–2132. https://doi.org/10.1007/s11869-023-01394-1

Touliabah HE-S, El-Sheekh MM, Ismail MM, El-Kassas H (2022) A review of microalgae- and cyanobacteria-based biodegradation of organic pollutants. Molecules 27(3):1141. https://doi.org/10.3390/molecules27031141

Tran HT, Lin C, Hoang HG, Bui XT, Le VG, Vu CT (2022) Soil washing for the remediation of dioxin-contaminated soil: a review. J Hazard Mater 421:126767. https://doi.org/10.1016/j.jhazmat.2021.126767

Tripathi M, Gaur R (2021) Bioactivity of soil microorganisms for agriculture development. Microbes Land Use Change Manage 197–220. https://doi.org/10.1016/b978-0-12-824448-7.00012-7

Tripathy B, Dash A, Das AP (2022) Detection of environmental microfiber pollutants through vibrational spectroscopic techniques: Recent advances of environmental monitoring and future prospects. Crit Rev Anal Chem, pp 1–11. https://doi.org/10.1080/10408347.2022.2144994

Tussipkan D, Manabayeva SA (2022) Alfalfa (*Medicago sativa* L.): genotypic diversity and transgenic alfalfa for phytoremediation. Front Environ Sci 10. https://doi.org/10.3389/fenvs.2022.828257

Ugya YA, Hasan DB, Tahir SM, Imam TS, Ari HA, Hua X (2021) Microalgae biofilm cultured in nutrient-rich water as a tool for the phycoremediation of petroleum-contaminated water. Int J Phytorem 23(11):1175–1183. https://doi.org/10.1080/15226514.2021.1882934

Uttaravalli AN, Dinda S, Kakara VR, Rao AVR, Daida T, Gidla BR (2022) Sustainable use of recycled soot (carbon black) for the cleaner production of value-added products: a compendium. Chem Eng J Adv 11:100324. https://doi.org/10.1016/j.ceja.2022.100324

Varjani SJ, Upasani VN (2017) Critical review on biosurfactant analysis, purification and characterization using rhamnolipid as a model biosurfactant. Biores Technol 232:389–397. https://doi.org/10.1016/j.biortech.2017.02.047

Vernez D, Oltramare C, Sauvaget B, Demougeot-Renard H, Aicher L, Roth N, Rossi I, Radaelli A, Lerch S, Marolf V, Berthet A (2023) Polychlorinated dibenzo-p-dioxins (PCDDs) and dibenzofurans (pcdfs) soil contamination in Lausanne, Switzerland: combining pollution mapping and human exposure assessment for targeted risk management. Environ Pollut 316:120441. https://doi.org/10.1016/j.envpol.2022.120441

Wagner A, Yarmak E (2017) Using frozen barriers for containment of contaminants. https://doi.org/10.21079/11681/23958

Wang D, Zhu S, Wang L, Zhen Q, Han F, Zhang X (2020) Distribution, origins and hazardous effects of polycyclic aromatic hydrocarbons in topsoil surrounding oil fields: a case study on the Loess Plateau, China. Int J Environ Res Public Health 17(4):1390. https://doi.org/10.3390/ijerph17041390

Wang W, Jariyasopit N, Schrlau J, Jia Y, Tao S, Yu TW, Dashwood RH, Zhang W, Wang X, Simonich SL (2011) Concentration and photochemistry of PAHs, NPAHs, and OPAHs and toxicity of PM2.5 during the Beijing Olympic Games. Environ Sci Technol 45(16):6887–6895. https://doi.org/10.1021/es201443z

Wang D, Wu S, Gong X, Ding T, Lei Y, Sun J, Shen Z (2023) Characterization and risk assessment of PM2.5-bound polycyclic aromatic hydrocarbons and their derivatives emitted from a typical pesticide factory in China. Toxics 11(7):637. https://doi.org/10.3390/toxics11070637

Wei Z, Wang JJ, Gaston LA, Li J, Fultz LM, DeLaune RD, Dodla SK (2020) Remediation of crude oil-contaminated coastal marsh soil: Integrated effect of biochar, RHAMNOLIPID biosurfactant and nitrogen application. J Hazard Mater 396:122595. https://doi.org/10.1016/j.jhazmat.2020.122595

Wulandari M, Effendi AJ (2018) Effect of frequency and ratio solid liquid on ultrasonic remediation of petroleum contaminated soil. AIP Conf Proc. https://doi.org/10.1063/1.5054524

Xiao W, Yao X, Zhang F (2019) Recycling of oily sludge as a roadbed material utilizing phosphogypsum-based cementitious materials. Adv Civil Eng 2019:1–10. https://doi.org/10.1155/2019/6280715

Xu M, Fu X, Gao Y, Duan L, Xu C, Sun W, Li Y, Meng X, Xiao X (2020) Characterization of a biosurfactant-producing bacteria isolated from marine environment: surface activity, chemical characterization and biodegradation. J Environ Chem Eng 8(5):104277. https://doi.org/10.1016/j.jece.2020.104277

Yousuf A, Manzoor SO, Youssouf M, Malik ZA, Khawaja KS (2020) Fly ash: production and utilization in India—an overview. J Mater Environ Sci 11(6):911–921

Zhang H-W, Kok VC, Chuang S-C, Tseng C-H, Lin C-T, Li T-C, Sung F-C, Wen CP, Hsiung CA, Hsu CY (2019) Long-term ambient hydrocarbons exposure and incidence of ischemic stroke. 4 ONE 14(12). https://doi.org/10.1371/journal.pone.0225363

Zhao C, Dong Y, Feng Y, Li Y, Dong Y (2019) Thermal desorption for remediation of contaminated soil: a review. Chemosphere 221:841–855. https://doi.org/10.1016/j.chemosphere.2019.01.079

Chapter 13
Recent Advances and Research in Biodiesel as a Sustainable, Eco-Friendly Approach Toward the Green Economy

Muheeb Yousuf, Alok Prasad Das, Prahlad Masurkar, Manas Kumar Bag, Debasis Mitra, and Anuprita Ray

Abstract The significant environmental threat posed by particulate matter is directly linked to the problems associated with fossil fuels, which are produced by harmful carbon dioxide emissions and climate change. Products made from biomass, such as biodiesel and bio-compressed natural gas (bio-CNG), have the potential to be more flexible and affordable. The recovery of high-quality glycerol as a by-product from the biodiesel through continuous transesterification of the spent cooking oils as the source material is one sustainable method of lowering the cost of biodiesel (glycerol). The basic processes for making biodiesel are transesterification, thermal cracking (pyrolysis), microemulsions, and direct usage and mixing. The method most frequently used is the transesterification of vegetable and animal fats. The transesterification process is influenced by a number of variables, including catalysts, the molar ratio of glycerides to alcohol, temperature and length of the reaction, and the water content of oils and fats. The hetero-catalytic transesterification process has become the most promising way to produce biodiesel because of its low cost and ease of use. The unique characteristics of seven different types of heterogeneous catalysts have been studied: hetero-polyacid; zeolite-based catalyst; layered-double-hydroxide-based catalyst; activated carbon; graphene; graphene oxide; activated carbon; biomass and non-biomass waste materials; and, lastly, graphene. Metal-

M. Yousuf · P. Masurkar · A. Ray (✉)
Molecular Biology and Genetic Engineering, School of Bioengineering and Biosciences, Lovely Professional University, Phagwara, Punjab, India
e-mail: Anuprita.ray88@gmail.com

A. P. Das
Department of Life Sciences, Rama Devi Women's University, Bhubaneswar, Odisha 751022, India

M. K. Bag
Crop Protection Division, ICAR-National Rice Research Institute, Cuttack 753006, India

D. Mitra
Department of Microbiology, Graphic Era (Deemed to be University), Dehradun 248002, India

I. D. Behera and A. P. Das (eds.), *Environmental Hydrocarbon Pollution and Zero Waste Approach Towards a Sustainable Waste Management*, Environmental Science and Engineering, https://doi.org/10.1007/978-3-031-81615-4_13

and metal-oxide-based catalysts are frequently used in the current biodiesel research. The history and recent progress of biodiesel are examined, focusing on the various types of biodiesels, their properties, manufacturing processes, and financial aspects of the biodiesel sector. This piece examines the utilization of biodiesel in the automotive sector, delving into the obstacles that impede its advancement, and evaluating pertinent regulations. It underlines the importance of organic resources in generating eco-friendly, renewable energy and investigates alternatives to depletable fuels, stressing the need for practical, eco-friendly replacements for fossil fuels.

Keywords Biodiesel · Transesterification · Pyrolysis · Lipids · Wastes · Sustainable development · Green economy

13.1 Introduction

Biodiesel production is a significant subject in the field of renewable energy, for it is a sustainable substitute for conventional petroleum-based fuels. Biodiesel is a liquid fuel made from biomass, like plant oils and animal fats, by the process known as transesterification (Bajpai et al. 2018; Jain 2019; Tyagi et al. 2019). The making of biodiesel has a long history, with the experiments that date back to the 1800s (Jain 2019). Lately, there has been an increasing interest in biodiesel as a renewable energy source because it reduces greenhouse gas emissions and dependency on fossil fuels (Brogan et al. 2017). The shift towards biodiesel also emphasizes the need to address environmental pollution, particularly the remediation of petroleum hydrocarbons in contaminated sites (Behera et al. 2022a, b; Kanungo et al. 2024). This chapter gives a brief of the production of biodiesel, its role as a renewable energy source, and the current research in the field. Environmental pollution is a major global concern today, with the increasing number of environmental pollutants like plastics (Tripathy et al. 2024; Mallick et al. 2023; Sahoo et al. 2022), heavy metals (Das et al. 2011; Prabhakar et al. 2019; Raj and Das 2023), biomedical waste (Khatun et al. 2023a, b; Sadasivuni et al. 2019) and electronic waste (Mishra et al. 2023; Mohanty et al. 2024) which marked by the harmful contamination of air, water, and soil. Key contributors to this issue include industrial processes, vehicle emissions, improper waste management, and the overuse of chemicals (Das and Mishra 2008; Kanungo et al. 2023; Singh et al. 2020).

Biodiesel is a crucial renewable energy source due to its several advantages over conventional petroleum-based fuels (Behera et al. 2020, 2021; Lahiri et al. 2021; Mohanta et al. 2024). For instance, biodiesel is eco-friendly, biodegradable, and non-toxic, making it a more sustainable and safer choice for the environment (Agarwal 2007; Huang et al. 2012). Moreover, biodiesel can be produced from a wide range of feedstocks, such as waste oils and fats, which offers a potential solution to the problem of waste disposal (Mishra et al. 2021; Chen et al. 2015). These advantages make biodiesel a viable alternative to conventional fuels and highlight its importance as a renewable energy source (Zhou and Thomson 2009; Prabhakar and Elder 2009; Huang et al. 2012).

Renewable energy is derived from natural resources that are constantly replenished, including sunlight, wind, rain, tides, and geothermal heat. Unlike fossil fuels, these energy sources are inexhaustible and have a much lower environmental impact, as they emit minimal to no greenhouse gases (Kumar et al. 2023). Biodiesel is a significant renewable energy source due to its numerous advantages over conventional petroleum-based fuels. For instance, biodiesel is eco-friendly, sustainable, and non-toxic, making it a safer and greener choice for the environment (Agarwal 2007; Huang et al. 2012). Additionally, biodiesel produces lower emissions than diesel, resulting in a reduction in greenhouse gas emissions and air pollution (Zhou and Thomson 2009; Prabhakar and Elder 2009). Moreover, biodiesel can be manufactured from various feedstocks, such as waste oils and fats, which helps to address the issue of waste disposal (Chen et al. 2015). These benefits make biodiesel a viable alternative to conventional fuels and highlight its importance as a renewable energy source (Zhou and Thomson 2009; Prabhakar and Elder 2009; Huang et al. 2012).

Fatty acid methyl esters (FAME), the chemical term for biodiesel, are created when oil or fat reacts with methanol in the presence of a catalyst during the transesterification process. (Pikula et al. 2020). Glycerol, a byproduct, can be repurposed. Biodiesel can replace or blend with petroleum-based diesel for transportation, power generation, and heating, sharing similar characteristics but offering advantages. It has a higher cetane number, enhancing ignition and combustion efficiency (Abdullah et al. 2019). Moreover, it contains less sulfur and pollutants, resulting in lower emissions of harmful substances like particulate matter and nitrogen oxides (Das et al. 2011, 2015, 2022).

When compared to regular diesel, biodiesel drastically lowers greenhouse gas emissions and is regarded as carbon neutral since the carbon dioxide generated during burning is balanced by the carbon absorbed by plants. Its production and use lower the overall carbon footprint and decrease dependence on finite fossil fuel resources (Angulo-Mosquera et al. 2021). Biodiesel production supports environmental protection and waste reduction, as it can be made from non-edible oils and fats, not affecting food production. Recycling used cooking oil and animal fats for biodiesel diverts waste from landfills and promotes a circular economy (Quevedo-Amador et al. 2024; Singh et al. 2022).

Despite the ongoing recognition and utilization of biodiesel as an alternative to conventional diesel fuel, it remains a highly appealing option in the renewable fuel sector due to its capacity to minimize greenhouse gas emissions, foster sustainable practices, and contribute to the development of a cleaner and more sustainable energy future (Das et al. 2024a; Dey et al. 2024a, b, c).

13.2 Commercial Biodiesel Production

Feedstock selection and preparation are crucial in commercial biodiesel production, comprising 75% of the total cost. The choice of feedstock depends on various factors, including source, price, and environmental impact (Dey et al., 2023a; John and Das 2024). Common feedstocks include vegetable oils, animal fats, and used cooking oil. Once selected, the feedstock must be purified by removing impurities and water to enhance conversion during transesterification, the predominant commercial biodiesel production method. This process involves heating the oil with alcohol and a catalyst, producing fatty acid methyl esters (FAMEs) and glycerol (Ferreira et al. 2018). Heterogeneous catalysts are preferred for their ease of separation, recyclability, and environmental benefits. The transesterification process yields fuel with a higher cetane number and improved combustion properties, making it a viable fossil fuel alternative. Efficiency can be further enhanced by regulating temperature, pressure, and catalyst concentration (Ferreira et al. 2018; Gnansounou and Escobar 2015). Table 13.1 shows the ideal temperatures and other experimental parameters for the conversion of biodiesel.

Biodiesel undergoes transesterification, refinement, and quality control to meet industry standards, removing impurities and testing properties like viscosity, flash point, and acid value (Da Costa Evangelista et al. 2016; Ferreira et al. 2020). Membrane technologies surpass traditional refining methods in producing high-quality fatty acid alkyl esters. These measures ensure biodiesel is eco-friendly, biodegradable, non-toxic, and carbon-neutral (Das et al. 2023, 2024b, c). Despite its advantages over petroleum-based diesel, such as smoother combustion and reduced sulfur oxide emissions, biodiesel's cold flow properties may limit its use in colder conditions. Its higher lubricant and oxygen content can enhance engine performance and emissions (Ferreira et al. 2020). Biodiesel standards and regulations ensure quality, safety, and compatibility with diesel engines, set emission thresholds, achieve environmental goals and maintain consistency in production and use within the industry.

Table 13.1 The ideal temperatures and other experimental parameters that provide the highest biodiesel conversion

Temperature (°C)	Residence time (min)	Alcohol to oil ratio	Catalyst wt%	Reference
55–65	60–120	6:1	1–3	Haas et al. (2006)
60–70	45–90	4–5:1	0.5–1.5	Karmee and Chadha (2005)
50–60	40–120	10:1	1–2	Nabi et al. (2012)
35–55	180–240	6:1	1–2	Demirbas (2006)
30–40	60–90	5–6:1	2–3	Thangalazhy-Gopakumar and Ghaziaskar (2008)
60–70	30–60	7–8:1	0.25–1	Mittelbach (1998)

Almost all vegetable oils used in producing biodiesel are: cinnabar oil, soy oil, palm oil, sunflower oil, and already used oil. Because they are in excessive supplies, relatively low cost and flexibility in undergoing a chemical process, these oils are recommended for transesterification to make biodiesel. With the highest yield and the more favourable fatty acid composition for biodiesel production, soybean oil leads the pack in this respect (Green et al. 2013). Canola oil, on the other hand, is marked with low levels of saturated fat and higher levels of monounsaturated fat.

13.3 Advantages of Using Vegetable Oils

The use of vegetable oils for biodiesel synthesis is notable for its low environmental impact and renewable nature. These oils reduce the carbon footprint in transportation by bypassing conventional engines, aligning with social ecology benefits. Biodiesel can be used without altering the diesel engine profile or fuel infrastructure (Gustavsson et al. 2010). Although vegetable oils appear greener than fossil fuels, their dual use for food creates competition, akin to catching every other word at a concert—both food supply and biodiesel production need attention, but resources are limited.

Common vegetable oils used in biodiesel production include:

13.3.1 Soybeans: A Versatile and Sustainable Option

Sustainable development is a comprehensive strategy for progress that aims to fulfill current needs without jeopardizing the capacity of future generations to meet theirs. It focuses on harmonizing economic, social, and environmental goals. This entails fostering economic development and growth while simultaneously safeguarding the environment and promoting social fairness (Paul et al. 2023; Pradhan et al. 2023). So, the world is now shifting towards a sustainable option or alternatives. Soybean oil is rich in fatty acids, with linoleic acid accounting for 51% of the total. This secret sauce facilitates the transformation of regular soybean oil into exceptional biodiesel through a process called transesterification (Gustavsson et al. 2010). Additionally, soybean oil possesses a high flashpoint, which minimizes the risk of fires during storage and transport, effectively functioning as an invisible safety shield. Further, when touching soybean leaves, one can appreciate their smooth and robust texture, highlighting the plant's sturdiness in nature.

Moreover, the utilization of soybeans does not involve deforestation, thus preserving both trees and the air, in contrast to alternatives such as palm oil (Zhang and Rozelle 2009). Moreover, soybeans exhibit impressive yield rates per hectare, making them a more planet-friendly choice overall.

13.3.2 Canola Oil

Canola oil is a widely known and commonly used vegetable oil for making biodiesel. It is derived from canola seeds, which contain a high amount of oil, sometimes up to 46%. Canola oil is rich in monounsaturated fatty acids, with oleic acid comprising 62% of its composition. This makes it a more stable oil than other vegetable oils, particularly in colder climates (Barnwal and Sharma 2005). Like soybean oil, canola oil has a relatively low impact on the environment due to its high yield per hectare and minimal land requirements. However, concerns have been raised about the use of fertilizers and pesticides in canola production, which could potentially harm the environment (Table 13.2). Therefore, it is essential to implement necessary regulations and sustainable farming practices to ensure the sustainability of canola oil for biodiesel production (Barnwal and Sharma 2005).

Table 13.2 Sources for biodiesel, composition and advantages and disadvantages

Vegetable oil	Source	Fatty acid composition	Advantages	Disadvantages	Reference
Soybean oil	Soybean	18% palmitic, 5% stearic, 22% oleic, 58% linoleic	• High yield per acre • Widely available	• Can be expensive • Requires high processing temperature	Demirbas (2005)
Canola oil	Rapeseed	6% palmitic, 4% stearic, 61% oleic, 23% linoleic	• Low processing temperature • Low viscosity	• High processing cost • Limited availability in some regions	Varma and Madras (2007)
Palm oil	Oil palm trees	44% palmitic, 5% stearic, 39% oleic, 10% linoleic	• High yield per acre • Can be produced sustainably	• Contains high levels of saturated fatty acids • Controversies surrounding deforestation for palm oil production	Lam and Lee (2011)
Sunflower oil	Sunflower seeds	6% palmitic, 4% stearic, 24% oleic, 75% linoleic	• Low processing temperature • High yield per acre	• Expensive compared to other vegetable oils • Limited availability in some regions	Thirumarimurugan et al. (2012)

13.3.3 Palm Oil

Palm oil, sourced from palm tree fruit, is the most widely produced vegetable oil globally, used in food, cosmetics, and biofuels. Its high saturated fatty acid content makes it more viscous and unsuitable as a biodiesel, but transesterification converts it into palm oil methyl ester (POME), the main component of palm-based biodiesel (Table 13.2) (Santos et al. 2011). Despite its ubiquity, palm oil production faces criticism for causing deforestation, biodiversity loss, and labour abuse (Table 13.2). Initiatives like the Roundtable on Sustainable Palm Oil (RSPO) aim to address these issues and promote sustainable practices, positioning palm oil as a renewable biodiesel resource.

13.3.4 Sunflower Oil

Sunflower oil is obtained from the seeds of the sunflower plant and it is known for its high oleic acid composition so that it can be used in biodiesel production. It is also rich in vitamin E an antioxidant which enhances the oxidative stability of biodiesel hence making it less degrade at storage. Sunflower oil a high flashpoint, means it less likely ignite or explode, and therefore safer to use than other oils (Table 13.2) (Joshi and Attri 2013).

Sunflower oil, extracted from sunflower seeds, is high in oleic acid, making it suitable for biodiesel production. It is rich in antioxidants, enhancing biodiesel's oxidative stability and reducing storage degradation. Its high flashpoint increases safety compared to other oils (Joshi and Attri 2013). Sunflower oil yields high per hectare with minimal land use but requires energy-intensive extraction, reducing its sustainability. New technologies like cold pressing aim to lower energy consumption (Moreond and Cantero 2011). Sustainable practices are vital for vegetable oils in biodiesel, including non-food oil sources and promoting oilseeds for marginal areas to minimize food production impact. Vegetable oils such as soybean, canola, palm, and sunflower are common in biodiesel due to their quantity and transesterification capacity. While they offer benefits like recyclability and emission reductions, sustainability concerns, food production impact, and environmental harm require careful management for long-term biodiesel production (Joshi and Attri 2013).

13.4 Animal Fats

Animal fats, such as tallow, lard, and yellow grease, can be used as feedstock in biodiesel production, primarily sourced from meat processing and rendering by-products (Keera et al. 2013). Tallow from cattle or sheep is favored due to its high saturated fat content and suitability for biodiesel conversion (Zhang and Huo 2015).

Lard from pigs and yellow grease from leftover cooking oils, which includes both animal fats and vegetable oils, can also serve as feedstocks (Chettri et al. 2021). The fatty acid profile of these fats is crucial for their viability in biodiesel synthesis. Consistent quality requires effective collection and processing. Unlike vegetable oil-based biodiesel, animal fat-derived biodiesel might need extra refining due to differing properties. Issues such as odor control, potential pollutants, proper storage, and handling are essential to prevent spoilage and maintain quality (Green and Kroschwitz 2012). Animal fats are composed of triglycerides, which are esters of fatty acids and glycerol. The fatty acids present in animal fats are typically palmitic, stearic, and oleic, with varying levels of saturation. These fatty acids contribute to the chemical composition and properties of animal fats, making them suitable for use in biodiesel production. Common animal fats used for biodiesel production include tallow (beef fat), poultry fat, lard (pig fat), and yellow grease (animal by-product).

13.4.1 Advantages of Using Animal Fats in Biodiesel Production

Animal fats are often chosen for producing biodiesel due to their abundance and low cost as by-products of the food industry. They offer a cost-effective alternative to virgin vegetable oils and have a higher energy content per unit weight, resulting in increased yields of biodiesel and reduced carbon emissions. Biodiesel produced from animal fats has properties that are similar to conventional diesel, which can enhance engine performance by providing better lubricity and reducing wear. Furthermore, it can help reduce emissions of sulfur, aromatics, and particulate matter, which can improve air quality and decrease health risks associated with traditional diesel fuels.

13.4.2 Challenges and Limitations

Animal fats present a combination of opportunities and obstacles in the production of biodiesel. While they have a higher melting point and viscosity than vegetable oils, particularly tallow, which makes processing more difficult and affects performance in cold weather, blending them with other feedstocks is common to enhance flow properties and improve cold-weather performance. However, another challenge is the inconsistency in quality and consistency due to differences in fatty acid composition from various animal sources. Nonetheless, animal fats are valued for their abundance, cost-effectiveness, and environmental benefits in biodiesel production. To fully harness their potential as a sustainable biodiesel feedstock, continued research and development are necessary to optimize the production process.

13.5 Waste Oils and By-Products

Utilizing waste oils and by-products for biodiesel synthesis is both environmentally friendly and cost-effective. Waste oils, including used cooking oil, animal fats, and other industrial by-products, can serve as significant biodiesel sources (Mahmud and Kadir 2014). Commonly discarded from food production, restaurants, and households, these materials provide a substantial feedstock for biodiesel. Animal fats from rendering plants and other industrial by-products also serve as viable biodiesel feedstocks, reducing waste and ensuring a sustainable supply of renewable fuels. The quality of waste oils influences their suitability for biodiesel production. Using these materials as biodiesel feedstock offers multiple benefits. Firstly, they are often disposed of through expensive and environmentally harmful methods such as landfilling or incineration. Biodiesel production provides a cost-effective waste management solution (Pakalin and Sarika 2020). Moreover, it reduces dependence on virgin vegetable oils, alleviating land use pressure for crop cultivation and enhancing food security by utilizing waste oils instead of food-grade oils for fuel production.

Biodiesel produced from waste oils and by-products reduces greenhouse gas emissions compared to fossil fuels, as it is considered carbon-neutral. However, the primary challenge in biodiesel production from waste oils and by-products is feedstock variability, which affects biodiesel quality and consistency due to contaminants such as water, free fatty acids, and impurities (Quaye 1996; Smith et al. 2023; Tan and Lee 2012; Yenesew et al. 2012). Research is needed to develop cost-effective technologies to remove these contaminants and produce high-quality biodiesel. Another challenge is the competition for feedstock between food and fuel industries, which may cause potential food shortages and price volatility. To address this, researchers are investigating non-edible feedstocks, like waste oils, and developing technologies to convert them into biodiesel, thereby reducing food-based feedstock competition and ensuring biodiesel industry sustainability. Despite the challenges, there have been significant developments in biodiesel production from waste oils and by-products. One example is the enzymatic conversion process, which uses enzymes to catalyze the transesterification reaction, resulting in a higher biodiesel yield with fewer impurities (Silva et al. 2019). This method has shown promising results in converting waste oils and by-products into high-quality biodiesel, making it one of the most promising technologies for the future of biodiesel production. Other technologies, such as supercritical methanol, ultrasound-assisted transesterification, and microwave-assisted transesterification, are also being researched and developed to improve the efficiency and cost-effectiveness of the biodiesel production process (Nurrochmad et al. 2016).

13.6 Transesterification Process

Transesterification converts fats or oils into methyl or ethyl esters, a key stage in biodiesel production (Fig. 13.1). This process has garnered attention due to the rising demand for sustainable energy (Mishra et al. 2020). The primary reaction transforms triglycerides in vegetable oils or fats into fatty acid methyl esters (FAMEs) and glycerol, catalyzed by bases, acids, or enzymes, with sodium or potassium hydroxide being the most common. The process includes esterification, transesterification, and neutralization. During esterification, triglycerides react with methanol in the presence of an acid catalyst to form diglycerides and monoglycerides, reducing viscosity and enhancing reaction efficiency. In the transesterification step, diglycerides and monoglycerides react with methanol and a base catalyst to yield FAMEs and glycerol. Finally, the excess base catalyst is neutralized with an acid, producing biodiesel and a salt (Gude 2016).

Various factors influence the transesterification process, including feedstock type, catalyst quantity and type, reaction temperature, and reactant ratios. Feedstock selection significantly impacts biodiesel yield and quality; for example, palm and soybean oils are preferred due to their high unsaturated fatty acid content. The catalyst type and amount are crucial; alkaline catalysts like sodium or potassium hydroxide are commonly used for their efficiency and low cost but can cause saponification, requiring careful neutralization to avoid product contamination (Mohanty et al. 2024b; Paul et al. 2024; Priyadarsini & Das 2024, 2025). Acidic catalysts, such as sulfuric acid, are less used due to slower reactions but avoid saponification. Recently, enzymatic catalysts have gained attention for reducing reaction time and producing high-quality biodiesel (Marchand et al. 2019).

Temperature and reactant ratios are crucial in transesterification. The optimal temperature is 60–65 °C to prevent methanol evaporation, and the best methanol-to-oil molar ratio is about 6:1, depending on the feedstock (Chhetri et al. 2008; Edeh 2020). This method is advantageous over other biodiesel production methods due

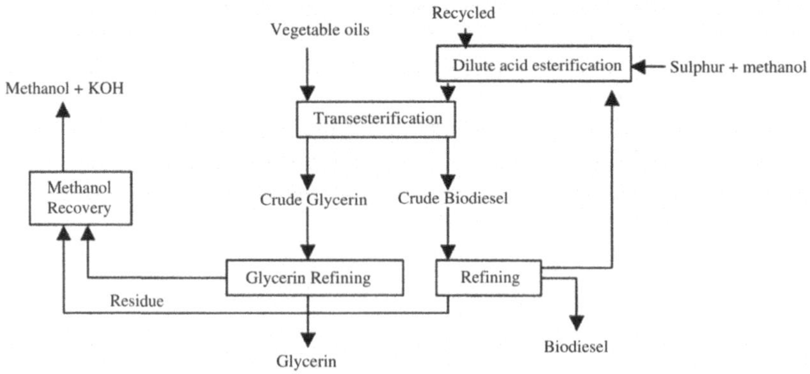

Fig. 13.1 Schematic flow chart of biodiesel production

to its simplicity, cost-effectiveness, scalability, high biodiesel yield, and valuable by-product glycerol (Bajpai and Tyagi 2018). Despite its benefits, further research is required to optimize the process and address limitations like saponification. As the demand for sustainable energy grows, transesterification is expected to significantly contribute to future energy needs. The process of transesterification, which produces glycerol and fatty acid methyl or ethyl esters (FAME or FAEE) from the reaction of triglycerides with alcohol (often methanol or ethanol) and a catalyst, is crucial to the synthesis of biodiesel (Khan 2019; Knothe et al. 2018). This process accommodates feedstocks like cooking oil, animal fats, and plant oils. During transesterification, triglycerides are converted into esters and glycerol, with alcohol replacing glycerol in the triglyceride molecule. Influencing factors include alcohol-to-oil ratio, catalyst type (e.g., sodium hydroxide or enzymes), temperature, pressure, water content, and reaction time. These factors are critical for optimizing conversion efficiency and biodiesel quality, with catalytic processes significantly enhancing transesterification kinetics and product yield.

13.6.1 Introduction to the Different Catalysts Used in Biodiesel Production

Solid acid catalysts, sulfuric acid (H_2SO_4), and alkali (basic) catalysts like potassium hydroxide (KOH) and sodium hydroxide (NaOH) are among the catalysts used in the production of biodiesel. Different manufacturing processes also make use of heterogeneous catalysts and enzymes. It is possible for the catalyst to be heterogeneous or homogeneous.

13.6.1.1 Homogeneous Catalysts

Typically, homogeneous catalysts take the form of acids, bases, or enzymes and are soluble in the reaction mixture. The most often used of these are potassium hydroxide (KOH) and sodium hydroxide (NaOH), which break the ester linkages in the feedstock and create new bonds with alcohol molecules to encourage the synthesis of FAME (Fig. 13.2).

13.6.1.2 Heterogeneous Catalysts

Heterogeneous catalysts are insoluble, allowing easy separation from the final product and thus streamlining and reducing purification costs. Common examples include acid resins, zeolites, and metal oxides like magnesium oxide (MgO) and calcium oxide (CaO), which have shown promising results in yield and selectivity compared to homogeneous catalysts (Rusydi et al. 2016). Enzymes are also

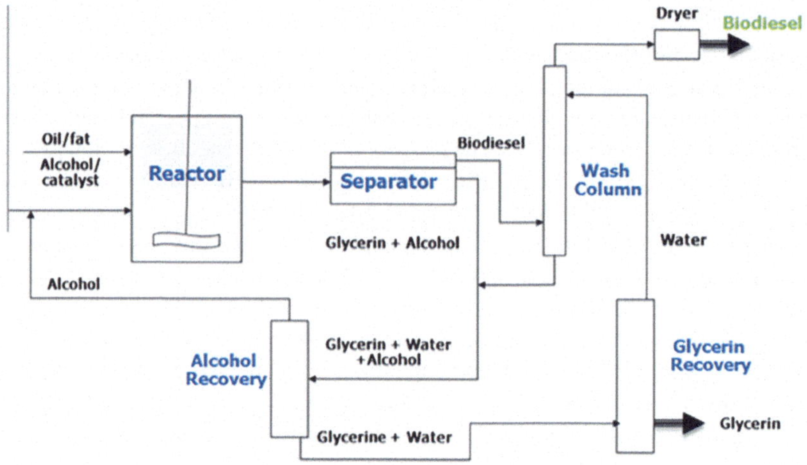

Fig. 13.2 Flow chart of homogeneous catalytic transesterification process

being explored as potential transesterification catalysts, offering advantages such as mild conditions, high specificity, and high-quality FAME production. However, their application is limited by high costs and specific reaction requirements (Marchand et al. 2019).

13.6.1.3 Interesterification

Interesterification is another catalytic process involving the rearrangement of fatty acid chains within triglyceride molecules. This process forms new ester bonds between fatty acids and glycerol, producing biodiesel with improved cold flow properties and stability (Bajpai and Tyagi 2018). The interesterification process is favored in producing biodiesel from feedstocks with high levels of polyunsaturated fatty acids, such as waste cooking oil, as it reduces the amount of unreacted feedstock and results in a higher biodiesel yield.

13.6.1.4 Hydrogenolysis

Hydrogenolysis is a catalytic process involving removing hydrogen atoms from unsaturated fatty acids. This produces saturated fatty acids, which have better fuel properties than unsaturated fatty acids. It is often used with transesterification to increase biodiesel yield (Pirselova et al. 2019). The catalysts used for this process are usually metal catalysts such as palladium or nickel, which are highly active and selective towards removing hydrogen atoms.

Catalytic Processes Pros and Cons

The following are some of the benefits of catalytic processes in biodiesel production:

Quicker reaction rates: Catalysts speed up the transesterification reaction, enhancing productivity.

- Increased yield: Catalysts frequently produce higher biodiesel yields, which improves process efficiency.
- Versatility: Different catalysts enable the use of diverse feedstocks, including low-grade oils and fats.

However, there are certain restrictions to consider:

- Catalyst cost: Some catalysts, such as enzymes, can be costly, which has an impact on manufacturing costs.
- Catalyst specificity: Some catalysts specialize in specific feedstocks or reactants, restricting their applicability to all biodiesel production situations.
- Waste management: Catalyst residues and by-products must be appropriately disposed of or treated, complicating manufacturing.

Factors Affecting Transesterification Process Optimization

Feedstock selection is crucial in optimizing biodiesel production through transesterification. Factors such as cost, quality, availability, and unsaturated fatty acid content must be considered. High impurity levels like free fatty acids and water in feedstocks can adversely affect transesterification, reducing biodiesel quality and efficiency; hence, careful evaluation of feedstock characteristics is essential. Additionally, reaction conditions, including the molar ratio of alcohol to oil and reaction temperature, significantly impact transesterification efficiency. The required molar ratio varies with feedstock and catalyst, with higher ratios needed for feedstocks with higher free fatty acid content. The optimal reaction temperature range is 50–60 °C, influencing reaction rates and by-product formation. Selecting the appropriate catalyst is also critical. Base catalysts like sodium hydroxide are cost-effective but corrosive and difficult to separate from the product. Acid catalysts, such as sulfuric acid, are effective but expensive and tend to form acidic impurities. Enzymatic catalysts are biodegradable and versatile with various feedstocks but are costly and sensitive to impurities. The choice of catalyst type affects overall process efficiency and biodiesel quality.

13.7 Factors Influencing Biodiesel Yield and Quality

Several factors critically influence biodiesel yield and quality. The alcohol-to-oil ratio in transesterification significantly affects conversion extent and yield. Catalyst type and concentration impact reaction kinetics and product quality. Reaction temperature and pressure influence rates, equilibrium conversion, and biodiesel properties. Feedstock quality, especially free fatty acid (FFA) and water levels, substantially affects the process and product quality. Reaction time is also crucial; longer durations can improve conversion but may produce unwanted by-products. Excessive water can cause soap formation and reduce catalyst efficiency, underscoring the need to control water levels for optimal production. Collectively, these factors determine biodiesel production efficiency and quality.

These American Society for Testing and Materials (ASTM) standards include requirements for purity, physical and chemical properties, and testing methods. Key quality aspects involve chemical composition, performance, and environmental impact (Dey et al. 2023b). Criteria like purity, viscosity, flashpoint, and cold flow properties ensure safe engine operation. Performance standards under ASTM D613 specify cetane numbers to match conventional diesel in power, fuel economy, and emission reductions. Environmental standards in ASTM D7467 limit sulfur, phosphorus, and alkali metals to reduce harmful emissions. Compliance with these standards ensures biodiesel's role in lowering carbon emissions and promoting environmental sustainability (Das, A. Prasad., & Ghosh, Shreya 2023).

13.8 Biodiesel Standards and Regulations

International standards for biodiesel production, such as ASTM D6751, EN 14214, and ISO 14214, play a vital role in regulating fuel quality, promoting market acceptance, and ensuring compliance with environmental and operational standards. These standards provide comprehensive guidelines for biodiesel quality and performance and enable uniformity and reliability in biodiesel production, facilitating its integration into global fuel markets. ASTM D6751, established by the American Society for Testing and Materials, defines specifications for B100 biodiesel, including physical and chemical characteristics, contaminants, and performance criteria. Similarly, EN 14214, adopted by the European Union, specifies requirements for biodiesel as a blend component in diesel fuel. ISO 14214, set by the International Organization for Standardization, focuses on biodiesel as a component in diesel fuel, addressing purity, oxidation stability, and cold flow properties. Adherence to these international standards ensures that biodiesel production maintains environmental and operational standards while facilitating its integration into global fuel markets.

13.8.1 Biodiesel Regulations

Regulations serve as the legal framework for biodiesel production, distribution, and use. In the United States, the EPA is the primary regulatory body for biodiesel, while other countries may have their governing bodies. The EPA's regulations for biodiesel are based on the Clean Air Act, which aims to reduce air pollution and protect public health.

The most notable regulation for biodiesel is the 2012 Renewable Fuel Standard (RFS2), which sets specific sustainability requirements for biodiesel production. In addition to fulfilling sustainability standards for land use, life cycle greenhouse gas emissions, and water quality and preservation, these requirements call for a minimum 50% reduction in greenhouse gas emissions when compared to petroleum diesel. Serious penalties and fines may be incurred for breaking these rules.

In addition, the EPA also has regulations for labeling and reporting of biodiesel. Biodiesel producers must adhere to proper labeling requirements, including indicating the biodiesel blend level and the ASTM standard followed. They must also report production and sales data to the EPA to ensure compliance with the RFS program.

13.8.2 Enforcement of Standards and Regulations

The responsibility of enforcing biodiesel standards and regulations lies primarily with the EPA. The agency conducts regular inspections and testing of biodiesel samples to ensure they meet the required ASTM standards and comply with the RFS2 program. In cases of non-compliance, the EPA has the authority to issue fines, revoke certification, or pursue legal actions against violators.

However, the responsibility of adhering to standards and regulations also falls upon biodiesel producers, who must consistently monitor their production processes and ensure the quality and sustainability of their products. To remain compliant, they must also stay current on any updates or changes to standards and regulations.

Biodiesel standards and regulations ensure this renewable fuel source's quality, safety, and sustainability. The development of ASTM D6751 and the implementation of the RFS2 program by the EPA have helped establish guidelines for biodiesel production and use, driving advancements in technology and sustainability. Both the EPA and biodiesel producers are responsible for enforcing and complying with these standards and regulations to reap the environmental and economic benefits of this renewable fuel source.

13.9 Biodiesel Production Strategies for Sustainable Land Use

Sustainable land use practices are critical for reducing the negative implications of biodiesel production on biodiversity and ecosystem services. Among the strategies are:

Utilization of Marginal Lands: Growing biodiesel feedstock on marginal or degraded lands can assist in reducing the burden on natural ecosystems and support landscape restoration.

Agroforestry and Integrated Land Management: Using agroforestry systems and integrated land management practices can help conserve biodiversity, improve soil health, and diversify farmer revenue streams.

Farming Practices That Protect Biodiversity: Crop rotation, cover cropping, and decreased tillage are examples of biodiversity-friendly farming practices that can assist in maintaining ecological balance and support biodiversity in biodiesel-producing settings.

Standards and certification: Adherence to sustainability standards and certification programs, such as the Round Table on Sustainable Palm Oil (RSPO) for palm oil-based biodiesel, can aid in conserving biodiversity and responsible land use practices.

In conclusion, assessing land use impacts, considering the effects on biodiversity and ecosystem services, and implementing sustainable land use strategies are critical to promote sustainable biodiesel production while minimizing negative environmental impacts.

13.10 Water Usage in Biodiesel Production

Biodiesel production is a process that necessitates water, which is required during various stages of production. The primary usage of water is during the extraction and processing of raw materials. The amount of water required for this step is dependent on the type of feedstock and the extraction method used. For example, traditional pressing methods can necessitate three to five times more water than solvent extraction methods (Qin 2019). Additionally, water is also required during the transesterification process to wash, purify, and neutralize catalyst residues. The quantity of water needed for this step is dependent on the type of catalyst and the impurities present. For example, using sodium hydroxide as a catalyst requires more water than using a solid acid catalyst (Chen et al. 2015). Apart from these direct requirements, water is also used for cooling, heating, and cleaning during production. The exact amount of water used in these steps is dependent on the size of the facility and the equipment used. Biodiesel production can consume significant amounts of water, which can be a concern in regions where water is scarce.

13.11 Waste Management in Biodiesel Production

The primary waste materials generated during biodiesel production are mainly composed of glycerol, wastewater, and spent catalysts. Glycerol is a by-product of the transesterification process, accounting for up to 10% of the biodiesel production volume (Nagaraja et al. 2019). Despite its value in various industries, such as cosmetics, pharmaceuticals, and food production, the excess glycerol generated in biodiesel production has resulted in its disposal as waste due to market oversupply (Mohanty et al. 2024a, b). Wastewater from biodiesel production contains impurities, including free fatty acids, methanol, and residual catalysts, which can be challenging to treat due to varying chemical compositions depending on the feedstock and production process (Chen et al. 2015). Moreover, the spent catalysts used to accelerate the transesterification process are hazardous due to their high alkalinity, and proper handling and treatment are required for their safe disposal.

13.12 Current Practices and Potential Solutions

Numerous methods are employed to handle water usage and waste in biodiesel manufacturing. These include conservation, recycling, and treatment of waste streams, as well as the creation of more sustainable production methods. One widely used method for conserving water and recycling it is the implementation of closed-loop water systems, where water is reused in various stages of production (Cantrell et al. 2013).

In terms of waste management, there is a focus on finding alternative uses for excess glycerol generated from biodiesel production. Some researchers propose converting it into biodegradable bioplastics (Nagaraja et al. 2019). Additionally, methods like filtration, coagulation, and activated sludge treatment have been explored to treat biodiesel wastewater (Jagadale et al. 2018).

The development of sustainable production processes is also a significant area of focus. For example, solid acid catalysts, like silica-supported heteropolyacids, are being investigated as a safer alternative to liquid catalysts (Chen et al. 2015). These alternative processes reduce water consumption, produce less waste, and have a smaller environmental impact.

13.13 Technological Advancements in Biodiesel Production

Biodiesel production faces various challenges, including high production costs, limited feedstock availability, and technical difficulties in the production process. To overcome these obstacles and promote the widespread use of biodiesel, technological advancements have been made in its production. For instance, enzymatic technology has emerged as a promising solution in biodiesel production, particularly in converting high-free fatty acid feedstock into biodiesel. The enzyme-catalyzed transesterification process is highly tolerant to impurities in feedstock, reducing the need for costly purification processes. The use of enzymes in biodiesel production could lead to reduced production costs, increased efficiency, and improved biodiesel quality. Additionally, enzymes are biodegradable, reducing the environmental impact of biodiesel production (Xie et al. 2020). Another technological advancement in biodiesel production is the use of supercritical transesterification, which allows for the use of unrefined feedstock. This reduces feedstock costs and increases the availability of biodiesel feedstock. Furthermore, using microalgae as a feedstock for biodiesel production has gained increasing attention due to advances in bioreactor design, genetic engineering, and harvesting techniques (Gong et al. 2020). Microalgae are photosynthetic microorganisms with high oil content and can be grown in various environments, including wastewater and non-arable land. As a result, their cultivation and processing into biodiesel have become more efficient and cost-effective. Additionally, using microalgae as a feedstock can help reduce competition for fertile land and food crops between biodiesel production and traditional agriculture, mitigating the issue of food security. Although technological advancements have significantly improved the efficiency and cost-effectiveness of biodiesel production, challenges remain to its widespread adoption. Additionally, the lack of standardized processes and regulations for these technologies can lead to uncertainties in their production and quality control (Brogan and Pratt 2017). However, further research and development are required to optimize these technologies and adapt them to different feedstocks and production scales. Government support and incentives are also necessary to promote the widespread adoption and utilization of these technologies in biodiesel production.

13.14 Prospects and Challenges in Biodiesel Production

The adoption of biodiesel across various industries such as transportation, power generation, and heating is a significant factor in this growth. One of the most promising aspects of biodiesel production is the use of waste materials as feedstock. Using waste oils and fats from restaurants, households, and industries can not only reduce production costs but also help with waste management by converting waste into a valuable resource. Non-food feedstock such as algae and Jatropha seeds are also gaining traction due to their high oil yield potential and ability to grow in marginal lands, reducing competition with food crops for arable land (Fernando and Gencoca 2017). Advancements in biodiesel production technologies are also promising. To increase the yield and quality of biodiesel, better manufacturing techniques including catalytic transesterification and supercritical transesterification are being explored. To improve the sustainability of the production process, various renewable energy sources like solar, wind, and geothermal energy are being investigated (Dornburg et al. 2010).

13.15 Challenges in Biodiesel Production

Despite the promising outlook, widespread biodiesel commercialization still needs improvement. One of the major obstacles is the availability and cost of feedstock. As global demand for biodiesel increases, competition for feedstock between food and non-food crops raises concerns about food security. This forces the industry to explore alternative feedstock sources. Furthermore, the fluctuating feedstock prices and its reliance on weather conditions pose challenges to the economic feasibility of biodiesel production (Wu and Zhao 2009). Another challenge in biodiesel production is the inconsistency in feedstock quality and composition, which significantly impacts the yield and quality of the final product. Variations in fatty acid composition and impurities in the feedstock can lead to impurities in biodiesel, resulting in engine performance issues (Dornburg et al. 2010). Hence, developing methods for consistent and reliable feedstock characterization is vital for efficient high-quality biodiesel production. The storage and distribution of biodiesel also pose challenges due to its chemical properties. Biodiesel has higher viscosity and lower volatility compared to conventional diesel, which can cause engine clogging in cold weather (Wu and Zhao 2009). Appropriate storage and blending with other fuels, such as petroleum diesel, are necessary to overcome these challenges.

13.16 Conclusion

Globally, large-scale production facilities are operating, utilizing a variety of feedstocks and production methods. The industry has grown significantly and complies with regulatory criteria, contributing to biodiesel's market viability as an alternative to conventional diesel fuel. Emerging technologies such as advanced feedstock utilization, novel process technologies, and catalyst research drive gains in biodiesel production efficiency, sustainability, and feedstock diversity. Renewable energy policies, global energy transition, and technology improvements impact the future market and demand for biodiesel. However, constraints and challenges, such as a lack of feedstock, regulatory uncertainty, and a competitive energy landscape, continue to hamper biodiesel adoption. Biodiesel, as a renewable alternative to conventional diesel fuel, provides environmental benefits, reduces greenhouse gas emissions, and has the potential to improve energy security. Its use helps to diversify the energy mix and reduces reliance on finite fossil fuel resources.

References

Abdullah B, Muhammad SAFS, Shokravi Z, Ismail S, Kassim KA, Mahmood AN, Aziz MMA (2019) Fourth generation biofuel: a review on risks and mitigation strategies. Renew Sustain Energy Rev 107:37–50. https://doi.org/10.1016/j.rser.2019.02.018

Agarwal AK (2007) Biofuels (alcohols and biodiesel) applications as fuels for internal combustion engines. Prog Energy Combust Sci 33:233–271

Angulo-Mosquera LS, Alvarado-Alvarado AA, Rivas-Arrieta MJ, Cattaneo CR, Rene ER, García-Depraect O (2021) Production of solid biofuels from organic waste in developing countries: a review from sustainability and economic feasibility perspectives. Sci Total Environ 795:148816. https://doi.org/10.1016/j.scitotenv.2021.148816

Bajpai S, Tyagi VK (2018) Manufacturing processes. Biodiesel: process, production, and analysis. Elsevier, Amsterdam, pp 91–112

Barnwal BK, Sharma MP (2005) Prospects of biodiesel production from vegetable oils in India. Renew Sustain Energy Rev 9(4):363–378

Behera ID, Basak G, Kumar RR, Sen R, Meikap BC (2020) Treatment of petroleum refinery sludge by petroleum degrading bacterium Stenotrophomonas pavanii IRB19 as an efficient novel technology. J Environ Sci Health Part A Toxic/Hazard Subst Environ Eng 56(2). https://doi.org/10.1080/10934529.2020.1866924

Behera ID, Nayak M, Biswas S, Meikap BC, Sen R (2021) Enhanced biodegradation of total petroleum hydrocarbons by implementing a novel two-step bioaugmentation strategy using indigenous bacterial consortium. J Environ Manag 292:112746. https://doi.org/10.1016/j.jenvman.2021.112746

Behera ID, Meikap BC, Sen R (2022a) Performance evaluation of bacterial consortium for biodegradation of total petroleum hydrocarbon: a comparative strategic biostimulation study. Res Square

Behera ID, Nayak M, Mishra A, Meikap BC, Sen R (2022b) Strategic implementation of integrated bioaugmentation and biostimulation for efficient mitigation of petroleum hydrocarbon pollutants from terrestrial and aquatic environment. Mar Pollut Bull 177:113492. https://doi.org/10.1016/j.marpolbul.2022.113492

Brogan JM, Pratt JE (2017) Technological advancements in biodiesel production. In: Sustainable technologies for a sustainable environment. Springer, Cham, pp 97–124

Cantrell DC, Jennings JC, He Z, Watson M, Sun Y (2013) Biochar benefits for biodiesel production and soil enhancement. BioResources 8(1):158–168

Chen J, Zhu Y, Aminabhavi TM, Bhatia S (2015) Biodiesel production using environmentally benign solid acid catalysts. Energy Convers Manag 105:774–788

Chhetri AB, Watts KC, Islam MA (2008) Waste cooking oil as an alternate feedstock for biodiesel production. Energies 1(3):178–189

Da Costa Evangelista P, Gondim AD, Di Souza L, Araujo AS (2016) Alumina-supported potassium compounds as heterogeneous catalysts for biodiesel production: a review. Renew Sustain Energy Rev 59:887–894. (1) (PDF) Heterogeneous catalysts for biodiesel production: a review. Available from: https://www.researchgate.net/publication/340132827_Heterogeneous_Catalysts_for_Biodiesel_Production_A_Review#fullTextFileContent. Accessed 06 July 2024

Das AP, Mishra S (2008) Hexavalent chromium (VI): health hazards and environmental pollutant. J Environ Res Dev 2(3):386–392

Das AP, Sukla LB, Pradhan N, Nayak S (2011) Manganese biomining: a review. Biores Technol 102(16):7381–7387. https://doi.org/10.1016/J.BIORTECH.2011.05.018

Das AP, Ghosh S, Mohanty S, Sukla LB (2015) Advances in manganese pollution and Its bioremediation. In: Sukla LB, Pradhan N, Panda S, Mishra BK (eds) Environmental microbe biotechnology, 1st edn. Springer, Cham, pp 313–328. https://doi.org/10.1007/978-3-319-19018-1_16

Das AP, Ghosh S (2022) Role of microorganisms in extenuation of mining and industrial wastes. Geomicrobiol J 39(3–5):173–175. https://doi.org/10.1080/01490451.2022.2038953

Das A. Prasad, Ghosh, Shreya (2023) Manganese mining microorganisms

Das AP, Dutta K, Khatun R, Behera ID, Singh S, Mishra S (2023) Microfiber pollution and its microbial mitigation: a review on current trends and future prospects. J Taiwan Inst Chem Eng 105104. https://doi.org/10.1016/J.JTICE.2023.105104

Das A, Ray SK, Mohanty M, Mohanty J, Dey S, Das AP (2024a) Ecotoxicity of microplastic wastes and their sustainable management: a review. Environ Chem Ecotoxicol 6:144–152. https://doi.org/10.1016/J.ENCECO.2024.05.003

Das A, Dey S, Bala K, Gnanasekaran R, Das AP (2024b) Microbial bioremediation technology for sustainable treatment and management of synthetic microfiber waste. In: Sustainable microbial technology for synthetic and cellulosic microfiber bioremediation, pp 221–238. https://doi.org/10.1007/978-3-031-62600-5_12

Das A, Dey S, Das AP (2024c) Ecotoxicological impacts of synthetic microfiber pollutants and development of sustainable mitigation strategies. Environ Chem Ecotoxicol 7:201–210. https://doi.org/10.1016/J.ENCECO.2024.12.003

Demirbas A (2006) Biodiesel production from vegetable oils via catalytic and non-catalytic supercritical methanol transesterification methods. Prog Energy Combust Sci 32(4):466–487. https://doi.org/10.1016/j.pecs.2006.03.002

Dey S, Tripathy B, Kumar MS, Das AP (2023a) Ecotoxicological consequences of manganese mining pollutants and their biological remediation. Environ Chem Ecotoxicol. https://doi.org/10.1016/J.ENCECO.2023.01.001

Dey S, Das A, Mallick K, Sahu A, Das AP (2023b) Environmental petroleum waste: pollution, toxicity, sustainable remediation. In: Behera ID, Das AP (eds) Impact of petroleum waste on environmental pollution and its sustainable management through circular economy. Environmental Science and Engineering. Springer, Cham. https://doi.org/10.1007/978-3-031-48220-5_7

Dey S, Bissoyi A, Koteswara Reddy G, Das AP (2024a) Synthetic microfibers: an urban emerging pollutant demanding for sustainable environmental solutions. In: Sustainable microbial technology for synthetic and cellulosic microfiber bioremediation, pp 107–121. https://doi.org/10.1007/978-3-031-62600-5_7

Dey S, Dash A, Das AP (2024b) Bioremediation as a sustainable and circular solution for mining waste management and bioenergy generation. In: Sustainable management of mining waste and tailings, pp 107–119. https://doi.org/10.1201/9781003442455-6

Dey S, Dash N, Ray SK, Priyadarsini S, Chhotaray C, Das AP (2024c) Occurrence and quantification of tire wear plastics pollutants from highway city dust of Bhubaneswar city of Odisha. In: Sustainable microbial technology for synthetic and cellulosic microfiber bioremediation, pp 77–89. https://doi.org/10.1007/978-3-031-62600-5_5

Dornburg V, Hermann B, Luksch T, Zech K, Faaij A (2010) Biodiesel production from different feedstocks and its environmental implications. Renew Sustain Energy Rev 14(2):682–692

Edeh I (2020) Biodiesel production as a renewable resource for the potential displacement of the petroleum diesel. In: Beschkov V (ed) Biorefinery concepts, energy and products. https://doi.org/10.5772/intechopen.93013

Fernando S, Gencoca F (2017) The prospects and challenges of biodiesel production using non-food crops. Renew Sustain Energy Rev 72:46–59

Ferreira AR et al (2018) Optimization of the transesterification of sunflower oil for the production of biodiesel: a response surface methodology (RSM) study. Renew Energy 130:744–755

Ferreira MFP, Oliveira BFH, Pinheiro WBS, Correa NF, França LF, Ribeiro NFP (2020) Generation of biofuels by slow pyrolysis of palm empty fruit bunches: optimization of process variables and characterization of physical-chemical products. Biomass Bioenergy 140:105707. https://doi.org/10.1016/j.biombioe.2020.105707

Gnansounou E, Escobar J (2015) Environmental and social issues in producing first- and second-generation biofuels: a review. Environ Manag 149:70–87

Gong Y, Jiang M, Chen Y, Tang X, Shang Z, Hu L, Wang X (2020) Cultivation and processing of microalgae for biofuels and high-value compounds. J Clean Prod 273:123140

Green PG, Kroschwitz JI (2012) Biodiesel production from animal fats. Biomass Convers Biorefinery 2(1):23–31

Gude VG (2016) Process technology for biodiesel production. In: Introduction to biodiesel production. Elsevier, Amsterdam, pp 161–190

Gustavsson M, Joelsson JM, Larsson EO (2010) Vegetable oils as diesel fuels: analysis of renewable energy options. Energy Fuels 24(1):596–604

Haas MJ, Scott KM, Foglia TA (2006) Biodiesel production: a review. Biores Technol 97(6):2246–2253. https://doi.org/10.1016/j.biortech.2005.09.013

Jagadale S, Chakrabarthy J, Kumar M (2018) Waste management from sustainable biodiesel production processes: hydrodynamic cavitation treatment of biodiesel effluent. J Clean Prod 194:896–903

John A, Das AP (2024) Government strategies for achieving circular economy for plastic waste management in Southeast Asia. In: Sustainable management of urban plastic waste through circular economic approaches, pp 72–82. https://doi.org/10.1201/9781003465447-6

Joshi V, Attri S (2013) Sunflower biodiesel: research, implementation and potential. Renew Sustain Energy Rev 22:414–425

Kanungo J, Sahoo T, Bal M, Behera ID (2023) Performance of bioremediation strategy in waste lubricating oil pollutants: a review. Geomicrobiol J. https://doi.org/10.1080/01490451.2023.2245395

Kanungo J, Sahoo T, Bal M, Behera ID (2024) Performance of bioremediation strategy in waste lubricating oil pollutants: a review. Geomicrobiol J 41(4):360–373. https://doi.org/10.1080/01490451.2023.2245395

Karmee SK, Chadha A (2005) Preparation of biodiesel from crude oil of *Pongamia pinnata*. Biores Technol 96(13):1425–1429. https://doi.org/10.1016/j.biortech.2004.11.003

Keera M, Rowe B, Nour DJ, Jamel MS (2013) Animal fats as feedstocks for biodiesel production: a review. Greener J Environ Manag Public Saf 2(2):68–77

Khan U et al (2019) Optimization of transesterification process for biodiesel production from sunflower oil using sodium hydroxide catalyst. In: Progress in biomass and bioenergy production. Springer, Singapore, pp 217–226

Khatun R, Singh S, Dubey NK, Prasad Das A (2023a) A review on marine-based phytochemicals and their application in biomedical research. Recent Front Phytochem Appl Food Pharm Cosmet Biotechnol 383–395. https://doi.org/10.1016/B978-0-443-19143-5.00013-X

Khatun R, Singh S, Dubey NK, Prasad Das A (2023b) A review on marine-based phytochemicals and their application in biomedical research. In: Recent frontiers of phytochemicals: applications in food, pharmacy, cosmetics, and biotechnology. Elsevier, Amsterdam, pp 383–395. https://doi.org/10.1016/B978-0-443-19143-5.00013-X

Knothe G et al (2018) Transesterification: a versatile and sustainable tool for the green production of biodiesel. Chem Rev 118(1):412–488. https://doi.org/10.1021/acs.chemrev.7b00447

Kumar M, Pandit S, Patel V, Khanna N, Nag M, Lahiri D, Ray RR, Das A, Das D (2023) Maximization of energy recovery from starch processing wastewater by thermophilic dark fermentation coupled with microbial fuel cell technology. Geomicrobiol J. https://doi.org/10.1080/01490451.2023.2209555

Lahiri D, Nag M, Dey A, Sarkar T, Joshi S, Pandit S, Das AP, Pati S, Pattanaik S, Tilak VK, Ray RR (2021) Biofilm mediated degradation of petroleum products. Geomicrobiol J 1–10. https://doi.org/10.1080/01490451.2021.1968979

Lam MK, Lee KT (2011) Alternative feedstocks and conversion processes. Biofuels 353–374

Mahmud N, Kadir FBA (2014) Recent advances in production and fuel properties of biodiesel from waste cooking oil. Energy Convers Manag 82:334–351

Mallick K, Sahu A, Dubey NK, Das AP (2023) Harvesting marine plastic pollutants-derived renewable energy: a comprehensive review on applied energy and sustainable approach. J Environ Manag 348:119371. https://doi.org/10.1016/J.JENVMAN.2023.119371

Marchand J, Dollet X, Tessèdre PL, Mouloungui Z (2019) Lipase-catalyzed biodiesel production. In Biodiesel: quality, emissions, and by-products. Royal Society of Chemistry, pp 119–151

Mishra S, Singh RP, Rath CC, Das AP (2020) Synthetic microfibers: source, transport and their remediation. J Water Proc Eng 38:101612. https://doi.org/10.1016/J.JWPE.2020.101612

Mishra A, Siddiqi H, Kumari U, Behera ID, Mukherjee S, Meikap BC (2021) Pyrolysis of waste lubricating oil/waste motor oil to generate high-grade fuel oil: a comprehensive review. Renew Sustain Energy Rev 150:111446. https://doi.org/10.1016/j.rser.2021.111446

Mishra S, Ghosh S, van Hullebusch ED, Singh S, Das AP (2023) A critical review on the recovery of base and critical elements from electronic waste-contaminated streams using microbial biotechnology. Appl Biochem Biotechnol 2023:1–30

Mittelbach M (1998) Production of biodiesel: prospects of finer chemical technology. Wear 231(2):304–313. https://doi.org/10.1016/s0043-1648(99)80032-4

Mohanta S, Pradhan B, Behera ID (2024) Impact and remediation of petroleum hydrocarbon pollutants on agricultural land: a review. Geomicrobiol J 41(4):345–359. https://doi.org/10.1080/01490451.2023.2243925

Mohanty M, Mohanty J, Dey S, Dutta K, Shah MP, Das AP (2024a) The face mask: a tale from protection to pollution and demanding sustainable solution. Emerg Contam 10(2):100298. https://doi.org/10.1016/J.EMCON.2023.100298

Mohanty M, Tripathy B, Koteswara Reddy G, Das AP (2024b) Greener technologies for establishing restraint over microplastic pollution. In: Sustainable microbial technology for synthetic and cellulosic microfiber bioremediation, pp 239–258. https://doi.org/10.1007/978-3-031-62600-5_13

Moreond AA, Cantero DA (2011) Cold pressing of sunflower oil. Chem Process Eng 32(1):15–22

Nabi MN, Rahman MM, Akhter MS (2012) Production of biodiesel from rapeseed oil: a review. Int J Sustain Energ 31(7):383–393. https://doi.org/10.1080/14786451.2010.506816

Nagaraja P, Reddy A, Sonar S (2019) Glycerol to bioplastic for sustainable resource management. In: Bio-plastics for a circular economy. Springer, Singapore, pp 243–261

Nurrochmad MH, Horiguchi T, Hirota Y, Shono K (2016) Synthesis of biodiesel from waste cooking oil by two-step process: microwave-assisted esterification followed by supercritical methanol reaction. Energy Convers Manag 126:574–580

Pakalin S, Sarika PR (2020) Biodiesel production from waste oils: a review. J Environ Chem Eng 8(2):104067

Paul A, Dey S, Ram DK, Das AP (2023) Hexavalent chromium pollution and its sustainable management through bioremediation. Geomicrobiol J 1–11. https://doi.org/10.1080/01490451.2023.2218377

Paul A, Priyadarsini S, Karakuş S, Behera JK, Gnanasekaran R, Das AP (2024) Fate and accumulation of synthetic microfibers in the environment. In: Sustainable Microbial Technology for Synthetic and Cellulosic Microfiber Bioremediation, pp 61–76. https://doi.org/10.1007/978-3-031-62600-5_4

Pikula K, Zakharenko A, Stratidakis A, Razgonova M, Nosyrev A, Mezhuev Y, Tsatsakis A, Golokhvast K (2020) The advances and limitations in biodiesel production: feedstocks, oil extraction methods, production, and environmental life cycle assessment. Green Chem Lett Rev 13(4):275–294. https://doi.org/10.1080/17518253.2020.1829099

Pirselova J, Waxy NR, Tirler WA, Devanathan R (2019) Biodiesel by hydrogenation and hydrogenolysis. In: Waste materials and by-products in concrete. Elsevier, Amsterdam, pp 209–227

Prabhakar SVRK, Elder M (2009) Biofuels and resource use efficiency in developing Asia: back to basics. Appl Energy 86:30–36

Prabhakar A, Mishra S, Das AP (2019) Isolation and identification of lead (Pb) solubilizing bacteria from automobile waste and its potential for recovery of lead from end of life waste batteries. Geomicrobiol J 36(10):894–903. https://doi.org/10.1080/01490451.2019.1654044

Pradhan G, Tripathy B, Ram DK, Digal AK, Das AP (2023) Bauxite mining waste pollution and its sustainable management through bioremediation. Geomicrobiol J 1–10. https://doi.org/10.1080/01490451.2023.2235353

Priyadarsini S, Das AP (2024) Analytical and structural characterization of waste lithium-ion batteries for their effective recycling strategy. Environ Chem Ecotoxicol. https://doi.org/10.1016/j.enceco.2024.12.004

Priyadarsini S, Das AP (2025) Lithium bioleaching: A review on microbial-assisted sustainable technology for lithium bio-circularity. J Water Process Eng, 69:106744. https://doi.org/10.1016/j.jwpe.2024.106744
Qin S (2019) Lifecycle assessment of biodiesel production: a rapid review. Energies 12(8):1386
Quevedo-Amador RA, Escalera-Velasco BP, Ramírez Arias AM, Reynel-Ávila HE, Moreno-Piraján JC, Giraldo L, Bonilla-Petriciolet A (2024) Application of waste biomass for the production of biofuels and catalysts: a review. Clean Technol Environ Policy 26:943–997. https://doi.org/10.1007/s10098-023-02728-4
Quaye EC (1996) Energy demands in the 21st century: the role of biofuels in a developing country. Renew Energy 9:1029–1032
Raj K, Das AP (2023) Lead pollution: impact on environment and human health and approach for a sustainable solution. Environ Chem Ecotoxicol 5:79–85. https://doi.org/10.1016/J.ENCECO.2023.02.001
Rusydi F, Hidayatulloh A, Wulandari PA (2016) Heterogeneous transesterification of waste cooking oil using membrane technology. Eur J Sustain Dev 5(1):191–198
Sadasivuni KK, Rattan S, Waseem S, Brahme SK, Kondawar SB, Ghosh S, Das AP, Chakraborty PK, Adhikari J, Saha P, Mazumdar P (2019) Silver nanoparticles and its polymer nanocomposites—synthesis, optimization, biomedical usage, and its various applications. In: Lecture notes in bioengineering, pp 331–373. https://doi.org/10.1007/978-3-030-04741-2_11
Sahoo PP, Singh S, Rout PK, Mishra S, Das AP (2022) Microbial remediation of plastic pollutants generated from discarded and abandoned marine fishing nets. Biotechnol Genet Eng Rev. https://doi.org/10.1080/02648725.2022.2152629
Santos DO, Hartmann LA, Blanco CW, Mattoon AT (2011) An overview of biodiesel production from palm oil. Renew Sustain Energy Rev 15(2):838–844
Silva CS, Pereira GR, Lopes MG, Meirelles AJA, Schwan RF, Rodrigues S (2019) Enzymatic biodiesel production from waste cooking oils and fats by immobilizing Candida antarctica lipase using rice husk silicas as support. Waste Manag 101:194–204
Singh RP, Mishra S, Das AP (2020) Synthetic microfibers: pollution toxicity and remediation. Chemosphere 257:127199. https://doi.org/10.1016/j.chemosphere.2020.127199
Singh AR, Singh SK, Jain S (2022) A review on bioenergy and biofuel production. Mater Today Proc 49:510–516. https://doi.org/10.1016/j.matpr.2021.03.212
Smith J, Johnson A, Brown C (2023) Production of biodiesel from vegetable oils: a review. Renew Sustain Energy Rev
Tan KT, Lee KT (2012) Palm oil: addressing issues and towards sustainable development. Renew Sustain Energy Rev 16(7):4276–4284
Thangalazhy-Gopakumar S, Ghaziaskar HS (2008) Factors influencing the production of biodiesel from soybean oil and its properties as a fuel. Biores Technol 99(11):4485–4491. https://doi.org/10.1016/j.biortech.2007.08.016
Thirumarimurugan M, Sivakumar VM, Xavier AM, Prabhakaran D, Kannadasan T (2012). Preparation of biodiesel from sunflower oil by transesterification. Int J Biosci Biochem Bioinf 2:443–446. https://doi.org/10.7763/IJBBB.2012.V2.151
Tripathy B, Sahoo PP, Sundaray H, Das AP (2024) In-vitro biodegradation of discarded marine microplastics across the eastern coast of the Bay of Bengal, India using *Exiguobacterium* sp. Environ Chem Ecotoxicol 6:236–247. https://doi.org/10.1016/j.enceco.2024.07.003
Tyagi RD, Surampalli RY, Zhang TC, Yan S, Zhang X (2019) Biodiesel production: technologies, challenges, and future prospects. https://doi.org/10.1061/9780784415344
Varma MN, Madras G (2007) Synthesis of biodiesel from castor oil and linseed oil in supercritical fluids. Ind Eng Chem Res 46(1):1–6. https://doi.org/10.1021/ie0607043
Wu J, Zhao J (2009) Biodiesel: prospects and challenges. Renew Sustain Energy Rev 13(9):2586–2592
Xie Y, Bian D, Du H, Li X, Jia F, Ma Y (2020) Advances in enzymatic process for biodiesel production. Ind Eng Chem Res 59(4):1281–1294

Yenesew MA, Hwang MH, Kim DK (2012) Analysis of cold flow properties of biodiesel and its blends by visible–near-infrared spectroscopy and chemometrics. Fuel 95(1):538–544

Zhang Y, Huo H (2015) Current production technologies for biodiesel and their by-products. In: Production of biofuels and chemicals from lignocellulosic biomass. Springer, Singapore, pp 139–188

Zhang Z, Rozelle S (2009) The emergence of the soybean oil supply chain and its role in the expansion of the oilseed rape and soymeal sectors in China. China Econ Rev 20(4):674–680

Zhou A, Thomson E (2009) The development of biofuels in Asia. Appl Energy 86:s11–s20Mohanty, M., Tripathy, B., Koteswara Reddy, G., & Das, A. P. (2024). Greener Technologies for Establishing Restraint Over Microplastic Pollution. Sustainable Microbial Technology for Synthetic and Cellulosic Microfiber Bioremediation, 239–258. https://doi.org/10.1007/978-3-031-62600-5_13

Chapter 14
Revolutionizing Solid Hydrocarbon Sourcing: Plastic's Journey from Waste to Upcycled Treasure

Akankshya Das, Sudeshna Dey, Amel Gacem, Krishna Kumar Yadav, Harjeet Nath, Deo Karan Ram, and Alok Prasad Das

Abstract Plastic is a material that is widely used in modern life due to its affordability, resilience, and adaptability. Nevertheless, a severe environmental issue has resulted from the yearly manufacturing of around 400 million tons of plastic and a meager 9% recycling rate. Around the world, a considerable quantity of plastic waste gets thrown away every year, contaminating the soil, water, and air along with generating a significant amount of waste in landfills. It is well known that plastic pollution in marine and coastal environments is a significant issue caused by humans. The growing amount of plastic pollution in aquatic environments has proven unprecedented and continuous due to anthropogenic causes, disrupting the ecosystem's structure, function, and ultimately its essential functions and values. Plastics may break down into micro- to nanoparticles, and the finer particles are more likely to travel via soil, water, and air. As a result, a variety of detrimental effects, including ingestion, entanglement, ulceration, decreased reproduction, and oxidative stress, affect both terrestrial and aquatic species. Reusing and upcycling

A. Das · S. Dey · A. P. Das (✉)
Department of Life Sciences, Rama Devi Women's University, Bhubaneswar, India
e-mail: alok1503@gmail.com

A. Gacem
Department of Physics, Faculty of Sciences, University 20 Août 1955, Skikda, Algeria

K. K. Yadav
Faculty of Science and Technology, Madhyanchal Professional University, Ratibad, Bhopal 462044, India

Environmental and Atmospheric Sciences Research Group, Scientific Research Center, Al-Ayen University, Dhi-Qar, Nasiriyah 64001, Iraq

H. Nath
Department of Chemical and Polymer Engineering, Tripura University (A Central University), Agartala 799022, India

D. K. Ram
Department of Chemical, Petroleum & Hydrogen Technology, Nims Institute of Engineering & Technology (NIET), Nims University, Jaipur, Rajasthan 303121, India

I. D. Behera and A. P. Das (eds.), *Environmental Hydrocarbon Pollution and Zero Waste Approach Towards a Sustainable Waste Management*, Environmental Science and Engineering, https://doi.org/10.1007/978-3-031-81615-4_14

waste materials to create fuels is a potential way to reduce the amount of garbage that would otherwise harm the environment and our reliance on fossil fuels. Conventional recycling methods for plastic trash either recover inefficient thermal energy or provide compounds with lesser value than the original plastic. Value-added products made from plastic trash may now be produced sustainably through upcycling or the valorisation strategy. Upcycling is a useful method of converting plastic waste into high-value products and has the potential to significantly lessen the negative environmental effects of plastic production and consumption. This study focuses, in particular, on the processes used to convert plastic trash into high-value products, such as pyrolysis, solvent extraction, hydrogenolysis, photo reforming, and biological upcycling.

Keywords Plastic pollution · Upcycling · Recycling · Value-added products · Valorisation

14.1 Introduction

The globe is overflowing with various environmental waste such as metal waste, and solid hydrocarbon waste, which includes various kinds of materials such as plastics food, animal waste, and woody biomass (Das et al. 2011 2015a; Das and Ghosh 2023c). The rapid development of industrialization and human activities has led to more and more serious environmental pollution in the world (Das and Mishra 2008; Mishra et al. 2020). It has aroused widespread concern due to which serious management of this waste is needed (Das and Ghosh 2022; Das and Mishra 2009; Khatun et al. 2023; Mallick et al. 2023; Mishra et al. 2019a). Plastic is just one of these solid hydrocarbon wastes that have been given a lot of consideration lately due to its large manufacturing volume and adverse. Plastics are frequently used in packaging due to their beneficial performance, low cost, compact size, and ease of processing (Zhao et al. 2022). With their extensive variety of characteristics, plastics are an assorted collection of artificial additives which might be widely utilized in the modern world. The structure of polymers is a critical feature that unites them aside. Long chains of repeated molecules make up polymers, and how these chains are structured offers plastics an extensive range of houses (Sahoo et al. 2022). For example, Polyethylene Terephthalate (PET), frequently utilized in food and drink containers, is renowned for its power, lightweight, and extraordinary recyclability. Milk jugs and pipelines made from excessive-density polyethylene (HDPE) offer similar advantages with prolonged sturdiness and sturdiness. Beyond those polyethylenes, the adaptability of polypropylene (PP) is noteworthy. Food boxes, garb, or even vehicle components are made from polypropylene (PP), which is robust, compact, and proof against chemicals and immoderate temperatures. Some polymers, even though, are less environmentally high-quality. Because it incorporates chlorine, polyvinyl chloride (PVC), which is frequently utilized in floors and pipes, isn't always recycled extensively. Comparably, the recyclability of Polystyrene (PS), which is applied in cups which is probably disposable and packaging, is restricted (Siddiqui and Pandey 2013). Plastics' great durability leads to significant garbage buildup in landfills and seas since most synthetic polymers prioritize lifetime and performance over recyclability and degradation. Plastics that are only used for sterilized packaging, storage, delivery, and disposable medical components contribute significantly to global trash. Globally, 380 metric tons of material made from plastic are manufactured yearly, with less than 20% recycled, leading to significant waste and environmental degradation (Das et al. 2015b; Dey et al. 2023b). Recycling post-consumer plastic trash can be challenging due to its unknown composition and the presence of organic and inorganic pollutants. The vast majority of trash made of plastic gets disposed of in landfills or marine environments, burning in thermal power plants, or deteriorating value (Zhao et al. 2022; Mallick et al. 2023).

Plastics are widely used in agriculture, housing, automobiles, packaging supplies, toys, electronics, and other industries, leading to enormous production volumes. Since 1990, the demand for polymers has climbed by 5% annually. As the usage of plastics rises, so does the amount of petroleum-based garbage created. Plastic garbage is divided into a minimum of two distinct groups: municipality and industrial. Industrial plastics are more uniform and devoid of contaminants, making them suitable for recycling and producing lower-grade plastic goods. Municipal plastics have a diverse composition and may incorporate other elements. Plastics account for approximately 10.675 weight per cent of municipal garbage (Kunwar et al. 2016). Plastic pollution in seas is expected to surpass marine species by 2050, with an estimated 8 million tons entering each year. One-third of global plastic production is either too complicated (e.g., complex structures, polymer mixtures) or too tiny for cost-effective recycling. Plastic garbage is chemically inert and takes around 250–500 years to disintegrate entirely. Plastic pollution may infiltrate the dietary chain utilizing crops and animals (Das et al., 2024b, c, Dey et al., 2024b). Crops may absorb and collect Nano plastics (e.g., < 6 nm) which penetrate their cell walls, such as the root. Plastic that are found in chickens can be a result of earthworm eating. This poses a significant health risk, the scale of which is unclear (Zhao et al. 2022).

When the pace of plastic pollution entry into a region exceeds the pace at which natural procedures for clearing or cleansing up, plastic builds up in the environment. In the environment, plastic persists at rates that range from decades to centuries due to natural clearance. In many places where plastic builds up in the environment, especially in isolated locales, cleanup efforts are impractical. Consequently, plastic meets the criteria for being a "poorly reversible pollutant" since its emissions are uncontrollable and persist in the environment for a considerable amount of time. One of the primary problems with inadequately recovered pollutants is that it is unlikely to be able to rapidly lower pollution levels underneath the threshold, thus if it accumulates up to levels above an effect threshold, this breach will have negative consequences that are extremely difficult to reverse. Even if there is a significant decline in plastic waste in the future, plastic emissions will still rise. The total amount of plastic debris dumped entering lakes, rivers, and the ocean has been estimated to be between 9 and 23 million metric tons per year globally as of 2016 (Jambeck et al. 2015). Between fourteen and twenty-five million metric tons were thought to be discharged towards the terrestrial realm each year. By 2025, these predicted 2016 greenhouse gas emissions will have about quadrupled under business-as-usual assumptions. Plastic emissions are predicted to increase annually even in cases where there is organized, cooperative global action, such as implementing the Basel standards aimed at preventing the movement of plastic waste to countries with not sufficient waste management practices or the European Union's ambition to start reuse and recycling greater quantities of plastic as an important aspect of the transition to a circular economy (MacLeod et al. 2021). India has inadequate infrastructure for managing solid waste, especially in metropolitan areas. The public's bad habit of littering has made the issue worse. Restrictions on the usage of thin plastic shopping bags have been put in place by the Union and various state government bodies to indirectly manage the trash problem. It is acknowledged, meanwhile, that several governmental notifications were not carried out efficiently. As a result, there was additional pressure on the regional authorities to enact stricter regulations, such as a total ban on plastic bags. It is understood that a total ban on plastic carry bags is not the answer; rather, it would promote the use of substitute materials, increasing real environmental pollution and ultimately contributing to climate change, the more dreaded reality that humanity faces today. The waste management issue may be resolved by the government authorities taking appropriate action in conjunction with the industry's and the general public's voluntary cooperation and self-regulation (Siddiqui and Pandey 2013).

Plastics likely have more opportunities to interact with different substances in the environment since they decompose slowly. Nowadays, it's widely acknowledged that plastics could function as "automobiles," for other harmful substances to the environment. Metals with toxic properties (Cd, Pb, Cu, Zn, etc.), tetracycline, dichloro diphenyl trichloroethane (DDT), aromatic polycyclic hydrocarbons (PAHs), and polychlorinated biphenyls (PCBs) are a few examples of compounds that have been detected on the surface of plastic garbage (Tripathy et al. 2022). As a result of these findings, environmental mixes including plastic pollutants were created. Apart from augmenting the increased dispersion of the pollutants, the formation of plastic-contaminant mixtures likely resulted in a novel environmental fate for the pollutants (Bala, Anshul et al. 2024; Bala, Dogra et al. 2024; A. Das, Dey, & Das 2024, Dey, Dash et al. 2024; Dey et al. 2023). It is currently unclear if plastics may serve as carriers of other pollutants, particularly primary or emergent pollutants, given the abundance of toxins found in the environment. Additionally, pathogens and germs may find a home in plastics, which likely speeds up their spread throughout the ecosystem and raises their dangers to living things. The combined impacts of plastics and other substances are influenced by the kinds and qualities of plastics (which include their dimensions, area of surface, frameworks, etc.), the properties of substances that have been absorbed, and environmental circumstances (Li et al. 2021).

Low rates of plastic recycling can be attributed to a number of factors, the main one being the absence of systems capable of effectively recycling plastic materials with different compositions. Because plastics have distinct chemical structures, there is variation in their content. This means that different recycling procedures are needed for different types of plastic.

The most recycled plastic is polyethylene terephthalate (PET), which is recycled by mechanical means (e.g., 20 and 58% of PET is recovered in the US and EU, respectively). PET is then cleaned, re-extruded, and flakes are produced for resale. Because this crushing and re-extrusion process reduces the molecular weight of recycled PET, recycled PET frequently has to be combined with virgin PET to produce a product with the right qualities for the intended end use. Furthermore, while PET recycling by mechanical means is effective for PET bottles, it is less effective for PET used as garment fabric or in clam-shell packaging (Rorrer et al. 2019). The second most recycled plastic is high-density polyethylene (HDPE); in the US and the EU, 10% and 10–15%, respectively, of HDPE is recycled. While the process of mechanical recycling of HDPE is comparable to that of PET, it necessitates a more thorough devolatilization of impurities and food aromas. Nevertheless, there is presently very little food-grade recycled HDPE on the market. The removal of pollutants and additives obtained during plastic processing (such as colorants, antioxidants, plasticizers, foils, and paper) as well as from plastic packaging material, such as those established by consumers (such as sugars and other foods), is another crucial challenge regarding the recycling or upcycling of plastics. Managing plastic garbage is a problem furthermore, plastics are hard to gather and sort because of their extremely low density. Reengineering packaging materials is being done to help with some of these issues, but even the most optimistic projections indicate that recycling will not be profitable for a minimum of 50% of plastic waste streams (Ragauskas et al. 2021). The upcycling and recycling of plastic is explained in Fig. 14.1.

Here, this chapter describes the techniques for repurposing garbage made of plastic. The challenges of controlling plastic waste and the increasing need for energy

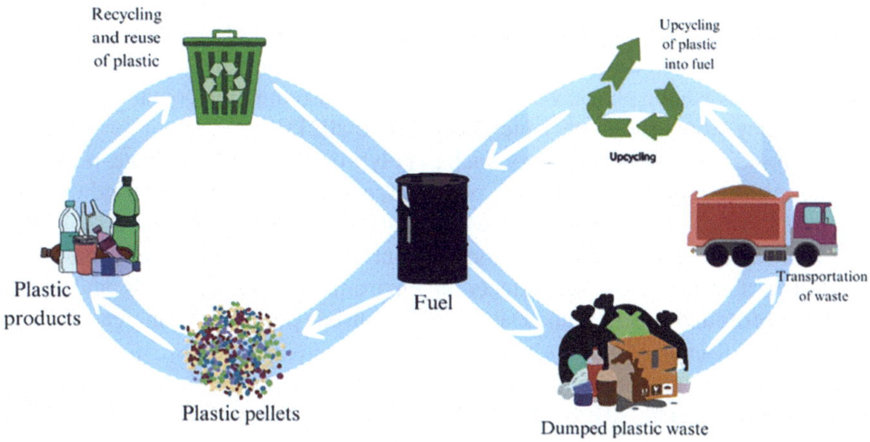

Fig. 14.1 The production of plastic from fuel and its upcycling and recycling

can be met simultaneously by producing bioenergy from plastics. Currently, several research teams are attempting to enhance this potential.

14.2 Origins of Plastic Pollution

We all rely on plastic products for our day-to-day activities. Everyday items like plastic shopping bags, food containers, and water bottles contribute significantly toward plastic pollution. These plastic items are not often discarded well into waterways and terrestrial landscapes. Synthetic fibers lost from garments and microbeads of plastic from cosmetics are two other factors that contribute to the issue, along with industrial operations.

Plastic packaging- Lightweight materials are needed by the packaging industry to minimize the quantity of product packaging, which lowers product weight and saves money on transportation. It also lessens the quantity of waste material generated by packaging that ends its useful life. Based on petroleum polymers, which include polypropylene (PP), polystyrene (PS), polyvinyl chloride (PVC), polyethylene (PE), and polyethylene terephthalate (PET), are widely used in packaging due to their lightweight, strong mechanical effectiveness, and good insulating qualities (Behera et al. 2020; Bal et al. 2020). A large number of them wind up in landfills, where it will take future generations for them to break down. Although they are not biodegradable, they generate a significant quantity of environmental contamination (Dey et al. 2023a; Mallick et al. 2024a; Sahu and Das 2024). Recyclable plastics are now being used in package designs by the industry as a way to cut down on trash that is dumped into the environment. Recycling companies must, however, keep pollutants in the regenerated plastic at levels that are appropriate for the final packaging's intended use. Because several sorts of plastic garbage need to be sorted and segregated from the collected debris, this has proven to be challenging, time-consuming, and expensive (Paul et al. 2024, Das, A. Prasad., & Ghosh, Shreya 2023). Subsequently, it might be necessary to thoroughly wash the plastic wastes to eliminate any remaining contaminants from the items they were used to package (Ncube et al. 2021).

Face masks - Synthetic polymers, such as polypropylene, polyurethane, polyacrylonitrile, polystyrene, polycarbonate, polyester, and polyethylene, are used to make face masks, which are used to cover the mouth and nose (Mishra and Das 2022). They serve as a defense against allergies, contaminants, and pathogenic pathogens. Waste pollution is a serious issue that affects the environment, human health, and global health in a profound way (Raj and Das 2023). The urban population's increased use of masks and its discharge of fibers and plastics into the environment are major contributors to the environmental contamination caused by synthetic plastic (Mishra and Das 2021; Mishra et al. 2019b). The COVID-19 pandemic, also known as the severe acute respiratory syndrome corona virusvirus 2 (SARS-CoV-2) global pandemic, has significantly changed the world environment. Several preventative steps are implemented to lessen the spread of the infection; among these, the usage of facemasks has been essential in controlling the outbreak (Mohanty et al. 2024a, b; Sahu et al. 2024).

Personal products and cosmetics- Thermoset polymers which include polyurethane (PU), polyethylene terephthalate (PET), and polymethyl methacrylate (PMMA), and thermoplastics which include polyethylene (PE), polypropylene (PP), polystyrene (PS), and polytetrafluoroethylene (PTFE), are the two main types of plastic used in cosmetics and personal hygiene products (Das et al. 2023a; Mishra et al. 2022a, b, c). Of all the microbeads in cosmetics, 93% are made of polyethylene plastic. Microbeads are present in a wide range of personal care and cosmetic goods, including shower gel, toothpaste, sunscreen, face cleansers, and hair dyes. Because of their tiny size, insoluble in water, and gradual breakdown, microbeads typically enter the drainage system along with wastewater washing (Mishra et al. 2022c; Mishra and Das 2021; Biswas et al. 2020). Plastic microbeads cannot currently be successfully removed by sewage treatment machinery. As a result, the sewage sludge that is commonly used in agriculture permits these microbeads to penetrate the environment (An et al. 2020).

Fishing nets- Some instances of plastic fishing debris include nets for fishing, buoys, floating containers, rods for fishing, aquariums, fishing lines, cables, and other objects. There are between 0.13 and 135,000 tons of commercial fishing gear that are abandoned annually, according to estimates. Large amounts of polystyrene foam have been found in marine plastic trash, for instance at Hofsgar Lake, Mongolia, and the Shandong Coast, China. The main contributors of this waste are the numerous discarded nets for fishing and foam pontoon boats that wind up in the ocean due to biological deterioration and normal wear from frequent use, as well as the fish farming industry's extensive and rapidly expanding consumption of Styrofoam floating devices (Moharana et al. 2024; Pradhan et al. 2024; Tripathy et al. 2024a). In addition, the quantity of polymers in the marine environment has increased as a result of the aquaculture industry's widespread feed waste bag abandonment. Additionally, more and more fishing boats are using plastic nets for fishing seafood, and as fishing gear is replaced, a sizable quantity of damaged plastic fishing nets are left floating randomly in the ocean. PP, LDPE, PA (nylon), and PE filament are the main materials found in nets for fishing and fishing ropes. It is possible to distinguish between nets and ropes made of these different materials by their diameter and transparency (An et al. 2020).

Vehicle tire wear- The synthetic material that tires wear discharges into the atmosphere is regarded to be one of the main sources of plastic in road dust. Tires used in motor vehicles go through three stages of life: manufacture, usage, and disposal. The usage stage has the highest carbon emissions among these three phases. Tires come in two varieties: natural rubber (NR) and synthetic rubber (SR). Most rubber that is synthetic is composed of butadiene rubber, or SBR (BR), because natural rubber is more durable than synthetic rubber, it is frequently utilized in modern automobile tires (Dey et al. 2023a; Sahu et al. 2024). The main source of natural rubber nowadays is the Brazilian rubber plant, or Hevea brasiliensis. The extraction and manufacturing of rubber using this tree is rather expensive. Synthetic rubber, on the other hand, comes from the petroleum sector (Dey et al. 2024a; Jagatee et al. 2024;

Mallick et al. 2024b). Rubber is typically treated with various kinds and concentrations of chemical additives by tire producers to satisfy the performance standards of motor vehicle tires. Tire debris from moving cars can wind up in a variety of places, including soil, surface water, sewage systems, and the atmosphere (Haque and Fan 2023). Increasing a motor vehicle tire's friction is typically thought to shorten the distance to stop the vehicle and hence improve driver safety. But more friction also means that the tire is releasing rubber particles more quickly. By not using winter-related tires in the summer, drivers may reduce the number of polymers generated by tire friction. On dry winter roads, winter tires release more tire fragments because they are often softer and more gripping than summer tires. Furthermore, to reduce the number of MPs generated by tire friction, drivers should make an effort to use low-emission tires, fix tires on time, and use public transportation (An et al. 2020).

14.3 Adverse Effects of Plastic Pollution

According to conventional ecotoxicological risk assessment, plastic now only affects a tiny but probably growing portion of the world's oceans. This is determined by comparing observed or expected quantities in the environment to toxicological impact thresholds calculated from routine testing (Ghosh et al. 2018; Nag et al. 2023a). The world's oceans now contain a very modest quantity of plastic, but this is expected to change shortly according to conventional ecotoxicological risk assessment. Toxicological effect levels derived from regular testing are compared to observed or anticipated environmental amounts (Nag et al. 2023b; Sadasivuni et al. 2019). Plastic pollution affects the cycle of carbon globally directly as well as indirectly (Mishra and Das 2023). The direct impact results from a small but considerable fraction of the 280 million to 350 million tons of fossil hydrocarbons expected to be converted into plastic yearly. The planet will continue to emit greenhouse gasses for millennia after fossil fuels are fully phased out due to waste management and plastic breakdown. One way that plastic may have an indirect effect on the carbon cycle that is bigger than the direct effects of greenhouse gas emissions is through the equilibrium state of the coastal carbon pump. An estimate from recently indicates that the seafloor currently gets 7.8 million metric tons of carbon derived from petroleum every year. Before they finally land on the seafloor, a large percentage of the polymers will first float about as neutrally buoyant nanoparticles in the water columns. The habitats of cyanobacteria and phytoplankton populations may be affected by increasing concentrations of hetero-aggregates and persistent plastic particles, which might affect the level of turbidity or food sources. Reduced populations of bacteria in the ecosystem would mean less carbon is taken up from the atmosphere. The unsequestered carbon may remain in the atmosphere and worsen global warming rather than being incorporated into the food chain in the oceans (Mishra et al. 2022b; Das et al. 2024a). Meanwhile, the increasingly significant amounts of carbon found in nonbuoyant polymers will sink; by 2050, some estimates suggest that in hotspot places, the amount of plastic carbon trapped in bottom sediments could surpass the amount of naturally existing organic carbon. The aqueous carbon pump is influenced by many variables that also affect the cycling of nutrients. It has been demonstrated that within aquatic systems, biofilms on plastics influence the cycle of phosphate and nitrogen (Das et al. 2023b). Similarly, the existence of plastic changed the nitrogen cycling in sediment, as shown by a microcosm experiment. Thus, plastic integrated into ocean particles may impact the transport of nutrients to the deep sea environments. Additionally, Earth system modeling indicates that zooplankton feeding on plastic may hasten the ocean's overall oxygen depletion. The detection of this waste is important for environmental protection to mitigate pollution and health hazards (Bhattacharjee et al. 2021; Das and Swain 2013; Mishra and Das 2008). Long-term alterations in soil characteristics, including water-holding capacity, microbial diversity and activity, availability of nutrients, and soil structure, may result from higher plastic loading (Mishra et al. 2021a, 2023; Tripathy et al. 2024a, b). In addition to possibly permanent soil deterioration, the buildup of polymers in soils can have an impact on plant variety and performance. Similar effects might result from the development of (micro)plastic hotspots on seabeds, which alter the structure and content of the sediment to the point that it affects both the marine carbon pump and sediment fertility.

There will only be more soils and sediments in the world in the future that are permanently impacted by plastic waste. There are multiple instances of wildlife getting in touch with trash that contains plastics. 914 marine megafaunal species, including 227 different species of seabirds, 86 different types of marine mammals, all species of sea turtles, and 430 different types of fish, were found to have been affected by entanglement or consumption, according to a recent analysis (Singh et al. 2020). It just takes a few contacts for population-level implications to be threatened for endangered animals (Bhuyan 2022; Zolotova et al. 2022). Entanglements and plastic ingestion pose a threat to thirteen percent of the 693 species encompassed under the International Union for Conservation of Nature Red List (MacLeod et al. 2021). After the deliberate killing, the most frequent cause of mortality in the northern Mediterranean was found to be the entanglement of 600–700 susceptible species of monk seal (Monachus monachus) in fishing gear (Karamanlidis et al. 2008). The invasion of plastic surfaces is a different type of interaction with living things. Six years following a single tsunami allowed over 300 living species to establish themselves on rafting debris, the species started propagating across the ocean (Mishra et al. 2021c). This suggests that during extreme weather events, animals may find it simpler to establish themselves due to plastic pollution. The complex and pervasive impacts of plastic pollution indicate that there is still much to learn about how it affects animals and ecosystems. Ingesting plastic has been connected to a number of adverse impacts, such as changed physiology, bodily injury, and lowered rates of eating, development, reproduction, and oxygen consumption, due to particle and chemical-related toxicity (Mishra et al. 2020). Over 0.5% concentrations of macro- and nanoplastic have been shown to affect the abundance of macroinvertebrates in sediments. Additives that leak plastic may have (eco)toxicological repercussions. One issue that has to be addressed is the presence of phthalate component esters in polyethylene mulches and their incorporation into grains used for human and animal consumption. Another is the recent discovery that coho salmon (Oncorhynchus kisutch) die quickly during stormwater overflow events as a consequence of a phototransformation brought on by a common antioxidant found in tire rubber (McIntyre et al. 2023). How plastic pollution interacts with other chemical, biological, and geophysical stresses to produce effects is a lesser-studied component of the problem. For example, the effects of plastics on carbon cycling, entanglement, ingestion, and toxicity may intensify the potential negative effects of overfishing and the impact of climate change on fisheries. Plastic is another stressor that aquatic animals must adjust to because of changes in temperature, nutrition availability, and chemical exposure, this may lead to a decline in biodiversity (Singh et al. 2020). In contemporary times, the focus of researchers has shifted towards bioremediation techniques centered around microorganisms, primarily due to their notable benefits such as high efficiency, cost-effectiveness, eco-friendliness, and absence of secondary contamination (Kanungo et al. 2024; Das and Ghosh 2023d, h, g; Behera et al. 2022a, b). The microbial approach to addressing pollution presents cost advantages compared to traditional physical and chemical methods owing to its minimal spatial requirements, uncomplicated infrastructure and procedures, minimal energy consumption, and absence of secondary pollution management (Das and Ghosh 2018, 2023a; Ghosh and Das 2017, 2020). Microorganisms possess the ability to combat the challenges posed by toxic environmental ions by swiftly adapting through mutation and evolutionary processes to develop resistance mechanisms (Kanungo et al. 2023; Behera et al. 2021, 2022a, b; Mishra et al. 2021b). Moreover, leveraging their swift reproduction rates, robust metabolic capabilities, and diverse species composition, microorganisms can effectively diminish the levels and harmful effects of heavy metals in the ecosystem through various mechanisms (Mohanta et al. 2024; Behera et al. 2022a, b; Bal et al. 2018; Ghosh et al. 2023, 2017; Ghosh and Das 2018; Mohanty et al. 2018).

14.4 Plastic as Solid Hydrocarbon and Its Upcycling

Any technique that makes use of plastic trash instead of discarding it is included in the category of plastic waste recycling. When recovered plastic trash is utilized again with no retaining its original qualities, it is referred to by the term downcycled plastic waste (Ghosh et al. 2016a, b; Sanket et al. 2016). Contrarily, plastic garbage is considered upcycled if it can be remanufactured without losing any of its functionality or performance, as shown in Table 14.1.

Developing techniques for upcycling and recycling plastic waste can help lower greenhouse gas emissions, the amount of waste that ends up in landfills and the ocean, and the need for non-renewable petroleum resources in the production of polymers—all while restoring the material's economic value (Zhao et al. 2022).

Pyrolysis- Pyrolysis is an intermediate recycling process that uses thermal breakdown to transform organic polymers into char, fumes, and liquid oil at high temperatures. Studies on pyrolysis were conducted at temperatures that ranged from 300 to 900 °C; however, 500–550 °C was generally the ideal range for plastic waste (Sipra et al. 2018). Additionally, pyrolysis of various plastic wastes was conducted at various

Table 14.1 Different upcycling procedures of different types of plastics and their end products

S. No	Type of upcycling process	Type of plastic	End products	References
1	Pyrolysis	PE	Biodiesel	Maqsood et al. (2021)
		HDPE	Biofuel and biogas	Sipra et al. (2018)
		LDPE	Biooil	Duan et al. (2021)
		PP	Biofuel and Biogas	Uzoejinwa et al. (2018)
		PS	Biofuel	Polymers (2021)
2	Photo reforming	PET	Hydrogen gas and Smaller organic molecule	Li and Zhang (2024)
		PS	Hydrogen gas and Hydrocarbons	Thiloka Edirisooriya et al. (2023)
3	Hydrogenolysis	PE	Liquid alkanes	Chen et al. (n.d.)
		PP	Liquid and gaseous alkanes	Chen et al. (n.d.)
		PS	Liquid alkanes and aromatic	Yao et al. (2022)
4	Solvent extraction	PET	High-quality polymer	Cejudo Bastante et al. (2018)
		PP	High-quality polymer	Cejudo Bastante et al. (2018)
5	Biological upcycling	PET	Bio-Based Monomers	Rorrer et al. (2019)
		PU	Bio-Based Monomers	Lee et al. (2023)

heating rates. Pyrolysis comes in two flavors: thermal and catalytic. Non-catalytic pyrolysis, another name for the thermal combustion process, is an endothermic technique that doesn't require a catalyst. Numerous plastics, such as PS, PP, and PE polymers, have undergone extensive thermal pyrolysis in the past (Polymers 2021; Mishra et al. 2021a, b, c). The catalytic pyrolysis is done with the aid of a catalyst (Uzoejinwa et al. 2018). The method showed great promise for producing liquid oil with higher standards at lower temperatures and quicker reactions when compared to thermal pyrolysis (Maqsood et al. 2021). These components have the potential to enhance the overall pyrolysis process and reduce the energy needed for the process (Duan et al. 2021). The pyrolysis of plastic waste is influenced by a number of process factors, such as temperature, retention time, feedstock composition, catalyst use, heating rate, and particle size. This process of pyrolysis helps in the generation of bioenergy (Miandad et al. 2016).

Chemical recycling- The growth of various recycling technologies and the reduction of the range of substances that can be disposed of in landfills are being driven by the increasing number of plastic products that are reaching the end of their useful lifespans and the rising demand over these products, along with the strengthening regulations regarding the environment in many nations. The environmental effects of the rise in plastic trash can be mitigated via chemical recycling of plastic garbage, which is a more environmentally friendly option than landfilling and incineration.

The chemical makeup of the plastic is one factor that influences the efficacy of various chemical processing methods. Upcycling acknowledges the following methods for handling plastic waste: photo-reforming, hydrogenolysis, and solvent extraction. These methods aim to transform recyclable plastics into goods.

Photo reforming Recovering all the recyclable plastic generated in a given year is projected to have saved roughly 3500 million gallons of oil, or greater in terms of 176 billion dollars, solely for the year 2017. This is because plastic is manufactured in large quantities and is sourced from fossil fuels. This estimate is based on the fact that these values increased significantly following the COVID-19 pandemic (Li and Zhang 2024). In addition, the materials' minuscule size poses a problem for the use of polymer recycling since small particles, or microplastics, are difficult to collect and recycle and wind up in drinking water, sediments, oceans, and almost every other ecosystem. It comprises of a straightforward, low-energy process that uses water, sunshine, and photocatalyst to convert waste plastic into H2 and other byproducts like acetic acid. Low performance and low selectivity are the outcomes of photo-reforming plastic trash at room temperature (Thiloka Edirisooriya et al. 2023). Photo-reform processes have been achieved for polymers such as polylactic acid, polyethene terephthalate (PET), and polyurethanes at room temperature with low-cost CdS/CdOx catalysts in an alkaline solution; nevertheless, this approach is dangerous. This strategy has not been widely adopted since it complicates performance. In general, this process uses the ingredients mentioned above together with an organic substrate that acts as a donor of electrons and becomes oxidized by the photocatalyst. The photogenerated electrons convert water to H2, which is subsequently transferred through the photocatalyst to a cocatalyst. This valuable product finds application in the chemical, medicinal, agricultural, and renewable energy sectors (Zúñiga et al. 2023).

Hydrogenolysis- Hydrogenation, or hydrogenolysis, is one of the numerous processes used to change organic molecules. It is an economical, ecologically friendly, and sustainable method since it uses renewable resources to make hydrogen, which is produced without waste. It explains using hydrogen for dissolving chemical bonds, particularly C–C bonds, and makes reference to a specific kind of depolymerization. The valuation of the various plastic wastes is made possible by the enzymatic hydroconversion of them, adding value (Chen et al. n.d.). Large chains may be broken down into monomers, oligomers, or additional goods that can be employed in a range of commercial contexts during the chemical decomposition process that produces this. Several methods that allow large polymers to be broken down into monomer have been studied up to this point. Trans-esterification or hydrolysis of PET is one technique for recycling conventional chemicals (Yao et al. 2022). The process of hydrogenation of PET on C/MoO2 and catalysts such as ruthenium in the presence of different solvents is an additional technique. An additional technique is the hydrogenolysis of aromatic compounds using Ru/Nb205 ruthenium/niobium catalysts, which produces excellent conversions yields to p-xylene and is used as a probe reaction to look at the C-O bond cleavage. Another method involves hydrogenating PET using ruthenium catalysts and DMSO or dimethyl sulfoxide as the solvent. The hydrogenation of polyolefins makes use of rhodium (Rh) catalysts because of its vacuum d orbital, which provides high catalytic efficiency, stability, and useful life (Das et al. 2023b; Kumar et al. 2023). However, petroleum derivatives and synthetic polymers like polylactic acid—which is produced by fermenting lactic acid—are also processed using this method. Research has been done on the development of solvent-free catalytic hydrogenolysis of PLA powder, yielding zero waste and conversion percentages of 99 and 100% (Lau et al. 2020). When this chemical gradually deteriorates in natural surroundings, it turns into solid waste. Because biofuels are produced, hydrogenolysis technologies enable the recycling of polymers, which gives them a significant additional value. It's a technique that can be used to produce this green technology because of its high-performance values (Zúñiga et al. 2023).

Solvent extraction- Chemolysis, another name for this process, is one of the most researched ways to depolymerize polymers in compounds containing carbonyl functional groups. Solvent molecules, which also act as reagents, depolymerize plastic chains when they are introduced in excess through nucleophilic substitutions. This process includes the breakdown of intramolecular bonds, and compounds with saturated carbon atoms coupled to groups of function that communicate with heteroatoms are more susceptible to it. Reports claim to have only discussed a small number of carbon fiber-based polymeric composites. While the majority of these methods rely on conventional pyrolysis, solvolysis research, and development seem to be an alternative since it has a variety of advantages, including reduced energy costs, which have a knock-on effect on economic expenses, and fewer emissions of greenhouse gases. The processes of ethylene glycol (glycolysis), freshwater (hydrolysis), methanol (methanolysis), amino acids (aminolysis), and ammonium (ammoniolysis) are the most often used solvents for this kind of procedure. There is a close relationship between these techniques. As a result, it is among the most studied techniques for depolymerizing waste plastic and creating synthetic, high-value chemical products. These items, which include textiles, plasticizing agents, adhesives, binding agents, and corrosion inhibitors, can be used to generate novel substances with either identical or distinct qualities from the originals, depending on the solvent used in the process. One excellent example of this type of recycling is PET waste (Cejudo Bastante et al. 2018). Additionally, this recycling technology may be used with other approaches like low-temperature pyrolysis (Das and Ghosh 2023b, f, e). In order to recover waste carbon fiber-reinforced polymer, they coupled low temperatures pyrolysis along with a solvolysis pretreatment, guaranteeing the process's maximum material usage and energy efficiency. Furthermore, it was shown that the solvolysis process reduced the consumption of heat during recycling, increasing the degree of retention of the mechanical qualities of the fibers to 90.53%. This percentage of retention was 10.21% greater than that produced in traditional pyrolysis procedures. With this knowledge, we can illustrate the many advantages of integrating upcycling with different methods. Water and alcohol are examples of supercritical and subcritical fluids that may partially oxidize and break down polymers, which makes them useful for use in solvolysis processes that recover or recycle waste chemicals. Additionally, the fluids' high solubility allows the polymer to break down (Zúñiga et al. 2023).

Biological upcycling- By using living things as an intermediary, bio-upcycling presents a viable strategy for managing plastic waste (Lee et al. 2023). This method uses enzymes produced by microorganisms to disassemble plastic polymers into their chemical constituents, as reported in a scientific study by Lomwongsopon and Varrone (2022). After being reabsorbed by the organisms, these monomers can be transformed into useful biobased products like biofuels or biodegradable polymers (Rorrer et al. 2019). The benefit of upcycling is its ability to break down plastics into new resources while simultaneously upcycling them, which encourages a more circular economy for plastic waste (Lomwongsopon and Varrone 2022).

14.5 Conclusion

Upcycling offers a different method of managing plastic waste that goes beyond downcycling. For instance, pyrolysis, chemical recycling, and biological upcycling can be used to convert plastic waste into useful resources. Through heat, pyrolysis breaks down plastics, thereby producing fuels and chemical feedstocks. Chemical recycling methods such as hydrogenolysis and solvent extraction target particular types of plastics, which makes it possible for them to create virgin-quality plastics or valuable chemicals. Photoreforming utilizes sunlight so that it can break plastics into reusable monomers. Biological upcycling involves the use of microbes which can degrade plastics and even upcycle them into bio-based products. Although every approach has got its advantages and disadvantages alone they have great potential when combined together too much good can come out from them hence the need for further studies while developing these processes to make them efficient enough in terms of cost scalability among others like this one should also address issues relating with mixed streams of plastic wastes during up-cycling as well ensuring minimum environmental impact during this stage would be important since through up-cycling only then can we achieve a closed loop system for materials but still we have more work ahead of us if want succeed in creating sustainable future where there no more pollution caused by single-use items made from various types polymers.

References

An L, Liu Q, Deng Y, Wennan W, Gao Y, Ling W (2020) Sources of microplastic in the environment. Handb Environ Chem 95:143–159. https://doi.org/10.1007/698_2020_449

Bala K, Anshul, Verma S, Dogra A, Das AP, Kumar R, Nandini (2024) Sustainable management for urban plastic waste generation. In: Sustainable Management of Urban Plastic Waste Through Circular Economic Approaches, pp 211–236. https://doi.org/10.1201/9781003465447-13

Bala K, Dogra A, Lata P, Das AP (2024) Discussing sources and biological-chemical remediation approaches for microplastic pollution. In: Sustainable Microbial Technology for Synthetic and Cellulosic Microfiber Bioremediation, pp 37–59. https://doi.org/10.1007/978-3-031-62600-5_3

Bal B, Ghosh S, Das AP (2018) Microbial recovery and recycling of manganese waste and their future application: a review. Geomicrobiol J 36(1):85–96. https://doi.org/10.1080/01490451.2018.1497731

Bal M, Behera ID, Kumari U, Biswas S, Meikap BC (2020) Hydrodynamic study and particulate matter removal in a self priming venturi scrubber. Environ Technol Innov 20:101167. https://doi.org/10.1016/j.eti.2020.101167

Behera ID, Basak G, Kumar RR, Sen R, Meikap BC (2020) Treatment of petroleum refinery sludge by petroleum degrading bacterium Stenotrophomonas pavanii IRB19 as an efficient novel technology. J Environ Sci Health Part A 56(2):226–239. https://doi.org/10.1080/10934529.2020.1866924

Behera ID, Nayak M, Biswas S, Meikap BC, Sen R (2021) Enhanced biodegradation of total petroleum hydrocarbons by implementing a novel two-step bioaugmentation strategy using indigenous bacterial consortium. J Environ Manag 292:112746. https://doi.org/10.1016/j.jenvman.2021.112746

Behera ID, Meikap BC, Sen R (2022a) Performance evaluation of bacterial consortium for biodegradation of total petroleum hydrocarbon: a comparative strategic biostimulation study. Research Square

Behera ID, Nayak M, Mishra A, Meikap BC, Sen R (2022b) Strategic implementation of integrated bioaugmentation and biostimulation for efficient mitigation of petroleum hydrocarbon pollutants from terrestrial and aquatic environment. Marine Pollution

Bhattacharjee J, Mishra S, Das AP (2021) Recent advances in sensor-based detection of toxic dyes for bioremediation application: a review. Appl Biochem Biotechnol 2021:1–20. https://doi.org/10.1007/S12010-021-03767-7

Bhuyan MS (2022) Effects of microplastics on fish and in human health. Front Environ Sci 10:827289. https://doi.org/10.3389/FENVS.2022.827289/BIBTEX

Biswas S, Diwakar RK, Behera ID, Meikap BC, Sen TK, Khiadani M (2020) Aqueous phase phenol removal from synthetic and real steel plant effluents through a batch and semifluidized bed column operation: experimental and model analysis. J Environ Chem Eng 8(5):104441. https://doi.org/10.1016/j.jece.2020.104441

Cejudo Bastante C, Casas Cardoso L, Fernández Ponce MT, Mantell Serrano C, Martínez de la Ossa-Fernández EJ (2018) Characterization of olive leaf extract polyphenols loaded by supercritical solvent impregnation into PET/PP food packaging films. J Supercrit Fluids 140:196–206. https://doi.org/10.1016/J.SUPFLU.2018.06.008

Chen L, Zhu Y, Meyer LC, Hale LV et al (n.d.) Effect of reaction conditions on the hydrogenolysis of polypropylene and polyethylene into gas and liquid alkanes. *Pubs.Rsc.Org*

Das A, Dey S, Das AP (2024) Ecotoxicological impacts of synthetic microfiber pollutants and development of sustainable mitigation strategies. Environ Chem Ecotoxicol. https://doi.org/10.1016/J.ENCECO.2024.12.003

Das AP, Ghosh S (2018) Bioleaching of manganese from mining waste materials. Mater Today Proc 5(1):2381–2390. https://doi.org/10.1016/J.MATPR.2017.11.459

Das A, Ghosh S (2022) Role of microorganisms in extenuation of mining and industrial wastes. Geomicrobiol J 39:173–175. https://doi.org/10.1080/01490451.2022.2038953

Das AP, Ghosh S (2023a) Cost-effectiveness of the biomining process. In: Manganese mining microorganisms, pp 143–151. https://doi.org/10.1016/B978-0-12-822147-1.00008-X

Das AP, Ghosh S (2023b) Evolution of biomining technology. In: Manganese mining microorganisms, pp 5–19. https://doi.org/10.1016/B978-0-12-822147-1.00012-1

Das AP, Ghosh S (2023c). Manganese and its application. In: Manganese mining microorganisms, pp 29–40. https://doi.org/10.1016/B978-0-12-822147-1.00007-8

Das AP, Ghosh S (2023d) Mechanism of manganese bioleaching. In: Manganese mining microorganisms, pp 87–98. https://doi.org/10.1016/B978-0-12-822147-1.00014-5

Das AP, Ghosh S (2023e) Metagenomic insights into unculturable microbial diversity. In: Manganese mining microorganisms, pp 63–76. https://doi.org/10.1016/B978-0-12-822147-1.00002-9

Das AP, Ghosh S (2023f) Microorganisms in manganese biomining processes. In: Manganese mining microorganisms, pp 55–61. https://doi.org/10.1016/B978-0-12-822147-1.00013-3

Das AP, Ghosh S (2023g) Mixed culture bioleaching: an insight into manganese biomining process and efficacy. In: Manganese mining microorganisms, pp 129–141. https://doi.org/10.1016/B978-0-12-822147-1.00001-7

Das AP, Ghosh S (2023h) Role of metabolites and metabolomics for understanding of manganese bioleaching. In: Manganese mining microorganisms, pp 117–127. https://doi.org/10.1016/B978-0-12-822147-1.00011-X

Das AP, Mishra S (2008) Hexavalent chromium (VI): health hazards and environmental pollutant. J Environ Res Dev 2(3):386–392

Das AP, Mishra S (2009) Hexavalent chromium [Cr (VI)]: yellow water pollution and its remediation. Sarovar Saurabh ENVIS Newsl Wetl Ecosyst 5(2):1–8

Das A. Prasad, Ghosh, Shreya (2023) Manganese mining microorganisms

Das AP, Swain S (2013) Algal biosorption of toxic dye methylene blue. In: A potential source of food, feed, biochemicals, biofuels and biofertilizers, international conference on algal biorefinery, Indian institute of technology, p 13

Das AP, Sukla LB, Pradhan N, Nayak S (2011) Manganese biomining: a review. Biores Technol 102(16):7381–7387. https://doi.org/10.1016/J.BIORTECH.2011.05.018

Das AP, Ghosh S, Mohanty S, Sukla LB (2015a) Advances in manganese pollution and its bioremediation. In: LB Sukla, N Pradhan, S. Panda, BK Mishra (eds) Environmental microbe biotechnology. Springer, Cham, Switzerland, pp 313–328

Das AP, Ghosh S, Mohanty S, Sukla LB (2015b) Consequences of manganese compounds: a review. Toxicol Environ Chem 96(7):981–997. https://doi.org/10.1080/02772248.2015.1005428

Das AP, Dutta K, Khatun R, Behera ID, Singh S, Mishra S (2023a) Microfiber pollution and its microbial mitigation: a review on current trends and future prospects. J Taiwan Inst Chem Eng 166:105104. https://doi.org/10.1016/J.JTICE.2023.105104

Das AP, Mohanty AK, van Hullebusch ED, Figueiredo G (2023b) Bioremediation as an emerging technology for the removal of synthetic microplastic pollutants from marine ecosystem. MarPB 194:115297. https://doi.org/10.1016/J.MARPOLBUL.2023.115297

Das AP, Behera ID, Bhanja D (2024a) Sustainable microbial technology for synthetic and cellulosic microfiber bioremediation. Springer Nature. https://doi.org/10.1007/978-3-031-62600-5

Das A, Dey S, Das AP (2024b) Ecotoxicological impacts of synthetic microfiber pollutants and development of sustainable mitigation strategies. Environ Chem Ecotoxicol. https://doi.org/10.1016/J.ENCECO.2024.12.003

Das A, Ray SK, Mohanty M, Mohanty J, Dey S, Das AP (2024c) Ecotoxicity of microplastic wastes and their sustainable management: a review. Environ Chem Ecotoxicol 6:144–152. https://doi.org/10.1016/J.ENCECO.2024.05.003

Dey S, Das A, Mallick K, Sahu A, Das AP (2023a) Environmental petroleum waste: pollution, toxicity, sustainable remediation, pp 159–175. https://doi.org/10.1007/978-3-031-48220-5_7

Dey S, Banismita Tripathy M, Kumar S, Das AP (2023b) Ecotoxicological consequences of manganese mining pollutants and their biological remediation. Environ Chem Ecotoxicol 5:55–61. https://doi.org/10.1016/J.ENCECO.2023.01.001

Dey S, Dash A, Das AP (2024a) Bioremediation as a sustainable and circular solution for mining waste management and bioenergy generation. In: Sustainable management of mining waste and tailings, pp 107–119. https://doi.org/10.1201/9781003442455-6/BIOREMEDIATION-SUSTAINABLE-CIRCULAR-SOLUTION-MINING-WASTE-MANAGEMENT-BIOENERGY-GENERATION-SUDESHNA-DEY-AKANSHYA-DASH-ALOK-PRASAD-DAS

Dey S, Bissoyi A, Koteswara Reddy G, Das AP (2024b) Synthetic microfibers: an urban emerging pollutant demanding for sustainable environmental solutions. In: Sustainable microbial technology for synthetic and cellulosic microfiber bioremediation, pp 107–121. https://doi.org/10.1007/978-3-031-62600-5_7

Dey S, Dash N, Ray SK, Priyadarsini S, Chhotaray C, Das AP (2024) Occurrence and quantification of tire wear plastics pollutants from highway city dust of Bhubaneswar City of Odisha. In:

Sustainable Microbial Technology for Synthetic and Cellulosic Microfiber Bioremediation, pp 77–89. https://doi.org/10.1007/978-3-031-62600-5_5

Dey S, Tripathy B, Kumar MS, Das AP (2023) Ecotoxicological consequences of manganese mining pollutants and their biological remediation. Environ Chem Ecotoxicol. https://doi.org/10.1016/J.ENCECO.2023.01.001

Duan D, Feng Z, Dong X, Chen X, Zhang Y, Wan K, Wang Y, Wang Q, Xiao G, Liu H, Ruan R (2021) Improving bio-oil quality from low-density polyethylene pyrolysis: effects of varying activation and pyrolysis parameters. Energy 232:121090. https://doi.org/10.1016/J.ENERGY.2021.121090

Ghosh S, Das AP (2017) Bioleaching of manganese from mining waste residues using acinetobacter sp. Geol Ecol Landsc 1(2):77–83. https://doi.org/10.1080/24749508.2017.1332847

Ghosh S, Das AP (2018) Metagenomic insights into the microbial diversity in manganese-contaminated mine tailings and their role in biogeochemical cycling of manganese. Sci Rep 8(1):1–12. https://doi.org/10.1038/s41598-018-26311-w

Ghosh S, Das AP (2020) Microbial metagenomics: current advances in investigating microbial ecology and population dynamics. In: Nayak SK, Mishra BB (eds) Frontiers in soil and environmental microbiology. CRC Press, Boca Raton, pp 247–254

Ghosh S, Mohanty S, Akcil A, Sukla LB, Das AP (2016a) A greener approach for resource recycling: manganese bioleaching. Chemosphere 154:628–639. https://doi.org/10.1016/J.CHEMOSPHERE.2016.04.028

Ghosh S, Mohanty S, Nayak S, Sukla LB, Das AP (2016b) Molecular identification of indigenous manganese solubilising bacterial biodiversity from manganese mining deposits. J Basic Microbiol 56(3):254–262. https://doi.org/10.1002/JOBM.201500477

Ghosh S, Bal B, Das AP (2017) Enhancing manganese recovery from low-grade ores by using mixed culture of indigenously isolated bacterial strains. Geomicrobiology 35(3):242–246. https://doi.org/10.1080/01490451.2017.1362080

Ghosh S, Kumar MS, Bal B, Das AP (2018) Application of bioengineering in revamping human health. In: Synthetic biology: omics tools and their applications, pp 21–37. https://doi.org/10.1007/978-981-10-8693-9_2

Ghosh S, Tripathy B, Dey S, Das AP (2023) Comparative investigation of fungal and bacterial manganese biomining mechanisms. In: Manganese mining microorganisms, pp 99–115. https://doi.org/10.1016/b978-0-12-822147-1.00016-9

Haque F, Fan C (2023) Fate and impacts of microplastics in the environment: hydrosphere, pedosphere, and atmosphere. Environ 10(5):70. https://doi.org/10.3390/ENVIRONMENTS10050070

Jagatee S, Priyadarshini S, Rath CC, Das AP (2024) Synthetic microfiber: an enduring environmental problem linked to sustainable development. Environ Sci Eng Part F2333:93–112. https://doi.org/10.1007/978-3-031-51792-1_6

Jambeck JR, Geyer R, Wilcox C, Siegler TR, Perryman M, Andrady A, Narayan R, Kara Lavender Law (2015) Marine pollution. Plastic waste inputs from land into the ocean. Sci 347(6223):768–771. https://doi.org/10.1126/SCIENCE.1260352

Kanungo J, Sahoo T, Swain LP, Behera ID (2023) Toxicity of persistent hydrocarbon pollutants, sources and sustainable remediation process. In: Behera ID, Das AP (eds) Impact of petroleum waste on environmental pollution and its sustainable management through circular economy. Springer Nature Switzerland, pp 39–65. https://doi.org/10.1007/978-3-031-48220-5_2

Kanungo J, Sahoo T, Bal M, Behera ID (2024) Performance of bioremediation strategy in waste lubricating oil pollutants: a review. Geomicrobiol J 41(4):360–373. https://doi.org/10.1080/01490451.2023.2245395

Karamanlidis AA, Androukaki E, Adamantopoulou S, Chatzispyrou A, Johnson WM, Kotomatas S, Papadopoulos A, Paravas V, Paximadis G, Pires R, Tounta E, Dendrinos P (2008) Assessing accidental entanglement as a threat to the mediterranean monk seal monachus monachus. Endanger Species Res 5(2–3):205–213. https://doi.org/10.3354/ESR00092

Khatun R, Singh S, Dubey NK, Das AP (2023) A review on marine-based phytochemicals and their application in biomedical research. In: Recent frontiers of phytochemicals: applications in food, pharmacy, cosmetics, and biotechnology, pp 383–395. https://doi.org/10.1016/B978-0-443-19143-5.00013-X

Kumar M, Pandit S, Patel V, Khanna N, Nag M, Lahiri D, Ray RR, Das A, Das D (2023) Maximization of energy recovery from starch processing wastewater by thermophilic dark fermentation coupled with microbial fuel cell technology. Geomicrobiol J. https://doi.org/10.1080/01490451.2023.2209555

Kunwar B, Cheng HN, Chandrashekaran SR, Sharma BK (2016) Plastics to fuel: a review. Renew Sustain Energy Rev 54:421–428. https://doi.org/10.1016/J.RSER.2015.10.015

Lau WWY, Shiran Y, Bailey RM, Cook E, Stuchtey MR, Koskella J, Velis CA, Godfrey L, Boucher J, Murphy MB, Thompson RC, Jankowska E, Castillo AC, Pilditch TD, Dixon B, Koerselman L, Kosior E, Favoino E, Gutberlet J, Baulch S, Atreya ME, Fischer D, He KK, Petit MM, Sumaila UR, Neil E, Bernhofen MV, Lawrence K, Palardy JE (2020) Evaluating scenarios toward zero plastic pollution. Sci 369(6510):1455–1461. https://doi.org/10.1126/SCIENCE.ABA9475

Lee S, Lee YR, Kim SJ, Lee JS, Min K (2023) Recent advances and challenges in the biotechnological upcycling of plastic wastes for constructing a circular bioeconomy. Chem Eng J 454:140470. https://doi.org/10.1016/J.CEJ.2022.140470

Li M, Zhang S (2024) Tandem chemical depolymerization and photoreforming of waste pet plastic to high-value-added chemicals. ACS Catal 14(5):2949–2958. https://doi.org/10.1021/ACSCATAL.3C05535/SUPPL_FILE/CS3C05535_SI_001.PDF

Li P, Wang X, Min S, Zou X, Duan L, Zhang H (2021) Characteristics of plastic pollution in the environment: a review. Bull Environ Contam Toxicol 107(4):577–584. https://doi.org/10.1007/S00128-020-02820-1/METRICS

Lomwongsopon P, Varrone C (2022) Critical review on the progress of plastic bioupcycling technology as a potential solution for sustainable plastic waste management. Polymers 14(22):4996. https://doi.org/10.3390/POLYM14224996

MacLeod M, Arp HPH, Tekman MB, Jahnke A (2021) The global threat from plastic pollution. Sci 373(6550):61–65. https://doi.org/10.1126/SCIENCE.ABG5433/ASSET/BB3AD5EB-E9B8-4B40-9244-9AD38C499B86/ASSETS/GRAPHIC/373_61_F3.JPEG

Mallick K, Sahu A, Dubey NK, Das AP (2023) Harvesting marine plastic pollutants-derived renewable energy: a comprehensive review on applied energy and sustainable approach. J Environ Manage 348:119371. https://doi.org/10.1016/J.JENVMAN.2023.119371

Mallick K, Sahu S, Sahu A, Dey S, Das AP (2024a) A critical review of marine microfiber pollution routes, toxicity, and its sustainable remediation, pp 189–211. https://doi.org/10.1007/978-3-031-51792-1_10

Mallick K, Sahu S, Sahu A, Dey S, Das AP (2024b) A critical review of marine microfiber pollution routes, toxicity, and its sustainable remediation. Environ Sci Eng Part F2333:189–211. https://doi.org/10.1007/978-3-031-51792-1_10

Maqsood T, Dai J, Zhang Y, Guang M, Li B (2021) Pyrolysis of plastic species: a review of resources and products. J Anal Appl Pyrol 159:105295. https://doi.org/10.1016/J.JAAP.2021.105295

McIntyre JK, Spromberg J, Cameron J, Incardona JP, Davis JW, Scholz NL (2023) Bioretention filtration prevents acute mortality and reduces chronic toxicity for early life stage coho salmon (Oncorhynchus kisutch) episodically exposed to urban stormwater runoff. Sci Total Environ 902:165759. https://doi.org/10.1016/J.SCITOTENV.2023.165759

Miandad R, Barakat MA, Aburiazaiza AS, Rehan M, Nizami AS (2016) Catalytic pyrolysis of plastic waste: a review. Process Saf Environ Prot 102:822–838. https://doi.org/10.1016/J.PSEP.2016.06.022

Mishra S, Das AP (2008) Hexavalent chromium reduction and 16S RDNA identification of bacteria isolated from a Cr (VI) contaminated site. Internet J Microbiol 7(1):1–6

Mishra S, Das AP (2021) Current treatment technologies for removal of microplastic and microfiber pollutants from wastewater. In: Wastewater treatment: cutting-edge molecular tools, techniques and applied aspects, 237–251. https://doi.org/10.1016/B978-0-12-821881-5.00011-8

Mishra S, Das AP (2022) Treatment of the wastewater polluted with synthetic microfiber released from washing machine. In: Das BB, Hettiarachchi H, Sahu PK, Nanda S (eds) Lecture notes in civil engineering, vol 207. Springer Science and Business Media Deutschland GmbH, pp 109–117

Mishra S, Das AP (2023) Microbial remediation of synthetic microfiber contaminated wastewater. In: Shah MP (ed) Microbial technologies in industrial wastewater treatment. Springer, Singapore, pp 337–350

Mishra S, Rout PK, Das AP (2019a) Solar photovoltaic panels as next generation waste: a review. Biointerface Res Appl Chem 9(6):4539–4546

Mishra S, Rath CC, Das AP (2019b) Marine microfiber pollution: a review on present status and future challenges. Mar Pollut Bull 140:188–197. https://doi.org/10.1016/J.MARPOLBUL.2019.01.039

Mishra S, Singh RP, Rath CC, Das AP (2020) Synthetic microfibers: source, transport and their remediation. J Water Process Eng 38:101612. https://doi.org/10.1016/J.JWPE.2020.101612

Mishra A, Siddiqi H, Kumari U, Behera ID, Mukherjee S, Meikap BC (2021a) Pyrolysis of waste lubricating oil/waste motor oil to generate high-grade fuel oil: A comprehensive review. Renew Sustain Energy Rev 150:111446. https://doi.org/10.1016/j.rser.2021.111446

Mishra S, Swain S, Sahoo M, Mishra S, Das AP (2021b) Microbial colonization and degradation of microplastics in aquatic ecosystem: a review. Geomicrobiol J. https://doi.org/10.1080/01490451.2021.1983670

Mishra S, Rout PK, Das AP (2021c) Emerging microfiber pollution and its remediation. In: Prasad R (ed) Environmental and microbial biotechnology. Springer, Singapore, pp 247–266

Mishra S, Dash D, Al-Tawaha ARMS, Das AP (2022a) A review on heavy metal ion adsorption on synthetic microfiber surface in aquatic environments. Appl Biochem Biotechnol 194(10):4639–4654. https://doi.org/10.1007/S12010-022-04029-W/METRICS

Mishra S, Dash D, Das AP (2022b) Detection, characterization and possible biofragmentation of synthetic microfibers released from domestic laundering wastewater as an emerging source of marine pollution. Mar Pollut Bull 185:114254. https://doi.org/10.1016/J.MARPOLBUL.2022.114254

Mishra S, Singh RP, Rout PK, Das AP (2022c) Membrane bioreactor (MBR) as an advanced wastewater treatment technology for removal of synthetic microplastics. In: Development in wastewater treatment research and processes, pp 45–60. https://doi.org/10.1016/B978-0-323-85583-9.00022-3

Mishra S, Dash D, Das AP (2023) Aquatic Microbial Diversity on Plastisphere: Colonization and Potential Role in Microplastic Biodegradation. Geomicrobiol J 41(4):312–323. https://doi.org/10.1080/01490451.2023.2209750

Mohanta S, Pradhan B, Behera ID (2024a) Impact and remediation of petroleum hydrocarbon pollutants on agricultural land: a review. Geomicrobiol J 41(4):345–359. https://doi.org/10.1080/01490451.2023.2243925

Mohanty M, Mohanty J, Dey S, Dutta K, Shah MP, Das AP (2024b) The face mask: a tale from protection to pollution and demanding sustainable solution. Emerg Contam 10(2):100298. https://doi.org/10.1016/J.EMCON.2023.100298

Moharana S, Dey S, Sailaja Priyadarsini M, Kumar S, Das AP (2024) Identification and characterization of microplastic pollutants from the marine sediments of Paradeep Coast of Bay of Bengal, India for their sustainable management. Environ Sci Eng Part F2333:57–73. https://doi.org/10.1007/978-3-031-51792-1_4

Nag M, Lahiri D, Ghosh S, Sarkar T, Pati S, Das AP, Ram DK, Bhattacharya D, Ray RR (2023a) Application of microorganisms in biotransformation and bioremediation of environmental contaminant: a review. Geomicrobiol J. https://doi.org/10.1080/01490451.2023.2261443

Nag M, Lahiri D, Ghosh S, Sarkar T, Pati S, Das AP, Ram DK, Bhattacharya D, Ray RR (2023b) Application of microorganisms in biotransformation and bioremediation of environmental contaminant: a review. Geomicrobiol J 41:1–18. https://doi.org/10.1080/01490451.2023.2261443

Ncube LK, Ude AU, Ogunmuyiwa EN, Zulkifli R, Beas IN (2021) An overview of plastic waste generation and management in food packaging industries. Recycling 6(1):12. https://doi.org/10.3390/RECYCLING6010012

Paul A, Priyadarsini S, Karakuş S, Behera JK, Gnanasekaran R, Das AP (2024) Fate and accumulation of synthetic microfibers in the environment. In: Sustainable Microbial Technology for Synthetic and Cellulosic Microfiber Bioremediation, pp 61–76. https://doi.org/10.1007/978-3-031-62600-5_4

Polymers, IM Maafa-, and undefined 2021 (2021) Pyrolysis of polystyrene waste: a review. Mdpi.Com. https://doi.org/10.3390/polym13020225

Pradhan G, Dey S, Sailaja Priyadarsini M, Kumar S, Das AP (2024) Characterization and quantification of microplastics pollutants in sediment samples from Daya River of Odisha state in India for their appropriate management. Environ Sci Eng Part F2333:75–92. https://doi.org/10.1007/978-3-031-51792-1_5

Ragauskas AJ, Huber GW, Wang J, Guss A, O'Neill HM, Lin CSK, Wang Y, Wurm FR, Meng X (2021) New technologies are needed to improve the recycling and upcycling of waste plastics. Chemsuschem 14(19):3982–3984. https://doi.org/10.1002/CSSC.202101872

Raj K, Das AP (2023) Lead pollution: impact on environment and human health and approach for a sustainable solution. Environ Chem Ecotoxicol 5:79–85. https://doi.org/10.1016/J.ENCECO.2023.02.001

Rorrer NA, Nicholson S, Carpenter A, Biddy MJ, Grundl NJ, Beckham GT (2019) Combining reclaimed PET with bio-based monomers enables plastics upcycling. Joule 3:1006–1027. https://doi.org/10.1016/j.joule.2019.01.018

Sadasivuni KK, Rattan S, Waseem S, Brahme SK, Kondawar SB, Ghosh S, Das AP, Chakraborty PK, Adhikari J, Saha P, Mazumdar P (2019) Silver nanoparticles and its polymer nanocomposites—synthesis, optimization, biomedical usage, and its various applications. Lect Notes Bioeng, 331–373. https://doi.org/10.1007/978-3-030-04741-2_11

Sahoo PP, Singh S, Rout PK, Mishra S, Das AP (2022) Microbial remediation of plastic pollutants generated from discarded and abandoned marine fishing nets. Biotechnol Genet Eng Rev. https://doi.org/10.1080/02648725.2022.2152629

Sahu S, Das AP (2024) Microplastics degradation mechanism by aquatic plastiomes: a review, pp 213–226. https://doi.org/10.1007/978-3-031-50840-0_12

Sahu A, Mallick K, Das AP (2024) Environmental impact of single-use synthetic face mask and its recycling: a sustainable approach, pp 197–212. https://doi.org/10.1007/978-3-031-50840-0_11

Sanket AS, Ghosh S, Sahoo R, Nayak S, Das AP (2016) Molecular identification of acidophilic manganese (Mn)-solubilizing bacteria from mining effluents and their application in mineral beneficiation. Geomicrobiol J 34(1):71–80. https://doi.org/10.1080/01490451.2016.1141340

Sansuta M, Ghosh S, Bal B, Das AP (2018) A review of biotechnology processes applied for manganese recovery from wastes. Rev Environ Sci Bio/technol 17(4):791–811. https://doi.org/10.1007/S11157-018-9482-1

Siddiqui J, Pandey G (2013) A review of plastic waste management strategies. Int Res J Environ Sci Int Sci Congr Assoc 2(12):84–88

Singh RP, Mishra S, Das AP (2020) Synthetic microfibers: pollution toxicity and remediation. Chemosphere 257:127199. https://doi.org/10.1016/j.chemosphere.2020.127199

Sipra AT, Gao N, Sarwar H (2018) Municipal solid waste (MSW) pyrolysis for bio-fuel production: a review of effects of MSW components and catalysts. Fuel Process Technol 175:131–147. https://doi.org/10.1016/J.FUPROC.2018.02.012

Thiloka Edirisooriya EMN, Senanayake PS, Pei X, Wang H (2023) Hydrogen production and value-added chemical recovery from the photo-reforming process using waste plastics. J Environ Chem Eng 11(6):111429. https://doi.org/10.1016/J.JECE.2023.111429

Tripathy B, Dash A, Das AP (2022) Detection of environmental microfiber pollutants through vibrational spectroscopic techniques: recent advances of environmental monitoring and future prospects. Crit Rev Anal Chem 54(7):1–11. https://doi.org/10.1080/10408347.2022.2144994

Tripathy B, Dey S, Dubey NK, Lo W-C, Das AP (2024a) An omics-based approach for managing the risk of COVID-19-generated wastes, pp 51–66. https://doi.org/10.1007/978-3-031-50840-0_4
Tripathy B, Sahoo PP, Sundaray H, Das AP (2024b) In-vitro biodegradation of discarded marine microplastics across the eastern coast of the Bay of Bengal, India using Exiguobacterium sp. Environ Chem Ecotoxicol 6:236–247. https://doi.org/10.1016/j.enceco.2024.07.003
Uzoejinwa BB, He X, Wang S, Abomohra A-F, Yamin H, Wang Q (2018) Co-pyrolysis of biomass and waste plastics as a thermochemical conversion technology for high-grade biofuel production: recent progress and future directions elsewhere worldwide. Energy Convers Manage 163:468–492. https://doi.org/10.1016/J.ENCONMAN.2018.02.004
Yao L, King J, Dezhen W, Ma J, Li J, Xie R, Chuang SSC, Miyoshi T, Peng Z (2022) Non-thermal plasma-assisted rapid hydrogenolysis of polystyrene to high yield ethylene. Nat Commun 13(1):1–8. https://doi.org/10.1038/s41467-022-28563-7
Zhao X, Korey M, Li K, Copenhaver K, Tekinalp H, Celik S, Kalaitzidou K, Ruan R, Ragauskas AJ, Ozcan S (2022) Plastic waste upcycling toward a circular economy. Chem Eng J 428:131928. https://doi.org/10.1016/J.CEJ.2021.131928
Zolotova N, Kosyreva A, Dzhalilova D, Fokichev N, Makarova O (2022) Harmful effects of the microplastic pollution on animal health: a literature review. PeerJ 10:13503. https://doi.org/10.7717/PEERJ.13503
ZúñigaD M, Aranda FL, Rivas BL (2023) Polymers recycling: Upcycling techniques. An overview. J Chil Chem Soc 68(2):5876–5886. https://doi.org/10.4067/S0717-97072023000205876

Chapter 15
Circularity and Sustainable Economy Model Towards Supply Chain Management and Zero Waste

Deeptimayee Pal and Sujit Sen

Abstract The integration of circularity and a sustainable economy framework in supply chain management represents a paradigm shift in waste reduction, especially regarding environmental hydrocarbon pollutants. Circularity fundamentally focuses on establishing closed-loop systems prioritizing items' durability, reusability, and recyclability. The transition above challenges conventional linear production methods, promoting a regenerative system in which materials are constantly recycled. When firms adopt sustainable supply chain management methods, they may effectively address the challenges of incorporating circular concepts across their whole value chain. The comprehensive approach endeavors to mitigate the ecological consequences of hydrocarbon pollutants and reevaluate the economic framework by harmonizing profitability with ecological accountability. This model aims to usher in a period when firms actively contribute to a circular and sustainable future by implementing waste reduction methods, innovative recycling programs, and improved treatment technologies.

Keywords Circular economy · Sustainable supply chain · Zero waste · Hydrocarbon pollutants · Environmental sustainability

15.1 Introduction

The concepts of a circular economy provide a revolutionary method for effectively regulating hydrocarbon contamination in the context of sustainable development (Chen et al. 2020; Paul et al. 2023; Raj and Das 2023). Circular economy concepts promote the continuous circulation of resources inside closed-loop systems, aiming to minimize waste creation, maximize resource efficiency, and support environmental sustainability, in contrast to the linear "take-make-dispose" approach (Mallick et al.

D. Pal (✉) · S. Sen
Department of Chemical Engineering, National Institute of Technology Rourkela, Rourkela, Odisha 769008, India
e-mail: deeptimayeenitrkl@outlook.com

I. D. Behera and A. P. Das (eds.), *Environmental Hydrocarbon Pollution and Zero Waste Approach Towards a Sustainable Waste Management*, Environmental Science and Engineering, https://doi.org/10.1007/978-3-031-81615-4_15

2023; Pradhan et al. 2023). Applying circular economy concepts to control hydrocarbon pollution provides creative techniques for minimizing the ecological consequences of hydrocarbon extraction, production, consumption, and waste disposal. Some of the fundamental principles that govern circular economy are as follows:

- The concepts of circular economy prioritize the optimal use of hydrocarbon resources over their entire life cycle, including extraction, processing, and end-of-life disposal. This entails reducing resource inputs, optimizing manufacturing processes, and promoting the reuse, recycling, and recovery of hydrocarbon-based products to extract the utmost value from limited resources.
- The fundamental concept of the circular economy is to minimize waste by eliminating or reducing the generation of hydrocarbon pollution at its origin (Heshmati 2017). This involves using cleaner production processes, reducing emissions and spills, and adopting closed-loop systems to reclaim and reuse hydrocarbon waste streams.
- The ideas of a circular economy promote the intentional design of goods and packaging to prioritize sustainability. This involves considering variables such as the choice of materials, the product's longevity, its capacity to be repaired, and its potential for recycling. Within the realm of hydrocarbon-based goods, this involves developing products that can be quickly taken apart, using components that are renewable or capable of being broken down by natural processes, and reducing the reliance on hazardous substances to promote the recovery and recycling of these products at the end of their lifespan.
- The concepts of circular economy prioritize extended producer responsibility, which means that producers are accountable for the environmental consequences of their goods from creation to disposal, including managing waste after consumers have used it. Hydrocarbon contamination may include implementing take-back systems, recycling programs sponsored by producers, and eco-labeling to promote responsible product stewardship and waste management.
- The concepts of circular economy strive to achieve the closure of material cycles by advocating for the reuse, recycling, and recovery of hydrocarbon-based materials as they reach the end of their useful lifespan (Wang et al. 2023). This entails the creation of cutting-edge technology and infrastructure to retrieve and treat hydrocarbon waste streams, including plastics, oils, and chemicals, to produce new goods or energy sources. This approach aims to decrease dependence on new resources and mitigate environmental degradation.
- The concepts of circular economy acknowledge the significance of cooperation and involvement of stakeholders in promoting systemic transformation towards more sustainable methods in managing hydrocarbons. This entails cultivating collaborations among the government, industry, academia, and civil society to develop and execute laws, legislation, and programs encouraging circularity, innovation, and exemplary methods to prevent and manage hydrocarbon pollution. Because hydrocarbon industries can lead to environmental damage in various ways (Pal and Sen 2023b), as listed in Table 15.1.

Table 15.1 Environmental challenges associated with hydrocarbon industries

Environmental challenge	Description
Water contamination	Surface and underground water sources are contaminated by hydrocarbons, chemicals, and generated water due to leaks, spills, and runoff
Air contamination	Industrial and combustion-related emissions of particulate matter, greenhouse gases, and volatile organic compounds (VOCs)
Soil contamination	Soil contamination may occur due to spills, leaks, and incorrect disposal of hydrocarbons, drilling liquids, and waste components
Biodiversity loss	Destruction of habitats, contamination of water sources, and disturbance of ecosystems caused by hydrocarbon activities are the leading causes of biodiversity loss and species extinction
Habitat destruction	Deforestation, land disturbance, and infrastructure development linked to hydrocarbon activities destroy ecosystems and habitats
Climate change	The generation and utilization of hydrocarbons contribute to climate change via the emissions of greenhouse gases, the destruction of forests, and the alteration of land use

Addressing hydrocarbon pollutants (Behera et al. 2020; Lahiri et al. 2021), heavy metals (Das et al. 2011; Raj and Das 2023), microplastics (Mishra and Das 2023; Singh et al. 2020), and electronics waste (Mishra et al. 2023) via sustainable supply chain management and circular economy techniques is crucial. Circular economy strategies prioritize the conservation of resources by encouraging the reuse, recycling, and recovery of materials derived from hydrocarbons within supply chains. Circular economy concepts contribute to reducing the environmental effect caused by hydrocarbon extraction and processing by completing material loops and minimizing the use of virgin materials. This approach helps conserve precious natural resources for the next generation. The ultimate goal of circular economy techniques is to reduce waste production by eliminating waste and contamination from the beginning and maximizing the utilization of goods and resources for as long as feasible (Desing et al. 2020). Implementing circular supply chain management strategies may effectively reduce emissions, spills, and other pollution types linked to hydrocarbon extraction, production, and disposal in the context of hydrocarbon pollutants. Implementing circular economy strategies may provide financial benefits for firms by enhancing resource efficiency, minimizing expenses related to waste disposal, and increasing chances for generating money by reclaiming and reusing hydrocarbon-based resources. Companies may improve their profitability and minimize their ecological impact by optimizing the movement of materials and maximizing the value derived from resources.

Adopting circular economy strategies may effectively reduce the potential negative impacts of hydrocarbon pollutants, including the challenges of adhering to regulations, managing environmental responsibilities, and protecting one's image.

Companies may reduce the risk of environmental mishaps and expensive remediation efforts by adopting closed-loop systems and ethical waste management procedures that can be mitigated with the aid of circular economy strategies. Adopting circular economy concepts promotes innovation and enhances competitiveness by motivating the advancement of sustainable goods, processes, and business models (Suárez-Eiroa et al. 2019). Companies that actively use circular supply chain management methods may gain a competitive advantage by distinguishing their offers, minimizing their environmental impact, and satisfying the increasing demand for sustainable goods and services. Consumers, investors, and regulatory agencies increasingly demand more openness and responsibility from corporations about their environmental performance and sustainability initiatives. By incorporating circular economy strategies into supply chain management, organizations may synchronize with stakeholder demands, amplify brand worth, and foster trust with consumers, investors, and other stakeholders. Circular economy strategies provide a means to achieve long-term sustainability by separating economic development from the use of resources and the deterioration of the environment. By adopting circular supply chains, enterprises can facilitate the transition toward a more resilient and environmentally friendly economy while assuring the long-term availability of hydrocarbon resources for the next generation (Lieder and Rashid 2016). A schematic representation of a circular supply chain is shown in Fig. 15.1.

Hence, the chapter aims to thoroughly comprehend circular economy ideas and their implementation in managing hydrocarbon pollutants. The aim is to investigate

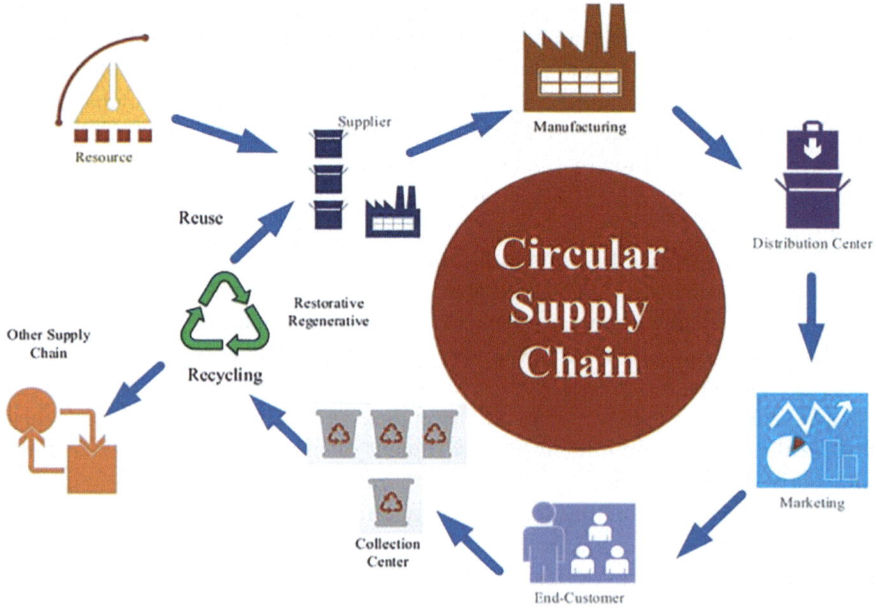

Fig. 15.1 Circular supply chain, as adapted from (Shahsavani and Goli 2023)

different techniques and methods of the circular economy that may be used to reduce hydrocarbon pollution along the whole supply chain, including extraction, production, consumption, and disposal. The chapter aims to showcase exemplary methods and real-life examples of firms and industries effectively using circular supply chain management and zero-waste programs to tackle hydrocarbon pollutants. The objective is to explore technical solutions and innovations that facilitate circular economy practices in managing hydrocarbon pollution, including waste reduction, recycling, and resource recovery technologies. This chapter aims to analyze the legislative and regulatory frameworks, both at the national and international levels, that encourage the concepts of circular economy and facilitate sustainable management practices for hydrocarbons. The objective is to identify the difficulties and possibilities linked to adopting circular economy practices in the hydrocarbon sector and to develop methods for overcoming obstacles and maximizing advantages. The objective is to identify the difficulties and possibilities linked to adopting circular economy methods in hydrocarbon sectors and to develop strategies for overcoming barriers and optimizing advantages. The chapter aims to offer explicit suggestions and instructions to businesses, policymakers, and other stakeholders on encouraging circularity and sustainability in hydrocarbon supply chains. This will help reduce hydrocarbon pollutants and facilitate the shift towards a more sustainable future.

15.2 Concept of Circular Economy and Hydrocarbon Contamination

A circular economy is designed to eliminate waste and optimize resource reuse, recycling, and recovery. The concept entails eliminating waste and pollution, maximizing the lifespan of goods and resources, and restoring natural systems (Das et al. 2023; Mohanty et al. 2024). A circular economy aims to reduce reliance on virgin hydrocarbon resources by maximizing existing resources via recycling and reuse rather than only depending on extraction (Jagaba et al. 2022). The concepts of circular economy promote the closure of material loops in the hydrocarbon value chain, which involves the ongoing recycling and repurposing of hydrocarbon-based goods and byproducts instead of disposing of them as waste (Mallick et al. 2023; Kanungo et al. 2023). The main objective of adopting a circular economy strategy for hydrocarbon pollution is to minimize waste and pollution at every stage of the lifespan of hydrocarbon-based goods, including extraction, manufacture, use, and disposal. The concepts of circular economy prioritize resource efficiency, promoting the optimal use of hydrocarbon resources while minimizing waste and emissions related to hydrocarbon extraction, processing, and utilization (Bala, Anshul et al. 2024; Bala, Dogra et al. 2024; A. Das, Dey, & Das 2024; A. Das, Dey, Bala et al. 2024; Dey, Bissoyi et al. 2024). To address hydrocarbon pollution using circular economy methods, it is necessary to use cutting-edge technology, establish new business models, and foster collaborative collaborations across many sectors. These efforts are crucial for shifting towards a more sustainable and circular economy (Ramakrishna et al. 2020).

15.2.1 Challenges of Supply Chain Hydrocarbon Pollution

Understanding the difficulties associated with hydrocarbon contamination in supply chains is essential to develop efficient ways to reduce environmental consequences. The main challenges include the fact that hydrocarbon supply chains are often multifaceted and interdependent, including many parties, procedures, and locations (Riera et al. 2023). Effectively managing pollution along the whole supply chain, including extraction, manufacturing, transportation, and distribution, poses substantial logistical and coordination complexities. Hydrocarbon contamination may arise from various sources, including petrochemical plants, refineries, extraction sites, transportation vessels, and end-user facilities (Behera et al. 2022a, b; Lahiri et al. 2021; Das et al. 2024). Finding and evaluating pollution sources within supply chains may take time since they vary in character and are subject to changing operating circumstances. Monitoring and quantifying hydrocarbon pollution levels in supply chains may pose challenges owing to the scattered distribution of pollution sources, restricted availability of monitoring equipment, and fluctuations in pollutant concentrations. Precise and timely data gathering is crucial for efficient pollution control and regulation adherence. Hydrocarbon pollution presents substantial environmental and health dangers, such as water, air, and soil contamination, deterioration of ecosystems, and threats to public health (Pal and Sen 2023a; Behera et al. 2021). To effectively handle these risks, it is necessary to take proactive steps to avoid pollution, reduce its impact, and guarantee the safety of workers, communities, and ecosystems across the supply chain.

Adherence to environmental legislation and standards that control hydrocarbon contamination is crucial for organizations that operate inside supply chains. However, negotiating intricate regulatory frameworks, acquiring permissions, and fulfilling compliance standards may be complex and demanding, especially in areas with strict environmental rules. Although pollution management technologies have made significant progress, some pollutants related to hydrocarbon activities, including VOCs, hazardous air pollutants (HAPs), and greenhouse gases, still need to improve inefficient collection, treatment, or mitigation. It is crucial to develop and implement pollution control systems that are both cost-effective and scalable to minimize the environmental impact. Incidents of hydrocarbon contamination, such as oil spills, leaks, and emissions, may lead to adverse publicity, harm to reputation, and erosion of stakeholder confidence. Effectively managing public image and stakeholder relationships requires clear and open communication, taking proactive steps to minimize risks, and promptly addressing environmental events. With the increasing worldwide need to decrease carbon emissions and shift towards renewable energy sources, companies that rely on hydrocarbons need help aligning their activities with sustainability objectives. Meeting the demand for hydrocarbon-based products while simultaneously addressing the need to decrease contamination and greenhouse gas emissions presents substantial strategic and operational obstacles for supply chains (Kaygusuz 2009).

15.2.2 Using Circular Economy Principles to Reduce Hydrocarbon Pollution

To address hydrocarbon pollution, using circular economy concepts requires implementing inventive approaches to minimize waste, enhance resource efficiency, and decrease environmental harm throughout the hydrocarbon value chain; various principles are employed. The various circular economy principles for managing waste are illustrated in Fig. 15.2. Implementing a design-for-circularity strategy entails creating goods, processes, and systems to reduce waste and optimize resource utilization. This concept involves developing goods and infrastructure that promote reuse, recycling, and recovery of hydrocarbon resources. This approach aims to minimize the production of waste and pollutants. Implementing closed-loop systems entails establishing systems in which resources and goods continuously circulate throughout the economy rather than being disposed of after use. This approach involves creating closed-loop systems to collect, recycle, and reuse hydrocarbon-based goods and byproducts to reduce the discharge of pollutants into the surroundings (Pan et al. 2015). Advocating for resource efficiency entails maximizing the value of resources while minimizing waste by optimizing their usage over their entire lifespan. This principle involves minimizing the utilization of original hydrocarbon resources by implementing tactics like recycling, reusing, and substituting them with renewable alternatives. This approach ultimately lessens the environmental consequences linked to the extraction and production of hydrocarbons.

Adopting extended producer responsibility entails ensuring that producers are held responsible for the environmental consequences of their goods throughout their entire lifespan, including their disposal at the end of their intended use. This concept involves imposing an obligation on producers to assume accountability for the gathering, reusing, and secure elimination of hydrocarbon-derived goods and packaging, thereby diminishing pollution and advocating for sustainable waste management methods. Emphasizing waste avoidance and reduction entails deliberately eliminating waste and pollution from the beginning and maximizing the lifespan of goods and resources. This concept involves the adoption of approaches aimed at minimizing the creation of waste. These tactics include diminishing disposable plastics' consumption, enhancing product longevity and repairability, and advocating for reusing and recycling materials derived from hydrocarbons. Promoting cooperation and innovation involves stakeholders along the hydrocarbon value chain to generate and execute inventive strategies for reducing pollution and advancing sustainability (Dey, Dash et al. 2024; Dey et al. 2023; John & Das 2024; Mishra et al. 2024; Mohanty et al. 2024). This principle involves working with business

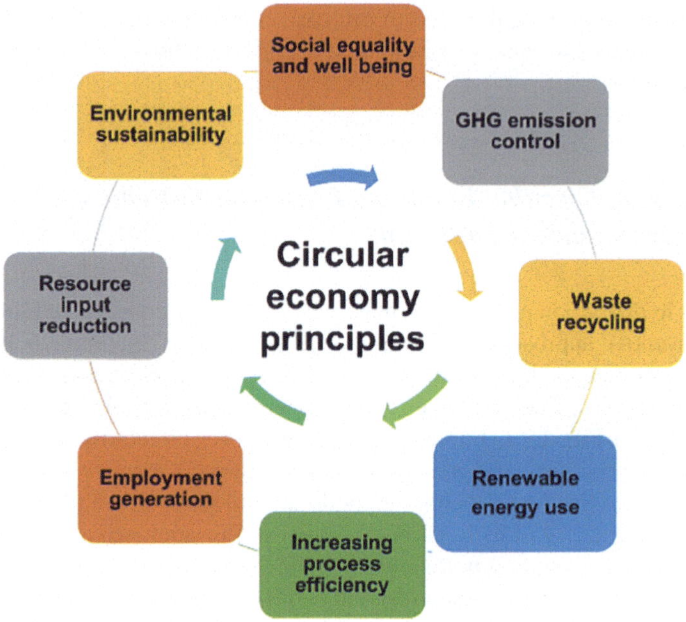

Fig. 15.2 Circular economy principles for waste management, as adapted from (Kumar et al. 2023)

associates, government agencies, research institutions, and civil society organizations to develop and implement advanced technologies, best practices, and policy frameworks that promote circular economy principles and minimize environmental harm (Iacovidou et al. 2017).

15.3 Hydrocarbon Pollution Regulation Using a Supply Chain in a Circularity Model

Effectively managing hydrocarbon pollution and advancing sustainability throughout the whole hydrocarbon value chain need the integration of circular economy concepts into supply chain management (Calzolari et al. 2021). These are some ways that supply chain management might reduce hydrocarbon contamination by using circular economy principles:

- Engaging in cooperation with suppliers to advance the concepts of a circular economy throughout the supply chain. Promoting sustainable practices among suppliers, including waste reduction, material recycling, and pollution minimization at the source.

- Executing life cycle assessments (LCA) to identify potential areas for minimizing environmental effects along the hydrocarbon value chain. Evaluating the ecological impact of goods, processes, and activities to determine which areas should be focused on for enhancement and to direct decision-making towards more environmentally friendly choices.
- Developing goods emphasizing circularity, prioritizing durability, reparability, and recyclability. Utilizing environmentally-friendly materials, reducing packaging waste, and building goods that can be quickly taken apart to promote recycling and reusing at the end of their lifespan.
- Utilizing closed-loop networks and reverse logistics techniques to recover, recycle, and reuse hydrocarbon-based resources and byproducts. Implementing take-back initiatives to gather and reprocess used goods and packaging materials, thus diminishing waste and pollution inside the distribution network.
- Incorporating circular economy standards into procurement procedures to encourage the acquisition of environmentally friendly materials and goods. Giving priority to suppliers that follow circular economy concepts, such as providing recycled materials, minimizing packaging waste, and adopting pollution protection measures (Park et al. 2010).
- Establishing waste management and recycling initiatives to reduce waste production and optimize the retrieval of resources. Allocating resources towards developing and enhancing the infrastructure necessary to collect, sort efficiently, and process hydrocarbon-based waste streams, including plastics, oils, and chemicals, to promote their reuse and recycling.
- Establishing strategic alliances and cooperative ventures with business stakeholders, research institutions, government agencies, and non-governmental organizations to foster the development of cutting-edge remedies for mitigating hydrocarbon contamination. Combining resources, exchanging the most effective methods, and cooperating on research and development initiatives to expedite the implementation of circular economy concepts in supply chain management.
- Promoting a culture that promotes ongoing enhancement and originality throughout the supply chain to propel sustainable performance. Motivating staff to explore waste reduction prospects, enhance resource efficiency, and adopt pollution control measures via innovation and technology (Pan et al. 2015).

15.3.1 Circular Supply Chains for Reducing Hydrocarbon Emissions and Waste Generation

To mitigate hydrocarbon emissions and waste production, the development of circular supply chains necessitates a reevaluation of conventional linear supply chain frameworks and the application of strategies that prioritize pollution prevention, resource conservation, and waste minimization. Circular supply chains can be designed in the following manner to accomplish many goals. It commences by delineating the hydrocarbon supply chain, pinpointing crucial procedures, participants, and possible emissions and waste production origins and utilizing data analytics and modeling methodologies to enhance the supply chain, identifying prospects for enhancing resource efficiency and minimizing waste and enforcing strategies to decrease the use of hydrocarbons and improve the efficiency of resources at all stages of the supply chain (Scheel et al. 2020). This may include enhancing industrial methods, reducing energy use, and applying strategies to minimize material waste and emissions throughout the extraction, processing, transportation, and storage and developing goods and packaging with a circular approach, prioritizing attributes such as longevity, ease of repair, and recyclability. Utilizing sustainable materials, reducing packaging waste, and manufacturing items that can be quickly taken apart to promote recycling and reusing at the end of their lifespan. Introducing closed-loop systems to reclaim, reprocess, and use hydrocarbon-based resources and byproducts throughout the supply chain and implementing take-back initiatives to establish systems for collecting and recycling old items and packaging, thereby minimizing waste and environmental contamination.

Implementing streamlined reverse logistics procedures to gather, categorize, and treat hydrocarbon-derived waste materials for recycling and repurposing. Developing transportation networks that enable the efficient return of discarded items and materials to recycling facilities or reverse supply chain partners. Establishing strategic alliances and cooperative ventures with suppliers, consumers, and other stakeholders to foster circularity across the supply chain. Combining resources, exchanging the most effective methods, and cooperating on research and development initiatives to develop inventive solutions for reducing waste and mitigating emissions and developing and deploying mechanisms to track and quantify hydrocarbon emissions and waste production at every stage of the supply chain and utilizing performance measurements and Key Performance Indicators (KPIs) to monitor and evaluate advancements made in waste reduction and emission reduction objectives and pinpoint areas that need improvement. It allocates resources towards implementing technology and solutions that facilitate waste reduction, emissions mitigation, and resource efficiency across the supply chain (Aljamal et al. 2024). This may include using more environmentally friendly industrial technology, sustainable energy sources, and state-of-the-art waste treatment systems and ensuring adherence to environmental legislation and standards that control the release of hydrocarbon emissions and waste handling.

15.3.2 Sustainable Hydrocarbon Sourcing via Circularity

Circular procurement methods are essential for improving sustainability in the hydrocarbon business. These policies encourage acquiring materials and products that are environmentally benign, resource-efficient, and support circularity. Procurement strategies might give priority to the acquisition of hydrocarbon-based commodities that have a high percentage of recycled content. This incentivizes suppliers to allocate resources towards developing recycling infrastructure and implementing efficient procedures, thereby closing material cycles and decreasing the need for new, unused resources. Circular procurement procedures may facilitate acquiring hydrocarbon-based goods from renewable feedstocks, such as bio-based polymers and biofuels. This reduces dependence on fossil fuels and lessens the environmental consequences linked to the mining and processing of hydrocarbons. Environmental impact assessments, considering carbon, water use, and ecological footprint, may provide valuable information to guide procurement choices (Jaeger and Upadhyay 2020). Suppliers may be obligated to provide ecological performance statistics for their goods, allowing customers to make well-informed choices based on sustainability standards. Procurement contracts may have product reuse and refurbishment clauses, incentivizing suppliers to provide items with prolonged lifespans and facilitating returning and refurbishing old products. This minimizes waste production and encourages the adoption of a circular economy.

Procurement contracts may stipulate that suppliers must establish take-back programs for hydrocarbon-based goods after they have reached the end of their intended use. This guarantees that items are appropriately disposed of or recycled, reducing the environmental consequences of waste disposal and encouraging the recovery of resources. Procurement practices may include actively involving suppliers to advance sustainability and circularity within the supply chain. Suppliers may be incentivized to embrace sustainable practices, including waste reduction, enhanced energy efficiency, and minimized emissions (Oghazi and Mostaghel 2018). Procurement rules may mandate that suppliers comply with established certifications and standards for sustainable procurement of hydrocarbon-based materials, such as ISO 14001 (Environmental Management Systems) and Forest Stewardship Council (FSC) accreditation for wood-based goods. Forming enduring alliances with suppliers may enhance cooperation on sustainability endeavors and foster ongoing enhancements in environmental performance. Collaborative efforts may be established to explore and execute creative ideas for sustainable sourcing and procurement. Procurement strategies should include systems for overseeing and documenting the environmental performance of suppliers and the sustainability of procured materials. Regular audits and performance evaluations may guarantee adherence to sustainable guidelines and promote ongoing improvement.

15.4 Zero Waste Strategies for Hydrocarbon Pollution

Zero waste has evolved as a transformational strategy within the hydrocarbon sector to effectively manage hydrocarbon contamination, aligning with the environmental sustainability goal. Zero waste strategies strive to minimize waste production, optimize the use of resources, and eliminate the idea of "waste" by observing wasted items as valuable resources that may be reused, recycled, or repurposed. Zero waste initiatives disrupt the conventional linear resource use and disposal model by reframing the notion of "waste." Zero waste principles acknowledge that all materials possess value and should be handled in a manner that optimizes their potential for reuse, recycling, or recovery, rather than observing waste as a mere byproduct to be disposed of. The primary objective of zero waste techniques is to reduce pollution and environmental damage throughout the whole hydrocarbon value chain. This entails implementing strategies to prevent pollution from occurring at its origin, decrease the release of harmful substances and waste generation, and encourage using sustainable alternatives to traditional waste disposal methods. Zero waste techniques prioritize resource recovery and recycling as crucial measures for effectively reducing hydrocarbon contamination (Awasthi et al. 2021).

Organizations may reduce the environmental impact of hydrocarbon activities and decrease the demand for extracting new resources by recovering and recycling hydrocarbon-based commodities and byproducts, such as plastics, oils, and chemicals. Zero waste techniques closely adhere to the concepts of a circular economy, prioritizing the ongoing circulation of resources within closed-loop systems. Zero waste initiatives facilitate the shift towards a sustainable and circular economy by increasing resource efficiency, waste reduction, and material reuse (Kerdlap et al. 2019). To successfully achieve zero waste targets, it is necessary to foster cooperation and active involvement among all participants in the hydrocarbon value chain, including industry stakeholders, government agencies, research institutions, and local communities. Through collaboration, stakeholders may foster the creation of inventive ideas, exchange exemplary methods, and surmount obstacles to the implementation of zero waste.

Zero waste strategies contribute to environmental protection by diminishing pollution, preserving natural resources, and minimizing the ecological impact of hydrocarbon activities. Zero waste techniques help maintain healthier air, water, and soil quality by reducing garbage from polluting landfills, rivers, and ecosystems. Zero waste initiatives encourage resource conservation and minimize the demand for extracting new resources by recovering and reusing valuable items (Paul et al. 2024; Priyadarsini & Das 2024, 2025; Ray et al. 2024; Tripathy et al. 2022). This practice not only preserves limited resources but also decreases energy use, greenhouse gas discharges, and other ecological consequences linked to the extraction and processing of resources. Implementing zero-waste strategies may benefit organizations by reducing waste disposal expenses, optimizing resource utilization, and discovering prospects for generating income by recovering and reutilizing materials. Organizations may enhance their financial performance and decrease environmental impact by optimizing resource use and minimizing waste production (Deselnicu et al. 2018).

15.4.1 Implementation of Zero Waste Strategies in Various Sectors

Implementing zero waste initiatives in oil and gas operations, refineries, and petrochemical sectors is particularly challenging. It offers distinct potential owing to the intricate nature of operations and the wide variety of waste streams. Nevertheless, by using strategic planning, innovation, and teamwork, these organizations have the potential to reduce waste production significantly, mitigate environmental consequences, and optimize resource utilization (Pietzsch et al. 2017). Below are the methods for implementing zero-waste plans in several sectors:

(i) **Oil and gas operations**: For example, strategies to minimize waste generation at its origin can be introduced by enhancing drilling and extraction techniques to decrease the formation of drilling fluids, cuttings, and other waste products. They are investigating possibilities for repurposing and reutilizing waste resources produced during oil and gas operations, such as using water for secondary recovery or reprocessing drilling fluids and mud. Utilizing waste-to-energy technology to transform organic waste materials, such as biomass and organic sludges, into sustainable energy sources reduces the need for landfill disposal and fossil fuel use (Mojarad et al. 2018; Behera et al. 2022a, b).

(ii) **Refineries**: Enhancing refining operations to minimize waste production and maximize the extraction of valuable hydrocarbon fractions and deploying sophisticated process control systems and monitoring technology to enhance operational efficiency and reduce emissions and enacting waste reduction tactics, such as decreasing the release of gases via flaring and venting, optimizing the use of catalysts, and minimizing spills and leaks by implementing improved maintenance and monitoring procedures and implementing circular economy concepts by examining possibilities for reusing materials, recycling, and recovering resources within the refinery processes and developing closed-loop systems that reclaim and reuse refinery wastes, such as exhausted catalysts and wastewater.

(iii) **Petrochemical industries**: Developing petrochemical goods and packaging that prioritize circularity by emphasizing recyclability, reusability, and biodegradability. Utilizing sustainable materials and packaging designs that promote efficient recycling and reusability at the end of their shelf life. Implementing strategies to enhance resource efficiency and minimize waste production in petrochemical manufacturing operations. Enhancing reaction conditions, minimizing energy use, and including solvent recovery devices to decrease waste and emissions. Engaging in cooperative efforts with suppliers and partners farther down the supply chain to advance sustainable sourcing methods, reduce waste production at every stage of the supply process, and enhance the efficiency of material movement to lessen the environmental consequences (Ropital 2009).

Through the implementation of strategic measures, the oil and gas sector, refineries, and petrochemical sectors can significantly decrease the production of waste, limit their adverse effects on the environment, and make a substantial contribution to a more sustainable and circular economy. It implements training and awareness initiatives for staff to promote waste reduction, recycling, and resource efficiency practices. It encourages staff involvement and active participation in zero-waste programs to cultivate a culture of sustainability inside the organization. Allocating funds towards cutting-edge technology and solutions that facilitate waste reduction, the recycling of materials, and the recovery of resources in oil and gas operations, refineries, and petrochemical sectors. It is investigating emerging technologies, including sophisticated waste management systems, sustainable energy technology, and eco-friendly chemical solutions, and guaranteeing adherence to environmental legislation and standards that regulate waste management and control of emissions. It establishes resilient monitoring, reporting, and verification mechanisms to effectively monitor advancements toward achieving zero waste objectives and showcase environmental efficacy to stakeholders and regulatory bodies (Lee et al. 2020).

15.5 Technological Innovations for Mitigating Hydrocarbon Pollution and Promoting the circular Economy

Technology is essential for implementing circular economy techniques in hydrocarbon management. It helps recover resources, reduce waste, and prevent contamination at every stage of the hydrocarbon value chain. Modern recycling technology allows for the effective recovery and recycling of materials and byproducts derived from hydrocarbons. Reducing the demand for virgin resources and minimizing waste, advanced recycling methods like pyrolysis, depolymerization, and chemical recycling enable the conversion of plastic waste and other hydrocarbon wastes into valuable feedstocks for new products. Data analytics and monitoring systems provide the continuous monitoring and analysis of hydrocarbon processes, enabling organizations to detect possibilities for minimizing waste, enhancing energy efficiency, and reducing emissions (Beier et al. 2022). Advanced sensor technologies and data analytics algorithms provide valuable insights into production processes, equipment performance, and resource utilization. This allows for informed decision-making and the optimization of resource usage. Blockchain technology and traceability solutions enhance transparency and accountability in the hydrocarbon supply chain by offering a secure and unchangeable record of transactions and product movements. Companies may ensure compliance with environmental rules and promote responsible sourcing practices by deploying blockchain-based systems to trace hydrocarbon products' origin, manufacturing, and distribution.

The deployment of the Internet of Things (IoT) devices and intelligent sensors throughout the hydrocarbon value chain facilitates the continuous real-time monitoring of equipment performance, ambient conditions, and product quality. By collecting and analyzing data from IoT devices, firms may enhance production processes, proactively mitigate equipment failures, and reduce energy consumption and emissions. This contributes to the optimization of resources and the avoidance of pollution. 3D printing, a kind of additive manufacturing, allows for developing personalized parts and components utilizing recycled and bio-based materials. Through 3D printing technology, organizations may minimize material waste, enhance product design, and enable manufacture as needed, thus encouraging efficiency and circularity in the production process. Artificial intelligence (AI) and machine learning algorithms facilitate the implementation of predictive maintenance, energy optimization, and process automation in hydrocarbon operations (Waqar et al. 2023). AI-driven systems use data analysis to anticipate equipment failures and maintenance requirements, prevent unexpected periods of inactivity, optimize resource allocation, and minimize waste production. This ultimately enhances operational efficiency and promotes sustainability. Renewable energy sources, including solar, wind, and bioenergy, provide sustainable alternatives to fossil fuels for powering hydrocarbon activities. By incorporating sustainable energy sources into their manufacturing processes, corporations may decrease the release of greenhouse gases, mitigate environmental harm, and shift towards an economy with reduced carbon emissions. Digital platforms and marketplaces support the interchange of recycled and recovered materials, allowing enterprises to acquire secondary raw materials and byproducts for manufacturing. Circular economy platforms connect buyers and sellers of recycled materials, promoting resource recovery, reducing waste, and fostering circularity in the hydrocarbon value chain.

15.5.1 Novel Approaches to the Treatment, Recycling, and Recovery of Hydrocarbons

Emerging technologies are fundamentally transforming the process of extracting, reusing, and treating hydrocarbons, providing inventive ways to tackle environmental issues and advance sustainability. Below are some significant emerging technologies for tacking hydrocarbon pollution:

(i) **Bioremediation**: Bioremediation uses enzymes, microbes, and other biological agents to break down soil, water, and air hydrocarbon pollutants. Emerging bioremediation methods use the inherent metabolic activities of microorganisms to degrade hydrocarbons into non-toxic substances, providing an environmentally benign and economically efficient approach to remediate pollutants (Mohanta et al. 2024; Salanitro 2001; Kanungo et al. 2024; Behera et al. 2022a, b).

(ii) **Chemical recycling**: Chemical recycling processes, such as gasification, pyrolysis, and depolymerization, facilitate the transformation of plastic waste and other hydrocarbon-based wastes into valuable raw materials for producing new goods (Mishra et al. 2021). These methods disassemble polymers into their molecular components, creating superior fuels, chemicals, and materials (Quicker et al. 2022).

(iii) **Electrochemical remediation**: Electrochemical remediation systems, such as electrochemical oxidation and electrokinetic, use electrical currents to mobilize and break down hydrocarbon pollutants in soil and groundwater. These technologies provide fast and efficient solutions for cleaning polluted areas without requiring excavation or expensive treatment processes.

(iv) **Carbon Capture and Utilization (CCU)**: CCU systems extract carbon dioxide emissions from hydrocarbon combustion and industrial operations and transform them into valuable resources such as chemicals, fuels, and construction materials. The emerging technologies for CCU have a double advantage: they help decrease greenhouse gas emissions and provide a financial incentive.

(v) **Advanced treatment methods**: Advanced water treatment methods, such as ultraviolet disinfection, electrocoagulation, and membrane filtration, are often used to treat water polluted with hydrocarbons from oil spills, industrial operations, and wastewater discharges. These methods eliminate hydrocarbons, suspended particles, and other contaminants, resulting in purified water that may be reused or safely released.

(vi) **Nanotechnology**: Nanotechnology focuses on improving material characteristics and performance via controlled manipulation of materials on a microscopic scale. Nanoporous membranes for water purification, oil refining catalysts enabled by nanotechnology, and environmental monitoring sensors on a nanoscale are all new nanotechnology uses in hydrocarbon recovery, recycling, and treatment (Kharisov et al. 2014).

(vii) **Hydrocarbon-resistant materials**: Modern infrastructure, tools, and safety gear are made from advanced materials that are more resistant to hydrocarbons. Hydrocarbon production, transportation, and storage activities benefit from these materials' enhanced dependability, safety, and durability, reducing the likelihood of accidents, spills, and leaks.

(viii) **AI and machine learning**: Artificial intelligence (AI) and machine learning algorithms are now used to enhance the efficiency of hydrocarbon recovery operations, accurately forecast equipment breakdowns, and optimize treatment efficiency. AI algorithms may improve operational efficiency and minimize environmental impact by scrutinizing extensive data to detect patterns, trends, and anomalies (Panja et al. 2018).

(ix) **Robotic technologies**: Automation and robotics are changing the face of the petroleum industry, from exploration to production to maintenance. Offshore platforms, pipelines, and refineries are just a few examples of complex settings that may be safely and efficiently navigated with the help of robotic drones, ROVs, and autonomous vehicles.

(x) **Decentralized and modular treatment systems**: Hydrocarbon recovery and treatment in underserved or remote regions may be addressed using scalable and adaptable decentralized and modular treatment systems. Rapid and economical deployment of these systems allows for treating polluted water, soil, or air, providing clean resources, and reducing the environmental pollution.

15.5.2 Optimizing Hydrocarbon Supply Chains with Data Analytics and Digitalization While Minimizing Environmental Impact

Digitalization and data analytics are essential for optimizing hydrocarbon supply chains and reducing environmental impact. This is achieved by acquiring valuable insights, improving efficiency, and facilitating proactive decision-making. Digitalization enables the immediate monitoring and observation of crucial factors throughout the hydrocarbon supply chain, including production, transportation, and distribution. Using IoT sensors, satellite images, and automation technologies enables the acquisition of up-to-the-minute data on inventory levels, equipment performance, and environmental conditions. This data facilitates enhanced planning, scheduling, and optimization of operations. Data analytics tools, such as predictive maintenance algorithms and condition monitoring systems, use historical and real-time data to forecast equipment failures and maintenance requirements in advance. Organizations may mitigate downtime, enhance asset utilization, and decrease the likelihood of environmental problems resulting from equipment failures by proactively resolving equipment concerns. Data analytics algorithms enhance routing and logistics operations by minimizing transit routes, fuel consumption, and emissions. Route optimization software uses many parameters, including traffic conditions, weather predictions, and vehicle capabilities, to determine the most optimal routes for moving hydrocarbon products. This minimizes the environmental effect and reduces operating expenses (Desai et al. 2021).

Digitalization facilitates monitoring energy consumption and greenhouse gas emissions across the supply chain using energy management systems and emissions monitoring platforms. Data analytics algorithms scrutinize energy use trends, detect inefficiencies, and propose optimization solutions to diminish energy consumption and emissions, promoting environmental sustainability and ensuring adherence to regulations. Blockchain and digital twin technologies are digitalization platforms that improve transparency and traceability in the hydrocarbon supply chain. Blockchain offers an unchangeable record of transactions and the movement of goods, allowing stakeholders to monitor hydrocarbon products' source, creation, and distribution. This ensures adherence to environmental legislation and sustainability criteria. Digital platforms and data analytics solutions aid in risk management and compliance endeavors by detecting possible environmental threats and instances of regulatory non-compliance. Advanced analytics methods use historical and real-time data to detect trends and irregularities that may signify environmental dangers or breaches of rules. This allows firms to reduce risks and guarantee proactive adherence to environmental standards. Data analytics tools facilitate scenario analysis and provide decision assistance for assessing the environmental consequences of various supply chain strategies and operational choices. By simulating different scenarios and evaluating their possible environmental effects, organizations may make well-informed choices to reduce environmental harm and maximize the efficient utilization of resources throughout the hydrocarbon supply chain (Bechtsis et al. 2017).

15.6 Policies and Regulations for Circular Economy and Hydrocarbon Pollution

Government policies and regulations are essential in supporting circular economy methods for hydrocarbon management. They define standards, offer incentives, and develop frameworks that promote sustainable behaviors and reduce environmental harm. Extended Producer Responsibility (EPR) legislation mandates that manufacturers assume accountability for properly handling their goods at the end of their life cycle, including collection, recycling, and disposal. Within hydrocarbon-based goods, EPR regulations incentivize manufacturers to generate long-lasting, easily repairable, and recyclable items. This approach encourages a circular economy and helps minimize waste. Waste management rules enforce stipulations for the management, containment, conveyance, and elimination of hydrocarbon waste products. These rules aim to reduce the amount of waste generated, encourage recycling and the recovery of resources, and avoid environmental degradation resulting from incorrect waste management methods. Certain governments require retrieving and reusing certain hydrocarbon-derived substances, such as plastics, oils, and chemicals. These laws necessitate that manufacturers and consumers of hydrocarbon products gather and repurpose waste materials, thereby advancing the ideals of resource preservation and circular economy (Pinheiro et al. 2021).

Government agencies may enforce green procurement rules that mandate the acquisition of ecologically friendly goods and services, including hydrocarbon-based materials and products. These rules promote the use of sustainable sourcing techniques, such as recycled materials, and reduce environmental harm over the whole lifespan of the product. Carbon pricing methods, such as carbon taxes or cap-and-trade systems, motivate corporations to decrease greenhouse gas emissions and encourage energy efficiency. By implementing a carbon pricing system, these methods incentivize corporations to allocate resources towards cleaner technology, reduce energy use, and mitigate environmental harm in hydrocarbon activities. Major hydrocarbon projects need Environmental Impact Assessments (EIAs) to evaluate the possible environmental consequences and hazards of project development, operation, and decommissioning. Governments use EIAs to assess the ecological ramifications of hydrocarbon operations, alleviate adverse effects, and guarantee adherence to environmental norms and standards. Several countries develop circular economy roadmaps and policies to formulate policies and encourage circularity in different areas, such as hydrocarbon management. These roadmaps include detailed objectives, benchmarks, and strategies for moving towards a more environmentally friendly and circular economy. They include measures to decrease waste, encourage recycling, and enhance the efficient use of resources in hydrocarbon activities. Governments may provide financial assistance for research and innovation initiatives to promote the development of circular economy techniques in the management of hydrocarbons. Funding initiatives promote cooperation among industry, academia, and research institutions to develop cutting-edge technologies, procedures, and remedies for minimizing waste, recovering resources, and ensuring environmental sustainability in the petroleum business (Kumar et al. 2021).

15.6.1 Sustainable Management of Hydrocarbon Resources: International Agreements and Initiatives

International agreements and initiatives are crucial in supporting the sustainable management of hydrocarbon resources by encouraging cooperation, establishing benchmarks, and organizing endeavors to tackle shared obstacles. The Paris Agreement, established under the framework of the United Nations Framework Convention on Climate Change (UNFCCC), seeks to restrict the rise in global temperatures to a level far below 2 °C over pre-industrial levels, with a specific focus on limiting the increase to 1.5 °C. The signatories of the agreement pledge to decrease the release of greenhouse gases, including those originating from hydrocarbon sources, and strengthen their ability to withstand the consequences of climate change. The Sustainable Development Goals (SDGs) provide a worldwide framework for sustainable development, specifically addressing Goal 7 (Affordable and Clean Energy) and Goal 13 (Climate Action), which directly pertain to the management of hydrocarbon resources. The Sustainable Development Goals (SDGs) advocate for enhanced availability of clean and cost-effective energy sources and strategies to alleviate climate change and its consequences. The Sustainable Development Solutions Network (SDSN), established by the United Nations Secretary-General in 2012, harnesses worldwide scientific and technology knowledge to advance sustainable development. The network prioritizes diverse domains such as environmental sustainability, climate change mitigation, and sustainable energy, all pertinent to managing hydrocarbon resources (Cohen et al. 2021).

The Extractive Industries Transparency Initiative (EITI) is a worldwide program that advocates for openness and responsibility in the extractive sectors, including oil, gas, and mining. Participating nations pledge to provide data on their extractive industry earnings, transactions, and agreements, fostering transparent government and mitigating the potential for corruption in the management of hydrocarbon resources. The Global Gas Flaring Reduction Partnership (GGFR) is a program headed by the World Bank to reduce gas flaring in oil-rich nations. This is achieved via collaborative efforts involving governments, oil firms, and other relevant parties. Flaring reduction projects aim to decrease greenhouse gas emissions, enhance air quality, and preserve natural resources in hydrocarbon production. The Oil and Gas Climate Initiative (OGCI) is an initiative by CEOs to unite significant oil and gas businesses to expedite the industry's efforts to address climate change. The OGCI members pledge to decrease methane emissions, enhance energy efficiency, and allocate resources to low-carbon technology to alleviate the environmental consequences of hydrocarbon production and use. The International Energy Agency (IEA) is an independent organization that strives to provide dependable, cost-effective, and environmentally friendly energy for its member nations. The IEA offers analysis, statistics, and policy suggestions to bolster sustainable energy policies, including initiatives to enhance energy efficiency, advocate for renewable energy sources, and mitigate greenhouse gas emissions from hydrocarbon utilization (Tükenmez and Demireli 2012). The Carbon Price Leadership Coalition (CPLC) is an international endeavor that unites governments, corporations, and non-governmental organizations to promote the development of carbon price policies and initiatives. Carbon pricing methods, such as carbon taxes and emissions trading systems, provide financial incentives for corporations to decrease the amount of greenhouse gas emissions produced by hydrocarbon sources and encourage them to invest in more environmentally friendly technology. A generalized sustainable operation management framework is illustrated in Fig. 15.3.

15.7 Implementing Circular Economy Policies to Mitigate Hydrocarbon Pollution: Challenges and Opportunities

The implementation of circular economy policies to reduce hydrocarbon pollution poses problems and possibilities, given the intricate structure of the petroleum industry and the need for significant and revolutionary change (Guerra-Rodríguez et al. 2020), as listed in Table 15.2.

15.7.1 Challenges in Implementing Circular Economy Strategies to Mitigate Hydrocarbon Pollution

The adoption of circular economy techniques for reducing hydrocarbon pollution is hindered by many constraints, which impede the shift towards more sustainable and circular activities. Various stakeholders, such as enterprises, governments, and consumers, may need more knowledge and comprehension of circular economy concepts and their potential advantages in reducing hydrocarbon pollution (Das, A. Prasad, & Ghosh, Shreya 2023). The absence of awareness about this matter might impede the implementation of circular practices and restrict the allocation of funds towards inventive solutions. Adopting circular economy techniques often requires substantial investment in new technology, infrastructural enhancements, and worker

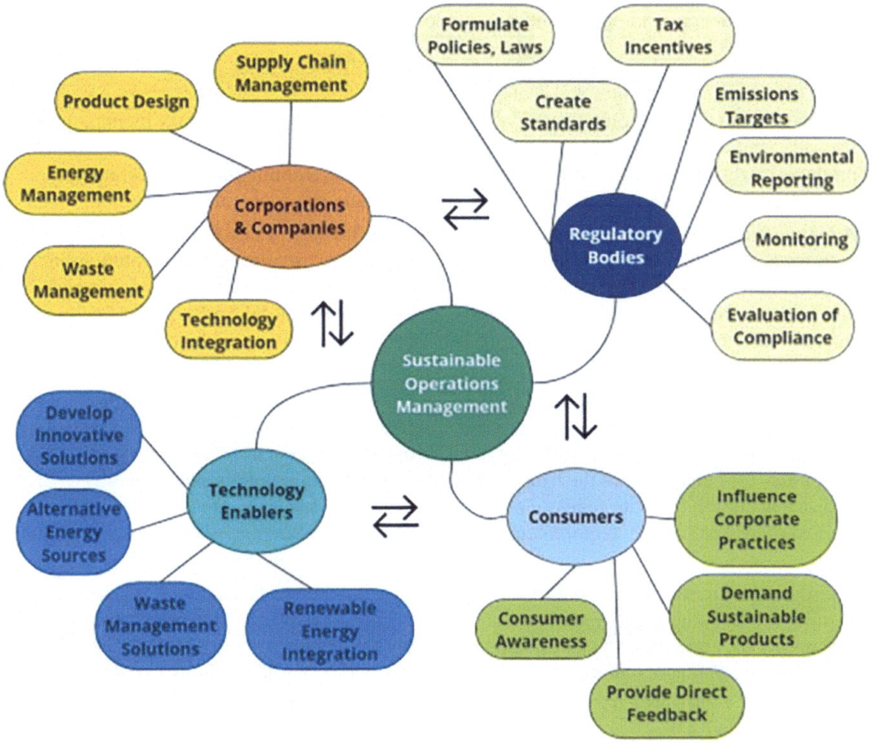

Fig. 15.3 Sustainable operation management framework, as adapted from (Alevizos et al. 2023)

education. The significant upfront expenses in adopting circular solutions might be an obstacle, especially for small and medium-sized firms (SMEs) with limited financial means. Hydrocarbon supply chains are intricate worldwide and include many stakeholders, a wide range of goods, and various environmental requirements. Coordinating circular economy activities across intricate supply chains may take time and require stakeholder cooperation and alignment. The current legislation and policies may not adequately promote or provide incentives for implementing circular economy activities to reduce hydrocarbon emissions. Regulatory impediments, such as the need for permits, laws on waste management, and standards for products, might hinder the implementation of circular solutions (Sheldon and Norton 2020).

Efficiently recycling, recovering, and reusing hydrocarbon-based products may be technically demanding, whether by adapting current technology or creating new ones. Technological constraints, such as the absence of economically viable recycling methods or the intricacy of segregating combined waste streams, might impede advancements toward achieving circularity. Businesses may hesitate to embrace circular economy methods owing to perceived risks, uncertainty, and inertia linked to altering existing business structures and procedures. Progress towards circularity may be hindered by the fear of failure, reluctance to embrace change, and the want

Table 15.2 Challenges and opportunities in implementing circular economy policies

Challenges	Opportunities
• The technological challenges and costs associated with enabling effective recycling, recovery, and reuse of hydrocarbon-based products might arise from adapting current technology or developing new ones	• Using circular economy strategies may optimize resource efficiency by maximizing the value derived from hydrocarbon-based resources, minimizing waste output, and lowering environmental impact
• Adopting circular economy techniques often requires a substantial initial investment in infrastructure, technological enhancements, and worker education, which may provide financial difficulties for organizations, especially smaller ones	• With the help of circular product design services, waste management solutions, and secondary markets for recycled materials, companies may take advantage of new possibilities presented by circular economy activities
• The current legislation and policies may need to adequately promote or encourage the use of circular economy practices, resulting in difficulties for enterprises in the hydrocarbon industry in complying with regulations and confusion over regulatory requirements	• Collaborative strategies that include industrial partners, civil society, academia, and government agencies may expedite the implementation of circular economy principles and overcome obstacles by sharing information, pooling resources, and engaging in collective action
• Hydrocarbon supply chains are intricate worldwide, including many participants, various goods, and differing environmental requirements. Managing circular economy projects across intricate supply networks may pose challenges	• Technological advancements, including chemical recycling, improved materials, and digitalization, provide prospects for developing novel solutions to reduce hydrocarbon pollution and promote circularity
• A shift in consumer and industrial behavior towards circularity necessitates cultural and mentality transformations, which may require considerable time and focused efforts	• Implementing circular economy techniques may save costs by optimizing resource utilization, minimizing waste disposal expenses, and improving operational efficiency

to preserve the existing state of affairs. The absence of robust market demand for recycled or reused hydrocarbon-based commodities might impede the implementation of a circular economy (Paletta et al. 2019). If there is more demand for circular goods and materials, firms can invest in recycling infrastructure or develop circular business models. The fragmentation and absence of collaboration within the hydrocarbon value chain might impede stakeholder cooperation and coordination. Fragmented value chains may lead to inefficiencies, redundant efforts, and obstacles to applying circular solutions across the value chain.

15.7.2 Strategies for Overcoming Challenges and Accelerating Circular Hydrocarbon Management

To address the challenges and expedite the shift towards circular hydrocarbon management, several approaches might be employed:

- Enlighten stakeholders, including enterprises, governments, and consumers, on the advantages of using circular economy methods to manage hydrocarbons. Deliver training and capacity-building initiatives to improve comprehension and expertise in circularity concepts and technology.
- Provide financial assistance, grants, subsidies, and tax benefits to promote and stimulate investment in circular economy projects. Establish finance tools and financing plans to assist firms in mitigating the substantial upfront expenses of adopting circular processes.
- Assess and modify current legislation and policies to conform with circular economy goals and encourage sustainable management of hydrocarbons. Implement and enforce additional legislation, standards, and incentives to promote the concept of circularity, minimize waste generation, and enhance the recovery of valuable resources (Hees et al. 2019).
- Encourage entities from different sectors, such as businesses, government, universities, and nonprofits, to work together and exchange information to improve business operations. Establish venues for teamwork, such as discussion boards, task forces, and research consortia, to ease the flow of information and promote coordinated efforts.
- Provide financial resources for research and development (R&D) projects that aim to develop novel technologies, techniques, and solutions for managing circular hydrocarbons. Facilitate cooperative research and development initiatives that unite business, academia, and research institutions to stimulate creativity and expedite the implementation of technologies (Khajuria et al. 2022).
- Facilitate the spread and adoption of circular economy technologies among enterprises by offering technical help, knowledge, and support. Establish technology transfer platforms, innovation centers, and demonstration projects to exhibit effective circular solutions and promote their replication (Bodar et al. 2018).
- Emphasize the creation of regenerative economic models that manage products through their entire lifespans while minimizing their impact on the environment. Help companies get more out of hydrocarbon resources with less waste and pollution by advocating for product-as-a-service, sharing economy platforms, and closed-loop technologies.
- Increase demand for circular goods and materials by educating buyers, encouraging eco-friendly shopping habits, and offering financial incentives for being green. Promote the use of hydrocarbon-based products that have been recycled, reused, or remanufactured as procurement priorities for both public and commercial organizations.
- Establish regional or sector-specific hubs and clusters for the circular economy to promote cooperation, innovation, and information exchange among enterprises,

research institutions, and other stakeholders. These hubs may function as centers of expertise for circularity, fostering economic growth and sustainability in the petroleum industry (Grim et al. 2020; Das et al. 2023).

- Implement monitoring and evaluation systems to measure advancements toward circular economy objectives and evaluate circular projects' environmental, social, and economic consequences. Utilize performance indicators, metrics, and benchmarks to assess results, pinpoint deficiencies, and guide decision-making for ongoing improvement.

15.8 Conclusions and Future Perspectives

The concepts of the circular economy provide a revolutionary method for managing hydrocarbon pollution, focusing on minimizing waste, recovering resources, and promoting sustainable resource utilization throughout the hydrocarbon value chain. Implementing circular economy methods in the petroleum business has several limitations, such as technical constraints, legal obstacles, and the intricacies of the supply chain. To overcome these obstacles, it is necessary to have a collaborative approach from all parties involved and implement favorable policies and incentives. Although there may be obstacles, there are substantial prospects for promoting circular economy concepts in controlling hydrocarbon contamination. Promising methods to minimize waste and pollution and maximize resource efficiency include advanced recycling technologies, bio-based feedstocks, digitalization, and circular supply networks. Implementing government laws and regulations is essential in fostering circular economy practices in the management of hydrocarbons. Regulatory assistance, monetary benefits, and cooperative endeavors may establish a conducive atmosphere for promoting circularity and minimizing pollution.

Technological advancements, including digitization, data analytics, and improved recycling technologies, play a crucial role in harnessing the benefits of circular economy concepts in the hydrocarbon business. Allocating resources to research and development and promoting collaborative innovation may accelerate progress toward achieving circularity. The increasing consumer consciousness about environmental problems and sustainability fuels the need for eco-friendly and sustainable goods in the hydrocarbon business. Companies that adopt circular economy ideas and exhibit environmental stewardship have the potential to earn a competitive advantage in the market. The cooperation and alliances between industry players, research institutions, and government agencies are crucial for promoting innovation and expediting the implementation of circular economy concepts. Collaborative endeavors, platforms for exchanging information, and groups of companies in the same sector may promote collective effort and expand the implementation of circular solutions.

Circular economy strategies prioritize resource efficiency via waste minimization, maximizing resource recovery, and encouraging material reuse and recycling. In the hydrocarbon business, where resources are limited and exploitation may have substantial environmental consequences, optimizing resource efficiency is crucial for ensuring long-term sustainability. Circular economy techniques aim to minimize waste creation and reduce contamination in the hydrocarbon supply chain by creating durable, repairable, and recyclable products. Circular techniques in hydrocarbon activities lessen the environmental impact and prevent pollution problems by effectively managing waste streams. Circular economy strategies facilitate the use of byproducts and waste streams produced during the extraction and processing of hydrocarbons. Rather than seeing these things as garbage, they may be reused, recycled, or converted into valuable resources, generating new sources of income and decreasing disposal expenses.

Circular economy techniques improve the resilience and security of hydrocarbon supply networks by fostering circular supply chains and closed-loop technologies. Circular practices enhance the stability and sustainability of hydrocarbon operations by diversifying feedstock sources, lowering reliance on virgin resources, and minimizing interruptions in the supply chain. Hydrocarbon firms may benefit from circular economy ideas since they can optimize resource utilization, decrease expenses related to waste disposal, and improve operational efficiency, resulting in cost savings. Companies may enhance their profitability and minimize their ecological footprint by embracing circular business models and implementing waste reduction methods. As rules on waste management, pollution prevention, and sustainability grow more strict, it is crucial to embrace circular economy ideas to comply with legislation and retain access to the market. Businesses that adopt circularity are in a more advantageous position to fulfill regulatory obligations, reduce compliance risks, and enter markets with strict environmental criteria. Amidst increasing environmental consciousness and stakeholder scrutiny, embracing circular economy strategies to meet stakeholder expectations and protect the business image is crucial. Hydrocarbon firms may increase their brand value and attract environmentally sensitive consumers, investors, and partners by showing a strong dedication to sustainability, conservation, and responsible resource management.

At this critical juncture between environmental sustainability and economic development, the need for action is more urgent than ever. Businesses, governments, and stakeholders need to come together in a collaborative endeavor to advance circular economy strategies to mitigate hydrocarbon pollution. Incorporating circular economy ideas as a fundamental aspect of organizational strategy. Allocating resources towards research and development to develop novel technologies and procedures that minimize waste, optimize resource efficiency, and diminish the environmental footprint in hydrocarbon activities. Implementing circular business models prioritizing sustainability, product stewardship, and value development from waste streams. Engaging in cooperation with supply chain partners, industry groups, and research institutes to stimulate innovation, exchange optimal methods, and expand the implementation of circular solutions.

Implementing favorable laws, regulations, and incentives to encourage using circular economy methods in the hydrocarbon business, providing carbon pricing systems, enacting extended producer responsibility (EPR) regulations, and developing circular economy roadmaps to provide a regulatory environment that promotes circularity. Allocating financial assistance, grants, and subsidies to enable investment in circular infrastructure, technological advancements, and workforce development. Promoting cooperation among government agencies, industry players, and civil society to initiate and execute circular economy projects. Promoting implementing circular economy principles among enterprises, policymakers, and the general public. Promoting understanding circularity's environmental and economic advantages and emphasizing the need to reduce hydrocarbon pollution. Participating in cooperative alliances, platforms for exchanging information, and industry forums to exchange ideas, share experiences, and collectively work towards achieving circularity. Endorsing enterprises that exhibit a dedication to sustainability and circularity by making informed choices when buying, investing, and advocating.

References

Alevizos V, Georgousis I, Kapodistria A (2023) Toward climate neutrality: a comprehensive overview of sustainable operations management, optimization, and wastewater treatment methods. Pollutants 3(4):521–543. https://doi.org/10.3390/pollutants3040036

Aljamal D, Salem A, Khanna N, Hegab H (2024) Towards sustainable manufacturing: a comprehensive analysis of circular economy key performance indicators in the manufacturing industry. Sustain Mater Technol: e00953. https://doi.org/10.1016/j.susmat.2024.e00953

Awasthi AK, Cheela VRS, D'Adamo I, Iacovidou E, Islam MR, Johnson M, Miller TR, Parajuly K, Parchomenko A, Radhakrishan L, Zhao M, Zhang C, Li J (2021) Zero waste approach towards a sustainable waste management. Resour Environ Sustain 3:100014. https://doi.org/10.1016/j.resenv.2021.100014

Bala K, Anshul, Verma S, Dogra A, Das AP, Kumar R, Nandini (2024) sustainable management for urban plastic waste generation. In: Sustainable Management of Urban Plastic Waste Through Circular Economic Approaches, pp 211–236. https://doi.org/10.1201/9781003465447-13

Bala K, Dogra A, Lata P, Das AP (2024) Discussing sources and biological-chemical remediation approaches for microplastic pollution. In: Sustainable Microbial Technology for Synthetic and Cellulosic Microfiber Bioremediation, pp 37–59. https://doi.org/10.1007/978-3-031-62600-5_3

Bechtsis D, Tsolakis N, Vlachos D, Iakovou E (2017) Sustainable supply chain management in the digitalisation era: the impact of automated guided vehicles. J Clean Prod 142:3970–3984. https://doi.org/10.1016/j.jclepro.2016.10.057

Behera ID, Basak G, Kumar RR, Sen R, Meikap BC (2020) Treatment of petroleum refinery sludge by petroleum degrading bacterium Stenotrophomonas pavanii IRB19 as an efficient novel technology. J Environ Sci Health Part A Toxic/Hazardous Subst Environ Eng 56(2). https://doi.org/10.1080/10934529.2020.1866924

Behera ID, Meikap BC, Sen R (2022) Performance evaluation of bacterial consortium for biodegradation of total petroleum hydrocarbon: a comparative strategic biostimulation study. Res Square

Behera ID, Nayak M, Biswas S, Meikap BC, Sen R (2021) Enhanced biodegradation of total petroleum hydrocarbons by implementing a novel two-step bioaugmentation strategy using indigenous bacterial consortium. J Environ Manage 292:112746. https://doi.org/10.1016/j.jenvman.2021.112746

Behera ID, Nayak M, Mishra A, Meikap BC, Sen R (2022) Strategic implementation of integrated bioaugmentation and biostimulation for efficient mitigation of petroleum hydrocarbon pollutants from terrestrial and aquatic environment. Marine Pollution

Beier G, Kiefer J, Knopf J (2022) Potentials of big data for corporate environmental management: a case study from the German automotive industry. J Ind Ecol 26(1):336–349. https://doi.org/10.1111/jiec.13062

Bodar C, Spijker J, Lijzen J, Waaijers-van der Loop S, Luit R, Heugens E, Janssen M, Wassenaar P, Traas T (2018) Risk management of hazardous substances in a circular economy. J Environ Manage 212:108–114. https://doi.org/10.1016/j.jenvman.2018.02.014

Calzolari T, Genovese A, Brint A (2021) The adoption of circular economy practices in supply chains—an assessment of European multi-national enterprises. J Clean Prod 312:127616. https://doi.org/10.1016/j.jclepro.2021.127616

Chen T-L, Kim H, Pan S-Y, Tseng P-C, Lin Y-P, Chiang P-C (2020) Implementation of green chemistry principles in circular economy system towards sustainable development goals: Challenges and perspectives. Sci Total Environ 716(1):136998. https://doi.org/10.1016/j.scitotenv.2020.136998

Cohen B, Cowie A, Babiker M, Leip A, Smith P (2021) Co-benefits and trade-offs of climate change mitigation actions and the sustainable development goals. Sustain Prod Consump 26:805–813. https://doi.org/10.1016/j.spc.2020.12.034

Das AP, Behera ID, Das NP (2024) Renewable energy generation and value addition from environmental microfiber pollution through advanced greener solution. Springer Nature

Das A, Dey S, Bala K, Gnanasekaran R, Das AP (2024) Microbial Bioremediation technology for sustainable treatment and management of synthetic microfiber waste. In: Sustainable Microbial Technology for Synthetic and Cellulosic Microfiber Bioremediation, pp 221–238. https://doi.org/10.1007/978-3-031-62600-5_12

Das A, Dey S, Das AP (2024) Ecotoxicological impacts of synthetic microfiber pollutants and development of sustainable mitigation strategies. Environ Chem Ecotoxicol. https://doi.org/10.1016/J.ENCECO.2024.12.003

Das AP, Dutta K, Khatun R, Behera ID, Singh S, Mishra S (2023) Microfiber pollution and its microbial mitigation: a review on current trends and future prospects. J Taiwan Inst Chem Eng 105104. https://doi.org/10.1016/J.JTICE.2023.105104

Das A. Prasad, Ghosh, Shreya (2023) Manganese mining microorganisms

Das AP, Sukla LB, Pradhan N, Nayak S (2011) Manganese biomining: a review. Biores Technol 102(16):7381–7387. https://doi.org/10.1016/J.BIORTECH.2011.05.018

Desai JN, Pandian S, Vij RK (2021) Big data analytics in upstream oil and gas industries for sustainable exploration and development: a review. Environ Technol Innov 21:101186. https://doi.org/10.1016/j.eti.2020.101186

Deselnicu DC, Militaru G, Deselnicu V, Zăinescu G, Albu L (2018) Towards a circular economy—a zero waste programme for Europe. In: ICAMS proceedings of the international conference on advanced materials and systems, 563–568. https://doi.org/10.24264/icams-2018.XI.4

Desing H, Brunner D, Takacs F, Nahrath S, Frankenberger K, Hischier R (2020) A circular economy within the planetary boundaries: towards a resource-based, systemic approach. Resour Conserv Recycl 155:104673. https://doi.org/10.1016/j.resconrec.2019.104673

Dey S, Bissoyi A, Koteswara Reddy G, Das AP (2024) Synthetic microfibers: An urban emerging pollutant demanding for sustainable environmental solutions. In: Sustainable Microbial Technology for Synthetic and Cellulosic Microfiber Bioremediation, pp 107–121. https://doi.org/10.1007/978-3-031-62600-5_7

Dey S, Dash N, Ray SK, Priyadarsini S, Chhotaray C, Das AP (2024) Occurrence and quantification of tire wear plastics pollutants from highway city dust of Bhubaneswar City of Odisha. In: Sustainable Microbial Technology for Synthetic and Cellulosic Microfiber Bioremediation, pp 77–89. https://doi.org/10.1007/978-3-031-62600-5_5

Dey S, Tripathy B, Kumar MS, Das AP (2023) Ecotoxicological consequences of manganese mining pollutants and their biological remediation. Environ Chem Ecotoxicol. https://doi.org/10.1016/J.ENCECO.2023.01.001

Grim RG, Huang Z, Guarnieri MT, Ferrell JR, Tao L, Schaidle JA (2020) Transforming the carbon economy: challenges and opportunities in the convergence of low-cost electricity and reductive CO_2 utilization. Energy Environ Sci 13(2):472–494. https://doi.org/10.1039/C9EE02410G
Guerra-Rodríguez S, Oulego P, Rodríguez E, Singh DN, Rodríguez-Chueca J (2020) Towards the implementation of circular economy in the wastewater sector: challenges and opportunities. Water 12(5):1431. https://doi.org/10.3390/w12051431
Hees T, Zhong F, Stürzel M, Mülhaupt R (2019) Tailoring hydrocarbon polymers and all-hydrocarbon composites for circular economy. Macromol Rapid Commun 40(1):1–18. https://doi.org/10.1002/marc.201800608
Heshmati A (2017) A review of the circular economy and its implementation. Int J Green Econ 11(3/4):251. https://doi.org/10.1504/IJGE.2017.089856
Iacovidou E, Velis CA, Purnell P, Zwirner O, Brown A, Hahladakis J, Millward-Hopkins J, Williams PT (2017) Metrics for optimising the multi-dimensional value of resources recovered from waste in a circular economy: a critical review. J Clean Prod 166:910–938. https://doi.org/10.1016/j.jclepro.2017.07.100
Jaeger B, Upadhyay A (2020) Understanding barriers to circular economy: cases from the manufacturing industry. J Enterp Inf Manag 33(4):729–745. https://doi.org/10.1108/JEIM-02-2019-0047
Jagaba AH, Kutty SRM, Lawal IM, Birniwa AH, Affam AC, Yaro NSA, Usman AK, Umaru I, Abubakar S, Noor A, Soja UB, Yakubu AS (2022) Circular economy potential and contributions of petroleum industry sludge utilization to environmental sustainability through engineered processes—a review. Clean Circ Bioecon 3:100029. https://doi.org/10.1016/j.clcb.2022.100029
John A, Das AP (2024) Government strategies for achieving circular economy for plastic waste management in Southeast Asia. In: Sustainable Management of Urban Plastic Waste Through Circular Economic Approaches, pp 72–82. https://doi.org/10.1201/9781003465447-6/GOVERNMENT-STRATEGIES-ACHIEVING-CIRCULAR-ECONOMY-PLASTIC-WASTE-MANAGEMENT-SOUTHEAST-ASIA-AKBAR-JOHN-ALOK-PRASAD-DAS
Kanungo J, Sahoo T, Bal M, Behera ID (2024) Performance of bioremediation strategy in waste lubricating oil pollutants: a review. Geomicrobiol J 41(4):360–373. https://doi.org/10.1080/01490451.2023.2245395
Kanungo J, Sahoo T, Swain LP, Behera ID (2023) Toxicity of persistent hydrocarbon pollutants, sources and sustainable remediation process. In: Behera ID, Das AP (eds) Impact of petroleum waste on environmental pollution and its sustainable management through circular economy. Springer Nature, Switzerland, pp 39–65. https://doi.org/10.1007/978-3-031-48220-5_2
Kaygusuz K (2009) Energy and environmental issues relating to greenhouse gas emissions for sustainable development in Turkey. Renew Sustain Energy Rev 13(1):253–270. https://doi.org/10.1016/j.rser.2007.07.009
Kerdlap P, Low JSC, Ramakrishna S (2019) Zero waste manufacturing: a framework and review of technology, research, and implementation barriers for enabling a circular economy transition in Singapore. Resour Conserv Recycl 151:104438. https://doi.org/10.1016/j.resconrec.2019.104438
Khajuria A, Atienza VA, Chavanich S, Henning W, Islam I, Kral U, Liu M, Liu X, Murthy IK, Oyedotun TDT, Verma P, Xu G, Zeng X, Li J (2022) Accelerating circular economy solutions to achieve the 2030 agenda for sustainable development goals. Circ Econ 1(1):100001. https://doi.org/10.1016/j.cec.2022.100001
Kharisov BI, Dias HVR, Kharissova OV (2014) Nanotechnology-based remediation of petroleum impurities from water. J Petrol Sci Eng 122:705–718. https://doi.org/10.1016/j.petrol.2014.09.013
Kumar A, Thakur AK, Gaurav GK, Klemeš JJ, Sandhwar VK, Pant KK, Kumar R (2023) A critical review on sustainable hazardous waste management strategies: a step towards a circular economy. Environ Sci Pollut Res 30(48):105030–105055. https://doi.org/10.1007/s11356-023-29511-8

Kumar R, Verma A, Shome A, Sinha R, Sinha S, Jha PK, Kumar R, Kumar P, Shubham DS, Sharma P, Vara Prasad PV (2021) Impacts of plastic pollution on ecosystem services, sustainable development goals, and need to focus on circular economy and policy interventions. Sustainability 13(17):9963.https://doi.org/10.3390/su13179963

Lahiri D, Nag M, Dey A, Sarkar T, Joshi S, Pandit S, Das AP, Pati S, Pattanaik S, Tilak VK, Ray RR (2021) Biofilm mediated degradation of petroleum products. Geomicrobiol J: 1–10. https://doi.org/10.1080/01490451.2021.1968979

Lee RP, Meyer B, Huang Q, Voss R (2020) Sustainable waste management for zero waste cities in China: potential, challenges and opportunities. Clean Energy 4(3):169–201. https://doi.org/10.1093/ce/zkaa013

Lieder M, Rashid A (2016) Towards circular economy implementation: a comprehensive review in context of manufacturing industry. J Clean Prod 115:36–51. https://doi.org/10.1016/j.jclepro.2015.12.042

Mallick K, Sahu A, Dubey NK, Das AP (2023) Harvesting marine plastic pollutants-derived renewable energy: a comprehensive review on applied energy and sustainable approach. J Environ Manage 348:119371. https://doi.org/10.1016/J.JENVMAN.2023.119371

Mishra A, Siddiqi H, Kumari U, Behera ID, Mukherjee S, Meikap BC (2021) Pyrolysis of waste lubricating oil/waste motor oil to generate high-grade fuel oil: a comprehensive review. Renew Sustain Energy Rev 150:111446. https://doi.org/10.1016/j.rser.2021.111446

Mishra S, Das AP (2023) Microbial remediation of synthetic microfiber contaminated wastewater. In: Shah MP (ed) Microbial technologies in industrial wastewater treatment. Springer, Singapore, pp 337–350. https://doi.org/10.1007/978-981-99-2435-6_17

Mishra S, Dash D, Das AP (2024) Microbial degradation of polyester microfibers using indigenously isolated bacterial strain Exiguobacterium Sp. CLEAN – Soil, Air, Water, 52(12):e202300343. https://doi.org/10.1002/CLEN.202300343

Mishra S, Ghosh S, van Hullebusch ED, Singh S, Das AP (2023) A critical review on the recovery of base and critical elements from electronic waste-contaminated streams using microbial biotechnology. Appl Biochem Biotechnol 2023:1–30. https://doi.org/10.1007/S12010-023-04440-X

Mohanta S, Pradhan B, Behera ID (2024) Impact and remediation of petroleum hydrocarbon pollutants on agricultural land: a review. Geomicrobiol J 41(4):345–359. https://doi.org/10.1080/01490451.2023.2243925

Mohanty M, Mohanty J, Dey S, Dutta K, Shah MP, Das AP (2024) The face mask: a tale from protection to pollution and demanding sustainable solution. Emerg Contaminants 10(2):100298. https://doi.org/10.1016/J.EMCON.2023.100298

Mohanty M, Tripathy B, Koteswara Reddy G, Das AP (2024) Greener technologies for establishing restraint over microplastic pollution. In: Sustainable Microbial Technology for Synthetic and Cellulosic Microfiber Bioremediation, pp 239–258. https://doi.org/10.1007/978-3-031-62600-5_13

Mojarad AAS, Atashbari V, Tantau A (2018) Challenges for sustainable development strategies in oil and gas industries. Proc Int Conf Bus Excellence 12(1):626–638. https://doi.org/10.2478/picbe-2018-0056

Oghazi P, Mostaghel R (2018) Circular business model challenges and lessons learned—an industrial perspective. Sustainability 10(3):739. https://doi.org/10.3390/su10030739

Pal D, Sen S (2023a) Emerging petroleum pollutants and their adverse effects on the environment. In: Impact of petroleum waste on environmental pollution and its sustainable management through circular economy, pp 103–137. https://doi.org/10.1007/978-3-031-48220-5_5

Pal D, Sen S (2023b) In-depth coverage of petroleum waste sources, characteristics, environmental impact, and sustainable remediation process. In: Impact of petroleum waste on environmental pollution and its sustainable management through circular economy, pp 1–38. https://doi.org/10.1007/978-3-031-48220-5_1

Paletta A, Leal Filho W, Balogun A-L, Foschi E, Bonoli A (2019) Barriers and challenges to plastics valorisation in the context of a circular economy: case studies from Italy. J Clean Prod 241:118149. https://doi.org/10.1016/j.jclepro.2019.118149

Pan SY, Du MA, Huang IT, Liu IH, Chang EE, Chiang PC (2015) Strategies on implementation of waste-to-energy (WTE) supply chain for circular economy system: a review. J Clean Prod 108:409–421. https://doi.org/10.1016/j.jclepro.2015.06.124

Panja P, Velasco R, Pathak M, Deo M (2018) Application of artificial intelligence to forecast hydrocarbon production from shales. Petroleum 4(1):75–89. https://doi.org/10.1016/j.petlm.2017.11.003

Park J, Sarkis J, Wu Z (2010) Creating integrated business and environmental value within the context of China's circular economy and ecological modernization. J Clean Prod 18(15):1494–1501. https://doi.org/10.1016/j.jclepro.2010.06.001

Paul A, Dey S, Ram DK, Das AP (2023) Hexavalent chromium pollution and its sustainable management through bioremediation. Geomicrobiol J: 1–11. https://doi.org/10.1080/01490451.2023.2218377

Paul A, Priyadarsini S, Karakuş S, Behera JK, Gnanasekaran R, Das AP (2024) Fate and accumulation of synthetic microfibers in the environment. In: Sustainable Microbial Technology for Synthetic and Cellulosic Microfiber Bioremediation, pp 61–76. https://doi.org/10.1007/978-3-031-62600-5_4

Pietzsch N, Ribeiro JLD, de Medeiros JF (2017) Benefits, challenges and critical factors of success for zero waste: a systematic literature review. Waste Manage 67:324–353. https://doi.org/10.1016/j.wasman.2017.05.004

Pinheiro CT, Quina MJ, Gando-Ferreira LM (2021) Management of waste lubricant oil in Europe: a circular economy approach. Crit Rev Environ Sci Technol 51(18):2015–2050. https://doi.org/10.1080/10643389.2020.1771887

Pradhan G, Tripathy B, Ram DK, Digal AK, Das AP (2023) Bauxite mining waste pollution and its sustainable management through bioremediation. Geomicrobiol J: 1–10. https://doi.org/10.1080/01490451.2023.2235353

Priyadarsini S, Das AP (2024) Analytical and structural characterization of waste lithium-ion batteries for their effective recycling strategy. Environ Chem Ecotoxicol. https://doi.org/10.1016/j.enceco.2024.12.004

Priyadarsini S, Das AP (2025) Lithium bioleaching: A review on microbial-assisted sustainable technology for lithium bio-circularity. J Water Process Eng, 69:106744. https://doi.org/10.1016/j.jwpe.2024.106744

Quicker P, Seitz M, Vogel J (2022) Chemical recycling: a critical assessment of potential process approaches. Waste Manage Res 40(10):1494–1504. https://doi.org/10.1177/0734242X221084044

Raj K, Das AP (2023) Lead pollution: impact on environment and human health and approach for a sustainable solution. Environ Chem Ecotoxicol 5:79–85. https://doi.org/10.1016/J.ENCECO.2023.02.001

Ramakrishna S, Ngowi A, De Jager H, Awuzie BO (2020) Emerging industrial revolution: symbiosis of industry 4.0 and circular economy: the role of universities. Sci Technol Soc 25(3):505–525. https://doi.org/10.1177/0971721820912918

Ray SK, Dey S, Das AP, John A (2024) Microplastic waste management through greener technologies: A review. In: Sustainable Management of Urban Plastic Waste Through Circular Economic Approaches, pp 237–261. https://doi.org/10.1201/9781003465447-14 RENEWABLE ENERGY GENERATION AND VALUE ADDITION FROM ENVIRONMENTAL. (2024)

Riera JA, Lima RM, Knio OM (2023) A review of hydrogen production and supply chain modeling and optimization. Int J Hydrogen Energy 48(37):13731–13755. https://doi.org/10.1016/j.ijhydene.2022.12.242

Ropital F (2009) Current and future corrosion challenges for a reliable and sustainable development of the chemical, refinery, and petrochemical industries. Mater Corros 60(7):495–500. https://doi.org/10.1002/maco.200805171

Salanitro JP (2001) Bioremediation of petroleum hydrocarbons in soil. Adv Agron 72:53–105. https://doi.org/10.1016/S0065-2113(01)72011-1

Scheel C, Aguiñaga E, Bello B (2020) Decoupling economic development from the consumption of finite resources using circular economy. A model for developing countries. Sustainability 12(4):1291. https://doi.org/10.3390/su12041291

Shahsavani I, Goli A (2023) A systematic literature review of circular supply chain network design: application of optimization models. Environ Dev Sustain: 0123456789. https://doi.org/10.1007/s10668-023-03362-2

Sheldon RA, Norton M (2020) Green chemistry and the plastic pollution challenge: towards a circular economy. Green Chem 22(19):6310–6322. https://doi.org/10.1039/D0GC02630A

Singh RP, Mishra S, Das AP (2020) Synthetic microfibers: pollution toxicity and remediation. Chemosphere 257:127199. https://doi.org/10.1016/j.chemosphere.2020.127199

Suárez-Eiroa B, Fernández E, Méndez-Martínez G, Soto-Oñate D (2019) Operational principles of circular economy for sustainable development: linking theory and practice. J Clean Prod 214:952–961. https://doi.org/10.1016/j.jclepro.2018.12.271

Tripathy B, Dash A, Das AP (2022) Detection of environmental microfiber pollutants through vibrational spectroscopic techniques: Recent advances of environmental monitoring and future prospects. Crit Rev Anal Chem, pp 1–11. https://doi.org/10.1080/10408347.2022.2144994

Tükenmez M, Demireli E (2012) Renewable energy policy in Turkey with the new legal regulations. Renew Energy 39(1):1–9. https://doi.org/10.1016/j.renene.2011.07.047

Wang M, Hossain MR, Si Mohammed K, Cifuentes-Faura J, Cai X (2023) Heterogenous effects of circular economy, green energy and globalization on CO_2 emissions: policy based analysis for sustainable development. Renew Energy 211:789–801. https://doi.org/10.1016/j.renene.2023.05.033

Waqar A, Othman I, Shafiq N, Mansoor MS (2023) Applications of AI in oil and gas projects towards sustainable development: a systematic literature review. Artif Intell Rev 56(11):12771–12798. https://doi.org/10.1007/s10462-023-10467-7

Chapter 16
Regulatory Framework and Policies to Achieve Zero Hydrocarbon Waste Management Through Circular Economy

Aishwarya Sahu, Laxmi Priya Swain, Ipsita Dipamitra Behera, and Geetanjali Basak

Abstract Hydrocarbon waste is generated from various industries such as oil and gas, petrochemicals, and transportation worldwide. It refers to a diverse range of byproducts, including crude oil spills, refinery sludges, and plastic pollution. These byproducts result from various activities such as extraction, production, use, and disposal of hydrocarbons. Recent trends have emphasised the imperative shift towards sustainable practices for managing hydrocarbon waste. Innovations such as advanced recycling technologies, bio-based alternatives, and circular supply chains have emerged to mitigate the environmental impact and promote resource efficiency. The regulatory frameworks and policies have a crucial impact on the direction of hydrocarbon waste management. The implementation of laws requiring pollution control, waste minimisation, and enhanced producer responsibility has stimulated comprehensive efforts across industries to reduce waste and promote recycling. Although there has been progress, there is still scope for enhancing regulatory frameworks in order to attain complete management of hydrocarbon waste with a zero-waste approach. Achieving this objective requires enhancing the effectiveness of enforcement of regulations, providing incentives for adopting sustainable practices, promoting collaborations across different sectors, and active involvement of all relevant parties within a circular economy framework. By coordinating policies with both environmental and economic goals, the adverse effects of hydrocarbon waste can be reduced, and we can move towards a more sustainable future.

Keywords Hydrocarbon waste · Sustainable waste management · Zero waste approach · Circular economy · Framework regulations and policies

A. Sahu · L. P. Swain · I. D. Behera (✉)
Department of Chemical Engineering, Indira Gandhi Institute of Technology, Sarang, Odisha 759146, India
e-mail: ipsitadbehera@gmail.com

G. Basak
Raniganj Girls' College, Affiliated to Kazi Nazrul University, Raniganj, West Bengal 713358, India

I. D. Behera and A. P. Das (eds.), *Environmental Hydrocarbon Pollution and Zero Waste Approach Towards a Sustainable Waste Management*, Environmental Science and Engineering, https://doi.org/10.1007/978-3-031-81615-4_16

16.1 Introduction

Hydrocarbons are the significant components that make up natural gas and oil. The majority of hydrocarbons on the planet are naturally occurring and largely present in crude oil. The hydrocarbons are organic molecules composed mostly of hydrogen and carbon, classified as aliphatic (straight-chain) and aromatic (cyclic) (Dey et al. 2023). They can exist in different forms, such as straight or branched chains, and can be either saturated or unsaturated. Also, these can also be classified into different groups based on their structure, including aliphatic (open-chain), alicyclic (cyclic), and aromatic compounds. Additionally, there are polyaromatic compounds. Halogenated hydrocarbons consist of halogen atoms, such as chlorine or fluorine, and can be either aliphatic or aromatic (Kanungo et al. 2024; Tormoehlen et al. 2014). Nonhydrocarbon compounds in petroleum include naphthenic acid, heterocyclic nitrogen, phenol, thiol, asphaltene sulfur, and metalloporphyrin compounds. The physical and chemical properties of petroleum extracted from different sources can vary significantly. (Rosner and Markowitz 2013). Petroleum hydrocarbons can be categorised into four separate fractions according to their chemical makeup (Behera et al. 2022a, b; Hazaimeh and Ahmed 2021). The first percentage is referred to as saturating compounds, often known as alkanes. The subsequent part consists of polyaromatic hydrocarbons (PAHs) and aromatics, such as benzene, toluene, ethylbenzene, and xylenes (commonly known as BTEX). The third fraction comprises resins, which are nitrogen, sulfur, and oxygen-containing molecules that are liquified in oil. Finally, the fourth portion consists of asphaltenes, which are intricate and sizable particles that are spread in crude oil in a colloidal fashion. The carbon atoms bond with each other to make the structure of the complex, and the hydrogen atoms can connect to them in several different ways.

Due to the growing population and urbanization, different pollutants like petroleum products (Lahiri et al. 2022), heavy metals (Bal et al. 2020; Das et al. 2011; Prabhakar et al. 2019; Pradhan et al. 2023; Raj and Das 2023), microplastics (Mallick et al. 2023; Mishra and Das 2022; Mishra et al. 2020, 2022; Tripathy et al. 2024), biomedical waste (Khatun et al. 2023; Sadasivuni et al. 2019; Sahu et al. 2024), and electronic waste (Mishra et al. 2023) increasing at a significant rate. All industrial sectors and refineries are calling for a significant global demand for crude oil. The significant growth and advancement in the production of crude oil have resulted in the widespread dissemination of various hydrocarbon-based products, including natural gas, fuel oil, lubricating oil, diesel, paraffin, kerosene and naphthene. The expansion has led to the production of petroleum waste byproducts as a consequence of inadequate disposal techniques. The exploration, extraction, and production of hydrocarbon products proceed in the creation of pollutants, including wastewater, solid waste, filter clay, and spent caustic, which contain significant amounts of heavy metals (Bala, Anshul et al. 2024; Bala, Dogra et al. 2024; A. Das, Dey, & Das 2024; A. Das, Dey, Bala et al. 2024; Dey, Bissoyi et al. 2024). Additionally, this process leads to the release of harmful gases such as hydrogen sulfide, carbon dioxide, nitrogen oxides, and carbon monoxide. The main contributors to pollution are the activities related to oil, such as exploration, followed by storage, then transportation, and its processing in refineries. Oil spills caused in the oceans are due to accidental or intentional release of oil into the water. They can also happen during the conveyance of petroleum hydrocarbons from production sites to distribution sites, as well as through leakages in pipelines used for gas exploration and crude oil transportation. Other sources of oil spills include retail service stations, mishaps occurring while storing petroleum and byproducts, and production refining and drilling activities (Aljamali et al. 2021; Behera 2022; Das et al. 2023; Kanungo et al. 2023; Mohanta et al. 2024). Leaking underground storage tanks is a common cause of groundwater contamination due to the discharge of fuels. Pollution from crude oil spills is common. Porphyrins, pyrrolic compounds, substantially correlate with crude oil metal content. Crude oil contains thousands of organic molecules, predominantly hydrocarbons, with a few nitrogen, sulphur, oxygen-containing compounds, and certain metals. Different petroleum fractions are created during refining, such as light oil, kerosene, naphtha, lube oil waxes, diesel, asphaltenes, etc. Heavy ends are high-pressure-distilled asphaltenes and lubricating oil, whereas light ends are at-atmosphere-pressure-distilled light fractions. Light ends have unsaturated and saturated hydrocarbons with lower molecular weight and fewer aromatic molecules than heavy ends (Manish et al. 2019). Organic substances that are toxic and generally detrimental to both the environment and other living things are the leading cause of hydrocarbon pollution or contamination. The aliphatic and aromatic hydrocarbons also pose a major risk to the health of humans and the environment. Since the hydrocarbons have low viscosity and surface tension, they enter the lungs quickly, giving rise to disorders such as pneumonia and affecting the alveoli, capillaries, and airways, inducing fever, atelectasis, and inflammation (Dey, Dash et al. 2024; Dey et al. 2023; John & Das 2024; Mishra et al. 2024; Mohanty et al. 2024). Hydrocarbons have an impact on microbial activity and soil fertility in addition to human health (NCBI). Hydrocarbon pollution has hazardous, fatal, and cataclysmic impacts on the atmosphere, marine life, and terrestrial life (Behera et al. 2021). Products that contain hydrocarbons, particularly oils, have detrimental effects on the growth of zooplankton and the fertility of fish. An estimated 35 million barrels of oil are shipped across the seas each year, despite the fact that oil spills and leaks are a worldwide threat to aquatic and marine life. Soil is contaminated by industrial waste. The discharge of hazardous wastes from refineries is one of the predominant reasons for groundwater pollution and overall reduced productivity of agrarian land (Varjani and Upasani 2016). We must mitigate the long-term repercussions that cause the ecosystem to be alarmed because of the dangerous environmental impact.

Physiochemical and biological techniques can lower the amount of contamination and the quantity of contaminants. In the first strategy, physiochemical techniques are used, which are frequently used in conventional engineering practices. Excavation and transportation of large amounts of contaminated material for treatment are expensive processes that contribute to the high costs of physicochemical decontamination techniques. A few crucial physiochemical techniques are microwave radiation, centrifugation, froth flotation, pyrolysis, solvent extraction, electrokinetic method, etc. Another technique is biological in nature, utilising bacteria to convert toxic constituents into less toxic ones. A few biological techniques include bioaugmentation, flocculation bio treatment, zero valent ion (ZVI) combined with/ the biological methods, phytoremediation, bioremediation, etc. (Behera et al. 2020; Mallick et al. 2024; Das et al. 2024a). When compared to the biological method, the physiochemical method has numerous drawbacks, including excessive chemical and energy consumption and the generation of secondary pollutants. The biological approach, on the other hand, is more economical and environmentally benign than the physiochemical method, which is why we chose it.

To develop successful bioremediation techniques for purifying polluted soil to a level that allows both safe disposal and reusing, the biodegradation and biotransformation processes of hydrocarbons in the soil must be thoroughly tested and characterised. This includes finding the precise communities of hydrocarbon-degrading microbes or consortia, understanding the environmental conditions that promote their activity, and assessing the degree and extent of hydrocarbon degradation (Das et al. 2024b, c). Researchers can design and apply biological remediation systems that effectively reduce contamination, allowing the soil to be removed or reused for various purposes. This is for the reason that petroleum hydrocarbons have the potential to pose a threat to both humans and the environment (Snežana et al. 2013). Pollutants are transformed by microorganisms into less hazardous and dangerous compounds that can subsequently be integrated into natural biogeochemical cycles through a type of metabolism called biodegeneration (Paul et al. 2024; Priyadarsini & Das 2024, 2025; Ray et al. 2024; Tripathy et al. 2022). The rate of petroleum degradation in soil is influenced by a number of factors, including soil chemical and physical features, pH value, nutrients, oxygen, its composition and concentration (Varjani and Upasani 2017).

Another green technique that accelerates the biological degradation method of pollutants or contaminants through the optimisation of limiting constrain is the bioremediation technique, which is eco-friendly, non-destructive, and cost-effective. The most effective technique is selected based on its biodegradation efficiency.

The United Nations Environmental Programme (UNEP) considers oil as hazardous waste. It is thus one of the restricted waste categories recognised under the Basel Convention on regulating the transboundary movement of hazardous waste and its disposal (Basel Convention—UN Environment Document Repository Home 2011). Minerals derived from crude oil or yielded as synthetic oils are termed waste oils. These fluids have several applications, including lubricating, cutting metal, and heat and power transmission (Mishra et al. 2021). The ecological consequences of waste oil on the ecosystem are significant and far-reaching when it is discharged into the environment. As stated by the Agency for Toxic Substances and Disease Registry (ATSDR) in 1991, when petroleum hydrocarbons are introduced into the water through spills or leaks, the hydrocarbon fractions will float on the water and create thin surface films. Any synthetic oil derived from petroleum is classified as waste oil if it comprises undesirable chemicals after being used. According to the United States Coast Guard, between 1991 and 2004, ships, tank vessels, and other sources were responsible for 35.7% of the oil leaks in the United States. Facilities and other non-vessel sources accounted for 27.6%, non-tank vessels for 19.9%, pipelines for 9.3%, and the origin of 7.4% of the spills was unclear (Mong'ina 2018).

Conversely, oil tankers were responsible for only 5% of the actual spills, while another vessel accounted for 51.8%. India is a large petroleum producer that produces a significant amount of petroleum waste (Aljamali et al. 2021). An oil spill, liquid petroleum hydrocarbon released into the land, ocean or coastal areas is a result of human activities. This can be caused by spills from drilling rigs, wells, and tankers during transportation and spills from offshore platforms containing distilled petroleum products for instance diesel and gasoline and their byproducts. It is also caused by spills of waste oils and heavier fuel usage by large ships, like bunker fuels or oil refuse. Cleaning up of these oil spills from various sources is quite tricky and mostly depends on different factors like temperature and weather at the spill point. Due to the lack of proper regulations and action bodies, it took weeks, months, or years to clean and restore the property.

This chapter focuses on waste release from different sources and its management and also emphasises the regulatory framework to achieve a cleaner environment. Apart from this, this highlights the policies made for properly managing Hydrocarbon waste at national and international levels.

16.2 Hydrocarbon Waste

A major source of hydrocarbon waste is the manufacture of various petrochemical products such as paints, fertilisers, plastics, and detergents. Hydrocarbon and its waste products encompass a variety of constituents, such as crude oil, emulsions, slurry and solids. The activities involved in the processing, exploration, and transportation of petroleum hydrocarbon wastes have the potential to cause significant harm to the environment, including aquatic and terrestrial biota, as well as pose risks to human health (Sattar et al. 2022). The most vulnerable elements of the environment—air, soil, and water—are key components of the biogeochemical ecosystem and are most prone to the detrimental impacts of hydrocarbons (Behera and Das 2023; Mitra et al. 2022). Hydrocarbon waste can manifest in various forms, including gaseous, liquid, and solid. Gaseous hydrocarbon waste, often referred to as flare gas or associated gas, is a common byproduct of oil and natural gas extraction (Arutyunov et al. 2023). While it can be combusted at the wellhead to enhance safety and reduce immediate emissions, flaring also releases greenhouse gases, exacerbating climate change. Consequently, there is a concerted effort to minimise flaring and repurpose this gas through methods like re-injection, using it as fuel, or converting it into chemical feedstock.

Liquid hydrocarbon waste, such as crude oil or refinery off-gases, is a byproduct of various industrial processes, including oil refining, petrochemical production, and transportation activities. These can be recycled, treated, and reused. These petroleum hydrocarbon compounds are often found in surface and subsurface water as a result of industrial processes, markedly those involving petrochemicals, accidental reservoir leaks, or improper waste disposal. Water streams are tainted by several processes, frequently with dissolved hydrocarbons such as BTEX (Mohammadi et al. 2020). Slop oils are a class of recovered petroleum hydrocarbons made up of complex waste streams extracted from petroleum crude oils during the refining process. Owing to their provenance from different phases of the refining process, these slop oils consist of a wide variety of water and petroleum hydrocarbons. For instance, waste crude oil can undergo treatment to remove impurities and contaminants, making it viable for reuse as fuel or feedstock (Shahbaz et al. 2023). Similarly, refinery off-gases can be captured and processed into valuable chemicals or fuels like ethylene or methanol.

Solid hydrocarbon waste, including drilling muds, oil spills, and industrial sludge, presents significant environmental challenges (Behera et al. 2022a, b; Biswas et al. 2020; da Silva et al. 2012). The gas and oil industry produces oily sludge in several regions, including cars, refining facilities, and storage vessels (Hossain et al. 2017). The primary methods for removing oil sludge include burning, solidification, oxidation, and biodegradation (Teng et al. 2021). On the other hand, meagre is recycled and transformed into by-products. There are three layers to the OS in the tank. Oil and water are present in the first layer, known as floating sludge, while solid or sludgy deposited materials make up above 50% of the sediments, which are the bottom layer (Egazar'Yants et al. 2015). Further, drilling mud containing water, clay, and hydrocarbons can be treated to extract and recover hydrocarbons for reuse as fuel or feedstock. Oil spills necessitate extensive cleanup and remediation efforts, often involving dispersants and in-situ burning to mitigate environmental damage. Industrial sludge, which may carry heavy metals and other pollutants, requires careful treatment and removal through techniques such as incineration or landfilling to ensure safety and environmental protection (Behera et al. 2022a, b; Johnson and Affam 2019) (Fig. 16.1).

16.3 Statistical Data on Hydrocarbon Waste Generation

In 2019–20, India consumed 194.2 million metric tons of petroleum products, indicating a corresponding increase in petroleum waste generation (Petroleum Planning and Analysis Cell, 2019–20). From 2020 to 2021, the total amount of hazardous waste generated by the petroleum sector in India was around 130,000 metric tons.

According to the Petroleum Planning and Analysis Cell, around 241,704 metric tonnes of crude oil were processed by different refineries in the year 2021–22 (Crude Oil Processed by Refineries 2024). Further, this amount gradually increases from 255,232 metric tonnes to 261,546 metric tonnes in the year 2022–23 and 2023–24, respectively.

According to the World Bank's report, in 2022, India's population will see a substantial increase, increasing from 361 million in 1951 to almost 1.4 billion in 2022. India presently has the third-largest oil user worldwide. Additionally, India is a prominent participant in the petroleum sector and produces a substantial quantity

Fig. 16.1 Hydrocarbon waste generation from various sources taken from Kanungo et al. (2023)

of petroleum waste (Central Pollution Control Board [CPCB] 2021; Ministry of Petroleum and Natural Gas, Government of India 2020). The oil refineries in India were the primary contributors to the 122,547.67 metric tons of lethal waste produced by the petroleum sector in 2020 and 2021. Over the same time period, the petroleum industry recycled 97.8% of its hazardous waste and sent 2.2% to a landfill (CPCB 2021) (Fig. 16.2).

To address these emerging issues, the government has proposed different policies and regulations for managing petrol and petroleum product wastes, along with establishing hazardous waste treatment facilities (National Green Tribunal, n.d.; Oil Industry Safety Directorate 2018). According to statistical data on petroleum waste in India, the sector is moving towards more sustainable waste management techniques, with a main emphasis on reprocessing and waste recovery. Nonetheless, further advances in wastewater treatment and management procedures are required to reduce environmental effects.

16.4 Recent Trends and Challenges in Hydrocarbon Waste Management

Hydrocarbon pollutants have a comprehensive variety of physical properties, chemical compositions, and environmental behaviours, complicating cleanup efforts. Different forms of hydrocarbons, such as aliphatic, aromatic, chlorinated, and volatile chemicals, necessitate cleanup strategies adapted to their distinct properties and

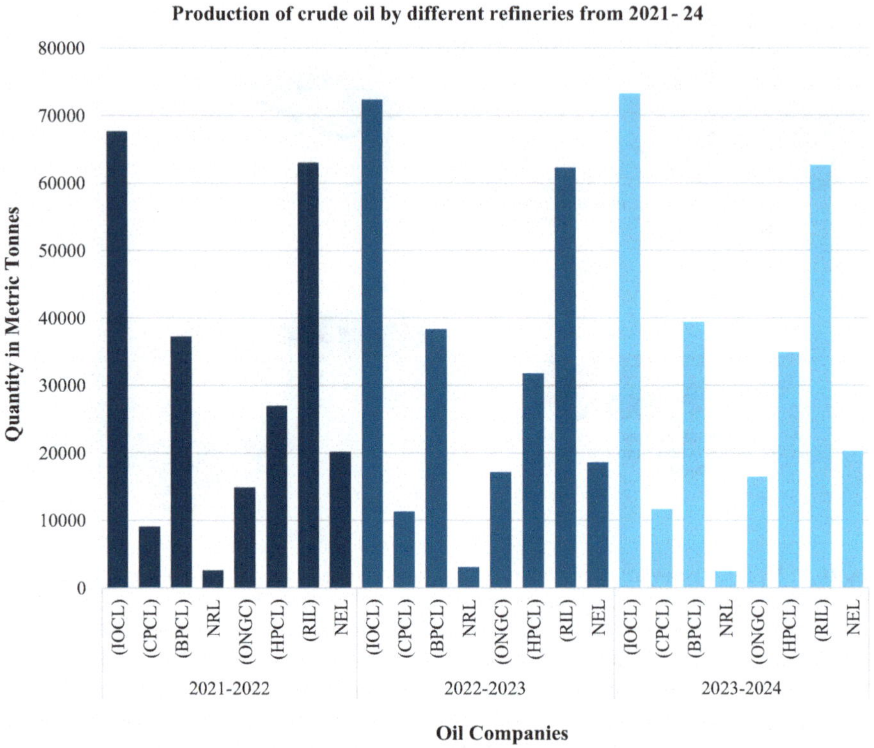

Fig. 16.2 Production of crude oil by different refineries from 2021 to 24

environmental interactions. Certain kinds of hydrocarbon pollutants endure in the environment, degrading slowly and persisting in soil, water, and sediments for long periods of time (Das et al. 2023; Mohanta 2016). Technological constraints in pollution treatment and cleanup provide substantial hurdles to efficiently addressing hydrocarbon contamination. Moreover, multifunctional technologies like incineration soil washing, and in-situ thermal desorption, solvent extraction, electrokinetic remediation, chemical oxidation, pyrolysis, bioreactors, composting/biopiles, bioremediation, vermiremediation, phytoremediation, biocatalyst-assisted remediation, nano remediation and so on., offer a range of possibilities for remediating petroleum hydrocarbon-contaminants, each with its own advantages and limitations (Kuppusamy et al. 2017). It is clear that, among all known remedial methods, bioremediation, such as Enzyme-mediated degradation and genetically modified organisms, followed by its integration with other treatments, has acquired widespread acceptance as a viable solution for treating petroleum-contaminated soils (Kumar et al. 2021).

Despite advances in remediation technology, multiple obstacles limit the efficiency, efficacy, and viability of pollution treatment projects. Various constraints can be widely categorised into technical and non-technical factors (Kuppusamy et al.

2017). Technical factors may include fluctuating soil and weather conditions, the presence and activity of specific microbial populations, the presence of other contaminants in the soil, and the cost of implementing green technologies for large-scale remediation projects (Behera et al. 2022a, b). Factors like limitations in time for remediating, public perception and regulatory acceptance, acquiescence with environmental laws and regulations and government funding can limit the use of certain remedial approaches and fall under non-technical factors. By solving these hurdles, stakeholders may improve the efficacy and cost-effectiveness of remediation operations while also supporting ecologically friendly and sustainable cleanup procedures. Implementing hydrocarbon remediation strategies is expensive to accomplish, involving major financial inputs, resource allocation, and specialist skills (Olabi et al. 2022). These financial consequences may present obstacles for organisations with limited resources, such as government agencies, businesses, and communities, restricting their ability to handle pollution effectively (Das, A. Prasad. & Ghosh, Shreya 2023). Henceforth, maintenance of the long-term success of remediation initiatives, regular monitoring, management, and evaluation of remediated sites is required. Establishing consistent funding, adequate staff, and institutional support is critical for ensuring continued remediation initiatives over time and overcoming any roadblocks that could jeopardise their effectiveness and longevity.

16.5 Regulatory Framework to Achieve Zero Waste

To achieve zero hydrocarbon waste within a regulatory framework, a comprehensive, multifaceted approach involving several interconnected policies and initiatives is essential. They establish standards, enforcement mechanisms, and guidelines to limit emissions, prevent pollution, and safeguard of human health and the biosphere (Awewomom et al. 2024). However, implementing and administering these standards can present several challenges and compliance issues. Governments must enact and rigorously enforce comprehensive regulations to curb hydrocarbon waste, including imposing strict emission limits, mandating the use of renewable energy and energy efficiency technologies, and incentivising or requiring the recycling and responsible disposal of hydrocarbons.

Industries frequently have to comply with the work on schedule in order to meet the requirements for reporting and monitoring emissions of hydrocarbon pollutants (Li 2024). This entails ongoing observation, data collection, and reporting of emissions information to regulatory organisations. Divergences in the monitoring, accuracy, and transparency of data can arise from different sectors adhering to different levels of responsibility to report and monitor. The extraction, production, and regulation of hydrocarbon emissions are advancing at a rapid pace, making it difficult for regulatory systems to keep up with new developments and industrial principles. Regulatory frameworks may face challenges in adapting to advanced technologies, innovative strategies, and shifting industrial standards, which could lead to breaches in regulation and compliance.

Further, the efficiency and credibility of these regulatory bodies on hydrocarbon pollution may decline due to insufficient involvement of stakeholders and the public in shaping and enforcing rules (Farombi and Klaus 2024). The lack of transparent information accessibility for the general public and opportunities for meaningful participation could shake off public trust in regulatory processes and decision-making. It's possible that non-compliance and illegal activities inside petroleum operations can be avoided with uniform implementation strategies, severe penalties, and adequate deterrence measures.

To guarantee benefits that are equal and inclusive for all, new policy systems, protocols, and governance frameworks are needed in the emerging digital era. Regulations that are both nature-led and nature-centred, as well as human-led and human-centred, are necessary for societies (Pathway, 2024). Public interests like human dignity and identity, trust, protecting the environment, and combating climate change ought to be equalized with private sector goals like profits and business disruption.

Implementing cap-and-trade systems is crucial, as it allows companies to trade carbon credits based on their historical emissions, thereby encouraging a financial incentive to spend money on green technology and reduce emissions and putting in place a carbon pricing mechanism, such as a cap-and-trade scheme or a carbon tax. Establishing a carbon pricing mechanism, whether through a carbon tax or cap-and-trade system, ensures that the environmental costs of carbon emissions are accurately reflected, further promoting waste reduction. Both governmental and private sectors ought to make research and development investments to innovate technologies that minimise hydrocarbon waste, improve extraction and utilisation methods, convert waste into usable products, and explore alternative energy sources. Effective collaboration between governments and businesses is also critical; public–private partnerships can drive research funding, technological adoption, and the implementation of beneficial policies and regulations. Moreover, encouraging consumers to adopt energy-efficient appliances and practices through incentives and education is vital, as awareness campaigns about reducing carbon footprints can significantly impact hydrocarbon waste reduction. On a global scale, international cooperation is necessary to tackle hydrocarbon waste effectively. Countries should work together to establish international standards, coordinate R&D efforts, and support the adoption of green technologies, particularly in developing nations. This holistic approach ensures a coordinated effort to reduce hydrocarbon waste, fostering sustainable development and environmental protection.

16.6 Policies in India for Hydrocarbon Waste Management

In India, the management of hydrocarbon waste is primarily overseen by the Ministry of Petroleum and Natural Gas (MoPNG) and the Central Pollution Control Board (CPCB), backed by a set of guidelines and rules intended to guarantee secure handling, storing, and disposal procedures (Hazardous Waste Management Rules 2022; Synthetic Fuels: Future Transport Fuels 2024). Regulations and policies of the government are essential to support circular economy practices for hydrocarbon waste management. These policies establish standards, provide reinforcement, and create frameworks that encourage sustainable behaviours and mitigate environmental damage. Extended Producer Responsibility (EPR) legislation, for instance, mandates that product makers be in charge of collecting, recycling, and disposing of their items at the end of their useful lives. EPR rules encourage businesses to build goods that are recyclable, easily repairable, and long-lasting when it comes to hydrocarbon-based products. This approach fosters a circular economy and helps to minimise waste.

Additionally, the Mission Circular Economy aims to promote sustainable waste management conventions and efficient use of resources. NITI Aayog, being a part of the initiative, constituted 11 committees to create action plans pertaining to the Circular Economy for a range of waste types, such as Li-ion batteries, electronic waste, hazardous and toxic industrial waste, tyres and rubber, end-of-life cars, gypsum, used oil, municipal solid garbage, and solar panels. Further, the Ministry of Environment, Forest and Climate Change (MoEFCC) implemented a number of rules as part of this aim, including the EPR, Battery Waste Management Rules, E-Waste Management Rules, and Waste Tyre Management Rules. In order to manage waste batteries, electronic waste, waste from solar panels, waste tyres, and waste plastic packaging, new or amended laws have been included, which form part of the EPR framework, which is an essential part of these regulations. The Oil Industry Safety Directorate (OISD), operating under MoPNG, has established comprehensive guidelines specifically targeting the oil and gas industry's hydrocarbon waste management (Oil Industry Safety Directorate 2024). Complementing this, the Hazardous Waste (Management, Handling and Transboundary Movement) Rules, 2008, amended in 2016, set forth detailed procedures encompassing the generation index, classification, consignment notes, transportation, and disposal of hazardous waste comprising hydrocarbon waste. The National Green Tribunal (NGT) has also played a significant role by issuing numerous orders and guidelines, notably directing the Central Ground Water Authority (CGWA) and the CPCB in 2015 to develop a national policy for the safe disposal of hydrocarbon waste. Additionally, the CPCB has provided guidelines for the treatment and discarding of generated water, a byproduct of oil and gas production laden with hydrocarbons and other contaminants. The Ministry of Environment, Forest and Climate Change (MoEFCC) contributes by issuing guidelines for managing industrial effluents covering hydrocarbon waste from various industries beyond the oil and gas sector (Policy on Waste Management 2023). Further reinforcing the regulatory landscape, the Ministry of Heavy Industries and Public Enterprises has established the National Oil Spill Disaster Response Centre (NOSDRC) to address oil spills and manage hydrocarbon waste during emergencies. Standards set by the Bureau of Indian Standards (BIS), such as IS: 14272:2010 and IS: 14273:2010, prescribe chemical and physical treatment methods for oil waste, respectively (Hazardous Waste Management Rules 2022). Another point to consider the measures, the Ministry of Petroleum and Natural Gas is actively developing a National Hydrocarbon Waste Management Policy aimed at creating a comprehensive framework for the effective management of hydrocarbon waste in the country.

Together, these initiatives and regulations form a robust system for managing hydrocarbon waste in India, reflecting a commitment to environmental safety and sustainability. Various organisations that govern the Oil & Natural Gas Industry in the country and its function are shown in Table 16.1.

16.7 International Agreements and Policies for Hydrocarbon Waste Management

International agreements and policies are key for the management of hydrocarbon waste and for promoting sustainable practices across the globe. Various frameworks have been implemented to address the environmental impacts of hydrocarbon extraction, production, and disposal. Sustainable hydrocarbon resource management is based on international agreements and initiatives that foster collaboration, establish standards, and coordinate solutions. Under the Vienna Convention, the objective of the Montreal Protocol is to phase out ozone-depleting substances, while the Basel Convention controls the transboundary movement of lethal waste. Other key international environmental conventions relevant to the petroleum industry include the Convention on Migratory Species, the Framework Convention on Climate Change, the Extractive Industries Transparency Initiative (EITI), the Biodiversity Convention, and the United Nations (UN) Law of the Sea. The International Convention for the Prevention of Pollution from Ships (MARPOL) and various regional sea conventions, such as the Barcelona Convention, Oslo Paris Commission (OSPAR), and the Kuwait Convention, play crucial roles in addressing marine pollution and environmental protection.

The Paris Agreement, established under the framework of the United Nations Framework Convention on Climate Change (UNFCCC), proposes to limit the global temperature increase to well below 2 °C above pre-industrial levels, with a particular emphasis on keeping it below 1.5 °C (The Paris Agreement 2015). The agreement commits the signatories to reducing the emission of greenhouse gases, including those from hydrocarbon sources, and enhancing their resilience to the impacts of climate change. The Paris Agreement and its long-term goals rely heavily on nationally determined contributions (NDCs) (Secretariat, 2020). NDCs reflect each country's efforts to reduce national emissions and adapt to the effects of climate change. when taken as a whole, these climate-related measures determine whether the world reaches the long-term objectives of the Paris Agreement, which include reaching the global peak of greenhouse gas (GHG) emissions as soon as feasible and then reducing them rapidly in accordance with the best available science to achieve a balance between anthropogenic emissions by sources and GHG removals by sinks in the second half of this century.

Table 16.1 Policies in India for hydrocarbon waste management

Various act	Its functions
The petroleum act of 1934	This act governs the importation of petroleum into India, as well as its circulation throughout the country, storing, production, refining, and blending, and it also contends with a significant number of midstream responsibilities
The petroleum and natural gas rules, 1959	These regulations establish a structure for granting exploration permits and mining leases. They, jointly with the Petroleum Act of 1934, govern the process of selling and distributing petroleum and petroleum products
The petroleum and natural gas regulatory board act, 2006	This legislation establishes the petroleum and natural gas regulatory board with the purpose of overseeing the refining, processing, storage, transportation, distribution, marketing, and sale of petroleum, petroleum products, and natural gas. It does not include the production of crude oil and natural gas
The oilfields (regulation and development) act, 1948	The act serves as the fundamental statute for the licensing and renting of petroleum and gas blocks by the Government of India. It grants the government extensive authority to establish rules for the regulation of oilfields and the development of mineral oil resources. This act also regulates the granting of production exploration licenses and mining leases
The mines act, 1952 (the mines act) and oil mines regulations, 2017	These regulations pertain to the health, safety, and well-being of workers in oil mines. This legislation also emphasises the responsibilities of owners, agents, and managers, as well as the consequences of rules of outbreaks of violence
The petroleum and natural gas (safety in offshore operations) rules, 2008 (the PNG safety rules)	The PNG safety rules provide the procedures for documenting information, the several forms of approval and notification needed for offshore facilities, the necessary precautions to ensure the health and safety of the public and the consequences for those who disobey these rules
Hydrogen exploration and licensing policy (HELP)	This policy seeks to boost distillate fuel oil and gas production by promoting discovery in sedimentary basins. It establishes various measures, such as a uniform license regime for both conventional and non-conventional hydrocarbons, an open acreage licensing policy, a revenue-sharing model, and flexibility in marketing and pricing (within certain limits)

The MARPOL Convention (International Convention for the Prevention of Pollution from Ships (MARPOL) 2023) was adopted on 2 November 1973 at the International Maritime Organization, the vital international convention covering the prevention of pollution of the marine environment by ships from operational or accidental discharge. It addresses oil pollution, including regulations on the discharge of oily waste and oil spill prevention measures.

The Extractive Industries Transparency Initiative (EITI) is a global program launched at the World Summit on Sustainable Development in Johannesburg in September 2002 (López-Cazar et al. 2021). The main objective is to "promote the open and accountable management of oil, gas and mineral resources". EITI promotes accountability, good governance, and transparency of public oil, gas, and mining revenue. EITI-implements countries progressively include taxes, levies, and other environmental payments in their EITI reports on the cause of the environmental loss by the extractive companies.

Air pollution emissions from refinery processes are also regulated. Figure 16.3 outlines the principal laws and regulations affecting the environmental impact of refineries in the United States (Khor et al. 2013).

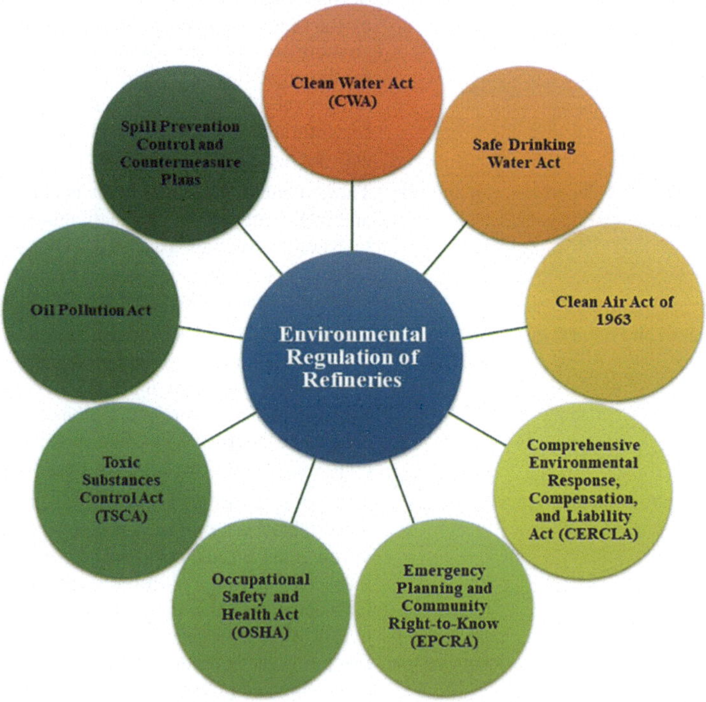

Fig. 16.3 U.S. federal and state legislation and regulations that affect the environmental effect of the refining industry

OGCI (Oil and Gas Climate Initiative 2024) is a CEO-led effort aimed at expediting action toward a net zero near future in accordance with the Paris Agreement. To attain its goals, OGCI's 12 members are aiming for net zero emissions in their particular operations while also working with firms and partners across industries and sectors to cut global GHG emissions urgently. The members commit to bringing down methane emissions, improving energy efficiency, and investing in low-carbon technologies to mitigate the environmental impacts of hydrocarbon production and use. The GGFR (Global Gas Flaring Reduction Partnership 2024), led by the World Bank, aims to reduce gas flaring in oil-producing countries. The GGFR collaboration works as a catalyst for reforming policies, facilitating stakeholder engagement, and implementing projects that reduce inefficient and undesired activities related to gas flaring and venting. It fosters collaboration among governments, oil companies, and other stakeholders; the initiative works to decrease greenhouse gas emissions and conserve valuable natural resources.

The United States Environmental Protection Agency's (EPA) New Source Performance Standards (NSPS) for the Oil and Gas Industry (2016) objective is to lessen volatile organic compound (VOC) and methane emissions from new and improved oil and gas facilities. It implements stricter controls on emissions from hydrocarbon production and waste management. The International Energy Agency (IEA) provides comprehensive policy advice and data to member countries to ensure reliable, affordable, and sustainable energy (International Energy Agency 2024). These policies support sustainable energy practices by advocating for the depletion of hydrocarbon emissions and the increased use of renewable energy sources. The IEA's initiatives include promoting energy efficiency, developing low-carbon technologies, and setting guidelines for the transition to sustainable energy systems. Through these efforts, the IEA aims to mitigate climate change, enhance energy security, and foster global collaboration in the energy sector.

The European Union's Circular Economy Action Plan (2020) aims to shift towards a circular economy, emphasising waste reduction and sustainable resource use (Circular Economy Action Plan 2024). It includes specific measures to manage hydrocarbon industry waste through recycling and reuse initiatives, Waste-to-Energy Projects, Extended Producer Responsibility (EPR), Hazardous Waste Management and Circular Supply Chains. By focusing on recycling, reuse, and sustainable practices, the plan aims to create a more resilient and environmentally friendly economy.

International agreements and recent policies play a vital role in managing hydrocarbon waste and promoting sustainable practices. Hence, by establishing regulations, encouraging transparency, and fostering international cooperation, these frameworks help mitigate the environmental impacts of hydrocarbon industries and advance global efforts toward sustainability.

16.8 Strategies for Hydrocarbon Waste Management Through Circular Economy

Hydrocarbon waste management is a critical issue globally, as the depletion of non-renewable energy sources and increasingly stringent environmental regulations underscore the need for sustainable waste disposal solutions. The circular economy is one strategy that is gaining popularity. It emphasises resource reuse, recycling, and repurposing in order to reduce waste and preserve natural resources.

Restoring natural systems, extending the life of products and resources, and removing trash and pollution are all part of the idea. Reducing dependency on virgin hydrocarbon resources is the goal of a circular economy, which maximises available resources through recycling and reuse as opposed to only relying on extraction. The ideas behind the circular economy encourage the continuous recycling and repurposing of products and byproducts derived from hydrocarbons rather than their disposal as waste, which closes material loops in the hydrocarbon value chain.

The goal of a circular economy is to maximise resource recovery, recycling, and reuse while removing waste. The idea comprises reducing pollution and waste, extending the life of products and resources, and repairing natural systems (Aithal and Aithal 2023). A circular economy aims to reduce reliance on virgin hydrocarbon resources by maximising existing resources via recycling and reuse rather than only depending on extraction. The concepts of circular economy promote the closure of material loops in the hydrocarbon value chain, which involves the ongoing recycling and repurposing of hydrocarbon-based goods and byproducts instead of disposing of them as waste (Mallick et al. 2023).

16.8.1 Benefits of the CE Model

- **Resource Efficiency**

The goal of a circular economy is to use resources as efficiently as possible. Materials are recycled and repurposed, and they have a long lifespan in product design. This preserves limited resources, including minerals, metals, and fossil fuels, by lowering the need for new raw materials.

- **Reduction in waste**

The circular economy seeks to drastically cut down on the quantity of debris that ends up in landfills and incinerators by promoting recycling, composting, and the construction of easily disassembled items. This reduces pollution to the environment and promotes sustainable waste management.

- **Reduced Carbon Emissions**

Energy-efficient procedures and renewable energy sources are common components of circular practices. The circular economy contributes to limiting emissions by decreasing the demand for new resource extraction and lowering energy usage through efficient manufacturing and recycling. This strategy works well to cut down on energy and resource usage since, according to research, over 80% of manufactured goods' total environmental effect is determined during the design stage (Diaz et al. 2021). Additionally, a major contribution to lowering trash output would come from the use of durable items that are upgradeable, repairable, and reusable. Right now, the packaging sector is expanding significantly. Taking on the issue of excessive packaging and improving its design to encourage recycling and reuse are the main objectives (Zhu et al., 2022).

- **Protects the environment**

The implementation of a Circular Economy (CE) can greatly decrease the use of natural resources, lessen the impact on landscapes and habitats, and support biodiversity preservation through practices such as product reuse and recycling. Additionally, adopting a CE structure can substantially reduce annual greenhouse gas emissions. To mitigate environmental effects, it is advised to focus on designing efficient and sustainable products from the outset.

- **Lessens Raw Material Dependence**

The global population is steadily increasing, which drives up the demand for raw materials. However, the supply of these essential raw materials is limited. This scarcity forces countries to depend on external resources for their raw material needs. Recycling raw materials mitigates risks related to supply issues, such as price volatility, scarcity, and reliance on imports. This is especially significant for critical raw materials needed to produce technologies like batteries and electric motors, which are essential for achieving climate-change targets (Malinauskaite et al. 2021).

- **Establishes employment and saves consumers money**

Adopting a CE model can boost competitiveness, stimulate innovation, promote economic growth, and create job opportunities. A press release from the European Commission (European Commission Adopts Circular Economy Action Plan 2020) indicates that implementing strong CE measures in Europe could increase the European Union's Gross Domestic Product by 0.5% by 2030 and create around 700,000 new jobs. Integrating CE principles into the design of materials and products can drive innovation across various economic sectors. Consumers will benefit from improved, durable products with innovative features, enhancing their quality of life and providing long-term cost savings.

In this context, hydrocarbon waste can be managed through circular economy strategies by recycling, repurposing, and reusing the materials in the waste stream. Recycling hydrocarbon waste involves techniques such as hydrotreating, distillation, and chemical and biological treatment to decompose waste components into reusable products (Das et al. 2024b). For instance, waste oil can be refined into diesel fuel or converted into chemicals or fuels for industrial applications, while natural gas can be processed for power generation or heating and cooling. Repurposing hydrocarbon waste is another sustainable strategy, where waste oil might be transformed into biofuels or chemicals for use in construction materials, plastics, and paints, and natural gas can be utilised to generate energy for homes or businesses or to develop new energy infrastructure. Additionally, reusing hydrocarbon waste offers effective management by repurposing waste gas for electricity generation, heating, and cooling, or for producing fertilisers and other chemicals used in agriculture, and waste oil for machinery lubrication or cosmetic products. Beyond these strategies, prevention and reduction of hydrocarbon waste are crucial and achievable through adopting more sustainable technologies and practices in the production, distribution, and consumption of hydrocarbon products. Energy efficiency measures can lower overall hydrocarbon demand, while technologies such as carbon capture and storage (CCS) can mitigate emissions from fossil fuel combustion. The circular economy approach thus presents a powerful framework for sustainable hydrocarbon waste management, reducing waste, conserving natural resources, and fostering sustainable development that benefits both the economy and the environment.

16.9 Conclusion

A strong regulatory framework and all-encompassing regulations are needed to achieve zero hydrocarbon waste management through a circular economy. A well-organized regulatory framework guarantees compliance, encourages innovation, and supports environmentally friendly business practices. Governments may pioneer the shift from a linear to a circular economy by enforcing resource efficiency standards, promoting recycling and reuse, and funding the advancement of sustainable technologies.

To create a regulatory environment that supports circularity, policies such as carbon pricing schemes, extended producer responsibility (EPR) laws, circular economy roadmaps, and favorable laws, regulations, and incentives should be put in place to encourage the use of circular economy methods in the hydrocarbon industry. distributing funds, grants, and subsidies to support workforce development, technology developments, and circular infrastructure investments. encouraging collaboration between governmental organizations, business leaders, and members of the public to plan and carry out circular economy initiatives.

Important policy initiatives should include financial incentives for companies to implement circular business models, stringent waste management rules, and investments in waste-to-energy and recycling infrastructure. Education initiatives and public awareness campaigns are also essential in encouraging consumers to adopt more environmentally friendly habits. The efficacy of these initiatives can be boosted by international cooperation and standardization, guaranteeing a unified and comprehensive strategy for the management of hydrocarbon waste. In eventuality, a circular economy model provides a workable way to get rid of hydrocarbon waste if it is backed by robust regulations and aggressive policies. This strategy not only builds social cohesion and fosters economic growth but also safeguards the environment, setting the stage for a sustainable future.

Acknowledgements The authors express their gratitude to the Department of Chemical Engineering, Indira Gandhi Institute of Technology, Sarang. The IDB expresses gratitude to the Department of Science and Technology for their support in the form of the Seed Fund Research Project (Project ID: *ST-BT-MISC-0005-2023-2427*/ST).

References

Aithal S, Aithal S (2023) Importance of circular economy for resource optimization in various industry sectors—a review-based opportunity analysis. Int J Appl Eng Manage Lett: 191–215. https://doi.org/10.47992/IJAEML.2581.7000.0182

Aljamali N, Najim M, Alabbasy A (2021) Treatment of oil spills and heavy metals in environment. 7:8–15. https://doi.org/10.37628/IJMA

Arutyunov V, Sedov I, Savchenko V, Nikitin A, Arutyunov A (2023) Utilization of waste hydrocarbon gases. Eurasian Chemico-Technol J 25:129–138. https://doi.org/10.18321/ectj1515

Awewomom J, Dzeble F, Takyi YD, Ashie WB, Ettey ENYO, Afua PE, Sackey LNA, Opoku F, Akoto O (2024) Addressing global environmental pollution using environmental control techniques: a focus on environmental policy and preventive environmental management. Discover Environ 2(1):8. https://doi.org/10.1007/s44274-024-00033-5

Bala K, Anshul, Verma S, Dogra A, Das AP, Kumar R, Nandini (2024) Sustainable management for urban plastic waste generation. In: Sustainable Management of Urban Plastic Waste Through Circular Economic Approaches, pp 211–236. https://doi.org/10.1201/9781003465447-13

Bala K, Dogra A, Lata P, Das AP (2024) Discussing sources and biological-chemical remediation approaches for microplastic pollution. In: Sustainable Microbial Technology for Synthetic and Cellulosic Microfiber Bioremediation, pp 37–59. https://doi.org/10.1007/978-3-031-62600-5_3

Bal M, Behera ID, Kumari U, Biswas S, Meikap BC (2020) Hydrodynamic study and particulate matter removal in a self priming venturi scrubber. Environ Technol Innov 20:101167. https://doi.org/10.1016/j.eti.2020.101167

Basel Convention—UN Environment Document Repository Home (2011) https://wedocs.unep.org/bitstream/handle/20.500.11822/8385/-BaselConventionontheControlofTransboundaryMovementsofHazardousWastes-20113644.pdf?sequence=2&isAllowed=y

Behera ID (2022) Biodegradation of petroleum sludge by indigenously developed bacterial consortium IIT Kharagpur

Behera ID, Basak G, Kumar RR, Sen R, Meikap BC (2020) Treatment of petroleum refinery sludge by petroleum degrading bacterium *Stenotrophomonas pavanii* IRB19 as an efficient novel technology. J Environ Sci Health Part A 56(2):226–239. https://doi.org/10.1080/10934529.2020.1866924

Behera ID, Das AP (2023) Impact of petroleum waste on environmental pollution and its sustainable management through circular economy. Springer

Behera ID, Meikap BC, Sen R (2022) Performance evaluation of bacterial consortium for biodegradation of total petroleum hydrocarbon: a comparative strategic biostimulation study. Res Square

Behera ID, Nayak M, Biswas S, Meikap BC, Sen R (2021) Enhanced biodegradation of total petroleum hydrocarbons by implementing a novel two-step bioaugmentation strategy using

indigenous bacterial consortium. J Environ Manage 292:112746. https://doi.org/10.1016/j.jenvman.2021.112746
Behera ID, Nayak M, Mishra A, Meikap BC, Sen R (2022) Strategic implementation of integrated bioaugmentation and biostimulation for efficient mitigation of petroleum hydrocarbon pollutants from terrestrial and aquatic environment. Mar Pollut Bull 177:113492. https://doi.org/10.1016/j.marpolbul.2022.113492
Biswas S, Diwakar RK, Behera ID, Meikap BC, Sen TK, Khiadani M (2020) Aqueous phase phenol removal from synthetic and real steel plant effluents through a batch and Semifluidized bed column operation: experimental and model analysis. J Environ Chem Eng 8(5):104441. https://doi.org/10.1016/j.jece.2020.104441
Central Pollution Control Board [CPCB] (2021). https://cpcb.nic.in/
Circular economy action plan (2024)
Crude Oil Processed by Refineries (2024) Petroleum planning & analysis cell, New Delhi. Retrieved from https://ppac.gov.in/governing-body
da Silva LJ, Alves FC, de França FP (2012) A review of the technological solutions for the treatment of oily sludges from petroleum refineries. Waste Manage Res 30(10):1016–1030
Das A, Dey S, Das AP (2024) Ecotoxicological impacts of synthetic microfiber pollutants and development of sustainable mitigation strategies. Environ Chem Ecotoxicol. https://doi.org/10.1016/J.ENCECO.2024.12.003
Das A, Dey S, Bala K, Gnanasekaran R, Das AP (2024) Microbial bioremediation technology for sustainable treatment and management of synthetic microfiber waste. In: Sustainable Microbial Technology for Synthetic and Cellulosic Microfiber Bioremediation, pp 221–238. https://doi.org/10.1007/978-3-031-62600-5_12
Das AP, Dutta K, Khatun R, Behera ID, Singh S, Mishra S (2023) Microfiber pollution and its microbial mitigation: a review on current trends and future prospects. J Taiwan Inst Chem Eng: 105104. https://doi.org/10.1016/j.jtice.2023.105104
Das AP, Behera ID, Das NP (2024a) Renewable energy generation and value addition from environmental microfiber pollution through advanced greener solution. Springer Nature
Das AP, Behera ID, Bhanja D (2024b) Sustainable microbial technology for synthetic and cellulosic microfiber bioremediation. Springer Nature. https://doi.org/10.1007/978-3-031-62600-5. Accessed 2024
Das AP, Nayak B, Behera ID (2024c) Applications of microorganisms in advanced geomicrobiological and biotechnological exploration. Geomicrobiol J 41(4):309–311. https://doi.org/10.1080/01490451.2024.2327059
Das A. Prasad, Ghosh, Shreya (2023) Manganese mining microorganisms
Das AP, Sukla LB, Pradhan N, Nayak S (2011) Manganese biomining: a review. Bioresour Technol 102(16):7381–7387. https://doi.org/10.1016/j.biortech.2011.05.018
Dey S, Bissoyi A, Koteswara Reddy G, Das AP (2024) Synthetic microfibers: an urban emerging pollutant demanding for sustainable environmental solutions. In: Sustainable Microbial Technology for Synthetic and Cellulosic Microfiber Bioremediation, pp 107–121. https://doi.org/10.1007/978-3-031-62600-5_7
Dey S, Das A, Mallick K, Sahu A, Das AP (2023) Environmental petroleum waste: pollution, toxicity, sustainable remediation. In: Behera ID, Das AP (eds) Impact of petroleum waste on environmental pollution and its sustainable management through circular economy. Springer Nature, Switzerland, pp 159–175. https://doi.org/10.1007/978-3-031-48220-5_7
Dey S, Dash N, Ray SK, Priyadarsini S, Chhotaray C, Das AP (2024) Occurrence and quantification of tire wear plastics pollutants from highway city dust of Bhubaneswar City of Odisha. In: Sustainable Microbial Technology for Synthetic and Cellulosic Microfiber Bioremediation, pp 77–89. https://doi.org/10.1007/978-3-031-62600-5_5
Dey S, Tripathy B, Kumar MS, Das AP (2023) Ecotoxicological consequences of manganese mining pollutants and their biological remediation. Environ Chem Ecotoxicol. https://doi.org/10.1016/J.ENCECO.2023.01.001

Diaz A, Schöggl JP, Reyes T, Baumgartner, RJ (2021) Sustainable product development in a circular economy: implications for products, actors, decision-making support and lifecycle information management. Sustain Prod Consum 26:1031–1045. https://doi.org/10.1016/j.spc.2020.12.044
Egazar'Yants S, Vinokurov V, Vutolkina A, Talanova MY, Frolov V, Karakhanov E (2015) Oil sludge treatment processes. Chem Technol Fuels Oils 51:506–515
European Commission Adopts Circular Economy Action Plan (2020). https://sdg.iisd.org/news/european-commission-adopts-circular-economy-action-plan/#:~:text=Applying%20circular%20economy%20measures%20in,including%20on%20their%20life%2Dspan
Farombi O, Klaus H (2024) Regulatory compliance and ethical considerations: compliance challenges and opportunities with the integration of big data and AI
Global Gas Flaring Reduction Partnership (2024) https://climateinitiativesplatform.org/index.php/Global_Gas_Flaring_Reduction_Partnership
Hazaimeh M, Ahmed E (2021) Bioremediation perspectives and progress in petroleum pollution in the marine environment: a review. Environ Sci Pollut Res: 28. https://doi.org/10.1007/s11356-021-15598-4
Hazardous Waste Management Rules (2022). https://cpcb.nic.in/rules/
Hossain ME, Al-Majed A, Adebayo AR, Apaleke AS, Rahman SM (2017) A critical review of drilling waste management towards sustainable solutions. Environ Eng Manage J (EEMJ) 16(7)
International Convention for the Prevention of Pollution from Ships (MARPOL) (2023). https://www.imo.org/en/about/Conventions/Pages/International-Convention-for-the-Prevention-of-Pollution-from-Ships-(MARPOL).aspx
International Energy Agency (2024). https://www.iea.org/
John A, Das AP (2024) Government strategies for achieving circular economy for plastic waste management in Southeast Asia. In: Sustainable Management of Urban Plastic Waste Through Circular Economic Approaches, pp 72–82. https://doi.org/10.1201/9781003465447-6/GOVERNMENT-STRATEGIES-ACHIEVING-CIRCULAR-ECONOMY-PLASTIC-WASTE-MANAGEMENT-SOUTHEAST-ASIA-AKBAR-JOHN-ALOK-PRASAD-DAS
Johnson OA, Affam AC (2019) Petroleum sludge treatment and disposal: a review. Environ Eng Res 24(2):191–201. https://doi.org/10.4491/eer.2018.134
Kanungo J, Sahoo T, Bal M, Behera ID (2024) Performance of bioremediation strategy in waste lubricating oil pollutants: a review. Geomicrobiol J 41(4):360–373. https://doi.org/10.1080/01490451.2023.2245395
Kanungo J, Sahoo T, Swain LP, Behera ID (2023) Toxicity of persistent hydrocarbon pollutants, sources and sustainable remediation process. In Behera ID, Das AP (eds) Impact of petroleum waste on environmental pollution and its sustainable management through circular economy. Springer Nature, Switzerland, pp 39–65. https://doi.org/10.1007/978-3-031-48220-5_2
Khatun R, Singh S, Dubey NK, Prasad Das A (2023) Chapter 22—a review on marine-based phytochemicals and their application in biomedical research. In: Pati S, Sarkar T, Lahiri D (eds) Recent frontiers of phytochemicals. Elsevier, pp 383–395. https://doi.org/10.1016/B978-0-443-19143-5.00013-X
Khor CS, Elkamel A, Riazi MR, Eser S, Agrawal SS, Peña Díez JL (2013) Environmental issues related to the petroleum refining industry. In: Petroleum refining and natural gas processing, vol MNL58-EB. ASTM International. https://doi.org/10.1520/mnl5820131214129
Kumar M, Bolan NS, Hoang SA, Sawarkar AD, Jasemizad T, Gao B, Keerthanan S, Padhye LP, Singh L, Kumar S, Vithanage M, Li Y, Zhang M, Kirkham MB, Vinu A, Rinklebe J (2021) Remediation of soils and sediments polluted with polycyclic aromatic hydrocarbons: to immobilize, mobilize, or degrade? J Hazardous Mater 420:126534. https://doi.org/10.1016/j.jhazmat.2021.126534
Kuppusamy S, Thavamani P, Venkateswarlu K, Lee YB, Naidu R, Megharaj M (2017) Remediation approaches for polycyclic aromatic hydrocarbons (PAHs) contaminated soils: technological constraints, emerging trends and future directions. Chemosphere 168:944–968. https://doi.org/10.1016/j.chemosphere.2016.10.115

Lahiri D, Nag M, Dey A, Sarkar T, Joshi S, Pandit S, Das AP, Pati S, Pattanaik S, Tilak VK, Ray RR (2022) Biofilm mediated degradation of petroleum products. Geomicrobiol J 39(3–5):389–398. https://doi.org/10.1080/01490451.2021.1968979

Li S (2024) Reviewing air pollutants generated during the pyrolysis of solid waste for biofuel and biochar production: toward cleaner production practices. Sustainability 16(3):1169. https://www.mdpi.com/2071-1050/16/3/1169

López-Cazar I, Papyrakis E, Pellegrini L (2021) The extractive industries transparency initiative (EITI) and corruption in Latin America: evidence from colombia, guatemala, honduras, peru, and trinidad and tobago. Resour Policy 70:101907. https://doi.org/10.1016/j.resourpol.2020.101907

Malinauskaite J, Anguilano L, Rivera XS (2021) Circular waste management of electric vehicle batteries: legal and technical perspectives from the EU and the UK post Brexit. Int J Thermofluids 10:100078. https://doi.org/10.1016/j.ijft.2021.100078

Mallick K, Sahu A, Dubey NK, Das AP (2023) Harvesting marine plastic pollutants-derived renewable energy: a comprehensive review on applied energy and sustainable approach. J Environ Manage 348:119371

Mallick K, Sahu S, Sahu A, Dey S, Das AP (2024) A critical review of marine microfiber pollution routes, toxicity, and its sustainable remediation. In: Das AP, Behera ID, Das NP (eds) Renewable energy generation and value addition from environmental microfiber pollution through advanced greener solution. Springer Nature, Switzerland, pp 189–211. https://doi.org/10.1007/978-3-031-51792-1_10

Manish S, Anamika S, Anjali Y, Varun R (2019) Source and control of hydrocarbon pollution. In: Muharrem I, Olcay Kaplan I (eds) Hydrocarbon pollution and its effect on the environment, Chap. 2. IntechOpen. https://doi.org/10.5772/intechopen.86487

Ministry of Petroleum and Natural Gas, Government of India (2020) https://mopng.gov.in/files/TableManagements/Indian-Petroleum--Natural-Gas_2020-21.pdf

Mishra A, Siddiqi H, Kumari U, Behera ID, Mukherjee S, Meikap BC (2021) Pyrolysis of waste lubricating oil/waste motor oil to generate high-grade fuel oil: a comprehensive review. Renew Sustain Energy Rev 150:111446. https://doi.org/10.1016/j.rser.2021.111446

Mishra S, Das AP (2022) Treatment of the wastewater polluted with synthetic microfiber released from washing machine. In: Das BB, Hettiarachchi H, Sahu PK, Nanda S (eds) Recent developments in sustainable infrastructure (ICRDSI-2020)—GEO-TRA-ENV-WRM: conference proceedings from ICRDSI-2020, vol 2. Springer, Singapore, pp 109–117. https://doi.org/10.1007/978-981-16-7509-6_9

Mishra S, Dash D, Das AP (2024) Microbial degradation of polyester microfibers using indigenously isolated bacterial strain Exiguobacterium Sp. CLEAN – Soil, Air, Water, 52(12):e202300343. https://doi.org/10.1002/CLEN.202300343

Mishra S, Ghosh S, van Hullebusch ED, Singh S, Das AP (2023) A critical review on the recovery of base and critical elements from electronic waste-contaminated streams using microbial biotechnology. Appl Biochem Biotechnol 195(12):7859–7888. https://doi.org/10.1007/s12010-023-04440-x

Mishra S, Singh RP, Rath CC, Das AP (2020) Synthetic microfibers: source, transport and their remediation. J Water Process Eng 38:101612. https://doi.org/10.1016/j.jwpe.2020.101612

Mishra S, Singh RP, Rout PK, Das AP (2022) Chapter 3—membrane bioreactor (MBR) as an advanced wastewater treatment technology for removal of synthetic microplastics. In: Shah M, Rodriguez-Couto S, Biswas J (eds) Development in wastewater treatment research and processes. Elsevier, pp 45–60. https://doi.org/10.1016/B978-0-323-85583-9.00022-3

Mitra S, Chakraborty AJ, Tareq AM, Emran TB, Nainu F, Khusro A, Idris AM, Khandaker MU, Osman H, Alhumaydhi FA, Simal-Gandara J (2022) Impact of heavy metals on the environment and human health: novel therapeutic insights to counter the toxicity. J King Saud Univ Sci 34(3):101865. https://doi.org/10.1016/j.jksus.2022.101865

Mohammadi L, Rahdar A, Bazrafshan E, Dahmardeh H, Susan MABH, Kyzas GZ (2020) Petroleum hydrocarbon removal from wastewaters: a review. Processes 8(4):447. https://www.mdpi.com/2227-9717/8/4/447

Mohanta S, Sahoo B, Behera ID, Pradhan S (2016) Effect of crushing on near-gravity material distribution in different size fractions of an Indian non-coking coal. J Southern African Inst Min Metall 116(2):209–213. https://doi.org/10.17159/2411-9717/2016/v116n2a12

Mohanta S, Pradhan B, Behera ID (2024) Impact and remediation of petroleum hydrocarbon pollutants on agricultural land: a review. Geomicrobiol J 41(4):345–359. https://doi.org/10.1080/01490451.2023.2243925

Mohanty M, Tripathy B, Koteswara Reddy G, Das AP (2024) Greener technologies for establishing restraint over microplastic pollution. In: Sustainable Microbial Technology for Synthetic and Cellulosic Microfiber Bioremediation, pp 239–258. https://doi.org/10.1007/978-3-031-62600-5_13

Mong'ina MK (2018) Waste oil management by petroleum business operators an analysis of the legal and institutional framework. University of Nairobi. http://erepository.uonbi.ac.ke/bitstream/handle/11295/105145/Mong%E2%80%99ina_Waste%20Oil%20Management%20By%20Petroleum%20Business%20Operators.pdf?sequence=1&isAllowed=y

Oil and Gas Climate Initiative (2024). https://www.ogci.com/

Oil Industry Safety Directorate (2018). https://www.oisd.gov.in/

Oil Industry Safety Directorate (2024). https://www.oisd.gov.in/

Olabi AG, Abdelkareem MA, Wilberforce T, Alkhalidi A, Salameh T, Abo-Khalil AG, Hassan MM, Sayed ET (2022) Battery electric vehicles: progress, power electronic converters, strength (S), weakness (W), opportunity (O), and threats (T). Int J Thermofluids 16:100212. https://doi.org/10.1016/j.ijft.2022.100212

Paul A, Priyadarsini S, Karakuş S, Behera JK, Gnanasekaran R, Das AP (2024) Fate and accumulation of synthetic microfibers in the environment. In: Sustainable Microbial Technology for Synthetic and Cellulosic Microfiber Bioremediation, pp 61–76. https://doi.org/10.1007/978-3-031-62600-5_4

Policy on Waste Management (2023). https://www.mospi.gov.in/sites/default/files/main_menu/Seminar/Policy%20on%20Waste%20Management%20-%20MOEFCC.pdf

Prabhakar A, Mishra S, Das AP (2019) Isolation and identification of lead (Pb) solubilizing bacteria from automobile waste and its potential for recovery of lead from end of life waste batteries. Geomicrobiol J 36(10):894–903. https://doi.org/10.1080/01490451.2019.1654044

Pradhan B, Behera ID, Sahoo K, Mohanta S (2023) Applicability of the teetered bed separator for beneficiating indian iron ore fines: an experimental study. J Mines Metals Fuels 71(11):1817–1825. https://doi.org/10.18311/jmmf/2023/31100

Priyadarsini S, Das AP (2024) Analytical and structural characterization of waste lithium-ion batteries for their effective recycling strategy. Environ Chem Ecotoxicol. https://doi.org/10.1016/j.enceco.2024.12.004

Priyadarsini S, Das AP (2025) Lithium bioleaching: A review on microbial-assisted sustainable technology for lithium bio-circularity. J Water Process Eng, 69:106744. https://doi.org/10.1016/j.jwpe.2024.106744

Raj K, Das AP (2023) Lead pollution: impact on environment and human health and approach for a sustainable solution. Environ Chem Ecotoxicol 5:79–85. https://doi.org/10.1016/j.enceco.2023.02.001

Ray SK, Dey S, Das AP, John A (2024) microplastic waste management through greener technologies: A review. In: Sustainable Management of Urban Plastic Waste Through Circular Economic Approaches, pp 237–261. https://doi.org/10.1201/9781003465447-14 RENEWABLE ENERGY GENERATION AND VALUE ADDITION FROM ENVIRONMENTAL. (2024)

Rosner D, Markowitz G (2013) Persistent pollutants: a brief history of the discovery of the widespread toxicity of chlorinated hydrocarbons. Environ Res 120:126–133. https://doi.org/10.1016/j.envres.2012.08.011

Sadasivuni KK, Rattan S, Waseem S, Brahme SK, Kondawar SB, Ghosh S, Das AP, Chakraborty PK, Adhikari J, Saha P, Mazumdar P (2019) Silver nanoparticles and its polymer nanocomposites—synthesis, optimization, biomedical usage, and its various applications. In: Sadasivuni KK, Ponnamma D, Rajan M, Ahmed B, Al-Maadeed MASA (eds) Polymer nanocomposites in biomedical engineering. Springer International Publishing, pp 331–373. https://doi.org/10.1007/978-3-030-04741-2_11

Sahu A, Mallick K, Das AP (2024) Environmental impact of single-use synthetic face mask and its recycling: a sustainable approach. In: Das AP, Mishra S (eds) Impact of COVID-19 waste on environmental pollution and its sustainable management: COVID-19 waste and its management. Springer Nature, Switzerland, pp 197–212. https://doi.org/10.1007/978-3-031-50840-0_11

Sattar S, Hussain R, Shah SM, Bibi S, Ahmad SR, Shahzad A, Zamir A, Rauf Z, Noshad A, Ahmad L (2022) Composition, impacts, and removal of liquid petroleum waste through bioremediation as an alternative clean-up technology: a review. Heliyon 8(10):e11101. https://doi.org/10.1016/j.heliyon.2022.e11101

Shahbaz M, Rashid N, Saleem J, Mackey H, McKay G, Al-Ansari T (2023) A review of waste management approaches to maximise sustainable value of waste from the oil and gas industry and potential for the State of Qatar. Fuel 332:126220. https://doi.org/10.1016/j.fuel.2022.126220

Snežana M, Božo D, Srđan R (2013) Petroleum hydrocarbon biodegradability in soil—implications for bioremediation. In: Vladimir K, Anton K (eds) Hydrocarbon, Chap. 3. IntechOpen. https://doi.org/10.5772/50108

Synthetic Fuels: Future Transport Fuels (2024) Policy on synthetic fuels, issue. https://mopng.gov.in/files/article/articlefiles/Policy-on-Synthetic-Fuels-Committiee-Report-March-2024.pdf

The Paris Agreement (2015). https://unfccc.int/process-and-meetings/the-paris-agreement

The Paris Agreement and NDCs (2020). https://unfccc.int/process-and-meetings/the-paris-agreement/nationally-determined-contributions-ndcs#Communications-received-from-Parties-in-relation-to-other-Parties-NDCs

Transformative technologies (AI) challenges and principles of regulation (2024). https://digitalregulation.org/3004297-2/#post-3004929-endnote-3

Teng Q, Zhang D, Yang C (2021) A review of the application of different treatment processes for oily sludge. Environ Sci Pollut Res 28:121–132

Tormoehlen LM, Tekulve KJ, Nañagas KA (2014) Hydrocarbon toxicity: a review. Clin Toxicol (Phila) 52(5):479–489. https://doi.org/10.3109/15563650.2014.923904

Tripathy B, Dash A, Das AP (2022) Detection of environmental microfiber pollutants through vibrational spectroscopic techniques: Recent advances of environmental monitoring and future prospects. Crit Rev Anal Chem, 1–11. https://doi.org/10.1080/10408347.2022.2144994

Tripathy B, Sahoo PP, Sundaray H, Das AP (2024) In-vitro biodegradation of discarded marine microplastics across the eastern coast of the Bay of Bengal, India using Exiguobacterium sp. Environ Chem Ecotoxicol 6:236–247. https://doi.org/10.1016/j.enceco.2024.07.003

Varjani SJ, Upasani VN (2016) Biodegradation of petroleum hydrocarbons by oleophilic strain of *Pseudomonas aeruginosa* NCIM 5514. Bioresour Technol 222:195–201. https://doi.org/10.1016/j.biortech.2016.10.006

Varjani SJ, Upasani VN (2017) A new look on factors affecting microbial degradation of petroleum hydrocarbon pollutants. Int Biodeterior Biodegradation 120:71–83. https://doi.org/10.1016/j.ibiod.2017.02.006

Zhu Z, Liu W, Ye S, Batista L (2022) Packaging design for the circular economy: a systematic review. Sustain Prod Consum 32:817–832. https://doi.org/10.1016/j.spc.2022.06.005

Chapter 17
Insights to Visual and Analytical Techniques for Identification and Characterization of Marine Micro Plastic as a Major Source of Petroleum Hydrocarbon Pollutants

Gao Pengqi, Akbar John, Demus Matheus Huang, and Alok Prasad Das

Abstract Marine microplastic pollution has become a major global environmental issue that is detrimental to the worldwide environment and ecology in the long term. Identifying and characterizing these microplastic pollutants are thus essential to fathom their potential environmental impacts, to determine their sources and pathways, and to devise effective governance measures. This chapter provides a comprehensive review on the current visual and analytical techniques that are popularly utilized in the identification and characterization of marine microplastics (i.e., optical and electron microscopy, FTIR, Raman spectroscopy, NMR, Py-GC-MS, DSC, and TGA). Several emerging microplastic identification and characterization technologies (i.e., hyperspectral imaging or HSI, artificial intelligence and machine learning, microfluidic devices) are also elucidated. Furthermore, the advantages and limitations of the techniques are highlighted to provide some insights on improving their efficiency in microplastics analysis in the future. In-depth research that explores other novel approaches which can be feasibly integrated into the current microplastic analytical technologies would be crucial as solution to address their limitations and to enhance their efficiency in microplastic analysis in the future.

Keywords Marine microplastic pollutants · Visual techniques · Analytical techniques · Identification · Characterization · Novel techniques

---

G. Pengqi · D. M. Huang
Environmental Technology Division, School of Industrial Technology, Universiti Sains Malaysia, 11800 Pulau Pinang, Malaysia

A. John (✉)
College of Marine Sciences and Aquatic Biology, University of Khorfakkan, Sharjah, United Arab Emirates
e-mail: akbarjohn50@gmail.com; Akbar.John@ukf.ac.ae

A. P. Das
Department of Life Sciences, Rama Devi Women's University, Bhubaneswar, Odisha, India

I. D. Behera and A. P. Das (eds.), *Environmental Hydrocarbon Pollution and Zero Waste Approach Towards a Sustainable Waste Management*, Environmental Science and Engineering, https://doi.org/10.1007/978-3-031-81615-4_17

17.1 Introduction

Marine microplastic pollution has emerged as a significant environmental issue, attracting attention from scientists, policymakers, and the public (Mishra et al. 2020; Singh et al. 2020). Microplastics, typically 0.05–5 mm in size, are prevalent in the marine environment due to their widespread use and persistence (Sarma et al. 2022; Das et al. 2023a). These pollutants mainly originate from various sources such as fragmentation of larger plastic debris, industrial processes, and consumer products (i.e., cosmetics and textiles) (Galloway et al. 2017; Mohanty et al. 2024). Additionally, the release of microplastics is often associated with industrial discharges, including those from the petroleum hydrocarbon and fly ash industries, which contribute to the contamination of marine ecosystems (Biswas et al. 2020; Bal et al. 2020; Behera et al. 2022a; Kanungo et al. 2023; Das et al. 2023b, 2024a; Pradhan et al. 2023). Petroleum hydrocarbon industries not only release microplastics but also other hazardous pollutants that can adsorb onto microplastic particles, further exacerbating environmental harm. These tiny particles have been detected globally, from surface waters to the deep sea and even in remote polar regions (Lusher et al. 2017; Mishra et al. 2021a, b, c; Mishra and Das 2023).

Identifying and characterizing microplastics is critical to understand their potential environmental and ecological impacts (Mishra et al. 2019, 2023). Accurate identification can provide feasible determination of the sources and pathways of microplastic pollution that is essential for developing effective mitigation strategies (Hidalgo-Ruz et al. 2012; Mallick et al. 2023). Besides, the characterization of microplastics by their size, shape, chemical composition, and surface properties is crucial as these factors influence their behavior, transport, and interaction with marine organisms (Hartmann et al. 2019; Sahoo et al. 2022) (Dey, Dash et al. 2024; Dey et al. 2023; John & Das 2024). Different polymers may have distinctive toxicological effects on marine life beings, which may further affect their feeding behavior, reproduction, and survival rates (Tripathy et al. 2024; Behera et al. 2020; Das et al. 2023b; Rochman et al. 2013). Additionally, understanding the characteristics of microplastics provides feasible assessment of their potential as vectors for harmful chemicals and pathogens that may pose detrimental risks to marine ecosystems and human health for long-term period (Ghosh and Das 2020; Koelmans et al. 2014). In the context of petroleum hydrocarbon pollutants, microbial degradation plays a vital role in breaking down these contaminants in marine environments. Bioaugmentation and biostimulation techniques improve natural degradation processes by adding certain microbial strains or providing necessary nutrients to increase microbial activity (Behera et al. 2020, 2021; Behera and Das 2023; Kanungo et al. 2024). These approaches are especially successful in dealing with oil spills, because hydrocarbon breakdown must occur quickly and efficiently in order to limit environmental harm. The use of microbial degrading techniques in conjunction with established cleanup procedures can considerably improve the remediation of petroleum hydrocarbon contaminants in marine ecosystems (Mohanta et al. 2024; Behera et al. 2022b; Das et al. 2024b, 2024c) (A. Das, Dey, & Das 2024; A. Das, Dey, Bala et al. 2024; Dey, Bissoyi et al. 2024).

This chapter provides a comprehensive review on the visual and analytical techniques used in the identification and characterization of marine microplastic contaminants. Firstly, we discuss current methodologies in visual identification, including optical and electron microscopy. Then, we explore several analytical techniques such as Fourier-transform infrared spectroscopy (FTIR), Raman spectroscopy, pyrolysis-gas chromatography-mass spectrometry (Py-GC-MS) that are mainly used to characterize the chemical composition and physical properties of microplastics. Eventually, the advantages and limitations of these techniques in marine environmental studies are evaluated. By examining the recent advancements and applications of the techniques as mentioned earlier, their critical roles could be elucidated to advance our understanding of marine microplastic pollution and supporting effective management strategies.

17.2 Visual Identification Methods

Visual identification methods, primarily using optical and electron microscopy, are fundamental physical analysis techniques in microplastic research (Song et al. 2015). Optical microscopy offers a rapid initial assessment of microplastic size, shape, and color, while electron microscopy (SEM and TEM) provides high-resolution imaging of surface morphology and internal structures, respectively. Combining these techniques provides a comprehensive understanding of microplastic physical properties, crucial for tracing their sources, fate, and potential impacts on marine ecosystems.

17.3 Optical Microscopy

Optical microscopy, a fundamental and widely accessible technique, is primarily used for the visual identification of microplastics larger than 20 μm. It employs visible light and lenses to magnify particles, allowing researchers to categorize them based on morphology (Murphy and Davidson 2012; Mallick et al. 2023). The process involves collecting samples, filtering to concentrate microplastics, and examining them under the microscope at magnifications ranging from 40× to 400× (Maes et al. 2017; Shim et al. 2017).

Optical microscopy has been effectively used to identify and quantify microplastics in various marine environments. This technique offers a simple, cost-effective, and rapid initial assessment of microplastic presence, making it ideal for large-scale surveys (Song et al. 2015; Tripathy et al. 2022). For instance, Lusher et al. (2016) used it to investigate microplastics in fish from the North Atlantic, finding 73% of examined fish contained microplastics, mainly fibers and fragments, highlighting their widespread distribution and potential impacts on marine organisms (Lusher et al. 2016). Similarly, Wang et al. (2017) (Mishra et al. 2022c) employed optical microscopy alongside SEM/EDS to analyze microplastics from ocean trawl samples and fish guts, effectively distinguishing plastics from other materials like fats and mineral particles by their texture and color (Bala, Anshul et al. 2024; Bala, Dogra et al. 2024). This pre-screening significantly reduced the number of particles needing advanced analysis. This study highlighted the importance of using optical microscopy as a preliminary step in the analysis of microplastics, enabling rapid and efficient screening before more detailed spectroscopic analysis (Mishra et al. 2021c; Wang et al. 2017).

Nevertheless, this technique possesses several limitations. Firstly, its resolution is limited to particles larger than 20 μm, making it unsuitable for nanoplastics. Additionally, visual identification can be subjective, necessitating the use of staining techniques for improved accuracy (Mishra et al. 2022a; Norén 2007). In summary, optical microscopy serves as an essential first step in visualizing microplastics, providing valuable insights into their abundance, size, shape, and color in marine environments (Mishra et al. 2024; Mohanty et al. 2024; Paul et al. 2024; Priyadarsini & Das 2024, 2025; Ray et al. 2024; Tripathy et al. 2022). While it has certain limitations, its role in the broader context of microplastic research remains indispensable.

Electron Microscopy

Electron microscopy provides higher resolution imaging than optical microscopy, making it suitable for detailed analysis of microplastic particles. This technique uses a beam of electrons to achieve magnifications far beyond optical microscopy, enabling visualization of fine structural details and comprehensive morphological and compositional data.

In general, there are two main types of electron microscopy used in microplastic analysis: scanning electron microscopy (SEM) and transmission electron microscopy (TEM), as shown in Fig. 17.1. SEM applies surface scanning of a sample with a focused electron beam that produces secondary electrons, backscattered electrons, and characteristic X-rays to create detailed images. Samples must be coated with a thin conductive layer, such as gold or platinum, to prevent charging, enhance resolution, and contrast (Duis and Coors 2016; Miserli et al. 2023). Whereas TEM utilizes transmitted electrons to generate high-resolution images of internal structures, requiring ultra-thin samples, typically less than 100 nm, prepared through ultramicrotomy (Hower et al. 2022).

A study by Miserli et al. (2023) employed SEM combined with energy-dispersive X-ray Spectroscopy (EDS) to analyze microplastics in aquaculture systems, specifically in fish, mussels, and water samples. SEM–EDS provided high-resolution images that revealed various morphological features such as cracks, cavities, and surface roughness, demonstrating its efficacy in identifying and characterizing microplastics in aquaculture systems (Miserli et al. 2023). Similarly, Caldwell et al. (2023) utilized a correlative workflow involving light microscopy, TEM, and Raman spectroscopy to detect and observe microplastic interactions with jellyfish (*Cassiopea andromeda*). TEM validated the presence and location of microplastics within the jellyfish tissues, highlighting TEM's utility in providing high-resolution images that confirm the ingestion and localization of microplastics in marine organisms, offering insights into their interactions and potential impacts (Caldwell et al. 2023).

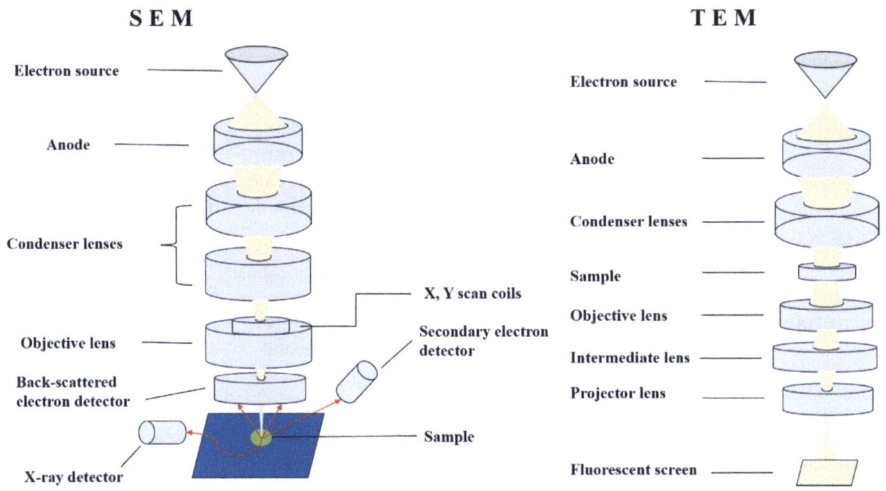

Fig. 17.1 Comparison of SEM and TEM working principles (Kannan 2018; Mohammed and Abdullah 2018)

Both SEM and TEM are powerful techniques for analyzing microplastics, each with distinct advantages. SEM excels at imaging surface morphology for particles ranging from 0.1 μm to several millimeters, revealing features like cracks and textures indicative of environmental degradation. Combined with Energy-Dispersive X-ray Spectroscopy (EDS), SEM enables rapid screening and elemental analysis, distinguishing plastics from other materials. However, SEM has some disadvantages. Sample preparation is complex, and it requires operation in a high-vacuum environment, which may alter the sample's morphology. Additionally, it is expensive to operate and maintain, and it can only provide information about surface morphology, not the chemical composition of the sample directly (Wang et al. 2017). In contrast, TEM is ideal for examining internal structures and crystallography at the nanoscale, typically for microplastics smaller than 100 nm as it offers higher resolution down to the atomic level. This feature allows TEM to effectively identify nanoplastics and observe internal features. However, TEM has drawbacks, including complex and limited sample preparation, expensive equipment and high operating and maintenance costs, the need for an ultra-high vacuum environment, and the ability to analyze only local areas (Caldwell et al. 2023). Although so, the combination of both SEM and TEM in microplastic analysis is considered highly effective. While SEM is practical for rapid surface and elemental analysis of larger microplastics, TEM provides detailed insights into the internal structures of smaller particles, making both techniques complementary in marine microplastic research (Table 17.1).

Table 17.1 Comparison of visual identification methods for marine microplastic analysis

Technique	Applicable size range (μm)	Advantages	Limitations
Optical microscopy	> 20	Accessible, widely available, low cost, rapid screening, can analyze large sample volumes	Limited resolution, subjective identification, cannot determine chemical composition, requires trained personnel
Scanning electron microscopy (SEM)	0.1 to several millimeters	High resolution, surface morphology imaging, elemental analysis (with EDS)	Requires sample preparation, expensive, vacuum environment, limited to surface analysis
Transmission electron microscopy (TEM)	< 100 nm (typically)	Highest resolution, internal structure imaging, crystallography	Complex sample preparation, time-consuming, expensive, limited sample volume

17.4 Analytical Identification and Characterization Techniques

In addition to visual identification, chemical characterization is essential to understanding the composition of microplastics, environmental impact, and sources and pathways. Analytical techniques provide detailed information about their chemical structure and composition, which helps to trace the source and assess toxicity. Chemical characterization mainly includes spectroscopic methods (i.e., Fourier-transform infrared spectroscopy (FTIR), Raman spectroscopy, nuclear magnetic resonance (NMR) spectroscopy, and pyrolysis-gas chromatography-mass spectrometry (Py-GC-MS)) along with thermal analysis techniques (i.e., differential scanning calorimetry (DSC) and thermogravimetric analysis (TGA)).

17.4.1 Spectroscopic Techniques

17.4.1.1 Fourier-Transform Infrared Spectroscopy (FTIR)

Fourier-transform infrared spectroscopy (FTIR) is a widely used technique for identifying the chemical composition of microplastics, typically ranging from 1 to 1000 μm, and down to the nanoscale with specialized methods. FTIR measures the absorption of infrared radiation, causing molecular vibrations. Each chemical bond absorbs infrared radiation at specific wavelengths, producing a characteristic spectrum that is compared to reference spectra for identification. This method is

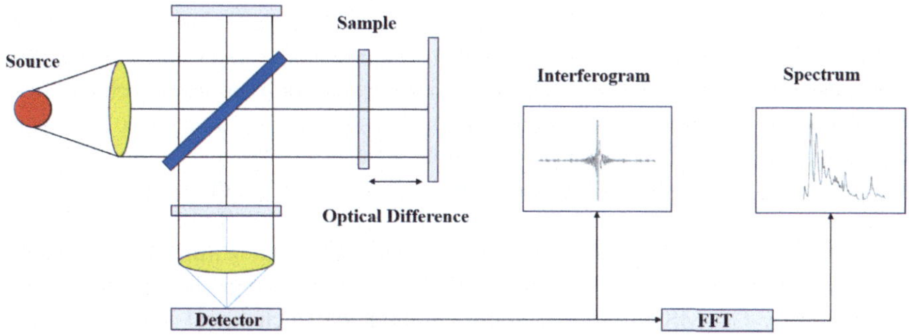

Fig. 17.2 Schematic diagram of the working principle of analyzing marine microplastics using FTIR (Xu et al. 2020)

particularly useful for detecting functional groups and identifying different plastic polymers and their additives (Harrison et al. 2012).

The general mechanism of FTIR is illustrated as in Fig. 17.2. The FTIR instrument includes an infrared light source, a beam splitter, a sample holder, and a detector. Infrared light is split into two beams: one passes through the sample, and the other serves as a reference. The beams are recombined and passed through the detector, which measures transmitted light intensity at different wavelengths. The data is processed using a Fourier transform algorithm to generate an absorbance spectrum, which is analyzed to identify sample components (Veerasingam et al. 2021).

FTIR has been extensively used in the analysis of microplastics from various environmental samples, including water, sediment, and biota. employed FTIR in their study of microplastics extracted from marine sediments. The researchers identified polymer types, with the majority being polyethylene and polypropylene. This study highlighted the utility of FTIR in environmental monitoring and the identification of plastic pollutants (Van Cauwenberghe et al. 2013). A study by Primpke et al. (2017) utilized FTIR spectroscopy to analyze microplastics in various environmental samples, including water, sediment, and biota. The researchers used an automated chemical imaging method, combining FTIR with focal plane array (FPA) detectors, to achieve high-throughput and high-resolution analysis. This approach allowed for the rapid identification and quantification of microplastics, providing detailed insights into their distribution and composition (Primpke et al. 2017).

FTIR offers significant advantages for microplastic analysis due to its non-destructive nature, preserving samples for further analyses. It provides rapid, accurate identification of polymer types by matching molecular fingerprints with reference databases. With focal plane array (FPA) detectors, FTIR imaging achieves high spatial resolution, enabling detailed mapping of microplastics in complex environmental matrices such as sediments, biota, and water samples. This versatile method can analyze both large and small particles, identifying a wide range of polymers, including PE, PP, PS, PET, and PVC, along with additives and degradation products.

Nevertheless, this instrument requires pure samples for microplastic analysis. Potential contaminants in the samples such as organic matter and salts may interfere with spectrum reading. Hence, effective sample preparation methods such as filtration, density separation, and enzymatic digestion should be conducted to eliminate those interfering contaminants and to promote accuracy of the measurement. While the application of traditional FTIR in measuring particles smaller than 20 μm is a challenging approach, microscopic FTIR can overcome this limitation. Additionally, FTIR mainly provides functional group information and is limited in analyzing complex polymer mixtures. Combining FTIR with techniques like Raman spectroscopy or Py-GC-MS allows for a more comprehensive chemical characterization of microplastics (Xu et al. 2019).

17.4.1.2 Raman Spectroscopy

Raman spectroscopy is a powerful technique for characterizing the chemical composition of microplastics, from a few micrometers to the nanoscale. Unlike FTIR, which measures infrared absorption, Raman spectroscopy measures light scattering by molecular vibrations, providing detailed molecular structure information. Raman microscopy, which combines Raman spectroscopy with optical microscopy, is essential for analyzing very small plastic particles (< 20 μm) that are challenging to detect with other methods (Araujo et al. 2018).

The general mechanism of Raman spectroscopy is illustrated as in Fig. 17.3. A typical Raman spectroscopy system includes a laser light source, a sample stage, a light collection system, a monochromator or spectrometer, a photomultiplier tube (PMT) or CCD detector, and a data acquisition and processing system. This instrument operates based on the concept of Raman effect, where monochromatic light from a laser is directed onto a sample that further interacts with molecular vibrations, scattering in various directions. Most scattered light remains at the incident wavelength (Rayleigh scattering), but a small fraction shifts wavelengths (Raman scattering) due to energy exchanges with vibrational modes. This shift produces a unique spectrum, serving as a molecular fingerprint to identify chemical bonds and molecular structure (Smith and Dent 2004).

Raman spectroscopy has been extensively applied to analyze microplastics as well as organic and inorganic particles in various environmental samples, including water, sediment, and biota (Ribeiro-Claro et al. 2017). The high feasibility of this powerful tool in analyzing these samples could be attributed to its non-destructive nature, minimal sample requirement, high-throughput capability, and rapid analysis (Primpke et al. 2017). Furthermore, it is particularly effective for identifying small particles and fibers. Ghosal et al. (2012) used Raman micro-spectroscopy to identify polymer residues and persistent organic pollutants (POPs) in samples from the Pacific Ocean trawl and fish stomachs, distinguishing polymer debris from natural particles (Picó and Barceló 2020). Donato et al. (2024) employed handheld Raman spectroscopy to monitor plastic pellet pollution along the Mediterranean coasts in

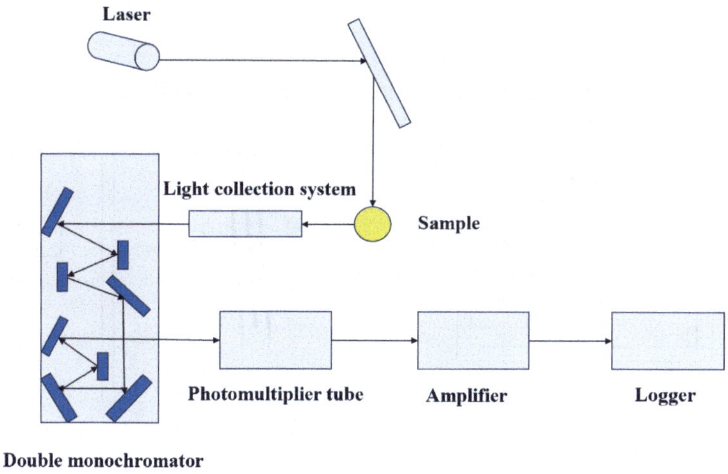

Fig. 17.3 Schematic diagram of the working principle of analyzing marine microplastics using Raman spectroscopy (Oßmann 2020)

Southern Italy, providing a rapid, effective means of identifying common polymers like PE, PP, and PS directly in the field (Donato et al. 2024).

Nevertheless, there are several limitations of Raman spectroscopy. Firstly, the operation of this instrument is susceptible to fluorescence interference, especially from organic or biological contaminants. Modern techniques such as near-infrared excitation, time-resolved, and frequency-shift excitation Raman spectroscopy help mitigate this issue. Despite these challenges, Raman spectroscopy remains essential for detailed chemical characterization of microplastics (Shim et al. 2017).

17.4.1.3 Nuclear Magnetic Resonance (NMR) Spectroscopy

Nuclear magnetic resonance (NMR) spectroscopy is a powerful technique that utilizes the magnetic properties of atomic nuclei, such as hydrogen (^{1}H) and carbon-13 (^{13}C), to elucidate the molecular structure and composition of samples. These nuclei possess a property called spin, which gives them a magnetic moment. When placed in a strong magnetic field, the nuclei align their spins, resulting in distinct energy levels. By applying radiofrequency pulses, NMR induces transitions between these energy levels, with the absorbed or emitted energy being characteristic of the surrounding chemical environment (Günther 2013), NMR is particularly effective for analyzing microplastics within the size range of 100 nm to 10 μm, as these particles can often be dissolved for solution-state NMR analysis (Ivleva 2021).

The general mechanism of an NMR spectrometer is illustrated as in Fig. 17.4. The basic components of NMR spectrometer include a powerful magnet, a radiofrequency transmitter, a probe containing the sample holder and radiofrequency coils, a

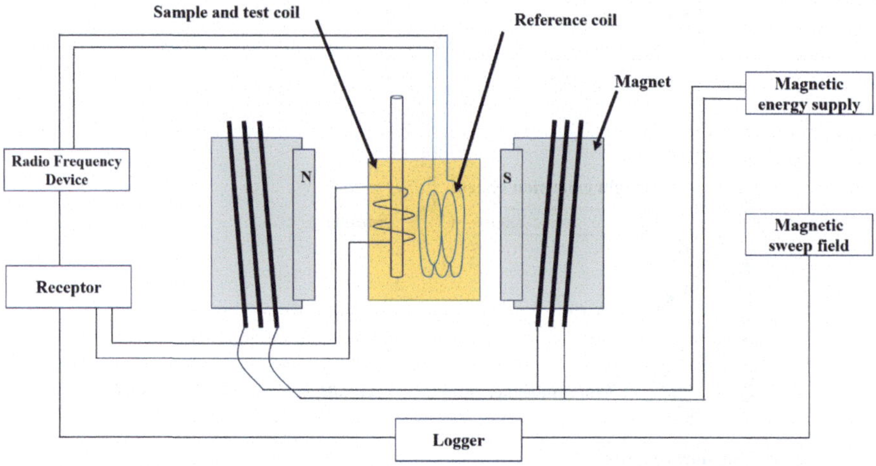

Fig. 17.4 Schematic diagram of the working principle of analyzing marine microplastics using NMR (Patel and Chudasama 2021)

receiver, and a computer to process the data and generate the NMR spectrum (Claridge 2016). Sample preparation involves extracting and purifying microplastics from environmental samples using techniques like filtration, density separation, or enzymatic digestion to remove contaminants (Hidalgo-Ruz et al. 2012). The microplastics are then dissolved in suitable solvents, such as deuterated chloroform ($CDCl_3$) or deuterated dimethyl sulfoxide (DMSO-d_6), to create a homogeneous solution. The solution is filtered to remove undissolved particles, transferred into an NMR tube, and may include an internal standard like tetramethylsilane (TMS) to calibrate chemical shifts. These steps ensure accurate and reliable NMR analysis, which can be adapted based on the polymer type and NMR technique used (e.g., solution-state NMR or solid-state NMR).

NMR is a powerful tool for determining the detailed chemical structure of microplastics, including identifying specific polymer types and additives. For instance, Materić et al. (2020) used NMR to analyze microplastics in glacial snow, identifying and quantifying polymers such as polyethylene (PE), polypropylene (PP), and polystyrene (PS). This study demonstrated NMR's effectiveness in identifying microplastics in complex environmental samples, highlighting its potential for marine pollution research (Materić et al. 2020). Additionally, NMR offers a non-destructive and highly detailed method for analyzing the chemical structure of microplastics, making it particularly useful for identifying additives and understanding polymer degradation processes.

However, there are several limitations of NMR application. Low sensitivity, long analysis times, the need for sophisticated instrumentation and expertise, and higher costs compared to other microplastic analytical techniques are among the common limitations that restrict the NMR performance efficiency (Peez and Imhof 2020).

Despite these challenges, NMR remains a valuable technique for the comprehensive chemical characterization of microplastics.

17.4.1.4 Pyrolysis-Gas Chromatography-Mass Spectrometry (Py-GC-MS)

Pyrolysis-gas chromatography-mass spectrometry (Py-GC-MS) is a powerful technique for detailed chemical characterization of microplastics, typically ranging from a few micrometers to tens of micrometers. This method combines pyrolysis, gas chromatography, and mass spectrometry to break down complex polymer structures into identifiable smaller molecules.

The general mechanism of Py-GC-MS is illustrated as in Fig. 17.5. The process is initiated by heating the sample to high temperatures in an inert atmosphere, breaking down the polymer into smaller, volatile fragments. These fragments are then separated by gas chromatography based on their volatility and interaction with the column material. Finally, mass spectrometry identifies and quantifies the separated molecules by measuring their mass-to-charge ratios. This enables precise identification of the original polymer types and any additives or degradation products present (Nuelle et al. 2014).

Py-GC-MS is highly effective for analyzing complex polymer mixtures and identifying chemical additives in microplastics due to high sensitivity, which may provide detailed information on their chemical structure and composition. For instance, Peters et al. (2018) used Py-GC-MS to analyze microplastics in the stomach contents of marine fish from the Texas Gulf Coast, identifying polymers such as PVC, PET, nylon, silicone, and epoxy resin (Peters et al. 2018). Similarly, Zhong et al. (2022) applied Py-GC-MS to analyze microplastics in aquatic shellfish, identifying and

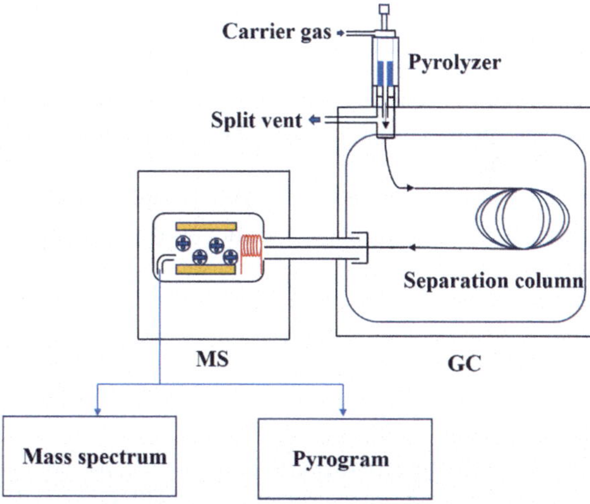

Fig. 17.5 Schematic diagram of the working principle of analyzing marine microplastics using Py-GC-MS (Wu et al. 2012)

quantifying microplastics like nylon 6 (PA6) and nylon 66 (PA66) with high precision. (Zhong et al. 2022). It shares benefits with FTIR and Raman spectroscopy, such as being non-destructive, requiring minimal sample amounts, and enabling high-throughput screening (Primpke et al. 2017).

However, Py-GC-MS possesses several limitations. Extensive sample preparation is often required, which can be time-consuming and costly. High temperatures used in pyrolysis can cause thermal degradation of some heat-sensitive additives, which may further complicate their identification (Mishra et al. 2021a, b, c). Also, high expertise is required in interpreting the mass spectra of some samples, especially the ones that involve complex mixtures. Additionally, Py-GC-MS measurement is a lengthy process, which may further restrict high-throughput applications. Lastly, quantitative analysis using this instrument can be difficult due to variations in pyrolysis efficiency and product distribution, and incomplete standard libraries limit identification capabilities (La Nasa et al. 2020). Despite these challenges, Py-GC-MS remains crucial for the detailed chemical characterization of microplastics, offering valuable insights into their composition, origin, and environmental impact. As technology advances and standard sample libraries improve, the applications of Py-GC-MS in microplastic analysis are expected to expand significantly.

17.4.2 Thermal Analysis Techniques

17.4.2.1 Differential Scanning Calorimetry (DSC)

Differential scanning calorimetry (DSC) is a thermo-analytical technique that measures heat flow associated with thermal transitions in a material as a function of temperature or time. It is particularly useful for analyzing microplastics within the size range of tens to hundreds of micrometers, providing valuable information about thermal properties such as melting temperature (Tm), crystallization temperature (Tc), and glass transition temperature (Tg). These properties are characteristic of different polymer types, allowing DSC to differentiate between various plastic polymers and provide insights into their composition and potential origins (Al-Bayaty et al. 2020).

The general mechanism of DSC is illustrated as in Fig. 17.6. A typical DSC system includes a furnace, thermocouples, a temperature controller, a differential thermal amplifier, a power compensation unit, and a data acquisition system (Paul Gabbott 2008). Sample preparation involves extracting and purifying microplastics from environmental samples, drying them, and accurately weighing a small amount into a DSC pan, which is then hermetically sealed. An empty DSC pan is used as a reference. Additional steps may be required for samples expected to undergo thermal decomposition (Dierkes et al. 2021).

DSC is effective for differentiating between types of plastics based on their thermal properties. For example, polyethylene (PE) and polypropylene (PP) have distinct melting points detectable by DSC, enabling their identification in environmental

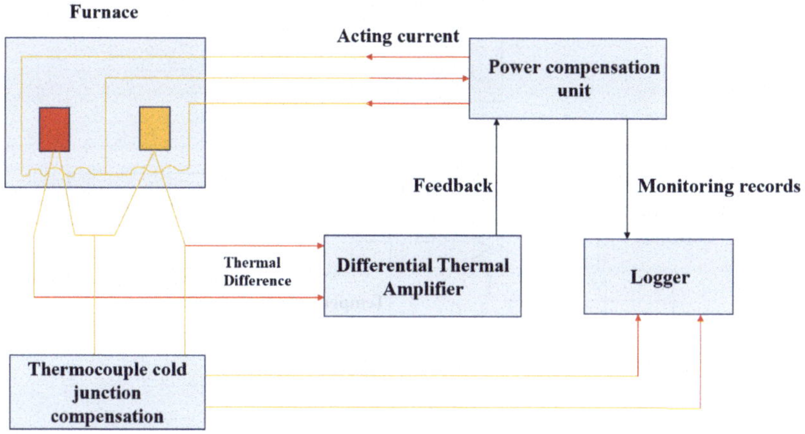

Fig. 17.6 Schematic diagram of the working principle of analyzing marine microplastics using DSC (Almoselhy 2020)

samples. Shabaka et al. (2019) used DSC to identify and quantify microplastics in sediment and water samples from the Eastern Harbor on the Mediterranean Coast of Egypt. The analysis revealed ten different polymers, including PE, PP, and polyethylene terephthalate (PET), with most being secondary microplastics from degraded larger items. This study demonstrated DSC's value in marine microplastic analysis, providing accurate identification and quantification of various polymers in complex samples (Shabaka et al. 2019). Furthermore, DSC is a powerful, non-destructive, quick, and reliable technique for identifying polymers based on their thermal properties. However, it requires relatively pure samples and is less effective with complex mixtures or heavily contaminated samples (Uurasjärvi et al. 2020).

17.4.2.2 Thermogravimetric Analysis (TGA)

Thermogravimetric Analysis (TGA) measures the mass change of a sample as a function of temperature or time under a controlled atmosphere, making it useful for investigating the thermal stability and composition of microplastics. By monitoring the weight loss profile as the temperature increases, TGA can identify the degradation temperatures of different polymers and provide insights into the presence of additives or fillers (Felix et al. 2022).

The general mechanism of TGA instrument is illustrated as in Fig. 17.7. A typical TGA instrument includes a precision balance, furnace, temperature controller, gas flow system, and data acquisition unit (Gabbott 2008). The sample, placed in a crucible, is heated or cooled according to a pre-set temperature program, and its mass is continuously measured. The gas flow system ensures a controlled atmosphere, which can be inert (e.g., nitrogen) or reactive (e.g., oxygen). Sample preparation involves drying the extracted and purified microplastics to remove residual solvents

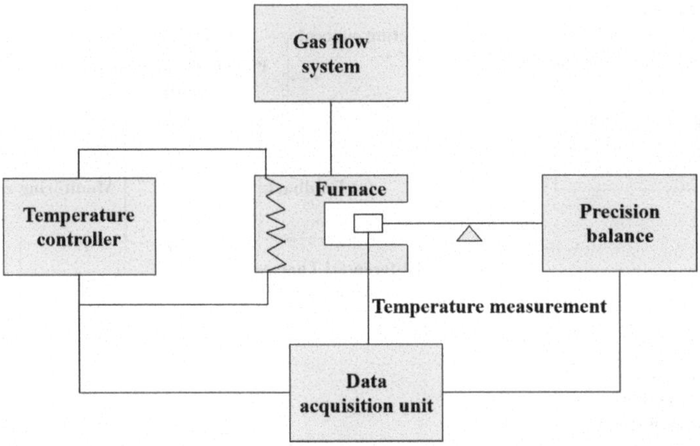

Fig. 17.7 Schematic diagram of the working principle of analyzing marine microplastics using TGA (Saadatkhah et al. 2020)

or moisture. A small amount of the dried sample (a few milligrams) is then weighed and placed in the TGA crucible. Additional steps like acid digestion or ashing may be needed if significant inorganic material is present (Hidalgo-Ruz et al. 2012).

TGA is widely used for thermal degradation analysis of microplastics, providing information about decomposition temperature, degradation kinetics, and the presence of volatile components. It can also determine filler content by analyzing the residual mass after the complete combustion of the organic polymer (Mansa and Zou 2021). For example, Yu et al. (2019) used TGA-FTIR to identify and quantify microplastics in mussel tissue, seawater, and soil samples in China. TGA decomposed the samples while monitoring weight loss, and the evolved gases were analyzed using FTIR to provide detailed chemical profiles. This method effectively distinguished different microplastics based on their thermal degradation profiles and characteristic infrared peaks, with recovery rates of 93% to 107%, demonstrating accuracy and reliability (Yu et al. 2019).

TGA offers several advantages for microplastic analysis. It provides valuable information on the thermal stability of polymers, aiding in their identification and differentiation. TGA can also quantify inorganic fillers or additives, as these components remain as residue after analysis. Additionally, TGA is a relatively simple and rapid technique requiring minimal sample preparation. However, TGA has limitations. It is not specific for polymer identification, as some polymers exhibit similar degradation profiles. Complementary techniques like FTIR or Py-GC-MS are often needed for unambiguous identification. Moreover, TGA cannot directly determine the chemical structure of microplastics, limiting its ability to identify complex polymer mixtures or specific additives (Mansa and Zou 2021; Table 17.2).

Table 17.2 Comparison of microplastic analytical identification and characterization techniques in marine environments

Technique	Applicable size range (μm)	Main applications	Advantages	Limitations
FTIR	1–1000	Polymer identification, quantification	Versatile and widely used; multiple modes for different microplastic sizes; provides chemical information	Cannot distinguish similar polymers; affected by sample purity; limited spatial resolution
Raman spectroscopy	1–1000 + (nanoscale)	Polymer identification, quantification, imaging	Advantageous for aqueous environments; can detect inorganic materials; faster analysis	Relatively low sensitivity; susceptible to fluorescence interference; longer measurement time for some samples; requires expertise for spectral interpretation
NMR	Dozens to hundreds	Polymer identification, quantification	Allows repeated analysis; provides rich structural information; non-destructive	Low sensitivity; time-consuming analysis; requires expertise for spectral interpretation; high sample purity requirement
Py-GC-MS	A few to dozens	Polymer identification, quantification	High sensitivity; accurate polymer identification; can distinguish additives and modifiers	Requires sample pretreatment; not suitable for large sample volumes; challenges in quantitative analysis; limited standard libraries

(continued)

Table 17.2 (continued)

Technique	Applicable size range (μm)	Main applications	Advantages	Limitations
DSC	Dozens to hundreds	Polymer identification, quantification	Non-destructive; can identify complex mixtures; provides thermodynamic information	Cannot distinguish similar polymers; affected by sample purity; quantitative analysis affected by sample homogeneity
Thermogravimetric analysis (TGA)	A few milligrams to tens of milligrams	Thermal stability, composition analysis (fillers, additives)	Simple, rapid, minimal sample preparation, quantifies inorganic content	Not specific to polymer identification, cannot determine the chemical structure of complex mixtures

17.5 Emerging Techniques and Future Directions for Microplastic Identification and Characterization

In recent years, as concerns about marine microplastic pollution have increased, researchers have continued to develop and apply new technologies to identify and characterize microplastics. The development trends of these technologies are mainly focused on improving detection accuracy, reducing analysis costs, and improving processing efficiency. In this chapter, we introduce several emerging technologies for microplastic analysis: hyperspectral imaging (HIS), artificial intelligence and machine learning, microfluidic devices, and other novel approaches that are currently being explored by several researchers.

17.5.1 Hyperspectral Imaging (HSI)

Hyperspectral imaging (HSI) is an emerging technique with immense potential in the field of marine microplastic analysis. As illustrated in Fig. 17.8, this technique involves the process of capturing images across a wide range of wavelengths, providing a spectral signature for each pixel in the image (Faltynkova et al. 2021). This wealth of spectral information allows for the simultaneous identification and quantification of different types of microplastics based on their unique spectral fingerprints.

HSI offers several advantages over traditional methods for microplastic analysis. Firstly, it enables rapid and non-destructive analysis of large sample areas without

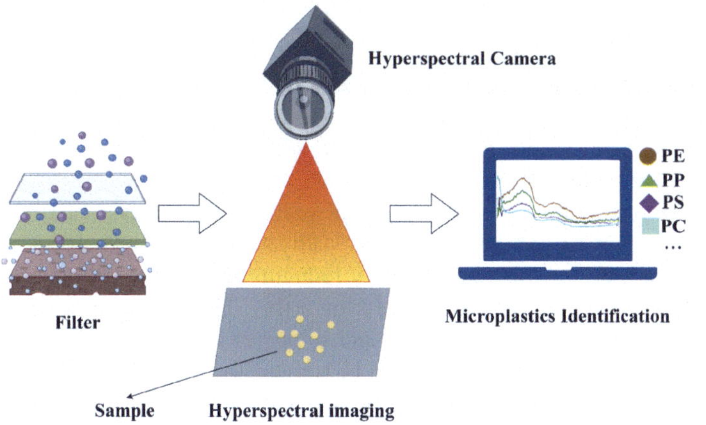

Fig. 17.8 Hyperspectral imaging system detects and characterizes marine microplastics (Shan et al. 2019) (by Figdraw)

the need for extensive sample preparation, making it a more efficient and less time-consuming method compared to traditional spectroscopic techniques (Primpke et al. 2017). Secondly, HSI boasts high spatial resolution, allowing for the identification and mapping of microplastics at a microscopic level, providing valuable insights into their abundance, size distribution, and spatial heterogeneity (Zhang et al. 2024). Thirdly, by analyzing the spectral signatures of microplastics, HSI can differentiate between various polymer types and even identify additives or other chemical components (Faltynkova et al. 2021). Lastly, HSI data analysis can be automated using machine learning algorithms, further enhancing the efficiency and objectivity of microplastic identification (Zhang et al. 2023b).

Despite its potential, HSI for microplastic analysis is still in its early stages of development, and thus may possess several limitations. One of the main challenges is the need for standardized spectral libraries and robust algorithms for accurate identification and quantification of microplastics in complex environmental matrices (Faltynkova et al. 2021). Further research is needed to optimize HSI systems and data analysis methods for specific applications in marine microplastic research. Additionally, the high cost of HSI instruments and the requirement for specialized expertise in data analysis may limit its widespread adoption.

17.5.2 Artificial Intelligence and Machine Learning in Microplastic Analysis

The application of artificial intelligence (AI) and machine learning (ML) in microplastic analysis has revolutionized the field by offering innovative solutions to address the challenges associated with traditional methods. Leveraging the power of algorithms and statistical models, AI and ML enable the analysis of vast datasets, automate laborious tasks, and enhance the accuracy and efficiency of microplastic identification and characterization (Mariani et al. 2023). This computational approach has the potential to transform microplastic research and monitoring, providing valuable insights into the sources, distribution, and impacts of these pervasive pollutants.

AI and ML offer numerous advantages in microplastic analysis. Firstly, they automate time-consuming and repetitive tasks such as image analysis, spectral interpretation, and data processing, significantly reducing the manual effort required and increasing throughput (Lönnstedt and Eklöv 2016). This automation not only saves time but also minimizes human error and improves the reproducibility of results. Secondly, machine learning models can be trained on large datasets of microplastic images and spectra, allowing them to recognize subtle patterns and features that might be overlooked by human analysts, thus improving the accuracy of microplastic identification and classification (Zhang et al. 2023a, b). Lastly, AI and ML facilitate the integration of data from multiple sources, such as microscopy images, spectroscopic data, and environmental parameters, to provide a more comprehensive understanding of microplastic distribution, sources, and potential impacts (Mariani et al. 2023).

The applications of AI and ML in microplastic analysis are diverse and rapidly expanding. In image analysis, AI-based algorithms can automatically detect, classify, and quantify microplastics in microscopic images, significantly reducing the time and labor required for manual analysis (Lönnstedt and Eklöv 2016). For spectral interpretation, machine learning models trained on extensive spectral libraries can accurately identify polymer types and even differentiate between different additives (Zhang et al. 2023a, b). Additionally, AI and ML can be used to develop predictive models that estimate microplastic concentrations in different environments based on various factors such as land use, population density, and waste management practices (Mariani et al. 2023). These models can aid in identifying pollution hotspots and assessing the effectiveness of mitigation strategies.

Despite the promising potential of AI and ML in microplastic analysis, several challenges remain to be addressed. One major challenge is the need for large and diverse training datasets to develop accurate and robust models. Acquiring and labeling such datasets can be time-consuming and resource-intensive. Additionally, the interpretability of complex ML models can be a concern, as it is important to understand the reasoning behind their predictions to ensure their reliability and trustworthiness (Mariani et al. 2023). Future research should focus on developing standardized protocols for data collection and analysis, expanding training datasets, and improving the interpretability of AI and ML models. Furthermore, the integration of AI and ML with other emerging technologies, such as hyperspectral imaging and microfluidics, holds great promise for further advancing the field of microplastic analysis and enhancing our understanding of the sources, fate, and impacts of these ubiquitous pollutants.

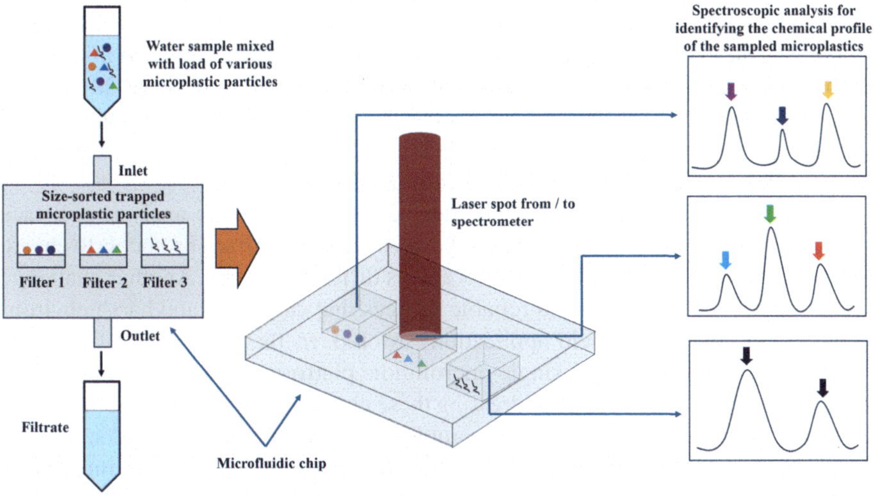

Fig. 17.9 Schematic diagram of the working principle of analyzing marine microplastics using microfluidic devices (Ece et al. 2023)

17.5.3 Microfluidic Devices

Microfluidic devices, often referred to as "labs-on-a-chip," are miniaturized systems that manipulate and analyze small fluid volumes within microchannels. In the realm of microplastic analysis, these devices have emerged as a promising tool due to their unique advantages over traditional methods. As shown in Fig. 17.9, the basic process involves injecting the sample into the microfluidic chip, using micropumps and microvalves to control fluid flow, employing laminar flow for precise sample delivery, efficiently separating and capturing microplastic particles within the microchannels, and finally detecting and analyzing the particles in real-time using integrated sensors (Zhang et al. 2023a, b).

The advantages of microfluidic devices in microplastic analysis are manifold. Firstly, their miniaturization and portability allow for on-site analysis in diverse environments, including remote locations and resource-limited settings (Zabihihesari et al. 2023). Secondly, they offer high throughput and efficiency, enabling the simultaneous analysis of multiple samples with minimal sample loss or contamination (Pollard et al. 2020). Thirdly, their small size and integrated design reduce reagent consumption and waste generation, making them a more sustainable option (Faramarzi et al. 2024). Lastly, microfluidic devices provide precise control over fluid flow, temperature, and other parameters, facilitating accurate manipulation and separation of microplastics based on their size, shape, density, and chemical properties (Mohammadimehr et al. 2024).

Microfluidic devices find diverse applications in microplastic analysis. They can be used for various sample preparation steps, such as filtration, extraction, and concentration of microplastics from environmental matrices (Faramarzi et al. 2024).

Additionally, they can be integrated with different separation techniques, like dielectrophoresis, inertial focusing, and magnetic separation, to isolate and sort microplastics based on their physical and chemical properties (Pollard et al. 2020). Moreover, these devices can incorporate diverse detection methods, including optical microscopy, spectroscopy, and electrochemical sensors, to identify and quantify microplastics in real-time (Zabihihesari et al. 2023).

Despite their potential, microfluidic devices for microplastic analysis still face several challenges. One key challenge is the need for standardized protocols to ensure consistency and comparability of results across different studies. Another challenge is the development of robust and reliable devices that can withstand harsh environmental conditions during field applications. Moreover, the integration of multiple analytical techniques within a single microfluidic platform remains a complex task. Future research should focus on addressing these challenges and exploring the potential of microfluidic devices for more complex microplastic analysis tasks, such as identifying polymer additives and assessing microplastic toxicity. The integration of artificial intelligence and machine learning algorithms with microfluidic devices could further enhance their capabilities and pave the way for automated, high-throughput microplastic analysis in various environmental settings (Zhang et al. 2023a).

17.6 Other Novel Approaches in Microplastic Analysis

Beyond the established and emerging techniques discussed above, several other novel approaches are being explored to enhance the identification and characterization of microplastics in marine environments. These approaches aim to address the limitations of existing methods and provide new insights into the complex issue of microplastic pollution.

One of these novel techniques is surface-enhanced Raman spectroscopy (SERS), which is a highly sensitive technique that amplifies Raman signals by several orders of magnitude, enabling the detection of trace amounts of microplastics. This method utilizes nanostructured metal surfaces to enhance the Raman scattering of molecules adsorbed on them, making it possible to identify and quantify even tiny microplastic particles (Mishra et al. 2022b; Mishra and Das 2022).

Another novel technique is stimulated Raman scattering (SRS) Microscopy, which is a label-free imaging technique that offers high sensitivity and fast acquisition speeds, making it suitable for analyzing microplastics in complex environmental samples. SRS microscopy can differentiate between different polymer types and provide information on their spatial distribution within a sample (Wang et al. 2017).

While fluorescence microscopy is not directly used for microplastic identification, it can be employed to visualize microplastics that have been labeled with fluorescent dyes or tags. This approach allows for the detection and tracking of microplastics in biological tissues and environmental samples (Aliro Villacorta et al. 2024).

Thermal extraction desorption gas chromatography-mass spectrometry (TED-GC-MS) is a technique that combines thermal extraction and gas chromatography-mass spectrometry to analyze volatile and semi-volatile organic compounds associated with microplastics. This method can provide valuable information on the additives, plasticizers, and other chemicals present in microplastics, which can help to identify their sources and assess their potential toxicity.

Lastly, bioanalytical assay is also regarded as a sustainable novel technique as it mainly utilizes the interactions between microplastics and biological systems to detect and quantify microplastics. Among of the examples of bioanalytical assay are enzyme-linked immunosorbent assays (ELISAs) and cell-based assays, which can detect specific types of microplastics based on their surface properties or chemical composition (Yu et al. 2022).

These novel approaches, along with ongoing advancements in existing techniques, are expected to play a crucial role in advancing our understanding of marine microplastic pollution. By providing more accurate, sensitive, and efficient methods for microplastic analysis, these approaches can help researchers to better assess the sources, fate, and impacts of microplastics in the marine environment, which may ultimately lead to more effective mitigation and prevention strategies.

17.7 Conclusion

The identification and characterization of marine microplastics, ranging in size from nanometers to millimeters and with diverse compositions, poses a great challenge due to their complex and varied composition and their presence in complex environmental matrices. Current visual and analytical techniques, including optical microscopy, electron microscopy, Fourier Transform Infrared Spectroscopy (FTIR), Raman Spectroscopy, Nuclear Magnetic Resonance (NMR) spectroscopy, Pyrolysis Gas Chromatography-Mass Spectrometry (Py-GC-MS), Differential Scanning Calorimetry (DSC), and Thermo-Gravimetric Analysis (TGA), have provided invaluable insights into the morphology, chemical composition and thermal properties of microplastics.

Nevertheless, there are still limitations in achieving standardized protocols, improving the sensitivity and selectivity of trace detection, and establishing comprehensive reference databases. Addressing these limitations may thus require a multifaceted approach. This approach should include improving existing techniques, exploring new methods such as surface-enhanced Raman spectroscopy (SERS) and stimulated Raman scattering (SRS) microscopy, and integrating multiple analytical tools. Future research should focus on developing more sensitive, high-throughput, and automated techniques that integrate artificial intelligence and machine learning algorithms to improve the accuracy and efficiency of microplastic analysis. Furthermore, research on the environmental behavior, fate, and toxic effects of microplastics needs to be strengthened to better assess their risks to marine ecosystems and human health. Interdisciplinary collaboration among scientists, engineers, policymakers, and stakeholders is essential to fully understand microplastic pollution and implement effective mitigation strategies to protect marine ecosystems and human health.

References

Al-Bayaty SA, Al-Uqaily RAH, Jubier NJ (2020) Using the coats-Redfern method during thermogravimetric analysis and differential scanning calorimetry analysis of the thermal stability of epoxy and epoxy/silica nanoparticle nanocomposites. J Southw Jiaotong Univ 55(4). https://doi.org/10.35741/issn.0258-2724.55.4.2

Almoselhy R (2020) Applications of differential scanning calorimetry (DSC) in oils and fats research. A review. Am Res J Agric 6:1–9. https://doi.org/10.21694/2378-9018.20002

Araujo CF, Nolasco MM, Ribeiro AMP, Ribeiro-Claro PJA (2018) Identification of microplastics using Raman spectroscopy: latest developments and future prospects. Water Res 142:426–440. https://doi.org/10.1016/j.watres.2018.05.060

Bala K, Anshul, Verma S, Dogra A, Das AP, Kumar R, Nandini (2024) Sustainable management for urban plastic waste generation. In: Sustainable Management of Urban Plastic Waste Through Circular Economic Approaches, pp 211–236. https://doi.org/10.1201/9781003465447-13

Bala K, Dogra A, Lata P, Das AP (2024) Discussing sources and biological-chemical remediation approaches for microplastic pollution. In: Sustainable Microbial Technology for Synthetic and Cellulosic Microfiber Bioremediation, pp 37–59. https://doi.org/10.1007/978-3-031-62600-5_3

Bal M, Behera ID, Kumari U, Biswas S, Meikap BC (2020) Hydrodynamic study and particulate matter removal in a self priming venturi scrubber. Environ Technol Innov 20:101167. https://doi.org/10.1016/j.eti.2020.101167

Behera ID, Das AP (2023) Impact of petroleum waste on environmental pollution and its sustainable management through circular economy. Springer

Behera ID, Basak G, Kumar RR, Sen R, Meikap BC (2020) Treatment of petroleum refinery sludge by petroleum degrading bacterium *Stenotrophomonas pavanii* IRB19 as an efficient novel technology. J Environ Sci Health Part A 56(2):226–239. https://doi.org/10.1080/10934529.2020.1866924

Behera ID, Nayak M, Biswas S, Meikap BC, Sen R (2021) Enhanced biodegradation of total petroleum hydrocarbons by implementing a novel two-step bioaugmentation strategy using indigenous bacterial consortium. J Environ Manag 292:112746. https://doi.org/10.1016/j.jenvman.2021.112746

Behera ID, Nayak M, Mishra A, Meikap BC, Sen R (2022b) Strategic implementation of integrated bioaugmentation and biostimulation for efficient mitigation of petroleum hydrocarbon pollutants from terrestrial and aquatic environment. Mar Pollut Bull 177:113492. https://doi.org/10.1016/j.marpolbul.2022.113492

Behera ID, Meikap BC, Sen R (2022a) Performance evaluation of bacterial consortium for biodegradation of total petroleum hydrocarbon: a comparative strategic biostimulation study. Research Square

Biswas S, Diwakar RK, Behera ID, Meikap BC, Sen TK, Khiadani M (2020) Aqueous phase phenol removal from synthetic and real steel plant effluents through a batch and semifluidized bed column operation: experimental and model analysis. J Environ Chem Eng 8(5):104441. https://doi.org/10.1016/j.jece.2020.104441

Caldwell J, Loussert-Fonta C, Toullec G, Heidelberg Lyndby N, Haenni B, Taladriz-Blanco P, Espiña B, Rothen-Rutishauser B, Petri-Fink A (2023) Correlative light, electron microscopy and Raman spectroscopy workflow to detect and observe microplastic interactions with whole jellyfish. Environ Sci Technol 57(16):6664–6672. https://doi.org/10.1021/acs.est.2c09233

Claridge TDW (2016) Experimental methods. In: High-resolution NMR techniques in organic chemistry. Elsevier, pp 457–498. https://doi.org/10.1016/B978-0-08-099986-9.00012-9

Das AP, Mohanty AK, van Hullebusch ED, Figueiredo G, Das AP, Mohanty AK, van Hullebusch ED, Figueiredo G (2023b) Bioremediation as an emerging technology for the removal of synthetic microplastic pollutants from marine ecosystem. MarPB 194:115297. https://doi.org/10.1016/J.MARPOLBUL.2023.115297

Das AP, Nayak B, Behera ID (2024c) Applications of microorganisms in advanced geomicrobiological and biotechnological exploration. Geomicrobiol J 41(4):309–311. https://doi.org/10.1080/01490451.2024.2327059

Das AP, Dutta K, Khatun R, Behera ID, Singh S, Mishra S (2023) Microfiber pollution and its microbial mitigation: a review on current trends and future prospects. J Taiwan Inst Chem Eng 105104. https://doi.org/10.1016/J.JTICE.2023.105104

Das AP, Behera ID, Das NP (2024a) Renewable energy generation and value addition from environmental microfiber pollution through advanced greener solution. Springer

Das AP, Behera ID, Bhanja D (2024c) Sustainable microbial technology for synthetic and cellulosic microfiber bioremediation. https://doi.org/10.1007/978-3-031-62600-5

Das A, Dey S, Bala K, Gnanasekaran R, Das AP (2024) Microbial bioremediation technology for sustainable treatment and management of synthetic microfiber waste. In: Sustainable Microbial Technology for Synthetic and Cellulosic Microfiber Bioremediation, pp 221–238. https://doi.org/10.1007/978-3-031-62600-5_12

Das A, Dey S, Das AP (2024) Ecotoxicological impacts of synthetic microfiber pollutants and development of sustainable mitigation strategies. Environ Chem Ecotoxicol. https://doi.org/10.1016/J.ENCECO.2024.12.003

Dey S, Bissoyi A, Koteswara Reddy G, Das AP (2024) Synthetic microfibers: An urban emerging pollutant demanding for sustainable environmental solutions. In: Sustainable Microbial Technology for Synthetic and Cellulosic Microfiber Bioremediation, pp 107–121. https://doi.org/10.1007/978-3-031-62600-5_7

Dey S, Dash N, Ray SK, Priyadarsini S, Chhotaray C, Das AP (2024) Occurrence and quantification of tire wear plastics pollutants from highway city dust of Bhubaneswar City of Odisha. In: Sustainable Microbial Technology for Synthetic and Cellulosic Microfiber Bioremediation, pp 77–89. https://doi.org/10.1007/978-3-031-62600-5_5

Dey S, Tripathy B, Kumar MS, Das AP (2023) Ecotoxicological consequences of manganese mining pollutants and their biological remediation. Environ Chem Ecotoxicol. https://doi.org/10.1016/J.ENCECO.2023.01.001

Dierkes G, Lauschke T, Földi C (2021) Analytical methods for plastic (microplastic) determination in environmental samples, pp 43–67. https://doi.org/10.1007/698_2021_744

Donato A, Spadaro D, Giuffrida D, Sabatino G, Di Bella M, Trusso S, Ponterio RC (2024) Monitoring plastic pellet pollution in coastal environments through handheld Raman spectroscopy: data from the Mediterranean coasts (Southern Italy). Mar Pollut Bull 202:116312. https://doi.org/10.1016/j.marpolbul.2024.116312

Duis K, Coors A (2016) Microplastics in the aquatic and terrestrial environment: sources (with a specific focus on personal care products), fate and effects. Environ Sci Eur 28(1):2. https://doi.org/10.1186/s12302-015-0069-y

Ece E, Hacıosmanoğlu N, Inci F (2023) Microfluidics as a ray of hope for microplastic pollution. Biosensors 13(3):332. https://doi.org/10.3390/bios13030332

Faltynkova A, Johnsen G, Wagner M (2021) Hyperspectral imaging as an emerging tool to analyze microplastics: a systematic review and recommendations for future development. Microplastics Nanoplastics 1(1):13. https://doi.org/10.1186/s43591-021-00014-y

Faramarzi P, Jang W, Oh D, Kim B, Kim JH, You JB (2024) Microfluidic detection and analysis of microplastics using surface nanodroplets. ACS Sens 9(3):1489–1498. https://doi.org/10.1021/acssensors.3c02627

Felix CB, Chen W-H, Ubando AT, Park Y-K, Lin K-YA, Pugazhendhi A, Nguyen T-B, Dong C-D (2022) A comprehensive review of thermogravimetric analysis in lignocellulosic and algal biomass gasification. Chem Eng J 445:136730. https://doi.org/10.1016/j.cej.2022.136730

Gabbott P (2008) In: Gabbott P (ed) Principles and applications of thermal analysis. Wiley. https://doi.org/10.1002/9780470697702

Galloway TS, Cole M, Lewis C (2017) Interactions of microplastic debris throughout the marine ecosystem. Nat Ecol Evol 1(5):0116. https://doi.org/10.1038/s41559-017-0116

Ghosal S, Janet MM, Ahmed K (2012) Raman Microspectroscopy-based identification of individual fungal spores as potential indicators of indoor contamination and moisture-related building damage. Environ Sci Technol 46(11):6088–6095

Ghosh S, Das AP (2020) Microbial metagenomics: current advances in investigating microbial ecology and population dynamics. In: Nayak SK, Mishra BB (eds) Frontiers in soil and environmental microbiology, 1st edn. CRC Press, pp 247–254

Günther H (2013) NMR spectroscopy: basic principles, concepts and applications in chemistry. https://api.semanticscholar.org/CorpusID:93021815

Harrison JP, Ojeda JJ, Romero-González ME (2012) The applicability of reflectance micro-Fourier-transform infrared spectroscopy for the detection of synthetic microplastics in marine sediments. Sci Total Environ 416:455–463. https://doi.org/10.1016/j.scitotenv.2011.11.078

Hartmann NB, Hüffer T, Thompson RC, Hassellöv M, Verschoor A, Daugaard AE, Rist S, Karlsson T, Brennholt N, Cole M, Herrling MP, Hess MC, Ivleva NP, Lusher AL, Wagner M (2019) Are we speaking the same language? Recommendations for a definition and categorization framework for plastic debris. Environ Sci Technol 53(3):1039–1047. https://doi.org/10.1021/acs.est.8b05297

Hidalgo-Ruz V, Gutow L, Thompson RC, Thiel M (2012) Microplastics in the marine environment: a review of the methods used for identification and quantification. Environ Sci Technol 46(6):3060–3075. https://doi.org/10.1021/es2031505

Hower JC, Berti D, Winkler CR, Qian D, Briot NJ (2022) High-resolution transmission electron microscopy study of a powder river basin coal-derived fly ash. Minerals 12(8):975. https://doi.org/10.3390/min12080975

Ivleva NP (2021) Chemical analysis of microplastics and nanoplastics: challenges, advanced methods, and perspectives. Chem Rev 121(19):11886–11936. https://doi.org/10.1021/acs.chemrev.1c00178

John A, Das AP (2024) Government strategies for achieving circular economy for plastic waste management in Southeast Asia. In: Sustainable Management of Urban Plastic Waste Through Circular Economic Approaches, pp 72–82. https://doi.org/10.1201/9781003465447-6/GOVERNMENT-STRATEGIES-ACHIEVING-CIRCULAR-ECONOMY-PLASTIC-WASTE-MANAGEMENT-SOUTHEAST-ASIA-AKBAR-JOHN-ALOK-PRASAD-DAS

Kannan M (2018) Transmission electron microscope—principle, components and applications. A textbook on fundamentals and applications of nanotechnology, pp 93–102

Kanungo J, Sahoo T, Bal M, Behera ID (2024) Performance of bioremediation strategy in waste lubricating oil pollutants: a review. Geomicrobiol J 41(4):360–373. https://doi.org/10.1080/01490451.2023.2245395

Kanungo J, Sahoo T, Swain LP, Behera ID (2023) Toxicity of persistent hydrocarbon pollutants, sources and sustainable remediation process. In: Behera ID, Das AP (eds) Impact of petroleum waste on environmental pollution and its sustainable management through circular economy. Springer, pp 39–65. https://doi.org/10.1007/978-3-031-48220-5_2

Koelmans AA, Besseling E, Foekema EM (2014) Leaching of plastic additives to marine organisms. Environ Pollut 187:49–54. https://doi.org/10.1016/j.envpol.2013.12.013

La Nasa J, Biale G, Fabbri D, Modugno F (2020) A review on challenges and developments of analytical pyrolysis and other thermoanalytical techniques for the quali-quantitative determination of microplastics. J Anal Appl Pyrol 149:104841. https://doi.org/10.1016/j.jaap.2020.104841

Lönnstedt OM, Eklöv P (2016) RETRACTED: environmentally relevant concentrations of microplastic particles influence larval fish ecology. Science 352(6290):1213–1216. https://doi.org/10.1126/science.aad8828
Lusher AL, O'Donnell C, Officer R, O'Connor I (2016) Microplastic interactions with North Atlantic mesopelagic fish. ICES J Mar Sci 73(4):1214–1225. https://doi.org/10.1093/icesjms/fsv241
Lusher A, Hollman P, Mendoza J (2017) Microplastics in fisheries and aquaculture: status of knowledge on their occurrence and implications for aquatic organisms and food safety
Maes T, Jessop R, Wellner N, Haupt K, Mayes AG (2017) A rapid-screening approach to detect and quantify microplastics based on fluorescent tagging with Nile Red. Sci Rep 7(1):44501. https://doi.org/10.1038/srep44501
Mallick K, Sahu A, Dubey NK, Das AP (2023) Harvesting marine plastic pollutants-derived renewable energy: a comprehensive review on applied energy and sustainable approach. J Environ Manag 348:119371. https://doi.org/10.1016/J.JENVMAN.2023.119371
Mansa R, Zou S (2021) Thermogravimetric analysis of microplastics: a mini review. Environ Adv 5:100117. https://doi.org/10.1016/j.envadv.2021.100117
Mariani MM, Machado I, Magrelli V, Dwivedi YK (2023) Artificial intelligence in innovation research: a systematic review, conceptual framework, and future research directions. Technovation 122:102623. https://doi.org/10.1016/j.technovation.2022.102623
Materić D, Kasper-Giebl A, Kau D, Anten M, Greilinger M, Ludewig E, van Sebille E, Röckmann T, Holzinger R (2020) Micro- and nanoplastics in alpine snow: a new method for chemical identification and (semi)quantification in the nanogram range. Environ Sci Technol 54(4):2353–2359. https://doi.org/10.1021/acs.est.9b07540
Miserli K, Lykos C, Kalampounias AG, Konstantinou I (2023) Screening of microplastics in aquaculture systems (fish, mussel, and water samples) by FTIR, scanning electron microscopy-energy dispersive spectroscopy and micro-Raman spectroscopies. Appl Sci 13(17):9705. https://doi.org/10.3390/app13179705
Mishra S, Dash D, Das AP (2024) Microbial degradation of polyester microfibers using indigenously isolated bacterial strain Exiguobacterium Sp. CLEAN – Soil, Air, Water, 52(12):e202300343 https://doi.org/10.1002/CLEN.202300343
Mishra S, Singh RP, Rath CC, Das AP (2020) Synthetic microfibers: source, transport and their remediation. J Water Process Eng 38:101612. https://doi.org/10.1016/J.JWPE.2020.101612
Mishra A, Siddiqi H, Kumari U, Behera ID, Mukherjee S, Meikap BC (2021a) Pyrolysis of waste lubricating oil/waste motor oil to generate high-grade fuel oil: a comprehensive review. Renew Sustain Energy Rev 150:111446. https://doi.org/10.1016/j.rser.2021.111446
Mishra S, Swain S, Sahoo M, Mishra S, Das AP (2021c) Microbial colonization and degradation of microplastics in aquatic ecosystem: a review. Geomicrobiol J. https://doi.org/10.1080/01490451.2021.1983670
Mishra S, Dash D, Das AP (2022a) Detection, characterization and possible biofragmentation of synthetic microfibers released from domestic laundering wastewater as an emerging source of marine pollution. Mar Pollut Bull 185:114254. https://doi.org/10.1016/J.MARPOLBUL.2022.114254
Mishra S, Dash D, Al-Tawaha ARMS, Das AP (2022b) A review on heavy metal ion adsorption on synthetic microfiber surface in aquatic environments. Appl Biochem Biotechnol 194(10):4639–4654. https://doi.org/10.1007/S12010-022-04029-W/METRICS
Mishra S, Das AP (2022) Treatment of the wastewater polluted with synthetic microfiber released from washing machine. In: Das BB, Hettiarachchi H, Sahu PK, Nanda S (eds) Lecture notes in civil engineering, vol 207. Springer, pp 109–117. https://doi.org/10.1007/978-981-16-7509-6_9/COVER
Mishra S, Das AP (2023) Microbial remediation of synthetic microfiber contaminated wastewater. In: Shah MP (ed) Microbial technologies in industrial wastewater treatment. Springer, Singapore, pp 337–350. https://doi.org/10.1007/978-981-99-2435-6_17

Mishra S, Rath CC, Das AP (2019) Marine microfiber pollution: a review on present status and future challenges. Mar Pollut Bull 140:188–197. https://doi.org/10.1016/J.MARPOLBUL.2019.01.039

Mishra S, Rout PK, Das AP (2021) Emerging microfiber pollution and its remediation. In: Prasad R (ed) Environmental and microbial biotechnology. Springer, Singapore, pp 247–266. https://doi.org/10.1007/978-981-15-5499-5_9

Mishra S, Singh RP, Rout PK, Das AP (2022) Membrane bioreactor (MBR) as an advanced wastewater treatment technology for removal of synthetic microplastics. Development in wastewater treatment research and processes, pp 45–60. https://doi.org/10.1016/B978-0-323-85583-9.00022-3

Mishra S, Dash D, Das AP (2023) Aquatic microbial diversity on plastisphere: colonization and potential role in microplastic biodegradation. Geomicrobiol J 1–12. https://doi.org/10.1080/01490451.2023.2209750

Mohammadimehr A, Rodríguez-Villarreal AI, Martínez JAL, Casals-Terré J (2024) Review: impact of microfluidic cell and particle separation techniques on microplastic removal strategies. J Agric Food Res 16:101–124. https://doi.org/10.1016/j.jafr.2024.101124

Mohammed A, Abdullah A (2018) Scanning electron microscopy (SEM): a review. In: Proceedings of the 2018 international conference on hydraulics and pneumatics—HERVEX, Băile Govora, Romania, 2018, pp 7–9

Mohanta S, Pradhan B, Behera ID (2024) Impact and remediation of petroleum hydrocarbon pollutants on agricultural land: a review. Geomicrobiol J 41(4):345–359. https://doi.org/10.1080/01490451.2023.2243925

Mohanty M, Mohanty J, Dey S, Dutta K, Shah MP, Das AP (2024) The face mask: a tale from protection to pollution and demanding sustainable solution. Emerg Contam 10(2):100298. https://doi.org/10.1016/J.EMCON.2023.100298

Mohanty M, Tripathy B, Koteswara Reddy G, Das AP (2024) Greener technologies for establishing restraint over microplastic pollution. In: Sustainable Microbial Technology for Synthetic and Cellulosic Microfiber Bioremediation, pp 239–258. https://doi.org/10.1007/978-3-031-62600-5_13

Murphy DB, Davidson MW (2012) Fundamentals of light microscopy and electronic imaging. Wiley. https://doi.org/10.1002/9781118382905

Norén F (2007) Small plastic particles in Coastal Swedish waters. KIMO Report, pp 1–11

Nuelle M-T, Dekiff JH, Remy D, Fries E (2014) A new analytical approach for monitoring microplastics in marine sediments. Environ Pollut 184:161–169. https://doi.org/10.1016/j.envpol.2013.07.027

Oßmann BE (2020) Microplastics characterization by Raman microscopy. In: Rocha-Santos T, Costa M, Mouneyrac C (eds) Handbook of microplastics in the environment. Springer, pp 1–28. https://doi.org/10.1007/978-3-030-10618-8_36-1

Patel GB, Chudasama D (2021) Nuclear magnetic resonance spectroscopy NMR. https://doi.org/10.37591/JoNET

Paul A, Priyadarsini S, Karakuş S, Behera JK, Gnanasekaran R, Das AP (2024) Fate and accumulation of synthetic microfibers in the environment. In: Sustainable Microbial Technology for Synthetic and Cellulosic Microfiber Bioremediation, pp 61–76. https://doi.org/10.1007/978-3-031-62600-5_4

Peez N, Imhof W (2020) Quantitative ^{1}H-NMR spectroscopy as an efficient method for identification and quantification of PVC, ABS and PA microparticles. Analyst 145(15):5363–5371. https://doi.org/10.1039/D0AN00879F

Peters CA, Hendrickson E, Minor EC, Schreiner K, Halbur J, Bratton SP (2018) Pyr-GC/MS analysis of microplastics extracted from the stomach content of benthivore fish from the Texas Gulf Coast. Mar Pollut Bull 137:91–95. https://doi.org/10.1016/j.marpolbul.2018.09.049

Picó Y, Barceló D (2020) Pyrolysis gas chromatography-mass spectrometry in environmental analysis: focus on organic matter and microplastics. TrAC Trends Anal Chem 130:115964. https://doi.org/10.1016/j.trac.2020.115964

Pollard M, Hunsicker E, Platt M (2020) A Tunable Three-Dimensional Printed Microfluidic Resistive Pulse Sensor for the Characterization of Algae and Microplastics. ACS Sensors 5(8):2578–2586. https://doi.org/10.1021/acssensors.0c00987

Pradhan B, Behera ID, Sahoo K, Mohanta S (2023) Applicability of the teetered bed separator for beneficiating Indian iron ore fines: an experimental study. J Mines Met Fuels 71(11):1817–1825. https://doi.org/10.18311/jmmf/2023/31100

Primpke S, Lorenz C, Rascher-Friesenhausen R, Gerdts G (2017) An automated approach for microplastics analysis using focal plane array (FPA) FTIR microscopy and image analysis. Anal Methods 9(9):1499–1511. https://doi.org/10.1039/C6AY02476A

Priyadarsini S, Das AP (2024) Analytical and structural characterization of waste lithium-ion batteries for their effective recycling strategy. Environ Chem Ecotoxicol. https://doi.org/10.1016/j.enceco.2024.12.004

Priyadarsini S, Das AP (2025) Lithium bioleaching: A review on microbial-assisted sustainable technology for lithium bio-circularity. J Water Process Eng, 69:106744. https://doi.org/10.1016/j.jwpe.2024.106744

Ray SK, Dey S, Das AP, John A (2024) microplastic waste management through greener technologies: A review. In: Sustainable Management of Urban Plastic Waste Through Circular Economic Approaches, pp 237–261. https://doi.org/10.1201/9781003465447-14 RENEWABLE ENERGY GENERATION AND VALUE ADDITION FROM ENVIRONMENTAL. (2024)

Ribeiro-Claro P, Nolasco MM, Araújo C (2017) Characterization of microplastics by Raman spectroscopy, pp 119–151. https://doi.org/10.1016/bs.coac.2016.10.001

Rochman CM, Hoh E, Kurobe T, Teh SJ (2013) Ingested plastic transfers hazardous chemicals to fish and induces hepatic stress. Sci Rep 3(1):3263. https://doi.org/10.1038/srep03263

Saadatkhah N, Carillo Garcia A, Ackermann S, Leclerc P, Latifi M, Samih S, Patience GS, Chaouki J (2020) Experimental methods in chemical engineering: thermogravimetric analysis—TGA. Can J Chem Eng 98(1):34–43. https://doi.org/10.1002/cjce.23673

Sahoo PP, Singh S, Rout PK, Mishra S, Das AP (2022) Microbial remediation of plastic pollutants generated from discarded and abandoned marine fishing nets. Biotechnol Genet Eng Rev. https://doi.org/10.1080/02648725.2022.2152629

Sarma H, Hazarika RP, Kumar V, Roy A, Pandit S, Prasad R (2022) Microplastics in marine and aquatic habitats: sources, impact, and sustainable remediation approaches. Environ Sustain 5(1):39–49. https://doi.org/10.1007/s42398-022-00219-8

Shabaka SH, Ghobashy M, Marey RS (2019) Identification of marine microplastics in Eastern Harbor, Mediterranean Coast of Egypt, using differential scanning calorimetry. Mar Pollut Bull 142:494–503. https://doi.org/10.1016/j.marpolbul.2019.03.062

Shan J, Zhao J, Zhang Y, Liu L, Wu F, Wang X (2019) Simple and rapid detection of microplastics in seawater using hyperspectral imaging technology. Anal Chim Acta 1050. https://doi.org/10.1016/j.aca.2018.11.008

Shim WJ, Hong SH, Eo SE (2017) Identification methods in microplastic analysis: a review. Anal Methods 9(9):1384–1391. https://doi.org/10.1039/C6AY02558G

Singh RP, Mishra S, Das AP (2020) Synthetic microfibers: pollution toxicity and remediation. Chemosphere 257:127199. https://doi.org/10.1016/j.chemosphere.2020.127199

Smith E, Dent G (2004) Modern Raman spectroscopy—a practical approach. Wiley. https://doi.org/10.1002/0470011831

Song YK, Hong SH, Jang M, Han GM, Rani M, Lee J, Shim WJ (2015) A comparison of microscopic and spectroscopic identification methods for analysis of microplastics in environmental samples. Mar Pollut Bull 93(1–2):202–209. https://doi.org/10.1016/j.marpolbul.2015.01.015

Tripathy B, Sahoo PP, Sundaray H, Das AP (2024) In-vitro biodegradation of discarded marine microplastics across the eastern coast of the Bay of Bengal, India using *Exiguobacterium* sp. Environ Chem Ecotoxicol 6:236–247. https://doi.org/10.1016/j.enceco.2024.07.003

Tripathy B, Dash A, Das AP (2022) Detection of environmental microfiber pollutants through vibrational spectroscopic techniques: recent advances of environmental monitoring and future prospects. Crit Rev Anal Chem 1–11. https://doi.org/10.1080/10408347.2022.2144994

Uurasjärvi E, Hartikainen S, Setälä O, Lehtiniemi M, Koistinen A (2020) Microplastic concentrations, size distribution, and polymer types in the surface waters of a northern European lake. Water Environ Res 92(1):149–156. https://doi.org/10.1002/wer.1229

Van Cauwenberghe L, Vanreusel A, Mees J, Janssen CR (2013) Microplastic pollution in deep-sea sediments. Environ Pollut 182:495–499. https://doi.org/10.1016/j.envpol.2013.08.013

Veerasingam S, Ranjani M, Venkatachalapathy R, Bagaev A, Mukhanov V, Litvinyuk D, Mugilarasan M, Gurumoorthi K, Guganathan L, Aboobacker VM, Vethamony P (2021) Contributions of Fourier transform infrared spectroscopy in microplastic pollution research: a review. Crit Rev Environ Sci Technol 51(22):2681–2743. https://doi.org/10.1080/10643389.2020.1807450

Villacorta A, Cazorla-Ares C, Fuentes-Cebrian V, Valido IH, Vela L, Carrillo-Navarrete F, Morataya-Reyes M, Mejia-Carmona K, Pastor S, Velázquez A, Arranz JA, Marcos R, López-Mesas M, Hernández A (2024) Fluorescent labeling of micro/nanoplastics for biological applications with a focus on "true-to-life" tracking. J Hazard Mater 476:135134. https://doi.org/10.1016/j.jhazmat.2024.135134

Wang Z-M, Wagner J, Ghosal S, Bedi G, Wall S (2017) SEM/EDS and optical microscopy analyses of microplastics in ocean trawl and fish guts. Sci Total Environ 603–604:616–626. https://doi.org/10.1016/j.scitotenv.2017.06.047

Wu S, Lyu G, Lou R (2012) Applications of chromatography hyphenated techniques in the field of lignin pyrolysis. https://doi.org/10.5772/32446

Xu J-L, Thomas KV, Luo Z, Gowen AA (2019) FTIR and Raman imaging for microplastics analysis: state of the art, challenges and prospects. TrAC, Trends Anal Chem 119:115629. https://doi.org/10.1016/j.trac.2019.115629

Xu J-L, Hassellöv M, Yu K, Gowen AA (2020) Microplastic characterization by infrared spectroscopy. In: Rocha-Santos T, Costa M, Mouneyrac C (eds) Handbook of microplastics in the environment. Springer, pp 1–33. https://doi.org/10.1007/978-3-030-10618-8_21-1

Yu J, Wang P, Ni F, Cizdziel J, Wu D, Zhao Q, Zhou Y (2019) Characterization of microplastics in environment by thermal gravimetric analysis coupled with Fourier transform infrared spectroscopy. Mar Pollut Bull 145:153–160. https://doi.org/10.1016/j.marpolbul.2019.05.037

Yu C, Takhistov P, Alocilja E et al (2022) Bioanalytical approaches for the detection, characterization, and risk assessment of micro/nanoplastics in agriculture and food systems. Anal Bioanal Chem 414:4591–4612. https://doi.org/10.1007/s00216-022-04069-5

Zabihihesari A, Khalili A, Farshchi-Heydari M-J, Eilaghi A, Rezai P (2023) Simple microfluidic device for simultaneous extraction and detection of microplastics in water using DC electrical signal. New J Chem 47(19):9050–9060. https://doi.org/10.1039/D2NJ06268B

Zhang M, Wang X, Zhang Y, Fan Y (2023a) Integrated sample processing and counting microfluidic device for microplastics analysis. Anal Chim Acta 1261:341237. https://doi.org/10.1016/j.aca.2023.341237

Zhang Y, Zhang D, Zhang Z (2023b) A critical review on artificial intelligence—based microplastics imaging technology: recent advances, hot-spots and challenges. Int J Environ Res Public Health 20(2):1150. https://doi.org/10.3390/ijerph20021150

Zhang N, Yang J, Liu S, Ma Y, Zhang J (2024) Hyperspectral remote sensing identification of marine oil spills and emulsions using feature bands and double-branch dual-attention mechanism network. J Oceanol Limnol. https://doi.org/10.1007/s00343-023-3122-5

Zhong Y, Bao Q, Yuan L, Liu J, Cai Y, Chen X (2022) Analysis of microplastics in aquatic shellfish by pyrolysis-gas chromatography/mass spectrometry after alkali digestion and solvent extraction. Polymers 14(18):3888. https://doi.org/10.3390/polym14183888

The manufacturer's authorised representative in the EU is Springer Nature Customer Service Centre GmbH, Europaplatz 3, 69115 Heidelberg, Germany. If you have any concerns regarding our products, please contact ProductSafety@springernature.com

Printed and bound by CPI Group (UK) Ltd, Croydon, CR0 4YY
30/09/2026
02225596-0015